PROGRESS IN
GRAPH THEORY

Academic Press Rapid Manuscript Reproduction

PROGRESS IN GRAPH THEORY

Edited by

J. ADRIAN BONDY
U. S. R. MURTY

Department of Combinatorics and Optimization
University of Waterloo
Waterloo, Ontario, Canada

1984

ACADEMIC PRESS
(Harcourt Brace Jovanovich, Publishers)
TORONTO   ORLANDO   SAN DIEGO   SAN FRANCISCO   NEW YORK
LONDON   MONTREAL   SYDNEY   TOKYO   SÃO PAULO

COPYRIGHT © 1984, BY ACADEMIC PRESS CANADA
ALL RIGHTS RESERVED.
NO PART OF THIS PUBLICATION MAY BE REPRODUCED OR
TRANSMITTED IN ANY FORM OR BY ANY MEANS, ELECTRONIC
OR MECHANICAL, INCLUDING PHOTOCOPY, RECORDING, OR ANY
INFORMATION STORAGE AND RETRIEVAL SYSTEM, WITHOUT
PERMISSION IN WRITING FROM THE PUBLISHER.

ACADEMIC PRESS CANADA
55 Barber Greene Road, Don Mills, Ontario   M3C 2A1

United States Edition published by
ACADEMIC PRESS, INC.
Orlando, Florida 32887

United Kingdom Edition published by
ACADEMIC PRESS, INC. (LONDON) LTD.
24/28 Oval Road, London NW1 7DX

Canadian Cataloguing in Publication Data

Main entry under title:

Progress in graph theory

Papers from the second week of the Silver Jubilee
Conference on Combinatorics, University of Waterloo,
June 14-July 2, 1982.
ISBN 0-12-114320-1

1. Graph theory—Congresses. 2. Combinatorial
analysis—Congresses. I. Bondy, J. A., 1944-
II. Murty, U. S. R. III. Silver Jubilee Conferences
on Combinatorics (1982: University of
Waterloo)

QA166.P76 1984   511'.5   C84-099462-1

Library of Congress Cataloging in Publication Data
Main entry under title:

Progress in graph theory.

   Proceedings of the second week of the Silver Jubilee
Conference on Combinatorics, held in 1982 at the University of Waterloo, organized by the Dept. of Combinatorics and Optimization.
   Includes index.
   1. Graph theory—Congresses. I. Bondy, J. A. (John
Adrian) II. Murty, U. S. R. III. Silver Jubilee Conference on Combinatorics (1982 : University of Waterloo)
IV. University of Waterloo. Dept. of Combinatorics and
Optimization.
QA166.P76 1984   511'.5   84-48434

ISBN 0-12-114320-1 (alk. paper)

PRINTED IN CANADA

84 85 86 87   9 8 7 6 5 4 3 2 1

CONTENTS

Contributors	ix
Preface	xiii

Lectures on Random Graphs   3
    B. Bollobás
Plane Representations of Graphs   43
    C. Thomassen
Marriage in Infinite Societies   71
    R. Aharoni, C. St. J. A. Nash-Williams, and S. Shelah
Dominating Cycles in Bipartite Graphs   81
    P. Ash and B. Jackson
Helms are Graceful   89
    J. Ayel and O. Favaron
Edge to Point Degree Lists and Extremal Triangle Free Point Degree Regular Graphs   93
    D. Bauer, G. Bloom, and F. Boesch
A Property of k-Optimal Path-Partitions   105
    C. Berge
Set-Labelled Graphs   109
    I. Z. Bouwer and B. Monson
Stability of Kings in Tournaments   117
    M. F. Bridgland and K. B. Reid
The Interchange Graph of Tournaments with the Same Score Vector   129
    R. A. Brualdi and Li Qiao
Linear Algorithms for Convex Drawings of Planar Graphs   153
    N. Chiba, T. Yamanouchi, and T. Nishizeki
Triangle-Free Graphs with Restricted Bandwidth   175
    F. R. K. Chung and W. T. Trotter, Jr.
On Clique Partition Numbers of Regular Graphs   191
    P. Eades, N. J. Pullman, and P. J. Robinson

Cube-Supersaturated Graphs and Related Problems P. Erdös and M. Simonovits	203
On the Complexity of Finding Sets of Edges in Graphs and Digraphs T. I. Fenner and A. M. Frieze	219
Cycle Decompositions, 2-Coverings, Removable Cycles and the Four-Color-Disease H. Fleischner	233
Strong Edge-Coloring of Cubic Planar Graphs J. L. Fouquet and J. L. Jolivet	247
Galois Theory of Tree Extensions G. Gati	265
Real Graph Polynomials C. D. Godsil	281
On the Minimal Cut Problem M. K. Goldberg and R. Gardner	295
On Isometric Embeddings of Graphs R. L. Graham	307
The Maximum Weighted Cycle-Packing Problem and its Relation to the Chinese Postman Problem M. Guan	323
Digraphs with Converse and Line Digraph Isomorphic R. L. Hemminger	327
Efficient Algorithms for the Maximum Weight Clique Problem on Circular-Arc Graphs and Circle Graphs Wen-Lian Hsu	335
On Some Algebraic Properties of Graphs F. Jaeger	347
The Tutte Polynomial of a Morphism of Matroids II. Activities of Orientations M. Las Vergnas	367
The Cycle-Chromatic Number of a Hypergraph and an Inequality of Lovász Li Wei-xuan	381
On k-Critically n-Connected Graphs W. Mader	389
Graph Width and Well-Quasi-Ordering: A Survey N. Robertson and P. D. Seymour	399
Cographic Subspaces of Cocycle Spaces for Line Graphs of Complete Graphs Ching-Hsien Shih	407
Extremal Graph Problems, Degenerate Extremal Problems, and Supersaturated Graphs M. Simonovits	419
Elements of a Decomposition Theory for Matroids K. Truemper	439

CONTENTS

Map-Colourings and Differential Equations *W. T. Tutte*	477
Combinatorial Classification of the Regular Polytopes *A. Vince*	487
Products of Highly Regular Graphs *P. M. Weichsel*	497
The Number of Complete Subgraphs in Graphs with Non-Majorizable Degree Sequences *D. B. West*	509
Unsolved Problems	525

CONTRIBUTORS

Numbers in parentheses indicate the pages on which the authors' contributions begin.

R. Aharoni (71), *Department of Mathematics, Technion, Israel Institute of Technology, Haifa, Israel*
P. Ash (81), *49 Felstead Road, Orpington, Kent BR6 9AA, England*
J. Ayel (89), *Université de Savoie, BP 1104, 73011 Chambrey Cédex, France*
D. Bauer (93), *Mathematics Department, Stevens Institute of Technology, Hoboken, New Jersey 07030, U.S.A.*
C. Berge (105), *10 rue Galvani, 75017 Paris, France*
G. Bloom (93), *Mathematics Department, Stevens Institute of Technology, Hoboken, New Jersey 07030, U.S.A.*
F. Boesch (93), *Mathematics Department, Stevens Institute of Technology, Hoboken, New Jersey 07030, U.S.A.*
B. Bollobás (3), *Department of Mathematics, University of Cambridge, 16 Mill Lane, Cambridge CB2 1SB, England*
I. Z. Bouwer (109), *Department of Mathematics and Statistics, University of New Brunswick, Fredericton, New Brunswick E3B 5A3, Canada*
M. F. Bridgland (117), *Department of Mathematics, Lawrence University, Box 599, Appleton, Wisconsin 54912, U.S.A.*
R. A. Brualdi (129), *Mathematics Department, University of Wisconsin, Madison, Wisconsin 53706, U.S.A.*
N. Chiba (153), *Department of Electrical Communications, Faculty of Engineering, Tohoku University, Sendai, Japan 980*
F. R. K. Chung (175), *Discrete Mathematics Research Group, Bell Communications Inc., 600 Mountain Ave., Murray Hill, New Jersey 07974, U.S.A.*
P. Duchet (525), *E. R. 175 Combinatoire, C. M. S., 54 boulevard Raspail, 75270 Paris Cédex, France*
P. Eades (191), *Department of Computer Science, University of Queensland, St. Lucia, Queensland 4067, Australia*
P. Erdös (203, 528, 532), *Mathematical Institute, Hungarian Academy of Sciences, Realtanoda utca 13-15, H-1053 Budapest, Hungary*

CONTRIBUTORS

O. Favaron (89), *Université de Paris-Sud, Centre d'Orsay, Laboratoire de Recherche en Informatique, Bâtiment 490, 91405 Orsay Cédex, France*

T. I. Fenner (219), *Department of Computer Science, Birkbeck College, Malet Street, London WC1E 7HX, England*

H. Fleischner (233, 537), *Institute for Information Processing, Austrian Academy of Sciences, Sonnenfelsgasse 19/2, A-1010 Vienna, Austria*

J. L. Fouquet (247), *Département de Mathématiques, Université du Maine, 72017 Le Mans Cédex, France*

A. M. Frieze (219), *Department of Computer Science and Statistics, Queen Mary College, Mile End Road, London E1 4NS, England*

R. Gardner (295), *Department of Mathematics and Computer Science, Clarkson College, Potsdam, New York, U.S.A.*

G. Gati (265), *Institut für Informatik, ETH-Zentrum, CH-8092 Zurich, Switzerland*

C. D. Godsil (281), *Department of Mathematics, Simon Fraser University, Burnaby, British Columbia V5A 1S6, Canada*

M. K. Goldberg (295), *Department of Mathematics and Computer Science, Clarkson College, Potsdam, New York, U.S.A.*

R. L. Graham (307), *Bell Laboratories, 600 Mountain Avenue, Murray Hill, New Jersey 07974, U.S.A*

M. Guan (323), *Department of Mathematics, Shandong Normal University, Jinan, Shandong, People's Republic of China*

R. L. Hemminger (327, 532), *Department of Mathematics, Vanderbilt University, Nashville, Tennessee 37235, U.S.A.*

D. A. Holton (532), *Department of Mathematics, University of Melbourne, Parkville, Victoria 3052, Australia*

Wen-Lian Hsu (335, 538), *Department of Industrial Engineering and Management Sciences, Northwestern University, Evanston, Illinois 60201, U.S.A.*

B. Jackson (81), *Goldsmiths' College, London SE8, England*

F. Jaeger (347), *IMAG, BP 53x, 38041 Grenoble Cédex, France*

J. L. Jolivet (247), *Département de Mathématiques, Université du Maine, 72017 Le Mans Cédex, France*

M. Las Vergnas (367), *E. R. 175 Combinatoire, C. M. S., 54 boulevard Raspail, 75270 Paris Cédex, France*

Li Qiao (129), *Mathematics Department, University of Science and Technology of China, Hefei, Anhui, People's Republic of China*

Li Wei-xuan (381), *Department of Fundamental Courses, Changsha Railway Institute, Changsha, Hunan, China*

W. Mader (389), *Institut für Mathematik, Universität Hannover, Welfengarten 1, 3000 Hannover, Federal Republic of Germany*

B. D. McKay (532), *Department of Computer Science, Australian National University, P.O. Box 4, Canberra, ACT 2600, Australia*

H. Meyniel (525), *E. R. 175 Combinatoire, C. M. S., 54 boulevard Raspail, 75270 Paris Cédex, France*

B. Monson (109), *Department of Mathematics and Statistics, University of New Brunswick, Fredericton, New Brunswick E3B 5A3, Canada*

C. St. J. A. Nash-Williams (71), *Department of Mathematics, University of Reading, Whiteknights, Reading, Berkshire RG6 2AX, England*

T. Nishizeki (153), *Department of Electrical Communications, Faculty of Engineering, Tohoku University, Sendai, Japan 980*

N. J. Pullman (191), *Department of Mathematics, Queen's University, Kingston, Ontario K7L 3N6, Canada*

K. B. Reid (117), *Department of Mathematics, Louisiana State University, Baton Rouge, Louisiana 70803, U.S.A.*

N. Robertson (399), *Department of Mathematics, Ohio State University, Columbus, Ohio 43210, U.S.A.*

P. J. Robinson (191), *Department of Computer Science, University of Queensland, St. Lucia, Queensland 4067, Australia*

P. D. Seymour (399), *Central Services Organization, Room 2C-120, Bell Laboratories, 600 Mountain Avenue, Murray Hill, New Jersey 07974, U.S.A.*

S. Shelah (71), *Department of Mathematics, Hebrew University, Jerusalem, Israel*

Ching-Hsien Shih (407), *Institute of Mathematics, Academia Sinica, Nankong, Taipei, Taiwan, Republic of China*

M. Simonovits (203, 419), *Eötvös Loránd University, H-1088 Budapest, Muzeum Krt 6-8, Hungary*

R. Stanley, (539), *Department of Mathematics, Rm 2-347, Massachussetts Institute of Technology, Cambridge, Massachussetts 02139, U.S.A.*

C. Thomassen (43), *Mathematics Institute, Bygning 303, Technical University of Denmark, DK-2800 Lyngby, Denmark*

W. T. Trotter, Jr. (175), *Department of Mathematics, University of South Carolina, Columbia, South Carolina 29208, U.S.A.*

K. Truemper (439), *School of Management and Administration, University of Texas, Dallas, Box 688, Richardson, Texas 75080, U.S.A.*

W. T. Tutte (477), *Department of Combinatorics and Optimization, University of Waterloo, Waterloo, Ontario N2L 3G1, Canada*

A. Vince (487), *Department of Mathematics, University of Florida, Gainesville, Florida 32611, U.S.A.*

P. M. Weichsel (497), *Department of Mathematics, University of Illinois at Urbana-Champaign, 1409 West Green Street, Urbana, Illinois 61801, U.S.A.*

D. B. West (509), *Department of Mathematics, University of Illinois at Urbana-Champaign, 1409 West Green Street, Urbana, Illinois 61801, U.S.A.*

T. Yamanouchi (153), *Department of Electrical Communications, Faculty of Engineering, Tohoku University, Sendai, Japan 980*

PREFACE

The University of Waterloo celebrated its twenty-fifth anniversary in 1982. To mark this occasion, the Department of Combinatorics and Optimization organized a conference on Combinatorics. The Silver Jubilee Conference on Combinatorics ran for three weeks, each week specializing in one of the areas of the subject. This volume is a record of the proceedings of the second week, which was devoted to Graph Theory. Instructional lectures by Professors Béla Bollobás and Carsten Thomassen were the main feature of this week. In addition, a number of invited and contributed papers were presented by participants from many parts of the world. This volume contains most of these papers.

The conference was made possible by generous grants from the Natural Sciences and Engineering Research Council of Canada, and the Faculty of Mathematics of the University of Waterloo. We are especially indebted to Dean Alan George for the encouragement and help he extended to us. We would like to express our gratitude to Kathy Knight and Joan Selwood for their secretarial assistance, Christy Folliott, Elinor Hawn, Archana Kundra, and Helen Warren for typesetting the manuscript, and to the staff of the Mathematics Faculty Computing Facility for their courteous service. Finally, we would like to thank the staff of Academic Press Canada for their cooperation in producing this volume.

J. A. Bondy
U. S. R. Murty

PROGRESS IN
GRAPH THEORY

Lectures on Random Graphs

Béla Bollobás

ABSTRACT

The study of random graphs was initiated by Erdős and Rényi in the sixties. They discovered that if a graph is viewed as a living organism which grows by acquiring more and more edges, then most properties appear rather suddenly. There is a vast literature devoted to the above theme and related applications of probabilistic methods to combinatorics. This paper contains a survey of the recent developments in random graph theory.

Introduction

Probabilistic arguments are the vogue in several branches of mathematics, perhaps nowhere more so than in graph theory. After some tentative and scattered results the study of random graphs got off to a flying start with two papers of Erdős [27], [28] concerning Ramsey numbers and graphs of large girth and chromatic number.

These papers contain applications of a method which has been found very successful in great many instances. In its simplest form this method says that in order to show the existence of a graph with certain seemingly surprising properties, we should construct a probability space of graphs in which the probability that a graph has all those properties is positive. Just a little later Erdős and Rényi [29], [30] began to consider random graphs for their own sake. Their main discovery was that if a graph is viewed as a living organism which develops by acquiring more and more edges, then most properties appear rather suddenly.

The last quarter of a century saw a proliferation of results concerning random graphs and by now there is a considerable literature of

---

Supported by NSF grant MCS - 8104854

probabilistic graph theory. A good selection of results can be found in Erdős and Spencer [34] and a substantial bibliography was collected by Karoński [52]. The recent reviews [13] and [45] emphasise various aspects of the theory. The aim of these lectures is to present some of the basic results for combinatorialists not primarily interested in random graphs. Due to lack of time no attempt will be made at an encyclopaedic treatment of the subject. A considerably more thorough account of random graphs will be given in the forthcoming book [21], on which these lectures are based.

1. Probability theoretic preliminaries

The aim of this section is to focus attention on the simple tools from probability theory one tends to apply in the study of random graphs. As customary, we shall denote by $E(X)$ the expectation of a random variable ($r.v.$) and by $\sigma^2(X)$ its variance: $\sigma^2(X) = E((X - E(X))^2)$. We shall often use the obvious fact that if X is a non-negative integer valued $r.v.$ then

$$P(X > 0) \leq E(X).$$

Another simple inequality one needs fairly often is Chebyshev's inequality: if X has expectation μ and variance σ^2 then for $d > 0$ we have

$$P(|X - \mu| \geq d) \leq \sigma^2/d^2.$$

In particular,

$$P(X = 0) \leq P(|X - \mu| \geq \mu) \leq \sigma^2/\mu^2.$$

Not surprisingly, one needs rather precise approximations of the binomial, hypergeometric and Poisson distributions. These approximations are often based on *Stirling's formula:*

$$n! = \left(\frac{n}{e}\right)^n \sqrt{2\pi n} \, e^{\alpha_n}, \tag{1}$$

where $1/(12n + 1) < \alpha_n < 1/(12n)$.

As customary, we shall adopt the convention that $0 < p < 1$ and $q = 1 - p$. We say that X is a *Bernouilli random variable with mean p* if X takes two values, 0 and 1, and $P(X = 1) = p$ and $P(X = 0) = q$. Clearly such a $r.v.$ arises as the outcome of tossing a biased coin. Let $X^{(1)}, X^{(2)}, \ldots$ be independent $r.vs$ with mean p. Then

$$S_{n,p} = \sum_{i=1}^{n} X^{(i)}$$

is a *binomial r.v. with parameters n and p*. The distribution of $S_{n,p}$ is given by

$$P(S_{n,p} = k) = b(k;n,p) = \binom{n}{k} p^k q^{n-k},$$

$k = 0, 1, \ldots, n$. By definition $b(k;n,p)$ is the probability that we get k heads when tossing a coin n times, provided the probability of getting a head is p. Since the expectation and variance of $X^{(i)}$ are $E(X^{(i)}) = p$ and $\sigma^2(X^{(i)}) = E((X^{(i)} - p)^2) = pq$, the binomial distribution with parameters n and p has mean $E(S_{n,p}) = pn$ and variance $\sigma^2(S_{n,p}) = pqn$.

It is easily seen that for fixed values of n and p the function $b(k;n,p)$ first increases with k and then it decreases. The maximum of $b(k;n,p)$ is attained when k is about pn, the expectation of $S_{n,p}$. Furthermore, by Stirling's formula (1) if p is fixed the $n \to \infty$ then

$$\max_k b(k;n,p) \sim \frac{1}{\sqrt{2\pi pqn}} = \frac{1}{\sigma\sqrt{2\pi}}.$$

Here and in what follows $A \sim B$ stands for the assertion "$A/B \to 1$ as $n \to \infty$".

A r.v. Y has Poisson distribution with mean $\lambda > 0$ if

$$P(Y = k) = p(k;\lambda) = e^{-\lambda}\lambda^k/k!, \quad k = 0, 1, \ldots.$$

This distribution, usually denoted by P_λ or $P(\lambda)$, is closely related to the binomial distribution. If λ is a positive constant and p depends on n in such a way that $pn \to \lambda$ as $n \to \infty$ then the binomial distribution with parameters n and p tends to the Poisson distribution with mean λ.

In the study of random variables occurring in combinatorics it is often profitable to calculate the factorial moments, which tend to arise naturally. The *rth factorial moment* of a r.v. X is $E_r(X) = E((X)_r)$, where $(x)_r = x(x-1) \cdots (x-r+1)$. The main reason why factorial moments occur frequently is that combinatorial random variables often count the number of certain objects like k-cycles, k-cliques, isolated vertices, etc. If X denotes the number of objects in a certain class then $E_r(X)$ is the expected number of *ordered r-tuples of elements of that class*.

The factorial moments of $S_{n,p}$ and P_λ are easily calculated:

$$E_r(S_{n,p}) = p^r(n)_r \text{ and } E_r(P_\lambda) = \lambda^r.$$

A random variable is *normally distributed with mean μ and variance σ^2* if it has density function

$$\phi(t) = \frac{1}{\sigma\sqrt{2\pi}} e^{-(t-\mu)^2/2\sigma^2}.$$

This distribution is usually denoted by $N(\mu,\sigma)$. Furthermore, $\Phi(x)$ is the distribution function of $N(0,1)$:

$$\Phi(x) = \frac{1}{\sqrt{2\pi}} \int_{-\infty}^{x} e^{-t^2/2} dt.$$

The distribution $N(0,1)$ is symmetric about 0 and if $x > 0$ and $l \geq 0$ then

$$1 + \sum_{m=1}^{2l+1} (-1)^m \frac{1 \cdot 3 \cdots (2m-1)}{x^{2m}} < (1-\Phi(x))/\{\frac{1}{x}\phi(x)\}$$
$$< 1 + \sum_{m=1}^{2l} (-1)^m \frac{1 \cdot 3 \cdots (2m-1)}{x^{2m}}. \qquad (2)$$

Relation (2) implies that as $x \to \infty$

$$1 - \Phi(x) \sim \frac{1}{x} \phi(x) = \frac{1}{\sqrt{2\pi}} \frac{1}{x} e^{-x^2/2}. \qquad (2')$$

In the theory of random graphs one often needs good estimates for the probability in the tail of the binomial distribution. The best known such approximation is the following classical De Moivre - Laplace theorem.

THEOREM 1. Suppose $0 < p < 1$ depends on n in such a way that $pqn = p(1-p)n \to \infty$.

(i) Suppose $h_1 < h_2$, $h_2 - h_1 \to \infty$ as $n \to \infty$ and $|h_1| + |h_2| = o((pqn)^{2/3})$. Put $h_i = x_i(pqn)^{1/2}$, $i = 1,2$. Then

$$P(pn+h_1 \leq S_{n,p} \leq pn+h_2) \sim \Phi(x_2) - \Phi(x_1).$$

(ii) If $0 < h = x(pqn)^{1/2} = o((pqn)^{2/3})$ then

$$P(S_{n,p} \leq pn+h) \sim 1 - \Phi(x).$$

In particular, if $x \to \infty$ then

$$P(S_{n,p} > pn+h) \sim \frac{1}{x\sqrt{2\pi}} e^{-x^2/2}. \qquad \square$$

If we are interested in the probability that $S_{n,p}$ is considerably larger than its mean then the result above is of not much use. In that case the following easy bounds may be applied.

THEOREM 2.

(i) Suppose $0 < p \leq 1/2$ and $0 < \epsilon < 1/20$. Then

$$P(|S_{n,p} - pn| \geq \epsilon pn) \leq (\epsilon^2 pn)^{-1/2} e^{-\epsilon^2 pn/3}.$$

(ii) If $u > e$ then

$$P(S_{n,p} \geq upn) < (e/u)^{upn}$$

and if $v > e^2$ then

$$P\left(S_{n,p} \geq \frac{2v}{\log v} pn\right) < e^{-vpn}. \quad \square$$

More often than not the random variables we have to consider do not have a well known and simple distribution. However, we may be able to estimate the factorial moments of our r.v. The question is then what information we can deduce about the distribution function from these estimates. Our main tool in answering this question is a set of inequalities due to Jordan [49], [50], [51], Bonferroni [23] and Fréchet [42]. A rather pleasant proof of these inequalities is based on the following theorem of Rényi [73].

THEOREM 3. Let $f_1, f_2, \ldots, f_k$ be Boolean polynomials in n variables $A_1, A_2, \ldots, A_n$ and let $b_1, b_2, \ldots, b_n$ be real numbers. Suppose

$$\sum_{i=1}^{k} b_i P(f_i(B_1, B_2, \ldots, B_n)) \geq 0 \tag{3}$$

whenever $B_1, B_2, \ldots, B_n$ are events in a probability space such that $P(B_i) = 0$ or 1. Then (3) holds for every choice of events $B_1, B_2, \cdots, B_n$ in every probability space. $\quad \square$

The result above is easily proved by expressing our Boolean polynomials as unions of complete products.

In order to state the Jordan-Bonferroni inequalities (Theorem 4) we introduce the following terminology and notation.

Let $A_1, A_2, \ldots, A_n$ be events in a probability space (Ω, P). For $J \subset N = \{1, 2, \ldots, n\}$ put

$$A_J = \bigcap_{j \in J} A_j \text{ and } A_\emptyset = \Omega.$$

Furthermore, set

$$P_J = P(A_J)$$

and

$$S_r = \sum_{|J|=r} P_J, \quad r = 0,1,\ldots,n.$$

The usual inclusion-exclusion formula, namely

$$P(\bigcup_{i=1}^{n} A_i) = \sum_{r=1}^{n}(-1)^{r+1}S_r, \tag{4}$$

is an easy consequence of Theorem 3. Indeed, by Theorem 3 relation (4) holds if and only if it holds whenever each A_i has probability 1 or 0. In the latter case relation (4) is easily checked.

Let us say that a sum

$$s = \sum_{k=1}^{n}(-1)^{k+1}x_k$$

satisfies the *alternating inequalities* if

$$(-1)^l\{s + \sum_{n=1}^{l}(-1)^k x_k\} \geq 0$$

for every l, $l = 0,1,\ldots,n$.

THEOREM 4. Let $A_1, A_2, \ldots, A_n$ be events in a probability space, let $S_1, S_2, \ldots, S_n$ be as above and denote by p_k the probability that exactly k of the A_i occur. Then

$$p_k = \sum_{r=k}^{n}(-1)^{r+k}\binom{r}{k}S_r \tag{5}$$

and the sum satisfies the alternating inequalities.

PROOF. By Theorem 3 it suffices to prove the assertions in the case when for some l, $0 \leq l \leq n$, exactly l of the sets A_i has probability 1 and all others have probability 0. Then $S_r = \binom{l}{r}$ and $p_k = \delta_{kl}$. If $l < k$ then all terms in (5) are 0 and if $l = k$ then (5) takes the form $1 = 1 - 0 + 0 - 0 + \cdots$.

Assume now that $l > k$. Then $p_k = 0$ and the right hand side of (5) is

$$\sum_{r=k}^{k+m}(-1)^{k+r}\binom{r}{k}\binom{l}{r} = \sum_{r=k}^{k+m}(-1)^{k+r}\frac{(r)_k}{k!}\frac{(l)_r}{r!}$$

$$= \sum_{r=k}^{k+m}(-1)^{r+k}\frac{(l)_k}{k!}\frac{(l-k)_{r-k}}{(r-k)!} = \binom{l}{k}\sum_{r=k}^{k+m}(-1)^{r+k}\binom{l-k}{r-k}.$$

If $k + m \geq l$ then the sum is 0 and if $0 \leq m < l-k$ then it is

$$\binom{l}{k}\sum_{i=0}^{m}(-1)^{i}\left\{\binom{l-k-1}{i-1}+\binom{l-k-1}{i}\right\} = \binom{l}{k}(-1)^{m}\binom{l-k-1}{m}.$$

Hence the sum is positive if m is even and negative if m is odd. This means exactly that (5) holds and satisfies the alternating inequalities. □

COROLLARY 5. Let X be a r.v. with values in $\{0,1,\ldots,n\}$ and let $E_r(X)$ be the rth factorial moment of X. Then

$$P(X = k) = \left\{\sum_{r=k}^{n}(-1)^{k+r}\frac{E_r(X)}{(r-k)!}\right\}/k!$$

and the sum satisfies the alternating inequalities.

PROOF. For $i = 1,\ldots,n$ denote by A_i the event $\{X \geq i\}$. Define the numbers $S_0, S_1, \ldots, S_n$ as above. Then

$$S_r = E_r(X)/r!$$

and $\{X = k\}$ is the event that exactly k of the A_i occur. Hence the assertion is a consequence of Theorem 4. □

COROLLARY 6. Let X be a non-negative integer valued r.v. whose rth factorial moment $E_r(X) = E((X)_r)$ is finite for $r = 1,\ldots,R$. Suppose $s,t \leq R$, $k + s$ is odd and $k + t$ is even. Then

$$\left\{\sum_{r=k}^{s}(-1)^{k+r}E_r(X)/(r-k)!\right\}/k! \leq P(X = k) \leq \left\{\sum_{r=k}^{t}(-1)^{k+r}E_r(X)/(r-k)!\right\}/k!.$$

PROOF. Set $X^{(n)} = \min\{X, n\}$. Then $\lim_{n\to\infty} E_r(X^{(n)}) = E_r(X)$ for $r = 1, 2, \ldots, R$, and for $n \geq k$ we have $P(X^{(n)} = k) = P(X = k)$. Hence our inequalities follow from Corollary 5. □

One often needs a third straightforward consequence of Theorem 4.

COROLLARY 7. Let X be a non-negative integer values r.v. with finite moments. If

$$\lim_{r\to\infty} E_r(X) r^m / r! = 0$$

for all m then

$$P(X = k) = \left\{\sum_{r=k}^{\infty}(-1)^{k+r} E_r(X)/(r-k)!\right\}/k!$$

and the sum satisfies the alternating inequalities. □

The results above can be extended to several sequences of events and to several random variables. Neither the statements nor the proofs hold any surprises so we shall not formulate these extensions.

In the theory of random graphs the inequalities above are especially important because they allow one to prove results about convergence in distribution. A sequence (X_n) of non-negative integer valued r.vs is said to *tend to a* r.v. X *in distribution* if $\lim_{n\to\infty} P(X_n = k) = P(X = k)$ for every $k \geq 0$. Similarly a sequence (X_n) of *r.vs tends to a r.v. X in distribution* if $\lim_{n\to\infty} P(X_n < x) = P(X < x)$ whenever x is a point of continuity of the distribution function $F(x) = P(X < x)$. Convergence in distribution is denoted by $X_n \xrightarrow{d} X$.

Corollaries 6 and 7 have the following important consequence.

THEOREM 8. *Let* $X_1, X_2, \ldots$ *and* X *be non-negative integer valued r.vs. Suppose*

$$\lim_{n\to\infty} E_r(X_n) = E_r(X), \quad r = 0, 1, \ldots,$$

and

$$\lim_{n\to\infty} E_r(X) r^m / r! = 0, \quad m = 0, 1, \ldots.$$

Then

$$X_n \xrightarrow{d} X. \quad \square$$

In many applications the r.v. X above is a Poisson r.v. with mean λ. In that case $E_r(X) = \lambda^r$ for every r so Theorem 8 implies the following result.

THEOREM 9. *Let* $\lambda > 0$ *and suppose the non-negative integer valued r.vs* $X_1, X_2, \ldots$ *are such that*

$$\lim_{n\to\infty} E_r(X_n) = \lambda^r, \quad r = 0, 1, \ldots.$$

Then

$$X_n \xrightarrow{d} P_\lambda. \quad \square$$

By the DeMoivre-Laplace theorem (Theorem 1 (ii)) if $pqn \to \infty$ then $(S_{p,n} - pn)/\sqrt{pqn} \xrightarrow{d} N(0,1)$. Also, if $\lambda > 0$ is a constant and $pn \to \lambda$ then $S_{p,n} \xrightarrow{d} P_\lambda$. Hence $(P_\lambda - \lambda)/\sqrt{\lambda} \xrightarrow{d} N(0,1)$ as $\lambda \to \infty$. By making use of the fact that if (X_n) is a sequence of real valued r.vs such that for every fixed r the rth moment of X_n tends to the rth moment of $N(0,1)$ then $X_n \xrightarrow{d} N(0,1)$, we can obtain the following simple result concerning normal convergence.

THEOREM 10. Let (X_n) be a sequence of r.vs and let $\lambda_n \to \infty$. Suppose
$$E_r(X_n) - \lambda_n^r = o(\lambda_n^{-s})$$
for all r and s, $1 \le r \le s$. Then
$$\frac{X_n - \lambda_n}{\sqrt{\lambda_n}} \xrightarrow{d} N(0,1). \qquad \square$$

Though the conditions in Theorem 10 are very restrictive, the result often enables us to conclude that our r.v. has asymptotically normal distribution.

2. Models of random graphs

Most results about random graphs concern labelled random graphs and in these lectures we shall also concentrate on labelled random graphs. This is partly because labelled graphs are easier to handle but also because, fortunately, most results about labelled random graphs can easily be translated to results about unlabelled graphs. This will be shown in Section 6.

We shall consider graphs of order n and the sake of simplicity we shall take $V = \{1, 2, \ldots, n\}$ as the vertex set. We write $N = \binom{n}{2}$ and we denote by $\mathbf{G}^n$ the set of all 2^N graphs with vertex set V.

When studying random graphs we tend to consider one of two models, $\mathbf{G}(n,M)$ and $\mathbf{G}(n,P(edge) = p)$. The first consists of all graphs with vertex set V which have M edges. In this model any two graphs have the same probability. Since $\mathbf{G}(n,M)$ has $\binom{N}{M}$ elements, every element has probability $\binom{N}{M}^{-1}$.

Almost always M is a function of n: $M = M(n)$. If we want to emphasise that the probability and expectation are taken in $\mathbf{G}(n,M(n))$ then we write $P_M(\mathbf{A}) = P_{M(n)}(\mathbf{A})$ and $E_M(X) = E_{M(n)}(X)$ instead of $P(\mathbf{A})$ and $E(\mathbf{A})$.

In the second model, $\mathbf{G}(n,P(edge) = p)$ or simply $\mathbf{G}(n,p)$ we have $0 < p < 1$, and the model consists of all graphs with vertex set V in which the edges are chosen independently and with probability p. Thus this model has 2^N points and if $G_0 \in \mathbf{G}^n$ has m edges then

$$P(\{G_0\}) = P(G = G_0) = p^m q^{N-m},$$

where G stands for a random element of $\mathbf{G}(n,p)$. Once again, we may emphasise the dependence on p by writing $P_p(\mathbf{A})$ and $E_p(\mathbf{A})$.

In the model $\mathbf{G}(n,p)$ the probability p may be a constant or it may depend on n. Note that $\mathbf{G}(n,1/2)$ is exactly $\mathbf{G}^n$ in which any two graphs are equiprobable.

It is rather convenient to abbreviate $\mathbf{G}(n,M(n))$ to $\mathbf{G}_M$, $\mathbf{G}(M)$ or $\mathbf{G}(M(n))$ and $\mathbf{G}(n,P(edge) = p)$ to $\mathbf{G}_p$, $\mathbf{G}(p)$ or $\mathbf{G}(p(n))$. Furthermore, one may write G_p or G_M for random graphs from $\mathbf{G}(n,p)$ or $\mathbf{G}(n,M)$. Thus $P(G_p$ has an isolated vertex) is the probability that a graph in $\mathbf{G}(n,p)$ has an isolated vertex. Finally, it is convenient to write '*r.g.*' instead of 'random graph'.

A subset Q of $\mathbf{G}^n$ is said to be a *property of graphs of order n* if $G \in Q$, $H \in \mathbf{G}^n$ and $G \simeq H$ imply that $H \in Q$. The statement "G has Q" is then equivalent to "$G \in Q$". Thus the property of being connected is identified with the set $\{G: G \in \mathbf{G}^n$ and G is connected $\}$. If Q is any property then $P_M(Q)$ (resp. $P_p(Q)$) denotes the probability that a graph of $\mathbf{G}(n,M(n))$ (resp. $\mathbf{G}(n,p)$) has Q.

A property Q is said to be *monotone increasing* or simply *monotone* if whenever $G \in Q$ and $G \subset H$ then also $H \in Q$. Thus the property of containing a certain subgraph, for instance a triangle, connected graph of order six or a Hamilton path, is monotone increasing. We call Q *convex* if $F \subset G \subset H$ and $F \in Q$, $H \in Q$ imply that $G \in Q$.

Let Ω_n be a model of random graphs of order n. (So usually $\Omega_n = \mathbf{G}(n,M(n))$ or $\Omega_n = \mathbf{G}(n,p)$.) We say that *almost every* (*a.e.*) graph in Ω_n has a given property Q if $P(Q) \to 1$ as $n \to \infty$. (Occasionally 'almost all' (*a.a.*) is used instead of 'almost every'.)

It is heuristically obvious that a monotone increasing property occurs with greater probability if our random graphs have (or are likely to have) more edges.

THEOREM 11. Suppose Q is a monotone increasing property, $0 \leq M_1 < M_2 \leq N$ and $0 \leq p_1 < p_2 \leq 1$. Then

$$P_{M_1}(Q) \leq P_{M_2}(Q) \text{ and } P_{p_1}(Q) \leq P_{p_2}(Q).$$

PROOF.

(i) Let us select M_2 edges one by one. The resulting graph will certainly have property Q if the graph formed by the first M_1 edges only has the property. Also, for any $G_0 \in \mathbf{G}(n,M_1(n))$ the probability that the first M_1 edges of a *r.g.* G_{M_2} form exactly G_0 is independent of G_0.

(ii) Put $p = (p_2-p_1)/(1-p_1)$. Choose independently $G_1 \in \mathbf{G}(n,p_1)$ and $G \in \mathbf{G}(n,p)$ and set $G_2 = G_1 \cup G$. Then the edges of G have been chosen independently and with probability $p_1 + p - p_1 p = p_2$, so G is exactly an element of $\mathbf{G}(n,p_2)$. As the property Q is monotone increasing, if G_1 has Q so has G_2. Hence $P_{p_1}(Q) \leq P_{p_2}(Q)$, as claimed. □

Usually it is more pleasant to work with the model $\mathbf{G}(n,P(edge) = p)$. On the other hand, results concerning $\mathbf{G}(n,M(n))$ tend to be sharper. Thus it is rather fortunate that in most investigations the models $\mathbf{G}(n,M(n))$ and $\mathbf{G}(n,P(edge) = p)$ are practically interchangeable, provided $M(n)$ is close to pN.

THEOREM 12.

(i) Let Q be any property and suppose $pqN \to \infty$. Then the following two assertions are equivalent.

  (a) Almost every graph in $\mathbf{G}(n,p)$ has Q.

  (b) Given $x > 0$ and $\epsilon > 0$, if n is sufficiently large, then there are $l \geq (1-\epsilon) 2x(pqN)^{1/2}$ integers $M_1, M_2, \ldots, M_l$,

$$pN - x(pqN)^{1/2} < M_1 < M_2 < \cdots < M_l < pN + x(pqN)^{1/2}$$

such that $P_{M_i}(Q) > 1 - \epsilon$ for every i, $i = 1, 2, \ldots, l$.

(ii) If Q is a convex property and $pqn \to \infty$ then a.e. graph in $\mathbf{G}(n,p)$ has Q iff for every fixed x a.e. graph in $\mathbf{G}(n,M)$ had Q, where $M = \lfloor pN + x(pqN)^{1/2} \rfloor$.

PROOF. By the De Moivre-Laplace theorem (Theorem 1) for every fixed $x > 0$ we have

$$P_p(|e(G) - pN| < x(pqN)^{1/2}) \sim \Phi(x) - \Phi(-x)$$

where, as customary, $e(G)$ denotes the number of edges of G. Since

$$P_p(e(G) = M) = b(M;p,N) < (pqN)^{-1/2}$$

for every M, the equivalence of (a) and (b) follows. Part (ii) is a consequence of (i) and Theorem 11. □

This result shows that if we know $P_M(Q)$ with a certain accuracy for every M close to pN then we know $P_p(Q)$ with a comparable accuracy. The converse is far from being true for a general property. However, by Theorem 12 if Q is a convex property and a.e. graph in $\mathbf{G}(n,p)$ has Q then a.e. graph in $\mathbf{G}(n, \lfloor pN \rfloor)$ has Q, provided $pqN \to \infty$.

Erdős and Rényi [30], [31], [32] discovered the important fact that most monotone properties appear rather suddenly: for some $M = M(n)$ almost no G_M has Q while for a 'slightly' larger M almost every G_M has Q. This phenomenon is usually described in terms of so called threshold functions. Given a monotone increasing property Q a function $M^*(n)$ is said to be a *threshold function* for Q if

$$M(n)/M^*(n) \to 0 \text{ implies that almost no } G_M \text{ has } Q$$

and

$$M(n)/M^*(n) \to \infty \text{ implies that almost every } G_M \text{ has } Q.$$

A priori there is no reason why a property should have a threshold function. Nevertheless, the properties commonly occurring in graph theory do have threshold functions. In fact, often one can do considerably better than determine a mere threshold function. For many a property Q one can determine an exact limiting probability distribution, that is for every x, $0 < x < 1$, one can find a function $M_x(n)$ such that $P(G_{M_x} \text{ has } Q) \to x$ as $n \to \infty$.

Clearly threshold functions are not unique. However, threshold functions are unique within factors $m(n)$ bounded away from 0 and ∞ ($0 < \underline{\lim} m(n) \le \overline{\lim} m(n) < \infty$). In other words, if M_1^* is a threshold function of Q then a function M_2^* is also a threshold function iff $M_2^* = 0(M_1^*)$ and $M_1^* = 0(M_2^*)$. In this sense we may speak of *the* threshold function of a property.

If M is neither too small nor too close to $N = \binom{n}{2}$ then *a.e.* G_M has the pleasant property that any fixed graph H can be embedded in it vertex by vertex. To be more precise, in a large range of $M = M(n)$ it is true that whenever $F \subset H$ are given graphs then *a.e.* G_M is such that if G_M has an induced subgraph F^* isomorphic to F then G_M also has an induced subgraph H^* containing F^* and for which there is an isomorphism $H^* \simeq H$ extending the isomorphism $F^* \simeq F$. This is an immediate consequence of the following result. Though this result is very simple, its proof (which will not be given here) is rather cumbersome because one is forced to work with the model $G(n,M)$ for the property we seek is not convex.

THEOREM 13. Suppose $M = M(n)$ is such that for every $\epsilon > 0$ we have

$$Mn^{-2+\epsilon} \to \infty \text{ and } (N-M)n^{-2+\epsilon} \to \infty. \tag{6}$$

Then for every natural number k almost every G_M is such that whenever W_1, W_2 are disjoint k-sets of vertices, there is a vertex

$z \in V - W_1 \cup W_2$ joined to every vertex in W_1 and to none in W_2. □

By Theorem 12 (ii) the assertion of Theorem 13 holds for the model $G(n,p)$ as well, provided

$$pn^\epsilon \to \infty \quad \text{and} \quad qn^\epsilon \to \infty \tag{7}$$

for every $\epsilon > 0$. However, a direct proof of this assertion is considerably easier than the proof of Theorem 13.

The simple result above has an interesting consequence concerning properties of random graphs defined in terms of first order sentences. In the theory of graphs there is only one relation, namely adjacency, and this binary relation is such that $\neg xRx$ and $xRy \to yRx$, where xRy means that the vertex x is adjacent to the vertex y. Consequently first order sentences (that is sentences involving $=, \vee, \wedge, \exists, \forall, \neg, \to$ and R) are particularly simple in this theory. The next result, which is rather striking, is a slight extension of a theorem of Fagin [38].

THEOREM 14. *Suppose M and p satisfy (6) and (7), and Q is a property of graphs given by a first order sentence. Then either Q holds for a.e. graph in $G(n,M)$ and $G(n,p)$ or it fails for a.e. graph in $G(n,M)$ and $G(n,p)$.*

PROOF. Let us say that a graph G has property P_k if G has more than $2k$ vertices and whenever W_1, W_2 are disjoint sets of at most k vertices each, there is a vertex $z \in V(G) - W_1 \cup W_2$ joined to every vertex in W_1 and none in W_2. Furthermore, given a graph H, let $C(H)$ be the property that H is an induced subgraph. Then, clearly P_k implies $C(H)$ if $k \geq H$.

Therefore it follows from Theorem 13 that in our range of p and M for every fixed k and every fixed H almost every graph has P_k and $C(H)$.

Let us assume that, contrary to the assertion of the theorem, $\overline{\lim}P(Q) > 0$ and $\overline{\lim}P(\neg Q) > Q$. We shall construct inductively a sequence of graphs $(G_k^*)_{k=0}^\infty$ as follows. Let G_0^* be a graph with vertex set $\{1,2,\ldots,m_0\}$ which has property Q. Suppose we have constructed $G_0^*, G_1^*, \ldots, G_{m_l}^*$, such that $V(G_j) = \{1,2,\ldots,m_j\}$, $m_0 < m_1 < \cdots < m_l$, G_{j-1}^* is an induced subgraph of G_j^* and G_j^* has properties P_j and Q, $j = 1,2,\ldots,l$. Since $\lim P(C(G_l^*) \wedge P_{l+1}) = 1$, we have

$$\overline{\lim}P(Q \wedge C(G_l^*) \wedge P_{l+1}) = \overline{\lim}P(Q) > 0.$$

Therefore there is a graph G_{l+1}^* with vertex set $\{1,2,\ldots,m_{l+1}\}$, $m_{l+1} > m_l$, which contains G_l^* as an induced subgraph and which has properties P_{l+1} and Q. Then the graph $G = \bigcup_{k=0}^{\infty} G_k^*$ has vertex set **N** and it has property Q since each G_k^* is an induced subgraph of G and has property Q. Furthermore, G has property P_k for every k since the graphs $G_k^*, G_{k+1}^*, \ldots$ all have property k.

Replacing Q by $\neg Q$, we can construct a graph H with vertex set **N** which does not have Q but has P_k for every k.

To complete the proof we shall show that if G and H are graphs with countable vertex sets and they both have P_k for every k then they are isomorphic, so if one has Q so has the other.

Let $V(G) = \{g_1, g_2, \ldots\}$ and $V(H) = \{h_1, h_2, \ldots\}$ and define 'partial isomorphisms' σ_n, $\sigma_n: \{g_{i_1}, \ldots, g_{i_n}\} \to \{h_{j_1}, \ldots, h_{j_n}\}$, $\sigma_n(g_{i_k}) = h_{i_k}$, inductively as follows. Let $g_{i_1} = g_1$ and $h_{j_1} = h_1$. If $n \geq 2$ is even, pick i_{n+1} to be the least $i \geq 1$ not in $\{i_1, \ldots, i_n\}$. Since H has P_n there is a vertex $h_{j_{n+1}} \notin \{h_{j_1}, \ldots, h_{j_n}\}$ such that for all s, $1 \leq s \leq n$, we have $h_{j_{n+1}} R h_{j_s}$ iff $g_{i_{n+1}} R g_{i_s}$. If $n \geq 2$ is odd, reverse the role of G and H above. The union of the partial isomorphisms is clearly an isomorphism between G and H. □

The morale of Theorem 14 is that for large ranges of M and p first order sentences are not too interesting when applied to random graphs in $\mathbf{G}(n,M)$ and $\mathbf{G}(n,p)$. Putting it a little more precisely, if Q is a property given by a first order sentence then Q should be studied in models $\mathbf{G}(n,M)$ and $\mathbf{G}(n,p)$ for which M and p do not satisfy (6) and (7). In fact, the range of M (and p) one considers should always depend on the property to be studied. For example, the statment 'almost every graph has diameter 2' (that is '$a.e.$ $G_{1/2}$ has diameter 2') is trivial while the assertion 'for $p = (2 \log n + \log \log n)^{1/2} n^{-1/2}$ $a.e.$ G_p has diameter 2' does have some genuine content. As another, somewhat artificial example, let Q be the property that whenever x_1, x_2, x_3 and x_4 are four distinct vertices of G_M then there is a fifth vertex y which is joined to x_1, x_2 and x_3 and which is not joined to x_4. Clearly Q is a property given by a first order sentence. It is not too hard to show that if $M = M(n) = n^{3/2}(\log n)^{1/2} - \frac{1}{8}(\log 6 + \log \log 2 + o(1))n^{3/2}(\log n)^{-1/2}$ then $P(G_M$ has $Q) \to 1/2$ as $n \to \infty$.

3. The degree sequence.

The degree of a vertex is one of the simplest graph invariants. This is especially so when we consider the model $G(n,p)$. If $x \in V$ is fixed then $d_G(x)$, as a function of $G \in G(n,p)$, has binomial distribution with parameters $n-1$ and p. Furthermore, if $x,y \in V$ and $x \neq y$ then $d_G(x)$ and $d_G(y)$ are rather close to being independent random variables on $G(n,p)$. In view of this it is to be expected that one can get rather detailed information about the degree sequence. Erdős and Rényi [29] were the first to study the distribution of the maximum (minimum) degree, and other aspects of the degree sequence were investigated by Ivchenko [48], Erdős and Wilson [36], Bollobás [10], [11], [16] and Cornuejols [26].

Let us write $(d_i)_1^n$ for the degree sequence of a graph $G \in G(n,p)$, with the degrees arranged in descending order: $d_1 \geq d_2 \geq \cdots \geq d_n$. Thus $d_1(G) = \Delta(G)$ is the maximum degree and $d_n(G) = \delta(G)$ is the minimum degree. Depending on the magnitude of $p = p(n)$, a number of fascinating phenomena may occur. For example in a certain range $a.e.$ graph has a unique vertex of minimum degree. In what follows we shall give a sample of these results.

The study of the degree sequence is based on the properties of some related random variables. Let us write $X_k = X_k(G)$ for the number of vertices of degree k, Y_k for the number of vertices of degree at least k and Z_k for the number of vertices of degree at most k. Thus

$$Y_k = \sum_{l \geq k} X_l \text{ and } Z_k = \sum_{l \leq k} X_l.$$

Clearly

$$d_r \geq k \text{ iff } Y_k \geq r$$

and

$$d_{n-r} \leq k \text{ iff } Z_k \geq r+1.$$

If $p = o(n^{-3/2})$ then

$$E(Y_2) = \sum_{j=2}^{n-1} E(X_j) \leq n \sum_{j=2}^{n-1} \binom{n}{j} p^j \leq \sum_{j=2}^{\infty} (pn)^j/j! = o(1),$$

so $a.e.$ $G_{n,p}$ consists of independent edges. Hence if also $pn^2 \to \infty$ then $a.e.$ $G_{n,p}$ has degree sequence $d_1 = d_2 = \cdots = d_{2M} = 1$, $d_{2M+1} = \cdots = d_n = 0$ where $M = e(G_{n,p}) \sim pn^2/2$. Therefore in the discussion of the degree sequence we may assume that neither $p = o(n^{-3/2})$ nor $1-p = o(n^{-3/2})$ holds.

The following theorem is the basis of a number of other results.

THEOREM 15. Let $\epsilon > 0$ be fixed, $\epsilon n^{-3/2} \leq p = p(n) \leq 1 - \epsilon n^{-3/2}$, let $k = k(n)$ be a natural number and set $\lambda_k = \lambda_k(n) = nb(k; n-1, p)$. Then the following assertions hold.

(i) If $\lim \lambda_k(n) = 0$ then $\lim P(X_k = 0) = 1$.
(ii) If $\lim \lambda_k(n) = \infty$ then $\lim P(X_k \geq t) = 1$ for every fixed t.
(iii) If $0 < \underline{\lim} \lambda_k(n) \leq \overline{\lim} \lambda_k(n) < \infty$ then X_k has asymptotically Poisson distribution with mean λ_k:

$$P(X_k = r) \sim e^{-\lambda_k} \lambda_k^r / r!$$

for every fixed r.

PROOF. We may assume without loss of generality that $p \geq 1/2$. Note that $\lambda_k = \lambda_k(n)$ is exactly the expectation of $X_k(G)$. Hence

$$P(X_k \geq 1) \leq E(X_k) = \lambda_k(n),$$

so the first assertion follows.

From now on we suppose that $\underline{\lim} \lambda_k(n) > 0$. The remaining assertions of the theorem will follow from the fact for every fixed $r \geq 1$ the rth factorial moment $E_r(X_r)$ of X_k is asymptotic to $\lambda_k(n)^r$.

Recall that $E_r(X_k)$ is the expected number or ordered r-tuples of vertices $(x_1, x_2, \ldots, x_r)$ such that each vertex x_i has degree k. What is the probability that r given vertices $x_1, x_2, \ldots, x_r$ all have degree k? Suppose we have chosen the edges joining the x_i's. If there are l such edges and vertex x_i is joined to $d_i \leq k$ vertices x_j, $1 \leq j \leq r$, then $\sum_{i=1}^{r} \tilde{d}_i = 2l$ and x_i has to be joined to $k - \tilde{d}_i$ vertices outside the set $\{x_1, x_2, \ldots, x_r\}$. The probability of this event is

$$\prod_{i=1}^{r} b(k - \tilde{d}_i; n-r, p).$$

Consequently with $R = \binom{r}{2}$ we have

$$E_r(X_k) \leq (n)_r \sum_{l=0}^{R} \binom{R}{l} p^l q^{R-l} \max_{\Sigma d_i = 2l} \prod_{i=1}^{r} b(k - d_i^*; n-r, p), \tag{8}$$

where the maximum is over all sequences $d_1^*, d_2^*, \ldots, d_r^*$ with $\sum_1^r d_i^* = 2l$ and $0 \leq d_i^* \leq \min\{r-1, k\}$.

Let us distinguish two cases according to the size of $p = p(n)$. Suppose first that $p(n)$ is bounded away from 0. Then it is easily seen that the function $k(n)$ has to be about pn, in fact

$$|k(n)-pn| = 0(n \log n)^{1/2}.$$

Furthermore, for every $r \in \mathbf{N}$ there is a function $\phi_r(n) \to 0$ such that

$$\frac{b(k-d;n-r,p)}{b(k;n-1,p)} \leq 1 + \phi_r(n)$$

whenever $0 \leq d \leq \min\{r-1,k\}$. Hence (8) implies

$$E_r(X_k) \leq (n)_r \sum_{l=0}^{R} \binom{R}{l} p^l q^{R-l} \{b(k;n-1,p)(1+\phi_r(n))\}^r$$

$$\leq (1 + \phi_r(n))^r \lambda_k^r. \qquad (9)$$

Now suppose that $p = o(1)$. Clearly if $0 \leq k \leq \min\{r-1,k\}$ and n is sufficiently large then

$$\frac{b(k-d;n-r,p)}{b(k;n-1,p)} \leq \frac{b(k-d;n-r,p)}{b(k;n-r,p)} \leq \frac{(k)_d}{(n-r-k)_d} p^{-d} < 2(\frac{k}{pn})^d.$$

Therefore (8) gives

$$E_r(X_k) \leq n^r b(k;n-r,p)^r \{1 + \sum_{l=1}^{R} \binom{R}{l} p^l 2^r (\frac{k}{pn})^{2l}\}$$

$$\leq n^r b(k;n-r,p)^r \{1 + 2^{r^2} \sum_{l=1}^{R} (\frac{k^2}{pn^2})^l\}. \qquad (10)$$

Our assumption that $\varliminf \lambda_k(n) > 0$ implies that $k^2 = o(pn^2)$. Indeed, if for some fixed $\eta > 0$ the inequality $k \geq \eta p^{\frac{1}{2}} n$ held for arbitrarily large n then we would have

$$\varliminf \lambda_k(n) = \varliminf n\binom{n}{k} p^k \leq \varliminf n(\frac{en}{k})^k p^k$$

$$= \varliminf n(\frac{epn}{k})^k \leq \varliminf n(\frac{ep^{1/2}}{\eta})^k \leq \varliminf n 2^{-k}$$

$$\leq \varliminf n 2^{-n^{1/5}} = 0,$$

contrary to our assumption. By $k^2 = o(pn^2)$ inequality (10) implies

$$E_r(X_k) \leq n^r b(k;n-r,p)^r (1+o(1)) = \lambda_k^r (1 + o(1)). \qquad (9')$$

By considering only independent r-tuples of vertices of degree k, we find that

$$E_r(X_r) \geq (n)_r b(k;n-r,p)^r = \lambda_k^r (1 + o(1)),$$

so (9) and (9') imply

$$E_r(X_k) = \lambda_k^r(1 + o(1)).\tag{11}$$

Thus the rth factorial moment of X_k is asymptotic to the rth factorial moment of $P(\lambda_k)$. Now if $\lim \lambda_k(n) = \infty$ then $E_2(X_k) \sim \lambda_k^2$ implies that $E(X_k^2) = \lambda_k^2(1 + o(1))$ so Chebyshev's inequality gives

$$\lim_{n \to \infty} P(X_k \geq t) = 1$$

for every fixed t.

Finally, if $\overline{\lim} \lambda_k < \infty$ then by Theorem 9 relation (11) implies that asymptotically X_k has Poisson distribution with mean λ_k. □

As an immediate consequence of Theorem 15 we see that if almost every G_p has a vertex of degree k then for every fixed t almost every G_p has at least t vertices of degree k.

The proof of Theorem 15 can be modified to give analogous results about Y_k and Z_k.

THEOREM 16. *Let $\epsilon > 0$ be fixed, $\epsilon n^{-3/2} \leq p = p(n) \leq 1 = \epsilon n^{-3/2}$, let $k = k(n)$ be a natural number and set*

$$\mu_k = nB(k; n-1, p) \quad \text{and} \quad \nu_k = \{1 - B(k+1; n-1, p)\},$$

where

$$B(l; m, p) = \sum_{j \geq l} b(j; m, p).$$

Then the assertions of Theorem 1 hold if we replace X_k and λ_k by Y_k and μ_k or Z_k and ν_k. □

This last result enables us to determine the distribution of the first few and the last few values in the degree sequence. First we present some results concerning the case when p is neither too large nor too small.

THEOREM 17. *Suppose $p = p(n)$ is such that $pqn/(\log n)^3 \to \infty$. Let c be a fixed positive constant, m a fixed natural number and let $x = x(n,c)$ be defined by*

$$\frac{1}{\sqrt{2\pi}} \frac{n}{x} e^{-\frac{1}{2}x^2} = c.\tag{12}$$

Then

$$\lim_{n \to \infty} P(d_m < pn + x(pqn)^{\frac{1}{2}}) = e^{-c} \sum_{k=0}^{m-1} \frac{c^k}{k!}.$$

PROOF. Put $K = \lceil pn + x(pqn)^{\frac{1}{2}} \rceil$. Then $d_m < pn + x(pqn)^{\frac{1}{2}}$ means that at most $m-1$ vertices have degree at least K, that is

$$Y_K \leq m - 1.$$

By Theorem 16 the distribution of Y_K tends to $P(\mu_K)$, where

$$\mu_K = nB(K; n-1, p).$$

Now by the definition of x we have

$$x(pqn)^{\frac{1}{2}} = o((pqn)^{2/3}),$$

so by Theorem 2

$$\mu_K \sim (c/n)n = c.$$

This implies the assertion of the theorem. □

With a little work we can rewrite the connection between x and c in Theorem 17 to yield the following result.

COROLLARY 18. *Suppose $pqn/(\log n)^3 \to \infty$ and y is a fixed real number. Then the maximal degree $\Delta = d_1$ of G_p satisfies*

$$\lim_{n\to\infty} P(\Delta < pn + (2pqn \log n)^{\frac{1}{2}}\{1 - \frac{\log\log n}{4\log n} - \frac{\log(2\pi^{\frac{1}{2}})}{2\log n} + \frac{y}{2\log n}\}) = e^{-e^{-y}}. \;\square$$

With the aid of Theorem 2 and Chebyshev's inequality one can locate the mth largest degree of G_p with fair accuracy.

THEOREM 19. *Suppose $0 < p < 1$ is fixed, $m = m(n) = o(n)$, $m(n) \to \infty$ and $\omega(n)$ tends to ∞ arbitrarily slowly. Define x by*

$$1 - \Phi(x) = \frac{m}{n}.$$

Then a.e. G_p satisfies

$$|d_m - pn - x(pqn)^{1/2}| \leq \omega(n)\left(\frac{pqn}{m \log(n/m)}\right)^{1/2}. \;\square$$

If m does not grow too fast then the function $x = x(m)$ appearing in Theorem 19 can be calculated rather precisely and one is led to the following corollary.

COROLLARY 20. *Let $0 < p < 1$ be fixed, let $\omega(n) \to \infty$ and $m = m(n) = 0((\log n)/(\log\log n)^4)$. Then a.e. G_p satisfies*

$$|d_m - (2pqn \log n)^{1/2} + (\log \log n)\left(\frac{pqn}{8 \log n}\right)^{1/2}$$
$$+ \log (2\pi^{1/2} m)\left(\frac{pqn}{2 \log n}\right)^{1/2} | \leq \omega(n)\left(\frac{pqn}{m \log n}\right)^{1/2}. \quad \square$$

It is rather interesting that, unless p is very close to 0 or 1, $a.e.$ G_p has a unique vertex of maximum (minimum) degree and, even more, the first several values of the degree sequence are strictly decreasing.

THEOREM 21. *If $pn/\log n \to \infty$ and $qn/\log n \to \infty$ then $a.e.$ G_p has a unique vertex of maximum degree and a unique vertex of minimum degree.* $\square$

THEOREM 22. *Suppose $m = o(pqn/\log n)^{1/4}$, $m \to \infty$ and $\alpha(n) \to 0$. Then $a.e.$ G_p is such that*
$$d_i - d_{i+1} \geq \frac{\alpha(n)}{m^2}\left(\frac{pqn}{\log n}\right)^{1/2}$$
for every $i < m$. $\square$

The phenomenon described in the last theorem has a beautiful and unexpected application to the graph isomorphism problem. The question is: how fast an algorithm can one give for deciding whether two given graphs are isomorphic or not. It is not known whether there is such an algorithm whose running time is polynomial in the number of vertices. However, Babai, Erdős, and Selkow [2] made use of the jumps in the degree sequence to give an algorithm which, for almost every graph G of order n, tests every graph H in time $0(n^2)$, whether G is isomorphic to H.

Let **K** be a set of graphs with vertex set $V = \{1, 2, \ldots, n\}$. A *canonical labelling algorithm* of **K** is an algorithm assigning to each $G \in \mathbf{K}$ a permutation π_G of V such that G and H in **K** are isomorphic if and only if $\pi_G(G)$ and $\pi_H(H)$ coincide. In other words, a canonical labelling algorithm relabels the vertices in such a way that two graphs are isomorphic if and only if the new labelled graphs coincide.

The next result is a slight extension of the result of Babai, Erdős and Selkow [2] mentioned above.

THEOREM 23. *Suppose $0 < p = p(n) \leq \frac{1}{2}$ is such that $p^5 n/(\log n)^5 \to \infty$. There is an algorithm which, for almost every graph G_p and every graph H, tests in $0(pn^2)$ time whether or not H is isomorphic to G_p.*

PROOF. First we give an algorithm which constructs a set $\mathbf{K} \subset \mathbf{G}(n,p)$ such that $P(\mathbf{K}) \to 1$. Then we construct a canonical labelling algorithm for $\mathbf{K}$. Both these algorithms will work in $0(pn^2)$ time.

Set $m = \lceil 3 \log_{(1/q)} n \rceil$. Compute the degrees of the vertices of G_p and then order the vertices as $x_1, x_2, \ldots, x_n$ so that $d(x_1) = d_1 \geq d(x_2) = d_2 \geq \cdots \geq d(x_n) = d_n$. If $e(G_p) \geq pn^2$ or $d_i \leq d_{i+1} + 2$ for some $i \leq m$ then put G_p into $\mathbf{G}(n,p) - \mathbf{K}$ and end the algorithm. Otherwise for $i = 1, 2, \ldots, n$ compute

$$f(x_i) = \sum_{j=1}^{m} a(i,j) 2^j,$$

where $a(i,j)$ is 1 if $x_i x_j \in E(G_p)$ or $i = j$ and 0 otherwise. If $f(x_i) = f(x_j)$ for some pair i, j, $1 \leq i < j \leq n$, then put G_p into $\mathbf{G}(n,p) - \mathbf{K}$. Otherwise put G_p into $\mathbf{K}$.

It is clear that this algorithm runs in $0(pn^2)$ time. Furthermore, straightforward calculations show that the conditions of Theorem 22 are satisfied with $\alpha(n) = 3m^2 (\log n/pqn)^{1/2}$. Consequently by Theorem 22 almost no graph is discarded because $d_i \leq d_{i+1} + 2$ for some $i \leq m$. What is the probability of discarding a graph G_p because of $f(x_i) = f(x_j)$? If $f(x_i) = f(x_j)$ then x_i and x_j are joined to exactly the same vertices in $V_{ij} = \{x_1, x_2, \ldots, x_m\} - \{x_i, x_j\}$. Let W_{ij} be the set of $m-2$ vertices of highest degree in $G_p - \{x_i, x_j\}$. Then $W_{ij} \subset V_{ij}$ since $d_1 \geq d_2 + 3 \geq d_2 \geq d_3 + 3 > \cdots \geq d_m \geq d_{m+1} + 3$. Since W_{ij} is independent of the edges incident with x_i and x_j,

$P(f(x_i) = f(x_j)) \leq P(x_i \text{ and } x_j \text{ are joined to the same vertices in } W_{ij}) \leq q^{m-2} = 0(n^{-3})$.

Consequently the probability of discarding a graph the second time round is not more than $\binom{n}{2} q^{m-2} = 0(n^{-1})$. Hence $P(\mathbf{K}) \to 1$ as claimed.

The required fast canonical labelling algorithm is easily constructed. Let $x_{\pi(1)}, x_{\pi(2)}, \ldots, x_{\pi(n)}$ be the order of the vertices according to their f-values: $f(x_{\pi(1)}) > f(x_{\pi(2)}) > \cdots > f(x_{\pi(n)})$ and label $x_{\pi(i)}$ by $\pi(i)$. Clearly two graphs in $\mathbf{K}$ are isomorphic iff the new labelled graphs coincide. □

In the theorem above we did not care about $1 - P(\mathbf{K})$, the probability of rejection. All we wanted to ensure was the $1 - P(\mathbf{K}) = o(1)$. Inspired by the above theorem of Babai, Erdős and Selkow, a number of fast algorithms were obtained by Lipton [58], Karp [53] and Babai and Kučera [3] which have exponentially small probability of rejection. A canonical labelling given by Babai and Kučera [3] runs in $0(n^2)$ time, with probability of rejection c^{-n} for some fixed $c > 1$. Furthermore, the rejected graphs are handled in such a way as to give a canonical labelling algorithm of all graphs with expected time $O(n^2)$.

4. Small subgraphs

The degree sequence of a graph with vertex set $V = \{1, 2, \ldots, n\}$ depends only on the number of edges in the sets $V_i^{(2)} = \{ij: 1 \leq j \leq n, j \neq i\} \subset V^{(2)}$, $i = 1, 2, \ldots, n$. The reason why the degree sequence of a random graph is almost determined is that there are only n of these sets and they close to being disjoint. The property of containing a fixed graph F has somewhat similar advantages in the context of random graphs: it depends on the distribution of edges in fairly few, fairly small and essentially independent subsets of $V^{(2)}$.

Erdős and Rényi [30], [31], [32] were the first to study the appearance of small subgraphs; further results were obtained by Schürger [77], Bollobás [15], Palka [68] and Barbour [8]. Here we shall present only a selection of the results.

The most complete results concern *balanced* and *strictly balanced* graphs. Loosely speaking these are the graphs which do not contain a denser subgraph.

The *average degree* or simply *degree* of a graph F of order k and size l is $d(F) = 2l/k$. The *maximum average degree* of a graph H is defined as

$$m(H) = \max\{d(F): F \subset H\}.$$

A graph H is balanced if $m(H) = d(H)$ and it is *strictly balanced* if $F \subset H$ and $d(F) = m(H)$ imply that $F = H$. Note that trees, cycles, complete graphs and complete bipartite graphs are strictly balanced, the union of at least two vertex disjoint complete graphs of order r is balanced but not strictly balanced, and the vertex disjoint union of a cycle and a tree is not balanced.

The significance of strictly balanced graphs is that it is fairly easy to determine the asymptotic distribution of the number of subgraphs of a random graph isomorphic to a given strictly balanced graph. The next results, from Bollobás [15], extend some results of Erdős and Rényi [30]. It will be convenient to call a graph F an H-graph if $F \simeq H$. Furthermore, we shall say that F is an H-subgraph of G if $F \subset G$ and $F \simeq H$.

THEOREM 24. Suppose H is a fixed strictly balanced graph with k vertices and $l \geq 2$ edges and with an automorphism having a elements. Let c be a positive constant and set $p = cn^{-k/l}$. Denote by $X = X(G^p)$ the number of H-subgraphs of a random graph G_p. Then $X \xrightarrow{d} P(c^l/a)$,

that is for $r = 0, 1, \ldots$ we have
$$\lim_{n \to \infty} P_p(X = r) = d^{-\lambda} \frac{\lambda^r}{r!},$$
where $\lambda = c^l/a$.

PROOF. Note first that
$$E(X) = \binom{n}{k} \frac{k!}{a} p^l \sim \frac{n^k}{a} p^l = \frac{c^l}{a} = \lambda. \tag{13}$$

Indeed, there are $\binom{n}{k}$ ways of selecting the k vertices of an H-graph from $V = \{1, 2, \ldots, n\}$ and then there are $k!/a$ ways of choosing the edges of an H-graph with a given vertex set. The probability of selecting l given edges is p^l.

By Theorem 9 it suffices to show that for every fixed r, $r = 0, 1, \ldots$, the rth factorial moment $E_r(X)$ of X tends to λ^r as $n \to \infty$.

Now let us estimate $E_r(X)$, the expected number of ordered r-tuples of H-graphs. Denote by $E_r'(X)$ the expected number of ordered r-tuples of H-graphs having disjoint vertex sets $V_1, V_2, \ldots, V_r$. Then clearly
$$E_r'(X) = \binom{n}{k}\binom{n-k}{k} \cdots \binom{n-(r-1)k}{k}\left(\frac{k!}{a}\right)^r p^{rl} = \frac{(n)_{rk}}{a^r} p^{rl} \sim \lambda^r. \tag{14}$$

Therefore all we need is that $E_r''(X) = E_r(X) - E_r'(X) = o(1)$. This will follow from the fact that H is strictly balanced.

Suppose the graphs A and B are such that B is an H-graph and it has exactly t vertices not belonging to A. If $0 < t < k$ then
$$e(A \cup B) \geq e(A) + tl/k + 1/k. \tag{15}$$
Indeed, since H is strictly balanced, fewer than $(k-t)l/k$ edges of B join its $k-t$ vertices in $V(A) \cap V(B)$.

Let $H_1, H_2, \ldots, H_r$ be H-graphs with $|\bigcup_{i=1}^r V(H_i)| = s < rk$. Suppose H_j has t_j vertices not belonging to $\bigcup_{i=1}^{j-1} H_i$, so that $\sum_{j=2}^r t_j = s - k$. We assume that $0 < t_2 < r$ so by (15)
$$e(H_1 \cup H_2) \geq l + t_2 l/k + 1/k.$$
Since also

$$e\left(\bigcup_{i=1}^{j} H_i\right) \geq e\left(\bigcup_{i=1}^{j-1} H_i\right) + t_j l/k,$$

we have

$$e\left(\bigcup_{i=1}^{r} H_i\right) \geq sl/k + 1/k.$$

Consequently, counting rather crudely, for $r \geq 2$ we have

$$E_r''(X) \leq \sum_{s=k}^{kr-1} \binom{n}{s}\binom{s}{k}\left(\frac{k!}{a}\right)^r p^{sl/k+1/k}$$

$$= \sum_{s=k}^{kr-1} 0(n^2 p^{sl/k+1/k}) = \sum_{s=k}^{kr-1} 0(n^{-1/l}). \tag{16}$$

In particular, $E_r''(X) = o(1)$, so the result follows. □

Because of Theorem 11 the result above implies the corresponding assertion about the model $G(n, M(n))$.

COROLLARY 25. *Let H, k, l, a, c and X be as in Theorem 24 and let $M(n) \sim \frac{c}{2} n^{2-k/l}$. Then*

$$\lim_{n \to \infty} P_m(X = r) = e^{-\lambda} \frac{\lambda^r}{r!}$$

for every r, $r = 0, 1, \ldots$.

PROOF. Let $0 < c_1 < c < c_2$ and $p_i = c_k n^{-k/l}$, $i = 1, 2$. Since $X(G) \leq X(G')$ whenever $G \subset G'$,

$$P_{p_1}(X \geq r) \leq P_M(X \geq r) \leq (1+o(1))P_{p_2}(X \geq r).$$

Hence the result follows from Theorem 1 if we recall that c_1 and c_2 are arbitrary numbers satisfying $0 < c_1 < c < c_2$. □

If our strictly balanced graph H is a tree then the subgraphs isomorphic to H turn out to be components of G_p, giving the following results.

COROLLARY 26. *Let H be a tree of order $k \geq 3$, let a be the order of the automorphism group of H and let $c > 0$. Then for $p = cn^{-k/(k-1)}$ a.e. G_p is such that every subgraph of G_p isomorphic to H is component of G_p. In particular, the number of components of G_p isomorphic to H tends to P_λ where $\lambda = c^{k-1}/a$.*

PROOF. Note first that the probability that G_p contains a cycle of length at most k is $0(n^{-k/(k-1)})$. Furthermore, since there are $(k+1)^{k-1}$ labelled trees of order $k+1$, the probability that G_p contains a tree of order $k+1$ is at most

$$\binom{n}{k+1}(k+1)^{k-1}p^k = 0(n^{-1/(k-1)}). \quad \square$$

COROLLARY 27. Let $k \geq 3$ be fixed, $c > 0$, $p = cn^{-k/(k-1)}$ and $\lambda = c^{k-1}k^{k-2}/k!$. Denote by X the number of trees of order k in G_p, by C the number of components of G_p with k vertices and by T the number of tree components of order k. Then $X \xrightarrow{d} P_\lambda$, $C \xrightarrow{d} P_\lambda$, $T \xrightarrow{d} P_\lambda$, $T \xrightarrow{d} P_\lambda$ and $P(X = C = T) \to 1$. $\quad \square$

If p is larger than in the range given in Theorem 24 then the expected number of H-subgraphs of G_p tends to infinity too fast then X is still an asymptotically Poisson random variable. To be precise, Barbour [8] proved the following beautiful result about Poisson approximation.

THEOREM 28. Suppose H is a fixed strictly balanced graph with k vertices, $l \geq 2$ edges and with a elements in its automorphism group. Denote by $X = X(G_p)$ the number of H-subgraphs of G_p. If

$$\lambda_n = \frac{(n)_k}{a}p^l \to \infty$$

and

$$p = o(n^{-k/\{l+1/(k-2)\}})$$

then

$$\sup\{|P(X \in A) - e^{-\lambda_n}\sum_{k \in A}\frac{\lambda_n^k}{k!}| : A \subset \{0,1,\ldots\}\}$$
$$= 0(n^{k-2}p^{(k-2)l/k+1/k}) = o(1). \quad \square$$

Now if p grows above the range covered in Theorem 28, then for a while X becomes asymptotically normally distributed. This follows from Theorem 10 and the fact that according to the proof of Theorem 24 if p is not too large then $E(X)^r$ is a fairly precise approximation of $E_r(X)$. A result of this type was proved by Schürger [77] for complete subgraphs.

The results concerning the appearance of a subgraph which is not strictly balanced are less sharp and the best results would take a little while to state. However, if one is interested only if the threshold function of a given fixed graph, then the situation is very simple for, as shown in Bollobás [15], the threshold function depends only on the maximum average degree.

THEOREM 29. The threshold function of the property of containing a fixed non-empty graph F is $n^{2-2/m}$, where $m = m(F)$. Equivalently,

$$\lim_{n\to\infty} P(G_M \supset F) = \begin{cases} 0 & \text{if } Mn^{-2+2/m} \to 0, \\ 1 & \text{if } Mn^{-2+2/m} \to \infty. \end{cases} \qquad \square$$

5. The evolution of sparse graphs

Following Erdős and Rényi [30], one often views a random graph G_M as a living organism, which develops by acquiring more and more edges in a random fashion. In this terminology, one's main aim is to determine at what stage of the development various properties are likely to appear and, in general, what does a 'typical' graph look like at a certain time in its evolution.

Let us give a precise meaning to the evolution of a random graph. A *graph process* with vertex set $V = \{1,2,\ldots,n\}$ is a sequence $(G_t)_0^N$ such that (i) each G_t is a graph on V with t edges, (ii) $G_0 \subset G_1 \subset \cdots \subset G_N$. Let $\tilde{\mathbf{G}}$ be the set of all graph processes. Clearly $|\tilde{\mathbf{G}}| = N!$. Turn $\tilde{\mathbf{G}}$ into a probability space by giving all members of it the same probability and write $\tilde{G}$ for random elements of $\tilde{\mathbf{G}}$. Furthermore, call G_t the *state* of the process $\tilde{G} = (G_t)_0^N$ at time t. The definition implies that a (random) graph process is a Markov chain whose states are graphs on V. This Markov chain is a precise model of the evolution of a random graph mentioned above.

The spaces $\tilde{\mathbf{G}}$ and $\mathbf{G}(n,M)$ are closely connected. Indeed, the map $\tilde{\mathbf{G}} \to \mathbf{G}(n,M)$, defined by

$$\tilde{G} = (G_t)_0^N \to G_M,$$

is measure preserving: if $H \in \mathbf{G}(n,M)$ then

$$P\{\tilde{G}: \tilde{G} = (G_t)_0^N, G_M = H\} = P_M(G_M = H).$$

In view of this it does not cause much difficulty that, strictly speaking, G_t stands for two different objects: a random element of $\mathbf{G}(n,t)$ and the state of a random graph process $\tilde{G} = (G_t)_0^N$ at time t.

At the first sight it is fairly surprising that for most values of t most random graphs G_t are rather similar. In other words, there *is* a *typical* random graph of size t. Erdős and Rényi [30] described the structure of a *typical* random graph G_t for most values of t. The most surprising phenomenon they discovered concerns the sudden change in the structure around time $t = n/2$. They proved that if $0 < c < 1/2$ is fixed and $t \sim cn$ then a.e. G_t is such that its largest component has $O(\log n)$ vertices; if $t \sim cn$ and $c > 1/2$ then the largest component of a.e. G_t has $1 - \alpha_c + o(1))n$ vertices, where $0 < \alpha_c < 1$; and if

$t = [n/2]$ then the maximal size of a component of $a.e$ G_t has order $n^{2/3}$. (In fact, the last statement was claimed by Erdős and Rényi for $t \sim n/2$, but it does not hold in that case.)

Recently Bollobás [19] proved considerably more precise results concerning the change of the structure of G_t in the neighbourhood of the critical value $t = n/2$.

It is convenient to divide the evolution of a random graph into six stages.

STAGE 1: $t = o(n)$.

If $t = o(n^{1/2})$ the structure of G_t is not too exciting: $a.e.$ G_t consists of t isolated edges. If $t \sim cn^{1/2}$ where c is a positive constant then paths of length 2 appear with positive probability but almost no G_t contains a triangle or a component of order more than 3. If $t/n^{1/2} \to \infty$ but still $t = o(n^{2/3})$ then $a.e$ G_t is the disjoint union of paths of length 2 and independent edges. If $t \sim cn^{2/3}$ for a positive constant c then trees of order 4 appear with positive probability, etc. In general, the following result if an immediate consequence of Theorem 29 and Corollary 26.

THEOREM 30. (i) Suppose $k \geq 2$ and $t = t(n)$ is such that $tn^{-(k-2)/(k-1)} \to \infty$ and $t = o(n^{(k-1)/k})$. Then $a.e.$ G_t is such that its components are trees of order at most k and every tree of order at most k occurs as a component of G_t. In particular, $a.e.$ G_t has maximum degree $k-1$.

(ii) Suppose $k \geq 3$, c is a positive constant and $t = t(n) \sim cn^{(k-1)/k} = cn^{1-1/k}$. Then $a.e.$ G_t is such that its components are trees of order at most $k+1$, every tree of order k occurs as a component and trees of order $k+1$ occur with probability tending to $1 - e^{-\lambda}$, where $\lambda = (2c)^k (k+1)^{k-2}/k!$. Furthermore, $P(\Delta(G_t) = k) \to 1 - e^{-\mu}$ and $P(\Delta(G_t) = k-1) \to e^{-\mu}$, where $\mu = (2c)^k/k!$. □

STAGE 2: $t \sim cn$ where $0 < c < 1/2$.

In this case $a.e.$ G_t is such that its components are trees or unicyclic connected graphs; every tree occurs as a component with probability tending to 1 and the number of cycles of G_t tends in distribution to the Poisson distribution with mean

$$\lambda = \sum_{k=3}^{\infty} (2c)^k/(2k)!.$$

In particular, $P(G_t$ contains a cycle$) \to 1 - e^{-\lambda}$. Furthermore, as the following theorem shows, one has rather precise information about the distribution of components.

THEOREM 31. Suppose $0 < c < 1/2$ is a constant and $t \sim cn$. Then the number of unicyclic components of order k tends to the Poisson distribution with mean

$$\frac{(2c\,e^{-2c})^k}{k!}(1 = k = \frac{k^2}{2!} + \cdots + \frac{k^{k-3}}{(k-3)!}).$$

Almost all vertices of G_t belong to trees in the sense that if $\omega(n) \to \infty$ then *a.e.* G_t has at most $\omega(n)$ vertices in its unicyclic components and almost no G_t contains a component with more than one cycle. Furthermore, *a.e.* G_t has at most $n - t + \omega(n)$ components and the largest component of *a.e.* G_t is a tree of order

$$\frac{1}{2c-1-\log 2c}(\log n - \frac{5}{2}\log \log n + o(\log \log n)). \qquad \square$$

STAGE 3: $t \sim n/2$.

This is the stage during which substantial changes take place. The larger components of G_t start to gel and within a short space of time they produce a so called *giant* component, a component very much larger than all other components. Even more, the essential change in the structure of G_t occurs very suddenly, when t is within $n^{2/3+\epsilon}$ ($\epsilon > 0$) of $n/2$. We shall not go into the detail and state only four results from Bollobás [19]. We start with a result concerning the process before but close to time $n/2$.

THEOREM 32. Let $0 < \gamma < 1/3$, $\omega(n) \to \infty$ and $\omega(n) = o(\log \log n)$, and set

$$k_1 = \{(2 - 6\gamma) \log n - 5 \log \log n\}n^2,$$

$$k_0 = k_1 - \omega(n)n^{2\gamma} \text{ and } k_2 = k_1 + \omega(n)n^{2\gamma}.$$

Then if $x \sim n^{1-\gamma}$ and $t = n/1 - s$ then *a.e.* G_t is such that its maximal component has at least k_0 and at most k_2 vertices. $\qquad \square$

Just after time $n/2$ almost every graph process has a unique largest component. Even more, all other components are considerably smaller, in fact, of less than half the size of the giant component.

THEOREM 33. *A.e.* graph process $\tilde{G} = (G_t)_0^N$ is such that for every $t \geq t_1 = \frac{n}{2} + (\log n)^{1/2}n^{2/3}$ the graph G_t has a unique component of order at least $n^{2/3}$. The other components of G_t have at most $n^{2/3}/2$ vertices each. $\qquad \square$

The significance of the large gap in the sequence of the components guaranteed by Theorem 33 is that when the addition unites two components of G_t to produce a component of G_{t+1} then either this new component is still much smaller than the giant component of G_t to another component of G_t. Thus from time t_1 on the giant component increases with leaps and bounds by swallowing more and more of the small components. Since every vertex not in a small component belongs to the giant component, we can get precise estimates of the order of the giant component and of the orders of the next largest components.

THEOREM 34. Let γ, $\omega(n)$, k_i and s be as in Theorem 32. Then for $t = \frac{n}{2} + s$ and $m \in \mathbb{N}$ a.e. G_t is such that its giant component has

$$4s + O\{\omega(n)n/s^{1/2} + (\log n)s^2/n\}$$

vertices and each of the next m largest components has at least k_0 and at least k_2 vertices. □

We see from this result that just after time $n/2$ the addition of one edges increases on average by 4 the order of the giant component.

STAGE 4: $t \sim cn$, $c > 1/2$.

The giant component, born during the previous stage, continues to grow during this stage. The speed of this growth, which was essentially 4 during the previous stage, eventually decreases to 0. The order of the second largest component continues to decrease. All components containing cycles melt into the giant component.

THEOREM 35. Let $c > 1/2$ be a constant and let $t = [cn/2]$, $\omega(n) \to \infty$. Then a.e. G_t is such that, with the exception of at most $\omega(n)$ vertices, all vertices of G_t belong to the giant component or to components with are trees. Furthermore, $L_1(G_t)$, the order of the giant component of G_t, satisfies

$$|L_1(G_t) - \alpha(2c)n| \leq \omega(n)n^{1/2} \log n,$$

where

$$\alpha(x) = \frac{1}{x}\sum_{i=1}^{\infty} \frac{k^{k-1}}{k!}(x e^{-x})^k.$$

The second largest component has at least

$$\frac{1}{\alpha}\{\log n - \frac{5}{2} \log \log n\} - \omega(n)$$

and at most

$$\frac{1}{\alpha}\{\log n - \frac{5}{2}\log\log n\} + \omega(n)$$

vertices, where $\alpha = 2c - 1 - \log(2c)$. □

The small tree-components of G_t are successively swallowed up by the giant component. The survival time of a tree of order k which is present in G_t has asymptotically exponential distribution with mean $n/(2k)$.

STAGE 5: $t \leq \frac{1}{2}n \log n(1 + o(1))$ then $t/n \to \infty$.

The graph continues to gel and during this stage eventually everything is swallowed up by the giant component. All components, except the giant one, become trees: for $\omega(n) \to \infty$ a.e. $\tilde{G}$ is such that for $t \geq \omega(n)n$ every component of G_t, with the exception of its giant component, is a tree. The distribution of small tree-components is easily determined, giving us firm control of w_t, *the number of components of* G_t.

THEOREM 36. Suppose $0 < \epsilon$, $\omega(n) \to \infty$ and $c(n) = \log n - \omega(n) \to \infty$. Then a.e. graph process $\tilde{G}$ is such that for every t satisfying $n \leq 2t \leq c(n)n$ we have

$$|w_t - n\beta(2t/n)| < \epsilon n\beta(2t/n),$$

where

$$\beta(x) = \frac{1}{x}\sum_{k=1}^{\infty}\frac{k^{k-2}}{k!}(x e^{-x})^k.$$ □

As the process continues, only smaller and smaller trees can survive the attraction of the giant component.

THEOREM 37. Suppose c is a constant, $k \in \mathbf{N}$ and

$$t = \frac{n}{2}\{\frac{\log n}{k} + \frac{k-1}{k}\log\log n + c + o(1)\}.$$

Then a.e. G_t consists of a giant component and a number of components which are trees of order at most k. A.e. G_t contains every tree of order less than k as a component and trees of order k appear with probability tending to $1 = \exp\{-\frac{e^{-2kc}}{k.k!}\}$. □

As a particular case of this result we see that if $\omega(n) \to \infty$ then

$$\lim_{n\to\infty} P(G_t \text{ is connected}) = \begin{cases} 0 & \text{if } t = \frac{n}{2}\{\log n - \omega(n)\}, \\ 1 & \text{if } t = \frac{n}{2}\{\log n + \omega(n)\}. \end{cases}$$

This relation says that $\frac{n}{2} \log n$ is a sharp threshold function of being connected.

STAGE 6: $t \le n^2/4$ and $t/n \log n \to \infty$.

During this stage the global structure of G_t becomes rather simple. In some sense G_t looks like a very pleasant regular graph: *a.e.* G_t has minimum degree $(2 + o(1))t/n$ and maximum degree $(2 + o(1))t/n$; furthermore, the vertex connectivity is equal to the minimum degree. On the other hand, *a.e.* G_t has a unique vertex of maximum degree and a unique vertex of minimum degree.

6. The automorphism group

Most models of random graphs, including the models $G(n, P(edge) = p)$ and $G(n,M)$, consists of *labelled* graphs. This is partly because labelled graphs are easier to work with but also because most results concerning labelled random graphs can easily be carried over to unlabelled graphs. The tool enabling one to do this is an important result due to Wright [81].

Let us examine the connection between labelled and unlabelled graphs. Denote by L_M the number of labelled graphs of order n, that is the number of graphs with vertex set $V = \{1, 2, \ldots, n\}$, and denote by U_M the number of unlabelled graphs of order n. Then trivially

$$L_M = |\mathbf{G}(n,M)| = \binom{N}{M},$$

but U_M is not easily determined. There are $n!$ ways of labelling a graph of order n, but these labellings need not be all different. (For example, all $n!$ labellings of K^n produce the same labelled graph, and there are $n!/(2n)$ labelled cycles of order n.) Consequently

$$L_M \le n!\, U_M. \tag{17}$$

Wright [81] showed that in a large range of M the two sides of (17) are asymptotically equal.

THEOREM 38. $U_M \sim L_M/n!$ if and only if

$$\min\{M,N-M\}/n - \frac{1}{2}\log n \to \infty \text{ as } n \to \infty. \quad \square$$

Since $U_M = U_{N-M}$ and $L_M = L_{N-M}$, when studying the connection between U_M and L_M we may assume that $M \leq N/2$. One of the implications in Theorem 38 is practically trivial. Indeed, suppose

$$M = M(n) \leq M_\circ = \frac{n}{2}\{\log n + c\}$$

for some fixed real number c. By Theorem 15, we find that $X_\circ = X_\circ(G_{M_\circ})$, The number of isolated vertices of $G_{M_\circ}$, tends to the Poisson distribution with mean

$$n(1 - \frac{\log n + c}{n})^n \sim e^{-c}.$$

Consequently with

$$P(M) = P(G_M \text{ has at least two isolated vertices})$$

we have

$$\underline{\lim} P(M) \geq 1 - e^{-e^{-c}}(1 + e^{-c}) = \epsilon(c) > 0.$$

A graph of order n with at least two isolated vertices has at most $n!/2$ different labellings so

$$U_M \geq \frac{2}{n!}P(M)L_M + \frac{1}{n!}(1 - P(M))L_M,$$

implying

$$\underline{\lim} n!U_M/L_M \geq \underline{\lim}(1 + P(M)) \geq 1 + \epsilon(c) > 1.$$

The beauty of Theorem 38 is that the above simple necessary condition for $U_M \sim L_M/n!$ is also sufficient. Though we shall not give the details of this proof, we shall indicate what it depends on.

Let S_n be the symmetric group acting on $V = \{1,2,\ldots,n\}$. For $\sigma \in S_n$ set

$$\mathbf{G}(\sigma) = \{G \in \mathbf{G}(n,M): G \text{ is invariant under } \sigma\}$$

and

$$I(\sigma) = |\mathbf{G}(\sigma)|.$$

Furthermore, denote by $a(G)$ the order of the automorphism group of a graph G. Clearly

$$\sum_{\sigma \in S_n} I(\sigma) = |\{(G,\sigma): G \in G(\sigma)\}| = \sum_{G \in G_M} a(G),$$

where for simplicity we put G_M instead of $G(n,M)$. Since G_M contains $n!/a(G)$ graphs isomorphic to a given graph $G \in G_M$,

$$U_M = \sum_{G \in G_M} (n!/a(G))^{-1} = \frac{1}{n!} \sum_{G \in G_M} a(G) = \frac{1}{n!} \sum_{\sigma \in S_n} I(\sigma).$$

As $I(1) = L_M$, we have

$$U_M \sim L_M/n!$$

if and only if

$$\sum_{\substack{\sigma \in S_n \\ \sigma \neq 1}} I(\sigma) = o(L_M).$$

Thus to prove Theorem 38 one has to find fairly good upper bounds for $I(\sigma)$, $\sigma \in S_n$, $\sigma \neq 1$.

Now $I(\sigma)$ depends only on M and the sizes of the orbits of σ acting on $V^{(2)}$, the set of unordered pairs of vertices. Indeed, $G \in G(\sigma)$ if and only if $E(G)$, the edge set of G, is the union of full orbits of σ acting on $V^{(2)}$. Thus if σ acting on $V^{(2)}$ has orbits of size $M_1, M_2, \ldots, M_l$, then $I(\sigma)$ is the number of ways M can be represented as a sum of some of these M_i's. The proof of the theorem hinges on the fact that one can estimate $I(\sigma)$ in terms of the number of fixed points of σ acting on V.

Let $U(n,M)$ be the probability space of all unlabelled graphs of order n and size M in which any two graphs have the same probability. Given a property Q of graphs, denote by $P_M^L(Q)$ the probability that a random graph from $U(n,m)$ has property Q. As mentioned earlier, one can easily translate results about $G(n,M)$ into results about $U(n,m)$, for Theorem 38 has the following immediate consequence.

COROLLARY 39. Suppose Q is a property of graphs and

$$\min\{M, N-M\}/n - \log n/2 \to \infty.$$

Then

$$P_M^L(Q) = P_M(Q) + o(1). \quad \square$$

According to this result, unless M is very small or very large, the models $G(n,M)$ and $U(n,M)$ are practically interchangeable.

A very simple consequence of Theorem 38 is that almost every graph (that is a.e. $G_{1/2}$) has no nontrivial automorphism group.

Given a finite group Γ, for $n \geq |\Gamma|$ let $a_n(\Gamma)$ be the ratio of the number of graphs $G \in \mathbf{G}^n$ whose automorphism group is Γ to the number of graphs whose automorphism group contains Γ:

$$a_n(\Gamma) = |\{G \in \mathbf{G}^n \colon Aut\, G = \Gamma\}|/|\{G \in \mathbf{G}^n \colon Aut\, G \geq \Gamma\}|.$$

According to the remark above if I is the trivial group of order 1 then $\lim_{n \to \infty} a_n(I) = 1$. Furthermore, by an old result of Frucht [43] if Γ is any finite group then $a_n(\Gamma) > 0$ whenever n is sufficiently large. Cameron [25] studied the sequence $a_n(\Gamma)$ for an arbitrary group Γ and proved a number of beautiful and surprising results, some of which are given in the following theorem.

THEOREM 40. (i) $a(\Gamma) = \lim_{n \to \infty} a_n(\Gamma)$ exists for every finite group Γ.

(ii) $a(\Gamma) = 1$ if and only if Γ is a direct product of symmetric groups.

(iii) The set $\{a(\Gamma) \colon \Gamma$ is a finite group $\}$ is dense in $[0,1]$. □

7. Other models

The models $\mathbf{G}(n,M)$ and $\mathbf{G}(n,P(edge) = p)$ are the most frequently investigated models of random graphs but by no means the only models considered in the literature. We shall mention briefly a number of other models.

Given a graph H and a number $0 < p < 1$, the space $\mathbf{G}(H,p)$ consists of those graphs which are obtained from H by deleting the edges of H independently of each other and with probability $q = 1 - p$. Thus $\mathbf{G}(K^n,p)$ is exactly $\mathbf{G}(n,P(edge) = p)$. Other examples of H for which $\mathbf{G}(H,p)$ is frequently investigated are the infinite lattice (or its finite portion) and the graph of the k-dimensional cube (see [20, 35, 37, 44, 46, 55, 56, 57, 59, 77, 78]).

The model $\mathbf{G}(n,p_{ij})$ is a refinement of $\mathbf{G}(H,p)$. Here $0 \leq p_{ij} = p_{ji} \leq 1$, $p_{ii} = 0$, $1 \leq i < j \leq n$, and space consists of all graphs with vertex set $V = \{1,2,\ldots,n\}$ in which the edges are present independently of each other, with p_{ij} for the probability of the edge ij.

Recently a considerable amount of effort has been devoted to random regular graphs, see e.g. [9, 12, 14, 17, 18, 22, 41, 79 and 80]. As expected, $\mathbf{G}(n,r-reg)$ is the space of all r-regular graphs with vertex set $V = \{1,2,\ldots,n\}$. The study of $\mathbf{G}(n,r-reg)$ is complicated by the fact that for $r \geq 3$ there is no simple explicit formula for the number of labelled r-regular graphs of order n.

A related model is $\mathbf{G}(n,M;\delta,\Delta)$, the set of all graphs of size M on $V = \{1,2,\ldots,n\}$ having minimum degree at least δ and maximum degree at most Δ.

A model which is perhaps not very natural, but can be used to shed light on other models is $G(n,k-out)$. Let $\vec{G}(n,k-out)$ be the probability space of all directed graphs on V in which every vertex has outdegree k. Given $\vec{G} \in \vec{G}(n,k-out)$, let $\pi(\vec{G})$ be the underlying graph of $\vec{G}$. The set of graphs in $G(n,k-out)$ is $\{\pi(\vec{G}): \vec{G} \in \vec{G}(n,k-out)\}$ and the probability on $\vec{G}(n,k-out)$. One usually chooses a fixed k and lets n tend to ∞. The significance of this model is that it consists of graphs with rather few edges, namely at most kn edges, but the minimum degree is at least k. As a consequence of this, almost every graph G_{k-out} turns out to be connected for $k \geq 2$ and for k large enough almost every G_{k-out} is Hamiltonian, as proved by Fenner and Frieze [39], [40]. The last result probably holds for $k = 3$ as well.

Several other models of random graphs have been studied in great detail, including random trees [32, 47, 60-64, 66, 75], order n and size $n + k$ [1, 5, 7, 54, 74, 82-84] and random tournaments [65].

References

[1] T.L. Austin, R.E. Fagen, W.F. Penney and J. Riordan, The number of components in random linear graphs, *Ann. Math. Statist.* 30 (1959) 747-754.

[2] L. Babai, P. Erdös and S.M. Selkow, Random graph isomorphisms, *SIAM Journal of Computing* 9 (1980) 628-635.

[3] L. Babai and L. Kučera, Canonical labelling of graphs in linear average time, *20th Annual IEEE Symposium on Foundations of Computational Science*, Puerto Rico, 1979, 39-46.

[4] G.N. Bagaev, A certain distribution in a random tree I, (in Russian), *Diskret. Analiz* 21 (1972) 3-9.

[5] G.N. Bagaev, Random graphs with degree of connectedness 2, (in Russian), *Diskret. Analiz* 22 (1973) 3-14.

[6] G.N. Bagaev, A certain distribution in a random tree II, (in Russian), *Diskret. Analiz* 24 (1974) 3-7.

[7] G.N. Bagaev, Distribution of the number of vertices in a component of a connected graph with randomly eliminated edges, (in Russian), *in III Vsesoy. Konf. Probl. Teor. Kiber. Novosibirsk*, 1974, 137-140.

[8] A.D. Barbour, Poisson convergence and random graphs, *Math. Proc. Cambridge Phil. Soc.* 92 (1982) 349-359.

[9] E.A. Bender and E.R. Canfield, The asymptotic number of labelled graphs with given degree sequences, *Journal of Combinatorial Theory (A)* 24 (1978) 296-307.

[10] B. Bollobás, The distribution of the maximum degree of a random graph, *Discrete Math.* 32 (1980) 201-203.

[11] B. Bollobás, Degree sequences of random graphs, *Discrete Math.* 33 (1981) 1-19.

[12] B. Bollobás, A probabilistic proof of an asymptotic formula for the number of labelled regular graphs, *European Journal of Combinatorics* 1 (1980) 311-316.

[13] B. Bollobás, Random graphs, in *Combinatorics* (H.N.V. Temperley, ed.),

London Math. Soc. Lecture Note Series 52, Cambridge University Press, 1981, 80-102.

[14] B. Bollobás, The independence ratio of regular graphs, *Proc. Amer. Math. Soc.* 83 (1981) 433-436.

[15] B. Bollobás, Threshold functions for small subgraphs, Math. Proc. Cambridge Phil. Soc. 90 (1981) 197-206.

[16] B. Bollobás, Vertices of given degree in a random graph, *Journal of Graph Theory* 6 (1982) 147-155.

[17] B. Bollobás, The asymptotic number of unlabelled regular graphs, *J. London Math. Soc.* 89 (1982) 201-206.

[18] B. Bollobás, Almost all regular graphs are Hamiltonian, *European Journal of Combinatorics*, 4 (1983) 97-106.

[19] B. Bollobás, The evolution of random graphs, Transactions Amer. Math. Soc., to appear.

[20] B. Bollobás, The evolution of the cube, in *Combinatorial Mathematics* (C. Berge, ed.), North-Holland, 91-97.

[21] B. Bollobás, *Random Graphs*, in preparation.

[22] B. Bollobás and W.F. de la Vega, The diameter of random regular graphs, *Combinatorica*, 2 (1982) 125-134.

[23] C.E. Bonferroni, Teorie statistiche delle classi e calcolo delle probabilita, *Publ. Inst. Sup. Sc. Ec. Comm. Firenze* 8 (1936) 1-62.

[24] Yu. D. Burtin, On the probability of connectedness of a random subgraph of an n-dimensional cube, *Problems of Information Transmission* 13 (1977) 90-95.

[25] P.J. Cameron, On graphs with given automorphism group, *European Journal of Combinatorics* 1 (1980) 91-96.

[26] G. Cornuejols, Degree sequences of random graphs, preprint.

[27] P. Erdös, Graph theory and probability, *Canadian Journal of Mathematics* 11 (1959) 34-38.

[28] P. Erdös, Graph theory and probability II, *Canadian Journal of Mathematics* 13 (1961) 346-352.

[29] P. Erdös and A. Rényi, On random graphs I, *Publ. Math. Debrecen* 6 (1959) 290-297.

[30] P. Erdös and A. Rényi, On the evolution of random graphs, *Publ. Math. Inst. Hungar. Acad. Sci.* 5 (1960) 17-61.

[31] P. Erdös and A. Rényi, On the strength of connectedness of a random graph, *Acta. Math. Acad. Sci. Hungar.* 12 (1961) 261-267.

[32] P. Erdös and A. Rényi, On the evolution of random graphs, *Bull. Inst. Internat. Statist. Tokyo* 38 (1961) 343-347.

[33] P. Erdös and A. Rényi, Asymmetric graphs, *Acta Math. Sci. Hungar.* 14 (1963) 295-315.

[34] P. Erdös and J. Spencer, *Probabilistic Methods in Combinatorics*, Academic Press, New York and London, 1974.

[35] P. Erdös and J. Spencer, Evolution of the n-cube, *Comput. Math. Appl.* 5 (1979) 33-39.

[36] P. Erdős and R.J. Wilson, On the chromatic index of almost all graphs, *Journal of Combinatorial Theory (B)* 23 (1977) 255-257.

[37] J.W. Essam, Graph theory and statistical physics, *Discrete Math.* 1 (1971) 83-112.

[38] R. Fagin, Probabilities on finite models, *Journal of Symbolic Logic* 41 (1976) 50-58.

[39] T.I. Fenner and A.M. Frieze, On the connectivity of random m-orientable graphs and digraphs.

[40] T.I. Fenner and A.M. Frieze, On the existence of hamiltonian cycles in a class of random graphs.

[41] T.I. Fenner and A.M. Frieze, Hamiltonian cycles in random regular graphs.

[42] M. Fréchet, Les Probabilités Associés á un Systéme d'Evénements Compatibles et Dépendants, *Actualité Scient. et Ind., Hermann, Partie I,* No. 859, Paris 1940.

[43] R. Frucht, Graphs of degree 3 with given abstract group, *Canadian Journal of Mathematics* 1 (1949) 365-378.

[44] G.R. Grimmett, On the differentiability of the number of clusters per vertex in the percolation model, *Journal of London Mathematical Society* 23 (1981), 372-384.

[45] G.R. Grimmett, Random graphs, in *Further Selected Topics in Graph Theory* (L. Beineke and R.J. Wilson, eds.), Academic Press, 1982.

[46] J.M. Hammersley and H.L. Frisch, Percolation processes and related topics, *J. SIAM* 11 (1963) 894-918.

[47] F. Harary and E.M. Palmer, The probability that a point of a tree is fixed, *Math. Proc. Cambridge Phil. Soc.* 85 (1979) 407-415.

[48] G.I. Ivchenko, On the asymptotic behaviour of the degrees of vertices in a random graph, *Theory of Probability and Applications* 18 (1973) 188-195.

[49] Ch. Jordan, Sur la probabilité des épreuves répetées, *Bull. Soc. Math. France* 54 (1926) 101-137.

[50] Ch. Jordan, Sur un cas généralisé de la probabilité des épreuves répetées, *Acta Sci. Math. (Szeged)* 3 (1927) 193-210.

[51] Ch. Jordan, La Théoréme de probabilité de Poincaré généralisé au cas de plusieurs variables indépendantes, *Acta Sci. Math. (Szeged)* 7 (1934) 103-111.

[52] M. Karoński, A review of random graphs, *Journal of Graph Theory* 6 (1982) 349-389.

[53] R.M. Karp, Probabilistic analysis of a canonical numbering algorithm for graphs, *Proc. Symposia in Pure Math.,* vol. 34, AMS, Providence, RI, 1979, 365-378.

[54] L. Katz, The probability of indecomposability of a random mapping function, *Ann. Math. Statist.* 26 (1955) 512-517.

[55] A.K. Kelmans, Comparison of graphs by their probability of connectedness , (in Russian), *in Kombinator. i Asimpt. Analiz,* R *Krasnoyarsk, 1977, 69-81*.

[56] A.K. Kelmans, The graph with the maximum probability of remaining connected depends on the edge-removal probability, *Graph Theory Newsletter* 9 (1) (1979) 2-3.

[57] A.K. Kelmans, On graphs with randomly deleted edges, *Acta Math. Sci. Hungar.,* to appear.

[58] R.J. Lipton, The beacon set approach to graph isomorphism, Yale University,

preprint.

[59] G.A. Margulis, Probabilistic properties of graphs with large connectivity, *Problems of Information Transmission* 10 (1974).

[60] A. Meir and J.W. Moon, On nodes of degree two in random trees, *Mathematika* 15 (1968) 188-192.

[61] A. Meir and J.W. Moon, Cutting down random trees, *J. Austral. Math. Soc.* 11 (1970) 313-324.

[62] A. Meir and J.W. Moon, Packing and covering constants for certain families of trees II, *Trans. Amer. Math. Soc.* 33 (1977) 167-178.

[64] J.W. Moon, On the maximum degree in a random tree, *Michigan J. Math.* 15 (1968) 429-432.

[65] J.W. Moon, *Topics on Tournaments*, Holt, New York, 1968.

[66] J.W. Moon, *Counting Labelled Trees*, Canad. Math. Congress, Montreal, 1970.

[67] J.W. Moon, A problem on random trees, *J. Combinatorial Theory (B)* 10 (1971) 201-205.

[68] Z. Palka, On the number of vertices of given degree in a random graph, *Journal of Graph Theory*, to appear.

[69] E.M. Palmer and A.J. Schwenk, On the number of trees in a random forest, *Journal of Combinatorial Theory (B)* 27 (1979) 109-121.

[70] Yu. L. Pavlov, The asymptotic distribution of maximum tree size in a random forest, *Theory Prob. Appl.* 22 (1977).

[71] Yu. L. Pavlov, Limit theorems for the number of trees of a given size in a random forest, *Math Sbornik* 32 (1977) 335-345.

[72] Yu. L. Pavlov, Limit distribution of a maximal order of a tree in a random forest - a special case, *Math. Notes* 23 (1979) 751-759.

[73] A. Rényi, Quelques remarques sur les probabilités des evenements dependents, *J. de Math. Pures et Appl.* 37 (1958) 393-398.

[74] A. Rényi, On connected graphs I, *Publ. Math. Inst. Hungar. Acad. Sci.* 4 (1959) 385-388.

[75] R.W. Robinson and A.J. Schwenk, The distribution of degrees in a large random tree, *Discrete Math.* 12 (1975) 359-372.

[76] K. Schürger, On the evolution of random graphs over expanding square lattices, *Acta Math. Sci. Hungar.* 27 (1976) 281-292.

[77] K. Schürger, Limit theorems for complete subgraphs of random graphs, *Per. Math. Hungar.* 10 (1979) 47-53.

[78] R.T. Smyth and J. Wierman, *First-Passage Percolation on the Square Lattice*, Lecture Notes in Math. No. 671, Springer Verlag, 1978.

[79] N.C. Wormald, The asymptotic connectivity of labelled regular graphs, *J. Combinatorial Theory (B)*, 31 (1981) 156-167.

[80] N.C. Wormald, The asymptotic distribution of short cycles in random regular graphs, *J. Combinatorial Theory (B)* 31 (1981) 168-182.

[81] E.M. Wright, Graphs on unlabelled nodes with a given number of edges, *Acta Math.* 126 (1971) 1-9.

[82] E.M. Wright, The number of connected sparsely edged graphs, *Journal of Graph*

Theory 1 (1977) 317-330.

[83] E.M. Wright, The number of connected sparsely edged graphs II, Smooth graphs and blocks, *Journal of Graph Theory* 2 (1978) 299-305.

[84] E.M. Wright, The number of connected sparsely edged graphs III, Asymptotic results, *Journal of Graph Theory* 4 (1980) 393-407.

Plane Representations of Graphs

C. Thomassen

ABSTRACT

This paper contains various results on representations of planar graphs. We extend a result of Duchet, Hamidoune, Las Vergnas, and Meyniel on representations of planar graphs such that the vertices are horizontal line segments and the edges are vertical line segments. We extend a result of Tutte on convex representations of 3-connected planar graphs in various directions. For example, the representation can always be chosen such that no polygon bounding a face has more than six corners, and this bound is sharp. Also, we extend a result of Ungar on rectangular representations. We present a short proof of a result of Mani-Levitska, Guigas and Klee on rectifiable doubly-periodic infinite planar graphs and prove a general result on convex representations of infinite finite-valent graphs.

1. Introduction

A *plane graph* is a graph represented in the Euclidean plane R^2 such that the vertices are points and the edges are polygonal arcs joining their ends. A *planar graph* is an abstract graph isomorphic to a plane graph.

In this paper we present various results on representations of planar graphs. We first describe some results on 3-connected abstract graphs which can be applied to planar graphs. For example, we obtain a partial solution to a problem of Melnikov [12] by showing that a 3-connected planar graph can always be represented in the plane such that the vertices are horizontal intervals and such that two vertices (i.e. intervals) are adjacent if and only if there is some vertical straight line segment that connects them and does not intersect any of the other intervals. The restriction of this result to maximal planar graphs was first found by Duchet, Hamidoune, Las Vergnas and Meyniel (private communication). Melnikov's more general problem (which is of practical interest, too) of characterizing all the graphs that can be represented in this way is unsolved. We also apply the results on 3-connected graphs to abstract graphs with convex representations.

Tutte [20] characterized the graphs that have convex representations with prescribed outer (strictly convex) polygon. The author [14] extended this result to the case where the outer polygon is not necessarily strictly convex. The motivation for that extension was an application to infinite graphs [14]: If an infinite 3-connected graph with no vertex of infinite degree has a plane representation with no vertex-accumulation-point, then it has a plane representation such that each face is convex. The above result on convex representations (and its proof) also plays an important role in the proof of the result in [16] on convex deformations of convex graphs which proves a conjecture of Grünbaum and Shephard [7] and extends an old result of Cairns [3,4].

In this paper we describe other applications and extensions. We show that each 3-connected planar graph has a 6-gon representation (meaning that every face is bounded by a k-gon where $k \leq 6$) and describe infinitely many 3-connected planar graphs that have no 5-gon representations. If we impose the additional restriction on the graph that it is cubic and cyclically 4-edge-connected, then the graph can be drawn such that all faces are rectangles as shown about 30 years ago by Ungar [23]. We extend Ungar's result by characterizing all cubic (not necessarily 4-edge-connected) planar graphs which have rectangular representations with prescribed outer rectangle. Our proof carries over to the infinite case as well.

We show how the result on convex representations [14] can also be used to give a short proof of the result of Mani-Levitska, Guigas and Klee [11] on convex representations of infinite doubly periodic 3-connected planar graphs which in [11] is proved by complicated arguments (this application was also noted by Grünbaum and Shephard [7]). Furthermore, we show that all 3-connected planar graphs with no vertex of infinite degree have convex representations. It is here essential that the vertices have finite degree. Indeed, we show by very simple examples that there are infinite 3-connected planar graphs with no convex representation and the examples indicate that perhaps it will be difficult to characterize all planar convex graphs. Grünbaum (private communication) conjectures that the above-mentioned result on 6-gon representations extends to infinite plane graphs with no vertices of infinite degree and no vertex-accumulation-points. We prove that all such graphs have 8-gon representations.

All graphs in this work are finite, except in sections 5 and 8,

Conjecture 7.2 and the definition of H-component (in section 2).

2. Terminology and preliminaries.

The terminology and notation are the same as in [14, 15]. For the sake of completeness we repeat the most important definitions. We denote abstract graphs by roman letters and plane Graphs by greek letters. If Γ is a plane graph we denote by Γ^* the point set of Γ. If p and q are points not in Γ^* we say that p and q are *related* if they can be joined by a polygonal arc not intersecting Γ^*. Clearly, this defines an equivalence relation and the points in an equivalence class are called a *face* of Γ. The *boundary* of a face is defined in the obvious way. Clearly, the boundary of a face is a subgraph of Γ and each edge is on at most two face boundaries. The basic tool for dealing with plane graphs is the Jordan Curve theorem in the restricted version (which has a very short proof, see e.g. [15]) saying that a cycle in the plane has precisely two faces each of which has the cycle as boundary.

Using this result it is easy to go to a step further and prove that any $K_{2,3}$ drawn in the plane has precisely three faces having the three cycles of $K_{2,3}$ as boundaries. An easy exercise shows that any 2-connected abstract graph can be obtained from a cycle by successively adding paths between distinct vertices. Combining this with the Jordan Curve Theorem and the above observation on $K_{2,3}$ we get, by induction, the result that each face of a 2-connected plane graph is bounded by a cycle. Also, we obtain Euler's formula for 2-connected plane graphs and using that it is easy to give a rigorous proof of the easy part of Kuratowski's theorem saying that a graph is planar iff it contains no subdivision of K_5 or $K_{3,3}$ (see [14, 15]). In [15] there are short proofs of the non-trivial part of Kuratowski's theorem based on characterizations of 3-connected graphs described in the next section.

If G is a graph and H a subgraph of G, then an $H-component$ (also called an $H-bridge$) is either a K_2 whose vertices (but not edge) are in H or its consists of a connected component of $G - V(H)$ together with the edges from that component to H and their ends. Those vertices of the H-component which are in H are called the *vertices of attachment*.

3. The 3-connected graphs.

Menger's theorem that, for each natural number k, any two nonadjacent vertices in a graph are either connected by k internally disjoint paths or separated by fewer than k vertices provides a good characterization of the k-connected graphs. However, for proofs on connectivity conducted by induction on the number of vertices or edges of a graph it is convenient also to have reduction results for k-connected graphs or, equivalently, methods for generating the k-connected graphs from simple ones. In the previous section we mentioned how to generate the 2-connected graphs from cycles and Tutte [21] proved an analogous result for 3-connected graphs:

THEOREM 3.1. (Tutte [21]). *Every minimally 3-connected graph which is not a wheel contains an edge e which is not in a K_3 such that G/e is 3-connected.*

Here G/e denotes the graph obtained from G by contracting e and replacing all double edges (if any) by single edges. A graph is *minimally k-connected* if it is k-connected and the deletion of any edge decreases the connectivity.

An easy exercise shows that, for any edge e in a 2-connected graph G of order >2, either $G - e$ or G/e is 2-connected. A analogous result holds for 3-connected graphs:

PROPOSITION 3.2. *If e is any edge in a 3-connected graph of order >4, then either G/e is 3-connected or $G - e$ is a subdivision of a 3-connected graph.*

PROOF. Suppose G/e is not 3-connected. Then G has a separating set $\{x, y, z\}$ such that x and y are the ends of e. Let x', y' be any vertices of C such that x, y, x', y' are all distinct. In G, x' and y' are joined by three internally disjoint paths. If one of these contains e, then x' and y' belong to the same component of $G - \{x, y, z\}$ or one of x', y' equals z. But then x' and y' are also connected by three internally disjoint paths in $G - e$ since e can be replaced by a path connecting x and y such that all intermediate vertices are in a connected component of $G - \{x, y, z\}$ which contains none of x', y'. So a separating set of $G - e$ consisting of two vertices must separate x or y from the rest of the graph. This proves the proposition.

The following two results show that the 3-connected graphs can be obtained from K_4 by a succession of edge-additions respectively vertex-splittings (where "edge-addition" here also includes the operation consisting of inserting a vertex of degree two on an edge and joining it to another vertex or inserting two vertices on distinct edges and adding the edge between them).

THEOREM 3.3. (Barnette and Grünbaum [2], Titov [18]). *Every 3-connected graph G of order >4 has an edge e such that $G - e$ is a subdivision of a 3-connected graph.*

THEOREM 3.4. (Thomassen [14]) *Every 3-connected graph of order >4 has an edge e such that G/e is 3-connected.*

If G is planar, then deletion of an edge in G corresponds to contraction of the corresponding edge in the dual graph G^*. So Theorems 3.3 and 3.4 are, in a sense, "dual statements" while Theorem 3.1 is "self-dual". Seymour [13] found a common generalization of Theorems 3.3 and 3.4 by showing that any 3-connected matroid with more than 6 edges contains an edge whose deletion leaves a subdivision of a 3-connected matroid and a further extension of this has been obtained by Truemper [19]. Theorems 3.3 and 3.4 have simple proofs (see e.g. [15]) and each of them can easily be obtained from the other. They can both easily be obtained from Theorem 3.1. Furthermore, Halin [8] gave a short proof of Theorem 3.1 using the fact that every minimally 3-connected graph has a vertex of degree 3 (a fact which follows immediately from Theorem 3.3).

We shall here indicate how to obtain Theorem 3.1 easily from Theorem 3.4: Let G be a minimally 3-connected graph such that, for each edge e not in a K_3, G/e is not 3-connected. We shall prove that G is a wheel. First we observe that G must contain a K_3. If all vertices of this K_3 have degree at least 4, then, using Menger's theorem, it is easy to find two vertices of the K_3 which are joined by four internally disjoint paths and hence the deletion of the edge between them does not decrease the connectivity. So the K_3 must have a vertex of degree 3. It must in fact have at least two such vertices. For if $V(K_3) = \{x, y, z\}$ and x and y have degree ≥ 4 and z has degree 3, then we let e denote the edge xy. Since $G - e$ is not 3-connected it has a separating set $\{z, z'\}$ such that x and y belong to distinct components, say G_x and G_y, of $G - \{z, z'\}$. Since x and y have degree at least 4, each of G_x and G_y has order at least 2 and hence z is joined to at least two vertices of both of them. But this contradicts the assumption that z has degree 3.

So any K_3 in G with vertices say x, y, z has at least two vertices, say x and y, of degree 3. If x' is the neighbour of x outside the K_3, then it is easy to see that G/xx' is 3-connected. Hence xx' must be in a K_3. The third vertex of this K_3 must be z (unless G is a K_4) and we conclude that x' has degree 3. We then consider the neighbour x'' of x' distinct from x and z and continuing in this way we see that G is a wheel with center z.

Barnette and Grünbaum [2] derived Theorem 3.3 in order to give a short proof of Steinitz' theorem that the 1-skeletons of the 3-dimensional polytopes are precisely the 3-connected planar graphs. In [15] it is shown how Theorems 3.3 and 3.4 provide simple proofs of Kuratowski's theorem as well as the result of Tutte [20] that 3-connected planar graphs have convex representations. In [14] it was shown how a slight extension of Theorem 3.4 also implies the result of

Tutte [22] that the induced non-separating cycles (in [22] called the peripheral polygons), i.e. the cycles which have no chords and whose deletion results in a connected graph, generate the cycle space of the graph. We shall here derive this from a slight extension of Theorem 3.3.

THEOREM 3.5. Let C be a cycle of a 3-connected graph G. Then G has an edge e not in C such that $G - e$ is a subdivision of a 3-connected graph unless G is isomorphic to a graph of the form $K_3 + \overline{K}_n$ i.e. a graph obtained from a K_3 by adding a set of independent vertices and joining them completely to the K_3.

PROOF. Suppose G is not of the form $K_3 + \overline{K}_n$. Since G is 3-connected we can extend C to a subdivision of K_4 in G. Now consider a subgraph H of G satisfying the following:

(i) $C \subseteq H \subsetneq G$,
(ii) if C is a K_3, then H contains all vertices that are joined completely to C,
(iii) H is a subdivision of a 3-connected graph, and
(iv) $|E(H)|$ is maximum subject to (i), (ii) and (iii).

We shall prove that $E(G)\setminus E(H)$ consists of a single edge. If H is not 3-connected, then H has two vertices x and y of degree at least 3 in H such that they in H are connected by a path P of least at least 2 all of whose intermediate vertices have degree 2 in H. Since G is 3-connected there is a path P' in $G - \{x, y\}$ such that P' has only its ends in common with H and such that precisely one end of P' is in P. Since $H \cup P'$ satisfies (ii) and (iii) we must have $H \cup P' = G$ and since G is 3-connected P' has length 1.

On the other hand, if H is 3-connected and $V(H) = V(G)$, then clearly $E(G)\setminus E(H)$ consists of one edge so assume that H is 3-connected and that there is a vertex x in $G - V(H)$. Let P_1, P_2, P_3 be paths from z to H which are disjoint except for z. Since $H \cup P_1 \cup P_2 \cup P_3$ satisfies (ii) and (iii) it must equal G and hence P_1, P_2, P_3 all have length 1. Since H satisfies (ii) we can assume that the ends z_1 and z_2 of P_1 and P_2, respectively, in H are not joined by an edge of C. But then we delete the edge $z_1 z_2$ from H (if it is present) and add instead $P_1 \cup P_2$ and obtain a contradiction to (iv). This proves the theorem.

Theorem 3.5 can be applied to convex graphs as we show in the next section. Here we mention the following

COROLLARY 3.6. (Tutte [22]). In a 3-connected graph G the induced nonseparating cycles generate the cycle space and each edge is in at least two such cycles.

PROOF. (by induction on $|V(G)|$). If G is of the form $K_3 + \overline{K}_n$, then the corollary is easily verified so assume this is not the case.

Let C be any cycle of G and, by Theorem 3.5, let e be an edge not in C such that $G - e$ is a subdivision of a 3-connected graph. By the induction hypothesis,

$$E(C) = E(C_1) + E(C_2) + \ldots + E(C_k)$$

where each C_i is an induced non-separating cycle of $G - e$. Then each C_i is also non-separating in G. If C_i has a chord in G this must be e and we get $E(C_i) = E(C_i') + E(C_i'')$ where C_i' and C_i'' are the two cycles of $C_i \cup \{e\}$ containing e. Since C_i' and C_i'' have no chords the proof of the first part of the corollary is complete. The second part is proved similarly.

Combined with MacLane's planarity criterion Corollary 3.6 proves the result of Tutte (and rediscovered by Kelmans [9]) that a 3-connected graph is planar if and only if each edge is contained in precisely two induced non-separating cycles.

As an application of Theorem 3.4 we provide a partial solution to the problem of Melnikov [12] mentioned in the introduction.

THEOREM 3.7. If G is a 3-connected planar graph, then G has a representation in the plane such that the vertices of G correspond to horizontal straight line segments and two vertices of G are adjacent if and only if the corresponding intervals can be connected by a vertical straight line segment not intersecting the other intervals.

PROOF. (by induction on $|V(G)|$). If $G = K_4$, the theorem is easily verified so we proceed to the induction step. By Theorem 3.4, G has an edge $e = xy$ such that the contraction of e (into a vertex z say) results in 3-connected graph. By the induction hypothesis, there exists a collection of horizontal intervals satisfying the statement of the theorem with G/e instead of G. An edge incident with z corresponds either to an edge incident with x or an edge incident with y or to two edges incident with x and y. In Figure 1 we illustrate this by letting the first (respectively second) type of edges by drawn in full (respectively broken) lines and the third type of edge is indicated by two vertical line segments close to each other.

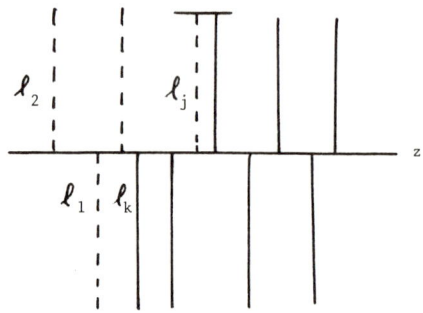

Figure 1. The edges incident with z.

Since G is planar we can label the vertical straight line segments incident with the interval representing z $l_1, l_2, \ldots, l_k$ such that they occur in clockwise order around z and such that $l_1, l_2, \ldots, l_j$ are the broken lines (as indicated in Figure 1). For otherwise, a close inspection of the proof of Kuratowski's theorem [15] shows that G contains a subdivision of K_5 or $K_{3,3}$, a contradiction. (Instead of referring to the proof of Kuratowski's theorem one can also appeal to Whitney's theorem that 3-connected planar graphs have unique plane embeddings).

In Figure 2 we show how to obtain the desired representation of G from the representation of G/e. In the proof it is essential that a horizontal interval may or may not include its ends.

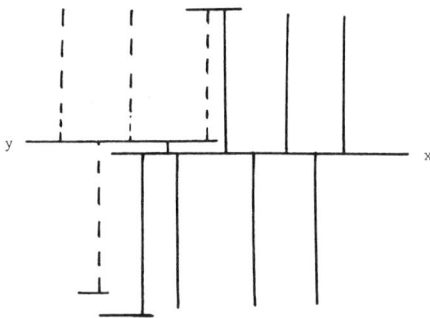

Figure 2. Vertex-splitting in the proof of Theorem 3.7.

We conclude this section with additional reduction results on 3-connected graphs and mention some unsolved problems.

Chartrand, Kaugars and Lick [5] proved a result which implies the following:

THEOREM 3.8. Every 3-connected graph of minimum degree at least 4 contains a vertex v such that $G - v$ is 3-connected.

By duality, every 3-connected planar graph of girth at least 4 has a cycle whose contraction results in a 3-connected graph.

Thomassen and Toft [17] investigated induced non-separating cycles in 3-connected graph and proved, among other things, the following results which are of interest in connection with Theorems 3.1 and 3.4.

THEOREM 3.9. Any 3-connected graph G of girth at least 4 contains a cycle C such that, for each edge e of C, G/e is 3-connected.

THEOREM 3.10. Any 3-connected graph of minimum degree at least 4 contains a cycle whose contraction results in a 3-connected graph.

As we have seen, 3-connectedness is very well understood in the sense that there are simple ways of generating the 3-connected graphs. For planar graphs, however additional problems may arise if one wants to prove statements involving, at the same time, a 3-connected graph and its dual. For example, Grünbaum [6] suggested the following extension of Barnette's result [1] that every 3-connected planar graph has a spanning tree of maximum degree at most 3.

CONJECTURE 3.11. Every 3-connected planar graph G has a

spanning tree T of maximum degree at most 3 such that the edges of G not in T correspond to a spanning tree of maximum degree at most 3 in the dual graph of G.

If true, the following conjecture provides a new planarity criterion for 3-connected graphs.

CONJECTURE 3.12 (Kelmans [10]). Any 3-connected non-planar graph of order at least 6 contains a cycle with three pairwise crossing chords. Another unsolved problem on 3-connected graphs has been proposed by the author: Is it true that every longest cycle in a 3-connected graph has a chord? This problem does not seem trivial even in the planar case.

4. Convex representations of graphs.

In [14] the author proved the following:

THEOREM 4.1 Let G be a 2-connected planar graph and let S be a cycle which is the face boundary of some plane representation of G. Let Σ be a convex polygon representing S. Then Σ can be extended to a graph isomorphic to G with convex facial cycles (and outer cycle Σ) if and only if

(i) each vertex x in $G - V(S)$ of degree at least 3 is joined to S by three paths that are disjoint except for x,

(ii) each cycle which is edge-disjoint from S has at least three vertices of degree at least 3, and

(iii) no S-component has all its vertices of attachment on a path of S corresponding to a straight line segment of Σ.

Figure 4 in [7] illustrates this theorem and Figure 3 below indicates the forbidden configurations for convex representations.

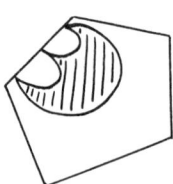 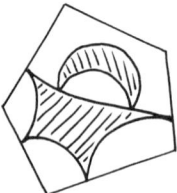

Figure 3. The forbidden configurations for convex representations.

If we restrict ourselves to the situation where Σ is strictly convex (i.e. its number of corners equals the length of S), then condition (iii) in Theorem 4.1 becomes redundant and we get the result of Tutte [20]. Vertices in $V(G)\backslash V(S)$ which have degree 2 must be on a straight line segment in any convex representation of G so it seems reasonable to restrict our attention to the case where all vertices (not in S) have degree at least 3. In this case condition (ii) of Theorem 4.1 can be omitted.

Barnette [1] defined an S-*circuit-graph* as a graph which consists of a cycle S of a 3-connected plane graph together with all edges and vertices in the interior of S and he proved that any such graph has a spanning tree of maximum degree at most 3. The circuit-graphs turn out to be interesting also in connection with convex graphs.

PROPOSITION 4.2. A graph G containing a cycle S such that all vertices not in S have degree ≥ 3 has a convex representation with outer cycle S if and only if G is an S-circuit-graph.

PROOF. Clearly an S-circuit graph satisfies condition (i) in Theorem 4.1. Conversely, if G and S satisfy the assumptions of Theorem 4.1 and Proposition 4.2, then the graph obtained from G by adding a vertex and joining it to all vertices of S is clearly planar and 3-connected and so G is an S-circuit-graph.

Combining Proposition 4.2 with results in Section 3 we get some reduction results for abstract graphs with convex representations analogous to those for 3-connected graphs.

PROPOSITION 4.3. Let G be a graph and S a cycle of G such that G has a convex representation with outer cycle S. Then for each edge e not in S either G/e or $G - e$ has a convex representation with outer cycle S.

PROOF. If an end of e has degree 2, then clearly G/e has the desired representation. So assume that both ends of e have degree at least 3. By Theorem 4.1 and Proposition 4.2, G is a subdivision of an S-circuit-graph and by Proposition 3.2, either G/e or $G - e$ is a subdivision of an S-circuit-graph. The result now follows from Proposition 4.2.

PROPOSITION 4.4. If G has a convex representation with outer cycle S and $G \neq S$, and all vertices not in S have degree at least 3, then G has an edge e not in S such that $G - e$ has a convex representation unless $G = K_4$.

PROOF. Let G_1 be the graph obtained from G by adding a vertex x and joining x to all vertices of S. Then G_1 is 3-connected. We delete successively edges from G_1 (and not in S) such that at each stage we have a subdivision of a 3-connected graph. By Theorem 3.5, some edge from $G - E(S)$ will be deleted unless $G = K_4$. If e is the first edge that is deleted from G, then $G - e$ is a subdivision of an S-circuit-graph and has a convex representation with outer cycle S.

There is also a contraction result for convex graphs analogous to Theorem 3.4:

PROPOSITION 4.5. If G has a convex representation with outer cycle S, and $V(G) \neq V(S)$, then G has an edge e not in S such that G/e has a convex representation with outer cycle S.

The proof of Proposition 4.4 is given in [16]. This result and Proposition 4.3 play an important role in the proof in [16] of the conjecture of Grünbaum and Shephard [7]: Two convex graphs which are isomorphic such that their outer cycles correspond to each other and have the same orientation can be continuously deformed into each other without losing convexity.

5. Doubly-periodic 3-connected plane graphs.

As pointed out by Grünbaum and Shephard [7], Theorem 4.1 implies easily the following result of Mani-Levitska, Guigas and Klee [11].

THEOREM 5.1. Let Γ be an infinite 3-connected plane graph such that the vertex set of Γ has no accumulation point. Assume further that Γ is doubly-periodic, i.e., the automorphism group of Γ includes two non-parallel translations. Then Γ is isomorphic to a doubly-periodic plane graph Γ' such that each face of Γ' is bounded by a convex polygon.

PROOF. Let t_1 and t_2 be non-parallel translations such that $t_1(\Gamma) = \Gamma$ and $t_2(\Gamma) = \Gamma$. Let Π be a path of Γ connecting vertices x and y such that $t_1(x) = y$. Assume Π is shortest possible. The minimality of Π easily implies that Π and $t_1(\Pi)$ have only y in common and hence $\Pi' = \bigcup_{k=-\infty}^{\infty} t_1^k(\Pi)$ is a two-way infinite path. Without loss of generality we can assume that $t_2(\Pi')$ and Π' are disjoint and we let Π'' be a (finite) path connecting them. It is no restriction to assume that $t_1(\Pi'')$ and Π'' are disjoint. Now the union of the paths $t_2^k(\Pi')$ and $t_1^p t_2^k(\Pi'')$ where k and p are integers) is a doubly-periodic subgraph Γ_1 of Γ and it is easy to redraw Γ_1 so as to obtain a doubly-periodic plane graph Γ_1' with convex faces (by tiling the plane into squares or hexagons). Now we can extend Γ_1' to a doubly-periodic convex graph representing Γ unless condition (iii) in Theorem 4.1 is violated. We consider a face of Γ_1 and if condition (iii) is violated, then Γ has a Γ_1-component say Ψ such that Ψ has all its vertices of attachment on a path of Γ_1 corresponding to a straight line segment of Γ_1'. We can assume that this path is Π'. Let Ψ' be the graph obtained from Ψ by adding the maximal segment of Π' connecting to vertices of attachment of Ψ. Then Ψ' is 2-connected and we call the number of vertices of $\Pi' - \Psi'$ (plus one) the *weight* of Ψ. We suppose that Γ_1 is chosen such that the number of the Γ_1-components in a face of Γ_1 violating (iii) in Theorem 4.1 is smallest possible and such that the smallest weight is smallest possible. We claim that under this assumption there is no Γ_1-component violating (iii). For if there were such a Γ_1-component Ψ, then we define Ψ' as above and instead of Π' we consider the path obtained from Π' by replacing $\Pi' \cap \Psi'$ by the other segment of the outer cycle of Ψ'. In this way we get rid of a Γ_1-component violating (iii). We may create a new one that violates (iii) but this will have smaller weight than Ψ. This contradiction shows that Γ_1' can be extended into the desired graph Γ'.

6. k-gon-representations.

If Γ is a 2-connected plane graph, we say that Γ is a *k-gon-representation* if each face of Γ is bounded by a convex polygon with at most k corners. Figure 4 shows a 3-gon-representation.

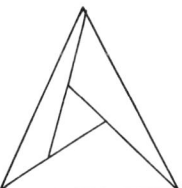

Figure 4. A 3-gon representation of the prism graph.

There are cubic 3-connected planar graphs with no 5-gon-representation. To see this we consider a cubic 3-connected plane graph with $2n$ vertices, $2n \geq 8$, and $n + 2$ faces and we replace each vertex by the configuration shown in Figure 5.

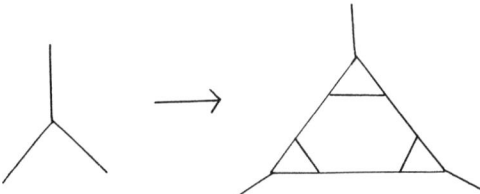

Figure 5. Replacement of a vertex.

In this way we obtain a cubic 3-connected plane graph with $18n$ vertices and $3n + 2$ faces that are not bounded by 3-cycles. Each vertex is a corner in at least one of these faces so the average number of corners of faces not bounded by 3-cycles is at least $\frac{18n}{3n + 2} > 5$. Conversely we have

THEOREM 6.1. *Every 3-connected planar graph has a 6-gon-representation.*

PROOF. Consider a 3-connected plane graph Γ. First we redraw the outer cycle of Γ as a triangle and we shall extend this triangle to a 6-gon-representation of Γ. Suppose we have already extended the triangle to a plane graph Δ which is a 6-gon-representation of a subgraph Γ' of Γ. Suppose further that Γ' can be extended to a convex graph isomorphic to Γ, i.e. each cycle Σ of Γ corresponding to a facial cycle of Γ' together with the subgraph in the interior of that cycle satisfies condition (iii) of Theorem 4.1 (conditions (i) and (ii) are trivially satisfied). For technical reasons we shall impose a further condition on Γ'. Consider again the cycle Σ above. The corresponding facial cycle in Γ' is a k-gon, $k \leq 6$. Let x be a vertex of Σ corresponding to a corner in Γ' and let Π and Π' be the paths of Σ corresponding to the two straight line segments of Γ' incident with x. Then we say that x is *blocked* in Σ if Γ has a path connecting a vertex of $\Pi - x$ to a vertex of $\Pi' - x$ such that all intermediate vertices are in the interior of Σ. The additional requirement of Γ' is that each cycle Σ of Γ corresponding to a facial k-gon of Γ' has at most $6 - k$ blocked corners.

We assume that Γ' is isomorphic to a proper subgraph of Γ and show that we can extend Γ' to a k-gon-representation of a larger subgraph of Γ satisfying the conditions given above. Suppose first that there is a blocked vertex x in a cycle Σ of Γ as described above, i.e., Γ has a path Π'' connecting Π and Π' and assume that the face in Γ' corresponding to Σ has more than three corners. Then we choose Π'' such that the ends of Π'' have largest possible distance from x (on Π and Π' respectively) and we represent Π'' by adding a straight line segment to Δ. The maximality property of Π'' ensures that the condition on the number of blocked corners in the resulting convex graph Δ' is satisfied. (We may create a facial $(k+1)$-gon but it has only $6 - k - 1$ blocked corners). The only problem is that Γ might have a $(\Gamma' \cup \Pi'')$-component with all vertices of attachment on Π''. But then, by the same argument as in the proof of Theorem 5.1, we can introduce and minimize the "weights" of such components and thereby modify Π'' so as to get another path with the same ends of Π'' such that (iii) is not violated. We may represent that path by the straight line segment of Δ' which was originally intended for Π''.

So we may assume that there are no blocked corners in cycles of Γ corresponding to a facial k-gon, $k \geq 4$ in Δ. We now consider the case where Γ has a cycle Σ corresponding to a facial k-gon, $k \geq 4$, in Δ such that the paths of Σ corresponding to straight line segments in Δ are $\Pi_1, \Pi_2, \ldots, \Pi_k$ in that order and such that Γ has a path Π'' connecting Π_1 with Π_3 and having all intermediate vertices in the interior of Σ. We choose Π'' such that the cycle of $\Sigma \cup \Pi''$ containing Π_2 and Π'' contains as many edges of Σ as possible and we extend Δ by representing Π'' by a straight line segment. The maximality property ensures that the polygon in the resulting graph containing Π'' and Π_4 has no blocked corners unless Π'' has an end (or both ends) in a corner of Σ in which case the polygon under consideration is a $(k-1)$-gon with at most one blocked corner or a $(k-2)$-gon with at most two blocked corners. Also, the 4-gon containing Π_2 and Π'' has at most two blocked corners. By choosing Π'' in an appropriate way, as in the previous case, we can make sure that condition (iii) in Theorem 4.1 is satisfied.

The next case to be considered is the case where Σ is a 6-gon with paths $\Pi_1, \Pi_2, \ldots, \Pi_6$ and Π'' connects a vertex of Π_1 with a vertex of Π_4. We proceed as in the previous cases except that we choose Π'' such that the cycle of $\Sigma \cup \Pi''$ containing Π_3 contains as few edges of Π_1 as possible and as many edges of Π_4 as possible. This ensures that the two 5-gons of $\Sigma \cup \Pi''$ have at most one blocked corner each.

Finally we consider the case where Σ is a cycle of Γ such that the corresponding cycle of Δ is a facial triangle and such that Γ has a Γ'-

component Γ'' inside Σ. If Γ'' has at most one corner of Σ as vertex of attachment we can dispose of this case as we did the first case. So assume Γ'' has at least two corners of Σ as vertices of attachment. Then Γ'' contains three paths connecting a vertex x not in Γ' to three vertices of Σ, two of which correspond to corners in Δ, such that the three paths are disjoint except for x. We now extend Δ by representing these three paths by straight line segments such that the interior of the triangle of Δ corresponding to Σ is partitioned into three triangular faces. By choosing the three paths appropriately we can ensure that condition (iii) in Theorem 4.1 is still satisfied.

So if Δ corresponds to a proper subgraph of Γ, then Δ can be extended to a larger 6-gon representation satisfying the conditions described in the beginning of the proof and so we obtain, in a finite number of steps, a 6-gon representation of Γ. This completes the proof.

As we have seen, Theorem 6.1 is in a sense best possible even for cubic 3-connected graphs. But maybe stronger results can be proved if we restrict ourselves to 4-connected or 5-connected planar graphs. For cubic cyclically 4-edge-connected graphs we can always get 4-gon representations as shown in the next section.

7. Rectangular representations.

In 1953 Ungar [23] proved that every cubic cyclically 4-edge-connected graph has a plane representation such that each face is bounded by a rectangle. We shall here extend this result.

THEOREM 7.1. Let Γ be a plane 2-connected graph such that precisely 4 vertices of the outer cycle Σ have degree 2 and all other vertices of Γ have degree 3. Let $\Pi_1, \Pi_2, \Pi_3, \Pi_4$ be the four paths of Σ with ends of degree 2 and all intermediate vertices of degree 3.

If Σ is redrawn such that Π_i is a vertical or horizontal straight line segment Π_i' for $i = 1, 2, 3, 4$ and the vertices of Π_1 and Π_2 are specified on the corresponding line segments, then the resulting rectangle $\Sigma' = \Pi_1' \cup \Pi_2' \cup \Pi_3' \cup \Pi_4'$ can be extended to a graph isomorphic to Γ such that all edges are vertical or horizontal if and only if

(i) for any vertex x of $\Gamma - V(\Sigma)$ and any set E of at most three edges, $\Gamma - E$ has a path from x to Σ unless E is the set of edges incident with x,

(ii) each Σ-component of Γ intersects two opposite paths Π_i, and

(iii) for any edge e not in Σ, each Σ-component of $\Gamma - e$

intersects at least two of the paths Π_i.

PROOF. It is easy to see that conditions (i), (ii) and (iii) are necessary for the rectangular representation. The three forbidden configurations are illustrated in Figure 6.

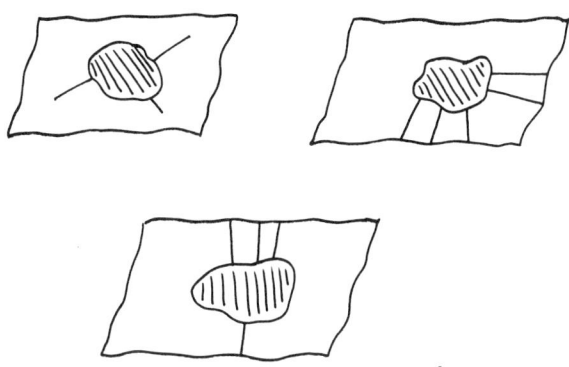

Figure 6. The forbidden configurations for rectangular representations.

Suppose conversely that Γ satisfies (i), (ii) and (iii). We prove, by induction on $|E(\Gamma)|$, that Γ has a representation as described in the theorem.

Suppose first that Γ has a Σ-component Ψ which intersects at most three of the paths Π_i. Without loss of generality we can assume that Ψ intersects Π_1 and Π_3 and possible Π_2. Now $\Sigma \cup \Psi$ is 2-connected and we let Π' be the path of Ψ which together with Π_4 and segments of Π_1 and Π_3 is the face boundary of a bounded face. We then add to Σ a vertical or horizontal straight line segment representing Π' and we apply the induction hypothesis to the two resulting rectangles. Clearly the two corresponding subgraphs of Γ satisfy (i). They also satisfy (ii) and (iii) because Γ satisfies (i), (ii) and (iii) and has maximum degree 3.

So we can assume that each Σ-component of Γ intersects each of $\Pi_1, \Pi_2, \Pi_3, \Pi_4$ and hence Γ has only one Σ-component. We consider a path Π in Γ from Π_1 to Π_3 such that all intermediate vertices are inside Σ. We then represent Π by a horizontal or vertical straight line segment connecting Π_1' with Π_3' and if possible we shall choose Π such that the induction hypothesis can be applied to the two subgraphs Γ_1

and Γ_2 of Γ whose outer cycles are the two cycles Σ_1 and Σ_2, respectively, of $\Sigma \cup \Pi$ distinct from Σ. Clearly Γ_1 and Γ_2 satisfy (i) and we shall show that Π can be chosen such that they satisfy (ii) as well. If Γ_1, say, has a Σ_1-component Γ_1' not satisfying (ii), then Γ_1' has all its vertices of attachment on Π and one of Π_1, Π_3 (say Π_1). We define the *weight* of Γ_1' as the length of the shortest subpath of Π from the vertex in Π_3 to a vertex of attachment of Γ_1 and we choose Π such that the sum of weights of the Σ_1- components of Γ_1 and the Σ_2- components of Γ_2 violating (ii) is least possible. Now if there is a Σ_1-component Γ_1' violating (ii) as above, then we replace the segment Π_1' of Π which connects a vertex of attachment of Γ_1' with Π_1 (and contains all vertices of attachment of Γ_1' on Π) with the path which together with Γ_1' and a segment of Π_1 forms a cycle outside which there is no edge of Γ_1'. Since Γ satisfies (iii), the total weight of the Σ_1- components of Γ_1 and the Σ_2-components of Γ_2 violating (ii) decreases, a contradiction. So we can assume that Γ_1 and Γ_2 both satisfy (i) and (ii). If Γ_1, say does not satisfy (iii), then Γ_1 has an edge e such that some Σ_1-component Γ_1' of $\Gamma_1 - e$ has all its vertices of attachment on one of the paths Π_1, Π_2, Π_3, Π. Since Γ satisfies (iii), Γ_1' has all its vertices of attachment on Π and since Γ_1 satisfies (ii), e must join Γ_1' to Π_2. We define the weight of Γ_1' as the number of edges in Π but not Π' where Π' is the smallest segment of Π containing all vertices of $\Gamma_1' \cap \Pi$. Furthermore, we choose Π such that Γ_1 and Γ_2 satisfy (ii) and such that the sum of weights of the Σ_1-components of Γ_1 and the Σ_2-components of Γ_2 violating (iii) is smallest possible. If this sum is zero we apply the induction hypothesis to Γ_1 and Γ_2 so assume that Γ_1 has an edge e such that $\Gamma_1 - e$ has a Σ_1-component Γ_1' as described above. Instead of Π we then consider the path Π'' obtained from Π by replacing Π' by the path which together with Π' forms the outer cycle of $\Gamma_1' \cup \Pi'$. Because of the minimality of Π, Γ_2 has an edge e' such that the graph obtained from Γ_2 by replacing Π by Π'' (and deleting e') has a $[(\Sigma_2 - \Pi) \cup \Pi'']$-component with the same weight as Γ_1'. This is illustrated in Figure 7.

PLANE REPRESENTATIONS OF GRAPHS 61

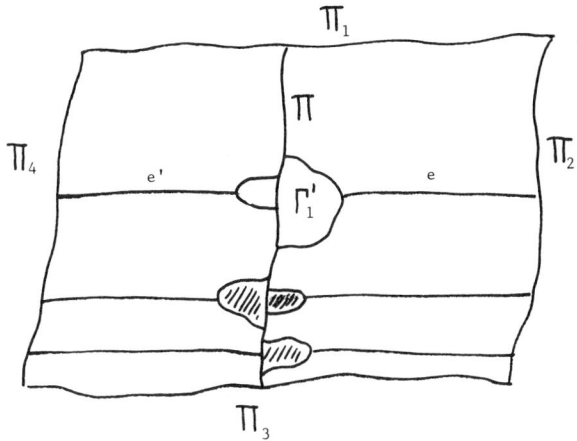

Figure 7. Components violating (iii).

There may be more than one such Σ_1-component or Σ_2-component violating (iii). In Figure 8 it is indicated how the paths in Figure 7 can be represented as vertical or horizontal straight line segments.

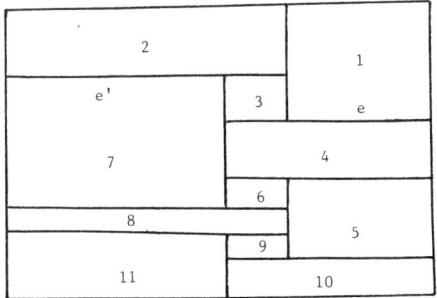

Figure 8. Rectangular representation of the part of Γ illustrated in Figure 7.

We now complete the proof by applying the induction hypothesis to each face in Figure 8 in the order indicated in the figure (so that at each face the vertices of only two consecutive sides of the rectangle are prescribed).

Using the same method as in the proof of Theorem 7.1 it is not difficult to show that any infinite 2-connected graph Γ having an outer cycle and satisfying (i), (ii) and (iii) and the same condition on the degrees as the graph in Theorem 7.1 can be redrawn such that all edges are horizontal or vertical (note that conditions (i), (ii) and (iii) are not necessary for such a representation when Γ is infinite). However, there is a more natural possible extension of Ungar's result [23] to infinite graphs which can probably also be proved with the method of Theorem 7.1 although the proof may be more complicated.

CONJECTURE 7.2. If Γ is an infinite cubic cyclically 4-edge-connected planar graph, then Γ has a plane representation such that each edge is a vertical or horizontal straight line segment. Moreover, this representation can be chosen such that the vertex set has no accumulation point provided Γ has a representation with no vertex-accumulation-point.

8. Infinite convex graphs.

In this section we describe convex representations of infinite *finite-valent* graphs (i.e. graphs with no vertex of infinite degree). Consider an edge e in a 2-connected finite-valent plane graph Γ. If we assign an orientation to e and walk from e in that direction turning sharp right at each vertex, then this walk results in either a cycle which we call a *facial cycle* or it results in an infinite path and in that case the sharp-left-turn-walk from e in the opposite direction also results in an infinite path and the union of these two paths is a two-way infinite path which we call a *facial path*. A facial cycle is the boundary of a face whereas a facial path need not be the boundary of a face.

We shall first extend Theorem 4.1 to the infinite case. For technical reasons we shall allow the outer "cycle" to contain infinite paths.

THEOREM 8.1. Let Σ be a convex polygon in the plane and let Γ be a 2-connected finite-valent plane graph such that all vertices and edges of Γ are on or inside Σ. Assume that each side of Σ is a finite or one-way infinite or two-way infinite path of Γ (such that Γ contains all corners of Σ). Suppose furthermore that

(i) each vertex in $\Gamma - V(\Sigma)$ of degree at least 3 is joined to Σ by three paths that are disjoint except for x,

(ii) each cycle which is edge-disjoint from Σ has at least three vertices of degree at least 3,

(iii) If Γ' is a Σ-component with all vertices of attachment on the same side of Σ, and Π' denote the minimal path of that side containing all vertices of attachment, then Γ' has no path Π'' such that $\Pi' \cup \Pi''$ is the outer cycle of $\Gamma' \cup \Pi'$ and

(iv) if some Σ-component has all its vertices of attachments on the same side of Σ, then this side is a finite path of Γ and at least one endvertex v of that path (which is a corner of Σ) satisfies the following; if v has degree 1 in Σ, then the other side containing v is a one-way infinite path all of whose vertices (except the endvertex) have degree 2 in Γ.

Then Γ can be redrawn such that Σ is unchanged and such that, for any two points p and q inside Σ and not in Γ^*, either p and q are separated by a closed polygon of Γ^* or else the straight line segment from p to q does not intersect Γ^*. In particular, all faces of the graph inside Σ are convex.

PROOF. Without loss of generality we can assume that all vertices not in Σ have degree at least 3. Since Γ is connected and finite-valent we can enumerate its vertices $\{x_1, x_2, \ldots \}$. Let x be the vertex which is not in Σ and which has the smallest label. Let Γ' be the Σ-component of Γ containing x. Suppose first that Γ' contains vertices of attachment on at least two different sides of Σ. Then we consider, by (i), three paths Π_1, Π_2, Π_3 from x to Σ which are pairwise disjoint except for x. We can assume that the ends of Π_1 and Π_2 are not on the same side of Σ. Now we draw Π_1, Π_2 and Π_3 as straight line segments and Γ is thereby partitioned into three parts each of which satisfies the condition of the theorem except possibly (iii). However, by choosing Π_1, Π_2, Π_3 such that appropriately defined "weights" of the $(\Sigma \cup \Pi_1 \cup \Pi_2 \cup \Pi_3)$-components of Γ violating (iii) are minimized (as in previous proofs) we can in fact assume that also condition (iii) is satisfied. Then we consider the vertex of Γ not in $\Sigma \cup \Pi_1 \cup \Pi_2 \cup \Pi_3$ of smallest label and continue as above with that vertex instead of x.

Assume next that Γ' has all its vertices of attachment on the same side of Γ and let Π' be defined as in condition (iii). Since Π'' in (iii) does not exist, the subgraph $\Gamma' \cup \Pi'$ has no outer cycle and so Π' together with two one-way infinite paths Π_1 and Π_2 of Γ' form a two-way infinite (outer) facial path of $\Gamma' \cup \Pi'$. We then select three paths Π_1', Π_2', Π_3' from x to $\Pi_1 \cup \Pi_2 \cup \Pi'$ which are pairwise disjoint (except for x). We can assume that the ends of Π_1' and Π_2' are not on the same of the paths Π_1, Π_2, Π'. We then consider the corner p of Σ on the same side as vertex v in condition (iv) and draw

Π_1, Π_2, Π_1', Π_2', Π_3' as indicated in Figure 9.

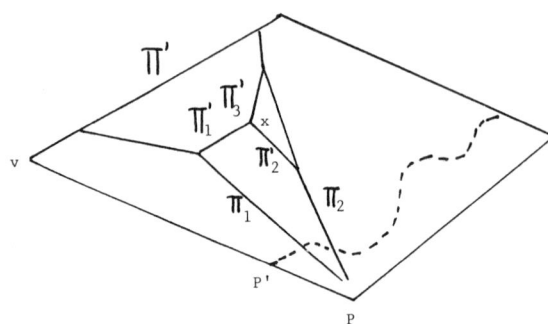

Figure 9. Convex representation of a subgraph containing x.

We can choose Π_1', Π_2' and Π_3' such that the resulting graph satisfies (iii). The only problem that might be left is that Γ has a path as indicated in dotted lines in Figure 9. If such a path exists, we first choose it such that (iii) is still satisfied and represent it as a straight line segment before drawing $\Pi_1, \ldots, \Pi_3'$ and we then proceed as above with p' instead of p. Since Γ is finite-valent and satisfies (iv) we get in this way, in a finite number of steps, a corner such that a path as the one indicated in dotted lines in Figure 9 does not exist.

So we can redraw a part of Γ containing x such that the interior of Σ is partitioned into a finite number of convex faces such that conditions (i) - (iv) are satisfied for the subgraphs of Γ inside each of these faces. We repeat the above procedure with the vertex of smallest label which has not yet been drawn instead of x, and we get thereby a representation of Γ with the desired properties.

In [14] it was shown that an infinite finite-valent 3-connected planar graph which has a plane representation with no vertex-accumulation- points also has a convex representation with no vertex-accumulation-points. We shall combine that result with Theorem 8.1 and obtain a general result on convex representations of finite-valent graphs. For this we need two lemmas.

LEMMA 8.2. Any infinite planar graph has a plane representation such that its vertex set is unbounded.

We shall only use Lemma 8.2 for 3-connected finite-valent graphs in which case the proof (which we leave for the reader) is very simple.

LEMMA 8.3. If G is an infinite finite-valent 3-connected planar graph, then G has an infinite subgraph H which is a subdivision of a 3-connected graph such that H has a plane representation which has no vertex- accumulation-point and which can be extended to a plane graph isomorphic to G such that each vertex of G not in H is in the interior of a facial cycle of H.

PROOF. By Lemma 8.2 there exists a planar graph Γ isomorphic to G and with unbounded vertex set. Since G is connected and finite-valent we can label its vertices $x_1, x_2, \ldots$. We now define a sequence $\Psi_1, \Psi_2, \ldots$ of finite subgraphs of Γ such that each Ψ_1 is a subdivision of a 3-connected graph. We let Ψ_1 be any subdivision of K_4 in Γ. Having already defined Ψ_k we define Ψ_{k+1} as follows: Let x be the vertex of smallest label in Γ outside the outer cycle of Ψ_k and let Ψ_k' be obtained by adding to Ψ_k three paths from x to Ψ_k which are pairwise disjoint (except for x). If Ψ_k' is a subdivision of a 3-connected graph we put $\Psi_{k+1} = \Psi_k'$. Otherwise, we can extend Ψ_k' to a subdivision Ψ_{k+1} of a finite 3-connected subgraph of Γ by adding paths outside Ψ_k' and inside the bounded face of Ψ_k that contains the ends of the three paths from x. Consider a bounded face of Ψ_k which is not a face in Ψ_{k+1}. Then the boundary of that face has at least two edges on the outer cycle of Ψ_k and the boundaries of the corresponding faces of Ψ_{k+1} each have fewer edges on the outer cycle of Ψ_{k+1}. From this follows that each cycle of $\Psi = \sum_{k=1}^{\infty} \Psi_k$ has only finitely many vertices of Ψ in its interior. Also, each vertex of Γ which is not in Ψ is in the interior of some cycle of Ψ. It is now easy to see that Ψ (if necessary) can be redrawn such that it has no vertex-accumulation-points and satisfies the conclusion of the lemma.

THEOREM 8.4. If G is an infinite finite-valent 3-connected planar graph, then G is isomorphic to a plane graph Γ such that, for any two points p and q not in Γ^*, either p and q are separated by a convex polygon of Γ^* or connected by a straight line segment disjoint from Γ^*.

PROOF. We consider a subgraph H of G satisfying the conclusion of Lemma 8.3. By [14, Theorem 8.6] H can be represented as a plane graph Θ such that each face is convex and such that Θ can be extended to a graph isomorphic to G by adding vertices and edges in the interior of facial cycles of Θ. We shall use Theorem 8.1 in order to obtain a graph Γ isomorphic to G and satisfying the assertion of the theorem. Clearly, condition (i) and (ii) and (iv) in Theorem 8.1 are satisfied. If condition (iii) is violated, we replace in H the segment Π' in condition (iii) of Theorem 8.1 by Π'' (without changing the point set of Θ). We may have to make infinitely many changes of this type but each vertex and edge of G is only involved in finitely many of these changes (compare with the proof of [14, Theorem 8.6]) and so we can assume that H and Θ also satisfy condition (iii) in Theorem 8.1. This completes the proof.

It is essential that the graphs in Theorem 8.4 are finite-valent. The graph G_0 obtained from a two-way infinite path by adding two vertices and joining them to all vertices of the path is 3-connected and planar but G_0 has no convex representation. However, if we add to G_0 two two-way infinite paths we obtain a graph which does have a convex representation as indicated in Figure 10.

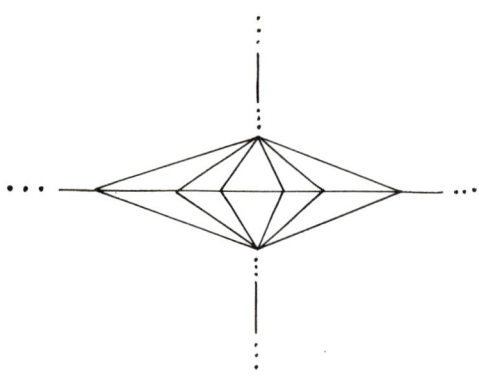

Figure 10. An infinite convex graph.

These examples indicate that it may be difficult to characterize all infinite graphs having convex representations.

B. Grünbaum (private communication) raised the problem if Theorem 6.1 can be extended to infinite finite-valent plane graphs with no vertex- accumulation-points. We conclude the paper by showing that all such graphs have 8-gon representations using the methods of Section 8 in [14]. We first prove a counterpart to [14, Theorem 8.4].

THEOREM 8.5 Let T be a finite-valent plane tree with no vertex of degree 1 and with no vertex-accumulation-point. Then T is isomorphic to a plane tree T' with no vertex- accumulation-point such that

  (i) each edge of T' is a straight line segment,
  (ii) each facial path in T corresponds to a facial path in T' and vice versa,

(iii) each face of T' is convex, and

(iv) each facial path of T' has at most three corners.

PROOF. We shall define a sequence $T_1, T_2, \ldots$ of subtrees of T with no vertex of degree 1 and a sequence $T_1', T_2', \ldots$ of plane trees such that $T_1 \subseteq T_2 \subseteq \cdots$ and $T_1' \subseteq T_2' \subseteq \cdots$ and for each natural number k, conditions (i) - (iv) are satisfied with T_k and T_k' instead of T and T'. In addition, we assume that any facial path Π of T_k which is not a facial path of T corresponds to a facial path Π' of T_k' consisting of two one-way infinite paths Π_1' and Π_2' that are straight non-parallel half-lines and possibly a finite path Π_3' which is a straight line segment (i.e. Π' has either one or two corners). Furthermore, we assume that the path of T_k corresponding to $\Pi_2' \cup \Pi_3'$ is part of a facial path of T.

Let $V(T) = \{x_1, x_2, \ldots\}$ and let T_1 consist of three one-way infinite paths each obtained by starting at x_1 and going sharp left each time. Suppose we have already defined T_k and T_k'. Assume $T_k \neq T$ and let m be the smallest natural number such that x_m is not in T_k. Let Π be the facial path in T_k which is the boundary of the face of T_k containing x_m. Let Π_1', Π_2' and Π_3' be the paths as above (see Figure 11(a)).

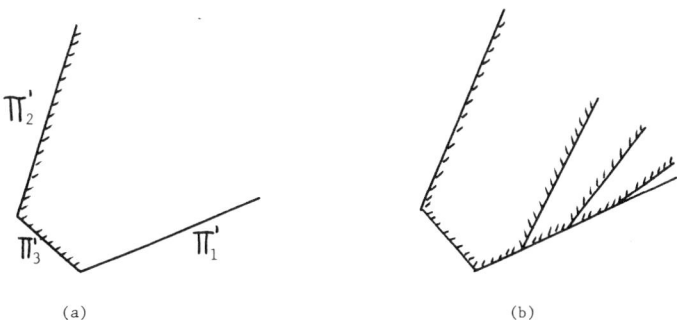

Figure 11. Convex representations of trees.

For each vertex of Π_1' we consider the corresponding vertex in T_k

and the one-way infinite path obtained from that vertex by walking sharp left in T. These one-way infinite paths are represented as straight half-lines as illustrated in Figure 11(b). The resulting trees are denoted T_{k+1} and T_{k+1}' and they clearly satisfy the conditions above. Also, the distance from x_m to T_{k+1} is smaller than the distance from x_m to T_k so $\bigcup_{k=1}^{\infty} T_k = T$ and hence $T' = \bigcup_{k=1}^{\infty} T_k'$ satisfies (i) - (iv).

THEOREM 8.6. Let Γ be a finite-valent 3-connected plane graph with no vertex-accumulation-points. Then Γ is isomorphic to a plane graph Γ' with no vertex-accumulation-point such that each face of Γ' is convex and each facial cycle (respectively path) has at most 8 (respectively 3) corners.

PROOF. By [14, Lemma 8.5] Γ contains a tree T which contains all facial paths of Γ and which has no vertex of degree 1. By Theorem 8.5, T has a "3-gon representation" T'. We extend T' to a convex graph Γ isomorphic to a subgraph of Γ by adding paths (represented by straight line segments) as in the proof of [14, Theorem 8.6] in such a way that Γ' can be extended to a convex graph isomorphic to Γ by adding vertices and edges inside the bounded faces of Γ'. These faces are bounded by 3-gons, 4-gons or 5-gons. If we choose the paths of Γ' which are added to T' such that the faces of the subgraph of Γ corresponding to Γ' have as few edges of Γ as possible, then the facial 5-gons of Γ' have at most three blocked corners and the facial 4-gons have at most two blocked corners. The proof of Theorem 6.1 can be modified so as to extend Γ' to an 8-gon representation of Γ.

References

[1] D. Barnette: Trees in polyhedral graphs, *Can. J. Math.* 18 (1966) 731-736.

[2] D.W. Barnette and B. Grünbaum: On Steinitz's theorem concerning convex 3-polytopes and on some properties of planar graphs. "The Many Facets of Graph Theory". *Lecture Notes in Mathematics* Vol. 110, Springer, Berlin 1969, 27-40.

[3] S.S. Cairns: Isotropic deformations of geodesic complexes on the 2-sphere and the plane. *Ann. of Math* (2) 45 (1944), 207-217.

[4] S.S. Cairns: Deformations of plane rectilinear complexes. *Amer. Math. Monthly* 51 (1944), 247-252.

[5] G. Chartrand, A. Kaugars and D.R. Lick: Critically n-connected graphs, *Proc. Amer. Math. Soc.* 32 (1972), 63-68.

[6] B. Grünbaum: Polytopes, Graphs, and Complexes, *Bull. Am. Math. Soc.* 76 (1970), 1131-1201.

[7] B. Grünbaum and G.C. Shephard: The geometry of planar graphs, *Proc. Eighth British Combinatorial Conf.* (1981), 124-150.

[8] R. Halin: Untersuchungen über minimale n-fach zusammenhängende Graphen, *Math. Ann.* 182 (1969), 175-188.

[9] A. K. Kelmans: A new planarity criterion for 3-connected graphs. *J. Graph Theory* 5 (1981), 259-267.

[10] A. K. Kelmans, Problem at the Sixth Hung. Coll. on Combinatorics, Eger 1981.

[11] P. Mani-Levitska, B. Guigas and V. Klee: Rectifiable n-periodic maps. *Geometriae Dedicata* 8 (1979), 127-137.

[12] L. A. Melnikov: Problem at the Sixth Hung. Coll. on Combinatorics, Eger 1981.

[13] P. D. Seymour: Decomposition of regular matroids, *J. Combinatorial Theory (B)* 28 (1980), 305-359.

[14] C. Thomassen: Planarity and duality of finite and infinite graphs. *J. Combinatorial Theory (B)* 29 (1980), 244-271.

[15] C. Thomassen: Kuratowski's theorem, *J. Graph Theory* 5 (1981), 225-241.

[16] C. Thomassen: Deformations of plane graphs, to appear in J. Combinatorial Theory (B).

[17] C. Thomassen and B. Toft: Induced non-separating cycles in graphs, *J. Combinatorial Theory (B)* 31 (1981), 191-224.

[18] V. K. Titov: A constructive description of some classes of graphs. Doctoral dissertation, Moscow 1975.

[19] K. Truemper, to appear.

[20] W. T. Tutte: Convex representations of graphs, *Proc. London Math. Soc.* 10 (1960), 304-320.

[21] W. T. Tutte: A theory of 3-connected graphs, *Nederl. Akad. Wetensch. Proc. Ser. A* 23 (1961), 441-455.

[22] W. T. Tutte: How to draw a graph, *Proc. London Math. Soc.* 10 (1960), 743-768.

[23] P. Ungar: On diagrams representing maps, *J. London Math. Soc.* 28 (1953), 336-342.

Marriage in Infinite Societies

R. Aharoni

C.St.J.A. Nash-Williams

S. Shelah

ABSTRACT

The question "Which societies have espousals ?" (which can also be formulated in various other ways, for instance in terms of systems of distinct representatives of families of sets) seems to be a fairly basic question of mathematics. For finite societies, it can fairly easily be answered by a classical theorem of P. Hall and D. König. In recent years, much work has been devoted to trying to answer the same question for infinite societies, which present substantially more complications. This is the subject of our paper.

A *society* is an ordered triple (G,M,W) such that G is a graph, M and W are sets, $M \cup W = V(G)$, $M \cap W = \emptyset$ and every edge of G joins a vertex in M to a vertex in W - so that G is bipartite and (M,W) is a bipartition [2, §1.2] of G. Elements of M and of W are called *men* and *women* respectively of the society. A man and woman are said to *know* each other if they are adjacent vertices in G. An *espousal* of (G,M,W) is a matching of M into W in G, i.e. a set $L \subseteq E(G)$ such that each element of M is incident with exactly one element of L and each element of W is incident with at most one element of L. Informally, an espousal of a society is a way of finding wives for all the men so that each marries a woman whom he knows. A society is *espousable* if it has an espousal, and *inespousable* if not.

The problem of trying to say, in some reasonable sense, which societies are espousable has been the theme of much research: see, for example, [10] and [11, Section 1]. The following classical theorem of König [7,8] and Hall [6] answers this question for finite societies (i.e. societies with only a finite number of men and women).

THEOREM 1. A finite society (G,M,W) is inespousable if and only if, for some r, there exist r men who know less than r women between them.

A society (G,M,W) will be called *countable* if the set $V(G)=M\cup W$ is countable, i.e. finite or denumerably infinite. Countable espousable societies were characterised in [3], and somewhat differently in [12]. We state the result of [12] after some required definitions.

Let Z^* denote a set whose elements are the integers and two further "numbers" ∞ and $-\infty$. For any set A, we define the *size* $||A||$ of A to be its cardinality $|A|$ if A is finite and to be ∞ if A is infinite: thus $||A||\in Z^*$ and all infinite sets, regardless of cardinality, have "size" ∞. We shall also need a few definitions relating to arithmetic in Z^*. The *sum* $a_1+a_2+\ldots+a_n$ of n numbers $a_1,\ldots,a_n\in Z^*$ has its usual meaning if $a_1,\ldots,a_n$ are integers, is defined to be ∞ if at least one a_i is ∞, and is defined to be $-\infty$ if no a_i is ∞ but at least one is $-\infty$. Expressions such as $a-b$, $a-b+c$ etc. denote $a+(-b)$, $a+(-b)+c$ etc. respectively, when $a,b,c\ldots\in Z^*$. The most important distinctive feature of these definitions is that $\infty-\infty$ is defined to be ∞, since we wish to think of $\infty-\infty$ as the *largest possible* value of $||A\setminus B||$ for sets A,B such that $B\subsetneq A$ and $||A||=||B||=\infty$. Inequalities between elements of Z^* are defined in the obvious way. The *infimum* inf S of a non-empty subset S of Z^* is the greatest $z\in Z^*$ such that $z\leq s$ for every $s\in S$, and the *supremum* sup S is analogously defined. If λ is a limit ordinal and a_θ is an element of Z^* for every ordinal $\theta<\lambda$, then $\lim_{\theta\to\lambda}\inf a_\theta$ is defined by analogy with the definition of $\lim_{n\to\infty}\inf u_n$ in analysis, that is, $\lim_{\theta\to\lambda}\inf a_\theta$ is sup $\{i_\phi:\phi<\lambda\}$, where i_ϕ denotes $\inf\{a_\theta:\phi\leq\theta<\lambda\}$.

We define a *transfinite sequence of distinct women* of a society (G,M,W) to be a transfinite sequence of the form $(w_\alpha)_{\alpha<\lambda}$ where λ is an ordinal number, $w_\alpha\in W$ for each ordinal $\alpha<\lambda$ and $w_\alpha\neq w_\beta$ whenever $\alpha<\beta<\lambda$. If s denotes this transfinite sequence, then the *demand set* $D(s)$ of s is defined to be the set of those men who know no women other than terms of s, that is,

$$D(s)=\{m\in M: N(m)\subseteq\{w_\alpha: \alpha<\lambda\}\},$$

where $N(m)$ is the set of those women whom m knows. (Thus $D(s)$ depends only on the set of terms of s and not on their order.) Roughly speaking, the men in $D(s)$ "demand" wives from amongst the terms of s. For each ordinal $\theta\leq\lambda$, s_θ will denote the transfinite sequence $(w_\alpha)_{\alpha<\theta}$ formed by the first θ terms of s. In particular, $s_\lambda=s$ and s_0 is the "empty" transfinite sequence, which has no terms at all.

If $s=(w_\alpha)_{\alpha<\lambda}$, we define a number $\mu(s)\in Z^*$ which, in some sense, measures (or bounds) "the number of women to spare" in s. To do this, we define an element $\mu(s_\gamma)$ of Z^* for each ordinal $\gamma\leq\lambda$ by transfinite induction on γ: then this will, in particular, define $\mu(s_\lambda)=\mu(s)$. We first define $\mu(s_0)$ to be $-||M_0||$, where M_0 is the set of those men who know no women at all. (Evidently, the society is not espousable if $M_0\neq\varnothing$.) Suppose now that $0<\gamma\leq\lambda$ and that $\mu(s_\theta)$ has been defined for every ordinal $\theta<\gamma$. Then we define $\mu(s_\gamma)$ to be

(i) $\mu(s_\beta)+1-||D(s_\gamma)\backslash D(S_\beta)||$ if γ is a successor ordinal $\beta+1$,

(ii) $\liminf\limits_{\theta\to\gamma}\mu(s_\theta)-||D(s_\gamma)\backslash\bigcup\limits_{\theta<\gamma}D(S_\theta)||$ if γ is a limit ordinal.

Some examples illustrating these ideas are discussed in [14, §4].

The following theorem was proved in [12].

THEOREM 2. A countable society is espousable if and only if $\mu(s)\geq 0$ for every transfinite sequence s of distinct women of the society.

(In fact, in Theorem 2, the hypothesis that the society is countable can be replaced by the weaker hypothesis that each woman knows only countably many men: see [13]).

If G is a graph and $X\subseteq V(G)$, then $G[X]$ denotes the subgraph of G such that $V(G[X])=X$ and $E(G[X])$ is the set of those edges of G which join two elements of X. Moreover $G-X$ denotes $G[V(G)\backslash X]$, i.e. the graph obtained from G by removing the vertices in X and their incident edges. If $\Gamma=(G,M,W)$ is a society and $A\subseteq M$, $X\subseteq W$, $P\subseteq M\cup W$ then $\Gamma[A,X]$, $\Gamma-P$ denote respectively the societies $(G[A\cup X],A,X)$, $(G-P,M\backslash P,W\backslash P)$. We say that Γ is *critical* if it is espousable but $\Gamma-\{w\}$ is inespousable for every $w\in W$. Thus critical societies are "only just espousable". A critical society has at least one espousal but every espousal of it makes use of *all* the women of the society.

The following theorem characterises critical societies using ideas like those in Theorem 2; but, unlike Theorem 2, it applies to both countable and uncountable societies.

THEOREM 3 [1, Corollary 2]. A society is critical if and only if it satisfies both of the conditions:

(i) $\mu(s)\geq 0$ for every transfinite sequence s of distinct women of the society;

(ii) $\mu(s)=0$ for some transfinite sequence s of distinct women such that EVERY woman of the society is a term of s.

Thus the structure of critical societies is fairly clearly known, and so it seems reasonable to use them in formulating a necessary and sufficient condition for an arbitrary society to be espousable.

To try to characterise espousable *uncountable* societies, one might begin by thinking what examples of *inespousable* ones are known. One might then try to generalize those examples in order to produce as wide a class of inespousable societies as possible, and then prove that all other societies are espousable. It has been realized for some while that inespousable uncountable societies can arise in ways which have no obvious counterpart in the countable case. For example [9], if (G, M, W) is a society such that $M = \{m_\alpha : \omega \leq \alpha < \omega_1\}$, $W = \{w_\alpha : \alpha < \omega_1\}$ (where ω is the first infinite ordinal and ω_1 is the first uncountable ordinal), $m_\alpha \neq m_\beta$ when $\alpha \neq \beta$, $w_\gamma \neq w_\delta$ when $\gamma \neq \delta$, and m_ϕ knows w_θ if and only if $\theta < \phi$, then (G, M, W) is inespousable. As a hint for proving this, recall that there is no infinite decreasing sequence $\alpha_1 > \alpha_2 > \alpha_3 > \ldots$ of ordinals. Thus, if we managed to marry off all the men in the required way, and formed a directed graph D whose vertices are the ordinals less than ω_1, with a directed edge from a vertex α to a vertex β if and only if $\alpha \geq \omega$ and m_α marries w_β, then every component of D must be a directed path with an ordinal less than ω as its terminal vertex, contradicting the uncountability of $V(D)$.

We might, in the first instance, generalize this example by letting $M = \{m_\alpha : \alpha \in S\}$, $W = \{w_\alpha : \alpha < \omega_1\}$, where S is some subset of the set of ordinals less than ω_1 but not necessarily the particular subset $\{\alpha : \omega \leq \alpha < \omega_1\}$. Again, we assume that $m_\alpha \neq m_\beta$ when $\alpha \neq \beta$, $w_\gamma \neq w_\delta$ when $\gamma \neq \delta$ and m_ϕ knows w_θ if and only if $\theta < \phi$. If S is a *reasonably large* subset of the set of ordinals less than ω_1, then perhaps the same sort of difficulty as before will arise when we try to marry off all of the men. In fact, this is easily seen to be so if S includes all but countably many of the ordinals less than ω_1; but we can say more. The notion of a "stationary" set of ordinals (defined below) has featured in the past literature of set theory, and it can be shown fairly easily that our society is inespousable if S is any stationary subset of the set of ordinals less than ω_1.

This construction for an inespousable society can be generalized much further, but the above discussion might suggest the clue that stationary sets have an important rôle to play in any such generalization and in the investigation of espousability, as indeed [15] demonstrated. We now define these sets.

We let ω_α denote the least ordinal whose cardinality is $\aleph_\alpha$. (In modern treatments of set theory, ω_α and $\aleph_\alpha$ are usually regarded as being the same object.) We assume ordinals to have been defined so that an ordinal is the set of all smaller ordinals.

DEFINITIONS. A subset S of ω_α is

(i) *closed* (in ω_α) if sup $S' \in S \cup \{\omega_\alpha\}$ for every non-empty subset S' of S.

(ii) *unbounded* (in ω_α) if sup $S = \omega_\alpha$.

(iii) ω_α-*stationary* (commonly abbreviated to *stationary* if the relevant ω_α is clear from the context) if $S \cap C$ is non-empty for every closed unbounded subset C of ω_α.

If $S \subseteq \omega_\alpha$ and $f: S \to \omega_\alpha$ is a function such that $f(\theta) < \theta$ for every $\theta \in S \setminus \{0\}$, then f is said to be *regressive*. An important alternative characterisation of stationary sets is given by the following proposition, provided that cf $\omega_\alpha > \omega$ (i.e. provided that ω_α is not the supremum of any countable set of ordinals).

PROPOSITION 4 (see [4], §3.7 (4), for example). *Suppose that* cf $\omega_\alpha > \omega$. *Then a subset S of ω_α is stationary if and only if, for every regressive function $f: S \to \omega_\alpha$, there exists $\phi < \omega_\alpha$ such that $\{\theta \in S: f(\theta) \leq \phi\}$ is unbounded in ω_α.*

This proposition could be used to prove inespousability of the society described above whose set of men was $\{m_\alpha: \alpha \in S\}$, where S is a stationary subset of ω_1. An espousal of such a society would give a regressive function $f: S \to \omega_1$ such that m_α married $w_{f(\alpha)}$ for each $\alpha \in S$, which is clearly incompatible with Proposition 4.

DEFINITIONS. A society (G, M, W) such that $M = \emptyset$ and $|W| = 1$ (i.e. a society containing only one woman and no men) will be called a *one-woman society*. If $\Gamma = (G, M, W)$ is a society, then $V(\Gamma)$ will mean $V(G)$. If A, X are subsets of M, W respectively then $\Gamma[A, X]$ will be called a *subsociety* of Γ. Thus, given the society Γ, a subsociety Δ of Γ is uniquely determined by the set $V(\Delta)$. If Γ_i is a subsociety of Γ for each $i \in I$ (where I is some set of indices), then the *join* $\vee \{\Gamma_i: i \in I\}$ of the subsocieties Γ_i ($i \in I$) is defined to be the subsociety Δ of Γ such that $V(\Delta) = \bigcup_{i \in I} V(\Gamma_i)$. The subsocieties Γ_i ($i \in I$) are *disjoint* if the sets $V(\Gamma_i)$ ($i \in I$) are disjoint. A subsociety $\Gamma[A, X]$ is *saturated* (in Γ) if, in Γ, no man in A knows any of the women in $W \setminus X$. Thus, if a man is in a saturated subsociety of Γ, then so are all the women whom he knows (in Γ).

We recall that an infinite cardinal κ is *singular* if it is the supremum of some set of fewer than κ cardinals less than κ, and is *regular* otherwise. For example, $\aleph_n$ is regular for every non-negative integer n, but $\aleph_\omega$ is singular because it is the supremum of the set $\{\aleph_n: n<\omega\}$ of $\aleph_0$ cardinals.

A κ-*subset* of a set S is a subset of S whose cardinality is κ.

For any cardinal κ such that $1\leq\kappa\leq\aleph_0$, we define a κ-*obstruction* in a society Γ to be a saturated subsociety $\Delta=\Gamma[A,X]$ of Γ such that $\Delta-L$ is critical for some κ-subset L of A. Notice that, if Γ has such a subsociety, then it cannot be espousable, because (to express it informally) $\Delta-L$ is "only just espousable" and so Δ has κ men too many to be espousable. Since Δ is saturated, any espousal of Γ would have to contain an espousal of Δ, and so the inespousability of Δ implies that Γ is inespousable. Thus Δ acts as an "obstruction" to finding an espousal of Γ.

By transfinite induction, we now define the meaning of "κ-obstruction" when κ is a regular cardinal greater than $\aleph_0$. Assume, therefore, that we have defined what is meant by saying that a subsociety of Γ is a μ-obstruction in Γ for every regular cardinal μ such that $\aleph_0<\mu<\kappa$. Of course, this has also been defined for cardinals μ such that $1\leq\mu\leq\aleph_0$. We shall say that a subsociety of Γ is a $(<\kappa)$-*obstruction* in Γ if it is a μ-obstruction in Γ for some $\mu<\kappa$ such that either $1\leq\mu<\aleph_0$ or μ is regular. Let ζ be the ordinal such that $\kappa=\aleph_\zeta$. We shall say that a subsociety Δ of Γ is a κ-*obstruction* in Γ if $\Delta=\vee\{\Delta_\alpha: \alpha<\omega_\zeta\}$ for some transfinite sequence $(\Delta_\alpha)_{\alpha<\omega_\zeta}$ of disjoint subsocieties of Γ such that

(i) for each $\alpha<\omega_\zeta$, Δ_α is either a one-woman society or a $(<\kappa)$-obstruction in $\Gamma-\bigcup_{\theta<\alpha}V(\Delta_\theta)$,

(ii) the set of those $\alpha<\omega_\zeta$ for which Δ_α is a $(<\kappa)$-obstruction in $\Gamma-\bigcup_{\theta<\alpha}V(\Delta_\theta)$ is ω_ζ-stationary.

The following is our main theorem, whose fairly lengthy proof will appear elsewhere.

THEOREM 5. A society Γ is inespousable if and only if there exists a κ-obstruction in Γ for some cardinal κ such that either $1\leq\kappa<\aleph_0$ or κ is regular.

Additional Remarks

I. The problem of characterising espousable societies can be formulated in other ways - e.g. given a family $(A_i: i\in I)$ of (not necessarily disjoint) sets, can we select $a_i\in A_i$ for each $i\in I$ so that $a_i\neq a_j$ whenever $i\neq j$? (This is equivalent to enquiring as to the espousability of a society with men $m_i(i\in I)$ and in which A_i is the set of women known by

m_i.) However it be formulated, the problem seems a fairly fundamental one, and related to a number of other questions in combinatorics and perhaps other areas of mathematics. Therefore one might hope that the ideas described in this paper will provide help in solving other problems and scope for further research. Some possible directions for such research were suggested in [14, §5]. We here mention just one of them.

The question answered by Theorem 5 can be generalized as follows. Let G be a graph and M, W be subsets of $V(G)$. We do not now assume that $M \cup W = V(G)$ and $M \cap W = \varnothing$ (although in fact $M \cap W = \varnothing$ could be assumed with no real loss of generality). A path from a vertex in M to a vertex in W will be called an MW-*path*, and a set of paths will be called *disjoint* if no two of them have a common vertex. For want of a better term, define an MW-*path-espousal* in G to be a set of disjoint MW-paths in G such that each element of M (but not necessarily each element of W) is an end-vertex of one of these paths. If (G, M, W) is a society (so that $M \cup W = V(G)$ and $M \cap W = \varnothing$), it is easily seen that an MW-path-espousal in G is essentially the same thing as an espousal of (G, M, W), the latter being a set L of edges and the former being a set of paths of length 1 whose edges are the elements of such a set L. Thus, when (G, M, W) is not necessarily a society, one might seek a generalization of Theorem 5 which states necessary and sufficient conditions for the existence of an MW-path-espousal in G.

In particular, such an investigation might cast light on the problem, which has (perhaps surprisingly) remained unsolved for nearly two decades, of proving or disproving the following conjecture of Erdős [5].

CONJECTURE. Let G be a graph and A, B be subsets of $V(G)$. Then there exist a subset S of $V(G)$ which separates A from B (in the sense that every AB-path includes at least one element of S), a set **P** of disjoint AB-paths in G and a bijection $\phi : \mathbf{P} \to S$ such that $\phi(P) \in V(P)$ for every $P \in \mathbf{P}$.

This conjecture has been proved for graphs in which there are no infinite paths: See [16].

Of course, these are really problems concerning infinite graphs, since Menger's Theorem provides the answers when G is finite.

II. Theorem 3 states that a society is *critical* if and only if it satisfies conditions (i) and (ii) of that theorem. Let us call a society *pseudo-critical* if it satisfies condition (ii) of Theorem 3. One could then define the term κ-*pseudo-obstruction* by simply re-writing the complete definition of "κ-obstruction" with the words "critical", "κ-obstruction", "μ-obstruction", "$(<\kappa)$-obstruction" changed to "pseudo-critical",

"κ-pseudo-obstruction", "μ-pseudo-obstruction" and "$(<\kappa)$-pseudo-obstruction" respectively.

We are reasonably sure that Theorem 5 remains true if, in its statement, the word "κ-obstruction" is changed to "κ-pseudo-obstruction" - and this may arguably be a more satisfactory form of the theorem since it somewhat more clearly provides an actual recipe for constructing all possible inespousable societies. Since every κ-obstruction is a κ-pseudo-obstruction, Theorem 5 immediately implies that, if a society Γ is inespousable, then there exists a κ-pseudo-obstruction in Γ for some κ such that either $1 \leq \kappa < \aleph_0$ or κ is regular. We believe that we also know how to prove the converse of this statement, although, at the time of this conference, the details of this proof have not yet been fully written down. Certainly, this converse seems intuitively very believable, since the power of a critical subsociety of Γ to help to obstruct espousability should surely lie in its satisfying condition (ii) of Theorem 3 and not condition (i).

References

[1] R. Aharoni, On two necessary conditions for the existence of transversals, *J. Combin. Theory Ser. A*, to appear.

[2] J. A. Bondy and U. S. R. Murty, *Graph theory with applications* (Macmillan, London, 1976).

[3] R. M. Damerell and E. C. Milner, Necessary and sufficient conditions for transversals of countable set systems, *J. Combin. Theory Ser. A* 17 (1974), 350-374.

[4] F. R. Drake, *Set theory - an introduction to large cardinals* (North-Holland/American Elsevier, 1974).

[5] P. Erdös, Problem 64, *Theory of graphs and its applications*, Proceedings of the Symposium held in Smolenice in June 1963 (edited by M. Fielder, Czechoslovak Academy of Sciences, Prague, 1964), 159.

[6] P. Hall, On representatives of subsets, *J. London Math. Soc.* 10 (1935), 26-30.

[7] D. König, Graphok és matrixok, *Mat. Fiz. Lapok* 38 (1931), 116-119 (Hungarian with German summary).

[8] D. König, Uber trennende Knotenpunkte in Graphen (nebst Anwendungen auf Determinanten und Matrizen), *Acta Litt. Sci. Sect. Sci. Math. (Szeged)*, 6 (1932 - 1934), 155-179.

[9] E. C. Milner and S. Shelah, Sufficiency conditions for the existence of transversals, *Canad. J. Math.* 26 (1974), 948-961.

[10] L. Mirsky, *Transversal theory* (Academic Press, New York, 1971).

[11] C. St. J. A. Nash-Williams, Marriage in denumerable societies, *J. Combin. Theory Ser. A* 19 (1974), 335-366.

[12] C. St. J. A. Nash-Williams, Another criterion for marriage in denumerable societies, *Advances in graph theory* (Proceedings of the Cambridge Combinatorial Conference in 1977, edited by B. Bollobás, North-Holland Publishing Company, Amsterdam), *Annals of Discrete Mathematics* 3 (1978), 165-179.

[13] C. St. J. A. Nash-Williams, Marriage in societies in which each woman knows countably many men, in *Proceedings of the Tenth Southeastern Conference on Combinatorics, Graph Theory and Computing, Volume 1* (edited by F. Hoffman, D. McCarthy, R. C. Mullin and R. G. Stanton, Congressus Numerantium XXIII, Utilitas Mathematica Publishing Inc, Winnipeg, 1979), 103-115.

[14] C. St. J. A. Nash-Williams, A glance at graph theory - Part I, *Bull. London Math. Soc.* 14 (1982), 177-212.

[15] S. Shelah, Notes on partition calculus, in *Infinite and finite sets* (Proceedings of Colloquium in 1973), Vol. III (edited by A. Hajnal, R. Rado and V. T. Sós, North-Holland Publishing Company, Amsterdam, 1975), 1257-1276.

[16] R. Aharoni, Menger's theorem for graphs containing no infinite paths, to appear.

Dominating Cycles in Bipartite Graphs

Pat Ash
Bill Jackson

ABSTRACT

We show that if G is a 2-connected bipartite graph of minimum degree k, and each set in the bipartition has at most 3k-3 vertices, then G contains a longest cycle which is also a dominating cycle.

Introduction

All graphs considered in this paper are assumed to be simple. A cycle C in a graph is a *dominating cycle* if every edge of G is incident with a vertex of C.

Nash-Williams [3] has shown that if G is a 2-connected graph of minimum degree k on at most 3k-2 vertices, then any longest cycle of G is a dominating cycle. We prove a related result for bipartite graphs.

THEOREM 1. If G is a 2-connected bipartite graph of minimum degree k, and each set in the bipartition has at most 3k-3 vertices, then G contains a longest cycle which is also a dominating cycle.

Suppose G is a 2-connected bipartite graph with bipartition (A,B). We shall use the following notation:

If C is a cycle in G and $u_1, u_2, \ldots, u_n$ are vertices of $G-C$, then we shall denote by $N_C(u_i)$ the set of vertices of C which are adjacent to u_i and put $N_C(u_1, u_2, \ldots, u_n) = \bigcup_{i=1}^{n} N_C(u_i)$. Labelling consecutive vertices of C by $c_0, c_1, \ldots, c_{s-1}$, then we define the set

$$N_C^+(u) = \{c_j : c_{j-1} \in N_C(u)\} \text{ and } N_C^-(u) = \{c_j : c_{j+1} \in N_C(u)\}$$

(working mod s on suffices). Similar definitions apply for the sets $N_C^{++}(u)$ and $N_C^{+++}(u)$. Clearly, if $u \in B$, $N_C(u) \subseteq A$ and $N_C^+(u) \subseteq B$. If $c_j, c_{j+2} \in N_C(u)$, we will call the edges uc_j, uc_{j+2} consecutive edges from u to C.

Let $[c_m, c_n]$ denote the set $\{c_m, c_{m+1}, \ldots, c_n\}$. We use a result of H. J. Voss and C. Zuluaga.

THEOREM 2 [4]. *A 2-connected bipartite graph, G, of minimum degree $k \geq 3$ contains a cycle of length at least $\min(4k-4, 2|A|, 2|B|)$.*

LEMMA 3. *Let G be a graph satisfying the conditions of Theorem 1 and such that $k \geq 3$. Let C be a longest cycle in G and H a component in $G - C$ having a longest path, P such that $|V(P)| \geq 2$. Then there are two consecutive edges from each end vertex of P to C.*

PROOF OF LEMMA 3. Let the end vertices of P be u and v and assume that no two edges from u to C are consecutive. If $\min(|A|, |B|) \leq 2k-2$ then by Theorem 2, a longest cycle contains all vertices of A or B. Therefore, every longest cycle is a dominating cycle. Hence, we have $\min(|A|, |B|) > 2k-2$, and Theorem 2 implies $|V(C)| \geq 4k-4$. Thus $|V(G-C)| \leq 2k-2$ and there are at most $k-1$ vertices of A and $k-1$ vertices of B in $G-C$. Since all vertices of G have minimum degree k, it follows that $|N_C(x)| \geq 1$ for all $x \in V(P)$.

Without loss of generality $u \in B$.

Case (i) where $|V(P)| = 2n \ (n \geq 1)$.

$N_C(u) \cap N_C^{++}(u) = \emptyset$ by assumption.

$N_C(u) \cap N_C^+(v) = N_C^+(v) \cap N_C^{++}(u) = \emptyset$ because of the maximality of C. Using the maximality of C and the fact that $|N_C(v)| \geq 1$ and $|N_C(u)| \geq 1$ we may choose two pairs of not necessarily distinct vertices $c_i, c_{j'} \in N_C(u)$ and $c_{i'}, c_j \in N_C(v)$ such that $N_C(u,v) \cap [c_{m+1}, c_{m'-1}] = \emptyset$ for $m = i, j$. Since $|[c_{m+1}, c_{m'-1}] \cap (N_C^{++}(u) \cup N_C^+(v))| = 1$ if $m = i, j$ and $|A \cap [c_{m+1}, c_{m'-1}]| \geq n$ for $m = i, j$ by maximality of C, we have

$$|V(C) \cap A| \geq |N_C(u)| + |N_C^{++}(u)| + |N_C^+(v)| + 2n - 2$$

$$\geq 3(k-n) + 2n - 2, \text{ by maximality of } P.$$

Hence

$$|A| \geq (3k-n-2) + n, \text{ since } |V(P) \cap A| = n$$

$$= 3k - 2.$$

Case (ii) where $|V(P)| = 3$.

Let t be the middle vertex of P, and suppose that there are m vertices of $B \setminus V(C)$ adjacent to t.

$N_C^+(u) \cap N_C^-(u) = \emptyset$ by assumption

$N_C^+(u) \cap N_C(t) = N_C^-(u) \cap N_C(t) = \emptyset$ because of the maximality of C.

Hence $|V(C) \cap B| \geq |N_C^+(u)| + |N_C^-(u)| + |N_C(t)|$

$\geq 2(k-1) + (k-m)$, by maximality of P.

Therefore $|B| \geq 2(k-1) + (k-m) + m$, by assumption

$= 3k - 2$

Case (iii) where $|V(P)| = 2n+1$, $(n>1)$.

$N_C^+(u) \cap N_C^{+++}(u) = \emptyset$ by assumption

$\left.\begin{array}{l} N_C^-(v) \cap N_C^+(u) \\ N_C^-(v) \cap N_C^{+++}(u) \end{array}\right\} = \emptyset$ by maximality of C.

We now consider 2 cases:

(a) $\underline{N_C(u) = N_C(v)}$

Since $|N_C(x)| \geq 1$ for all $x \in V(P)$ there exists at least one edge from u' to C where u' is the neighbour of u on P.

Choose $c_r \in N_C(u')$. Then we can find $c_j, c_i \in N_C(v)$ such that $c_r \in [c_j, c_i]$ and $N_C(v) \cap [c_{j+1}, c_{i-1}] = \emptyset$ (the argument at the beginning of the proof of the Lemma and $|V(P)| \equiv 1 \pmod{2}$ implies $|N_C(v)| > 1$).

$\left.\begin{array}{l} |B \cap [c_{j+1}, c_{r-1}]| \\ |B \cap [c_{r+1}, c_{i-1}]| \end{array}\right\} \geq n$, by maximality of C

then $|B \cap [c_{j+1}, c_{i-1}]| \geq 2n + 1$, since $c_r \in B$.

Further $\left.\begin{array}{l} |N_C^+(u) \cap [c_{j+1}, c_{i-1}]| \\ |N_C^{+++}(u) \cap [c_{j+1}, c_{i-1}]| \\ |N_C^-(v) \cap [c_{j+1}, c_{i-1}]| \end{array}\right\} = 1$ (by the choice of c_j, c_i).

Therefore we have

$|V(C) \cap B| \geq |N_C^+(u)| + |N_C^{+++}(u)| + |N_C^-(v)| + (2n+1) - 3$

$$\geq 3(k-n) + 2n - 2$$
$$= 3k - n - 2.$$
Hence $|B| \geq (3k - n - 2) + (n + 1)$, since $|V(P) \cap B| = n+1$
$$= 3k - 1.$$

(b) $\underline{N_C(u) \neq N_C(v)}$

Since $|N_C(u)| \geq 1$ and $|N_C(v)| \geq 1$ we may choose pairs of not necessarily distinct vertices $c_i, c_j \in N_C(u)$ and $c_{i'}, c_j \in N_C(v)$ such that

$$N_C(u,v) \cap [c_{m+1}, c_{m'-1}] = \emptyset \text{ for } m = i, j \quad \text{(i)}$$
$$\{c_j, c_{j'}\} \not\subseteq N_C(u) \cap N_C(v). \quad \text{(ii)}$$

Since $|B \cap [c_{m+1}, c_{m'-1}]| \geq n+1$ for $m=i,j$ by maximality of C, and because of (i)

$$|\{B \cap [c_{i+1}, c_{i'-1}]\} \setminus \{N_C^+(u) \cup N_C^{++}(u) \cup N_C^-(v)\}| \geq n-2$$

and further, by (ii)

$$|\{B \cap [c_{j+1}, c_{j'-1}]\} \setminus \{N_C^+(u) \cup N_C^{++}(u) \cup N_C^-(v)\}| \geq n-1.$$

Hence
$$|V(C) \cap B| \geq |N_C^+(u)| + |N_C^{++}(u)| + |N_C^-(v)| + 2n-3$$
$$\geq 3k - n - 3.$$

Therefore
$$|B| \geq (3k - n - 3) + (n + 1)$$
$$= 3k - 2.$$

Since all possible cases contradict the hypothesis that $|A|$ and $|B|$ are at most $3k-3$, this completes the proof of Lemma 3.

PROOF OF THEOREM 1. For $k=1$, Theorem 1 is vacuously true; for $k=2$, the validity of Theorem 1 is easily checked; hence we assume $k \geq 3$. Let $C = c_0 c_1 \cdots c_n c_0$ be a longest cycle in G such that $|E(G-C)|$ is as small as possible, H_0 a component in $G-C$ and P_0 a longest path in H_0 such that $|V(P_0)| \geq 2$. If no such component exists then C must be a dominating cycle and Theorem 1 holds. Let u_0 be an end vertex of P_0 and without loss of generality let $u_0 \in B$. By Lemma 3 there are 2 consecutive edges from u_0 to C. Let the end vertices of these edges be $x, \bar{x}$ and let y be the vertex $c_h \in C$ where $c_{h-1} = x$ and

$c_{h+1} = \bar{x}$. Now let C_1 be the longest cycle $u_0[c_{h+1}, c_{h-1}]u_0$ and u_0' be the neighbour of u_0 on P_0. Then $u_0 u_0' \in E(G-C)$, but $u_0 u_0' \notin E(G-C_1)$. Since $|E(G-C)|$ is as small as possible, there exists at least one edge from y to a component H_1 of $G-C$. By the maximality of C, $H_1 \neq H_0$. Since $|V(H_1) \cap A| \geq 1$ we may choose $u_1 \in A \cap V(H_1)$. Let $|V(H_1) \cap B| = m$ and P^* be a path from y to u_1 in $G[V(H_1) \cup \{y\}]$.

(i) $|V(P_0)| = 2n$, $n \geq 1$

$N_C(u_1) \cap N_C^-(u_0) \subseteq \{y\}$ since if $c_i \in N_C(u_1) \cap N_C^-(u_0) \setminus \{y\}$ then the cycle $[c_{i+1}, y] P^* u_1 c_i c_{i-1} \cdots \bar{x} u_0 c_{i+1}$ contradicts the maximality of C. Let v_0 be the other end vertex of P_0. $N_C^-(u_0) \cap N_C^{--}(v_0) = \emptyset$ since if $c_i \in N_C^-(u_0) \cap N_C^{--}(v_0)$ then the cycle $c_{i+1} u_0 P_0 v_0 [c_{i+2}, c_{i+1}]$ contradicts the maximality of C. Hence $y \notin N_C^{--}(v_0)$.

$N_C(u_1) \cap N_C^{--}(v_0) = \emptyset$ since if $c_i \in N_C(u_1) \cap N_C^{--}(v_0)$ then the cycle $[c_{i+2}, y] P^* u_1 c_i c_{i-1} \cdots \bar{x} u_0 P_0 v_0 c_{i+2}$ contradict the maximality of C.

Since $|N_C(u_l)| \geq 1$ for $l = 0, 1$ and $|N_C(v_0)| \geq 1$ we may choose $c_i \in N_C(u_0, u_1) \setminus \{y\}$ and $c_j \in N_C(v_0)$ such that $N_C(u_0, u_1, v_0) \cap [c_{i+1}, c_{j-1}] = \emptyset$. Since

$$|[c_{i+1}, c_{j-1}] \cap \{N_C(u_1) \cup N_C^-(u_0) \cup N_C^{--}(v_0)\}| = 1$$

and $|B \cap [c_{i+1}, c_{j-1}]| \geq n$ by the maximality of C we have

$$|V(C) \cap B| \geq |N_C(u_1)| + |N_C^-(u_0)| + |N_C^{--}(v_0)| + n - 2$$
$$\geq (k - m) + 2(k - n) + (n - 2)$$
$$= 3k - n - m - 2$$

Hence $|B| \geq (3k - n - m - 2) + (n + m) = 3k - 2$, since $|V(P_0) \cap B| = n$ and $|V(H_1) \cap B|| = m$. This contradicts the assumption on $|B|$.

For the case when $|V(P_0)| = 2n + 1$, $n \geq 1$ we obtain a similar contradiction by considering the sets $N_C(u_1)$, $N_C^+(v_0)$ and $N_C^-(u_0)$. Hence there does not exist a component H_0 in $G - C$ having a longest path P_0 such that $|V(P_0)| \geq 2$. It follows that C must be a dominating cycle.

This completes the proof of Theorem 1.

The following graph shows that the bound of $3k - 3$ in the statement of Theorem 1 is best possible:

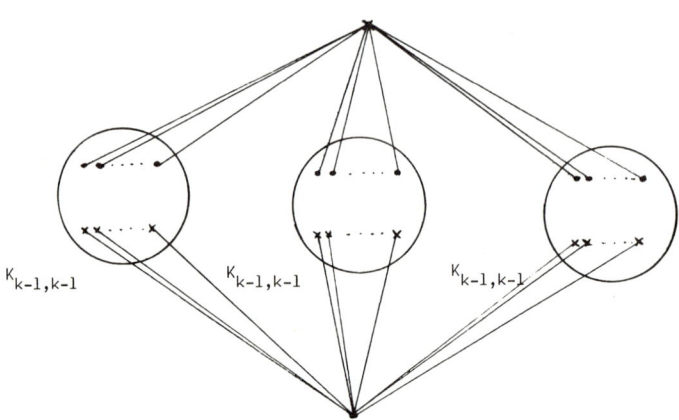

Every longest cycle in this graph omits a subgraph $K_{k-1,k-1}$. Moreover the conclusion of Theorem 1 cannot be strengthened to show that "every longest cycle is a dominating cycle", as the following graph illustrates:

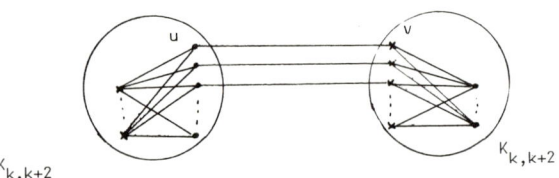

A longest cycle has length $4k+2$. However there exist longest cycles which avoid both u and v.

We hope that Theorem 1 may be useful in attacking the following conjecture due to R. Häggkvist. [2]

CONJECTURE All 2-connected, k-regular bipartite graphs on at most $6k-6$ vertices are hamiltonian.

A final problem concerning dominating cycles, is a strong form of a conjecture of Fleischner [1].

CONJECTURE Every 3-regular, cyclically 4-edge connected graph has a dominating cycle.

References

[1] H. Fleischner, Cycle Decomposition, 2-Coverings, Removable Cycles, and The Four-Color-Disease, in "Proceedings, Silver Jubilee Conference on Combinatorics, University of Waterloo 1982".

[2] R. Häggkvist, unsolved problem, in "Proceedings, Fifth Hungarian Colloquium on Combinatorics, 1976".

[3] C. St. J. A. Nash-Williams, Edge-disjoint hamiltonian circuits in graphs with vertices of large degree in "Studies in Pure Mathematics", papers presented to Richard Rado, Academic Press, London/New York, 1971.

[4] H. J. Voss and C. Zuluaga, Maximale gerade and ungerade Kreise in Graphen I. Wiss Z. Techn. Hochsch. Ilmenau 23.4 (1977), 57-70.

Helms are Graceful

Jacqueline Ayel

Odile Favaron

ABSTRACT

Let a helm be a graph obtained from a wheel by adding a pendent edge at each vertex on the rim of the wheel. In this paper, we prove that the helms are graceful thus answering a conjecture of Koh, Rogers, Teo, Yap.

Introduction

A graph $G=(X,E)$ is *numbered* if each vertex v is assigned a non-negative integer $\phi(v)$, and each edge (v,w) is assigned the absolute value of the difference of the numbers at its endpoints, that is $|\phi(v)-\phi(w)|$.

The numbering is called *graceful* is furthermore the vertices are labelled with distinct integers of $\{0,1,\ldots,m\}$ and the edges with the integers, from 1 to m.

A graph which admits a graceful numbering is said to be *graceful*. (see [1] for a survey on graceful graphs).

The *helm* H_n, $n \geq 3$, is the graph obtained from the wheel W_n with n spokes, by adding a pendent edge at each vertex on the wheel's rim. In [2] Koh, Rogers, Teo, Yap prove that the helms are graceful for n odd and conjectured that they are also graceful for n even.

Here we prove this conjecture.

THEOREM: the helms are graceful.

PROOF: By the results of [2] it suffices to consider the case n even. However our methods applied in the case n odd give graceful numberings different from those of [2].

Let H_n a helm, the number m of edges is equal to $3n$ and the number of vertices is $2n+1$.

We give a numbering which depends on the residue of n modulo

4. The center w of the wheel is always labelled with O.

Let w_i, $i=1,\ldots,n$ be the vertices of the wheel's rim in a cyclic order and u_i, $i=1,\ldots,n$, the pendent vertex adjacent to w_i.

CASE 1: $n=4p$: Define ϕ on the vertices of H_n for $p\geq 1$ by:

$\phi(w)=0$

$$\phi(w_i)=\begin{cases} 10p+k, & i=2k; \quad 1\leq k\leq 2p; \\ 10p-k, & i=2k+1; \quad 0\leq k\leq p-1; \\ 10p-k-1, & i=2k+1; \quad p\leq k\leq 2p-1. \end{cases}$$

For $p>1$,

$$\phi(u_i)=\begin{cases} 4p+2k-1, & i=2k \quad ; \quad 1\leq k\leq 2p \\ 4p-2k-1, & i=2k+1 \; ; \; 0\leq k\leq 2p-1, k\neq p-2 \text{ and } k\neq p-1 \\ 2, & i=2p-3 ; \\ 2p+2, & i=2p-1. \end{cases}$$

For $p=1$, the numbering of the pendent vertices is the same, but the formulation is different: $\phi(u_1)=4$, $\phi(u_2)=2$, $\phi(u_3)=1$, $\phi(u_4)=7$. One can verify that ϕ is a graceful numbering of H_n. Indeed the edges of the cycle are numbered with all the integers from 1 to $4p$; the edges of the spokes of the wheel with all the integers from $8p$ to $12p$ except $9p$ and the pendent edges with the remaining values. (see figure 1).

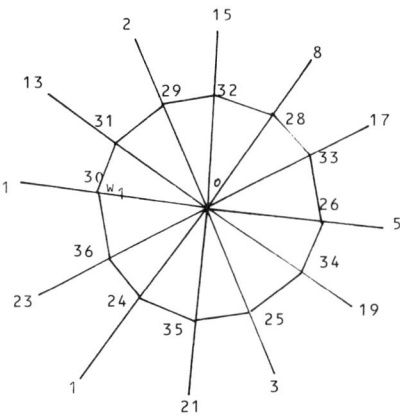

Figure 1 ($n=12$)

CASE 2: $n=4p+2$: Define ϕ on the vertices of H_n for $p \geq 1$ by:

$\phi(w)=0$

$$\phi(w_i)=\begin{cases} 10p+k+5, & i=2k \quad ; \quad 1\leq k\leq 2p+1 \\ 6p+5, & i=1 \\ 10p-k+6, & i=2k+1 \quad ; \quad 1\leq k\leq 2p. \end{cases}$$

$$\phi(u_i)=\begin{cases} 2, & i=1 \\ 4p+2, & i=2 \\ 4p+5, & i=4 \\ 4p+2k+2, & i=2k \; ; \; 3\leq k\leq 2p+1 \\ 4p-2k+1, & i=2k+1 \; ; \; 1\leq k\leq 2p \end{cases}$$

One can verify that ϕ is a graceful numbering of H_n. Indeed the edges of the cycle are numbered with all the integers from 1 to $4p+1$, and $6p+1$; the edges of the spokes of the wheel with all the integers from $8p+6$ to $12p+6$, and $6p+5$; the pendent edges with the remaining values. (see figure 2).

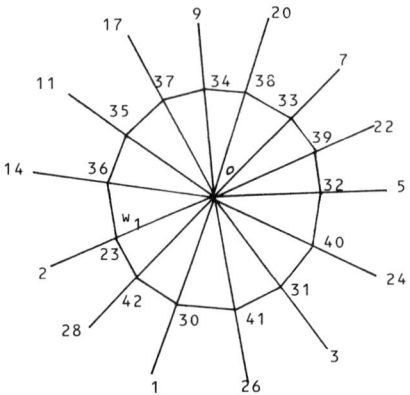

Figure 2 ($n = 14$)

References

[1] J.C. Bermond: Graceful graphs, radio antennae and French windmills. One day combinatorics conference, open university 1978. Pitman 1979, 18-37.

[2] K.M. Koh, D.G. Rogers, H.K. Teo and K.Y. Yap: Graceful graphs: some further results and problems, in Proc. 11th S.E. Conference on Combinatorics, Graphs theory and Computing, Congressus Numerantium Utilitas Math. Vol 29(1980), pp 559-571.

Edge to Point Degree Lists
and
Extremal Triangle Free
Point Degree Regular Graphs

D. Bauer

G. Bloom

F. Boesch[*]

ABSTRACT

The well known concept of the degree of a point of a graph may be viewed as the number of points in the neighborhood of a given point. The dual concept of defining the number of edges in the neighborhood of a given edge as the edge degree has already appeared in the literature. In this work we introduce a new concept of degree called the edge-to-point degree, defined as the number of points in the neighborhood of a given edge. We then investigate the realizability question for edge-to-point degree lists. The study of realizability of such lists when all terms are equal leads to our two main results.

First we solve a realizability problem involving triangle free, point degree regular graphs. Next we consider regular lists of length q, i.e., all of whose terms are equal to a single number r. We completely determine those regular lists which are realizable as edge-to-point degree lists when $r \geq 6$ and $q \leq 3r-3$.

---

[*]This author's work was supported in part by the National Science Foundation under Grant Number ECS-8100652.

1. Introduction

In this paper we introduce the notion of an edge-to-point degree list. First we define some elementary concepts for the convenience of the reader. Notation and terminology not introduced here are as in the book by Harary [11]. A *graph* G is an ordered pair of sets (V,X) where elements of X, called *edges*, are two element subsets of non-empty set of points V. The elements of an edge are called its two *endpoints*. The orders of V and X are denoted by p and q respectively. Two points are *adjacent points* if they are endpoints of an edge. Two edges are said to be *adjacent edges* if they share an endpoint. It is assumed that a point is not adjacent to itself; likewise an edge is not adjacent to itself. An edge is said to be *incident* at its endpoints. The total number of edges incident at a point v of a graph G is called the *degree of the point* v and is denoted by $\deg v$. If the degree of each point of a graph is placed in a list, then the list is called the *degree list of G and is denoted by* $\pi(G)$.

An arbitrary list is said to be *realizable* if there is some graph G which has this list as its degree list. An explicit set of inequality conditions for realizability are due to Erdős and Gallai [8], and a simple recursive algorithm for testing the realizability of a list is due to Havel [12] and Hakimi [9]. Both of the preceding conditions deal with general realizability, i.e., to determine if there is any graph with the given list as its degrees. If one restricts the realizability to specific types of graphs, however, the conditions often become quite simple. For example, tree realizability is well known. Likewise, simple conditions for realizability of a graph containing only one cycle (a *unicycle*) were given by Boesch and Harary [5].

Another special case is *point degree regular* graphs, i.e., those graphs whose points each have the same degree r. Any list of p terms all equal to r we call a regular list $(r)^p$, where $0 \le r \le p-1$. Such a list is realizable provided that r and p are not both odd. We mention one of the many possible proofs here because a particularly interesting class of graphs realize all point degree regular lists. Let C_p denote a cycle on p points, and let C_p^k be formed by adding edges between pairs of points on C_p whenever they are endpoints of a path on C_p of length not exceeding k. It is easy to see that for each $k \le \lfloor p/2 \rfloor$, C_p^k is point degree regular of degree $2k$. Thus we can realize any even degree r. The case of odd r is handled by adding edges between opposite pairs of points on C_p^k.

The graphs C_p^k form a subset of a more general class of graphs which are defined as follows. Let E denote a $p \times p$ circulant matrix where $e_{i,i+1}=1$, $e_{p,1}=1$ and $e_{ij}=0$ otherwise. We define a *circulant* graph $C_p<n_1,n_2,\ldots,n_k>$ for any sequence of integers where $1\le n_1<n_2<,\ldots,n_k<p/2$ as a graph whose adjacency matrix A may be expressed, using $()^t$ to denote transpose, as $A=\sum_{i=1}^{k}E^{n_i} + (\sum_{i=1}^{k}E^{n_i})^t$. Clearly $C_p<n_1,n_2,\ldots,n_k>$ is a generalization of C_p^k in that $C_p^k=C_p<1,2,\ldots,k>$.

A wide variety of realizability problems were recently surveyed by Hakimi and Schmeichel [10]. At the end of their survey they discuss the concept of a degree for an edge of a graph. This idea was introduced by Harary [11] in the section on symmetric graphs. In [10,11] the authors all define an unordered pair of integers representing the degrees of the endpoints of an edge to be an edge degree pair. Necessary and sufficient conditions for a list of integer pairs to be a list of edge degree pairs of a graph are derived by Hakimi and Patrinos in [13]. Now if a dual concept to point degree is desired for edges, then it is natural to associate a single number with each edge as its edge degree. One such possibility is to define the *degree of an edge x of graph G*, $dl_x(G)$, as the total number of edges that x is adjacent to in G. For the purpose of this work, we shall assume that the term edge degree always has this meaning. This is equivalent to defining $dl_x(G)$ as the degree of the point corresponding to x in the line graph $L(G)^*$; hence it represents a legitimate dual concept. It is easy to see that if $x = \{u,v\}$ then $dl_x(G) = \deg u + \deg v - 2$. This definition likewise gives rise to a question of realizability of the obviously associated degree list, which we denote by $\pi_l(G)$ and call the *edge degree list* or *line graph degree list*. The general case of determining explicit necessary and sufficient conditions for realizability is an unsolved problem, and there are only results for some special cases, c.f., Bauer [2,3]. The only other published results on realizing edge degree lists are in Rao [14]. Therein he characterizes lists that are *forcibly line-graphic*, i.e., every realization is a line graph.

For edge degree lists it happens that the edge degree regular case can again be solved, and the conditions for realizability are straightforward, albeit nonobvious [2]. We digress a bit at this point to comment on some related results. A recent work of Caccetta and Häggkvist [7] shows that $dl_x(G)$ has a role to play in the study of "optimal diameter graphs". More specifically, let a shortest path between two points be called a *geodesic*, and denote the length of the longest geodesic in a graph G by the *diameter*, $d(G)$. In [7], G is called *diameter k-critical* if $d(G)=k$ and for each edge x of G, $d(G-x) > k$. They then obtain an

---
*c.f. [11,p71] for the definition of a line graph.

upper bound on the value of $(dl_x(G)+2)$ averaged over all edges x of a diameter 2 critical graph. This result is of interest in applications of graph theory to the design of reliable minimum cost computer communication networks. The connection between networks and graph diameter was observed by Bollobás [6].

Herein we now introduce yet another way to associate a single number with each edge x of G and define the *edge-to-point degree* $dp_x(G)$ of edge x in G as the total number of distinct points in $G-x$ adjacent to the endpoints of x. Notice that $dl_x = dp_x$ if and only if x is not contained in any triangle of G. However in general they are not equal. Also note that an attempt to define a dual concept of a point-to-edge degree does not yield a new invariant as the number of edges incident to a point v is equal to the number of points adjacent to v. Since the numbers $dp_x(G)$ are new, they lead to new realizability questions for the q term list of edge-to-point degrees, which we call the *edge-to-point degree list* $\pi_e(G)$. For simplicity we use e.p.d. for edge-to-point degree and refer to dp_x as the *e.p.d.* of x.

It is interesting to note that the e.p.d. of x concept also plays a role in the study of diameter two graphs. We define the complement $\overline{G}$ of G to be the graph having the same points as G, with two points of $\overline{G}$ being adjacent if and only if they are not adjacent in G. Bloom, Kennedy, and Quintas [4] show that a necessary and sufficient condition for a graph to have diameter two is that the maximum value of $dp_x(G)$ over all edges x in $\overline{G}$ not exceed $p-3$.

In the remainder of this work we initiate a study of the realizability of e.p.d. lists by first investigating the question for regular e.p.d. lists. This problem appears to be more difficult than it was for regular lists of the form $\pi(G)$ or $\pi_l(G)$. For instance it is easy to see that the circulants C_p^k, for $k \geq 2$, are point degree regular and edge degree regular but are not e.p.d. regular. Nevertheless we begin to investigate the regular case by noting that a triangle free graph which is point degree regular will also be e.p.d. regular. This naturally leads to the question of realizing triangle free point degree regular graphs. The solution of this problem is Theorem 1 in the next section.

In section 3 we characterize those regular e.p.d. lists that can be realized by a connected triangle free graph. Our main result in section 4 states precisely which regular e.p.d. lists of the form $\pi_e = (r)^q$ can be realized by a connected graph if $r \geq 6$ and if $q \leq 3r - 3$. Finally we conclude with some remarks concerning future work.

2. Extremal Triangle Free Regular Graphs

As noted in the introduction, a graph G is e.p.d. regular if G is point degree regular and triangle free. For simplicity we will henceforth refer to point degree regular as simply regular. Thus if we determine the maximum allowable degree of a regular triangle free p point graph, we can then investigate realizability of regular e.p.d. lists by considering regular triangle free graphs of degree less than or equal to this maximum.

Now if p is even, the above extremal problem may be solved by invoking the famous Turán Theorem [17]. Specifically the Turán result states that the maximum number of edges in any triangle free p point graph is $\lfloor p^2/4 \rfloor$. Now when p is even the extremal graph for the Turán result is the complete bigraph $K_{p/2, p/2}$. As this graph is regular of degree $p/2$ and any regular graph of larger degree would have more than $\lfloor p^2/4 \rfloor$ edges, it follows that $p/2$ is the maximum degree of regular triangle free graphs in this case. However, when p is odd the Turán extremal graph is not regular. If p is odd, the solution to the extremal problem follows from work of both Sheehan [15,16], and Andrásfai, Erdős, and Sós [1]. The realizability question is answered below in Theorem 1.

THEOREM 1. There exists a regular triangle free graph of degree r on p points if and only if p is even and $p \geq 2r$; or p is odd, r is even and $p \geq 5r/2$.

PROOF. First we dispense with the case of p even. As noted above the maximum value of r is $p/2$. Furthermore, the following construction yields a regular bigraph of degree r for $1 \leq r \leq p/2$. Let the points in the two parts of a bigraph be number $1,2,\ldots,p/2$ and $1',2',\ldots,(p/2)'$ and connect each point i to $(i+k)'$ mod $p/2$ for $0 \leq k \leq r-1$.

Now suppose p is odd. To show how to realize triangle free regular graphs when $p \geq 5r/2$, suppose first that $p \geq 3r-1$. Then the circulant graph $C_p <1,3,5,\ldots,r-1>$ has the required properties. Now if $5r/2 \leq p \leq 3r-1$, form G as in Figure 1.

The number of points in each of the five induced subgraphs of G, denoted A, B, C, D, and E, is indicated in Figure 1. Furthermore, each induced subgraph contains no internal edges, e.g., $A = (r/2)K_1$. The solid lines joining pairs of the induced subgraphs indicate that every point in one of the subgraphs is adjacent to every point in the other subgraph. The dashed lines joining D and E indicate that the induced subgraph on the points in D and E form a regular bigraph of degree $r/2$. Note that this is possible since $|D| = |E| \geq r/2$.

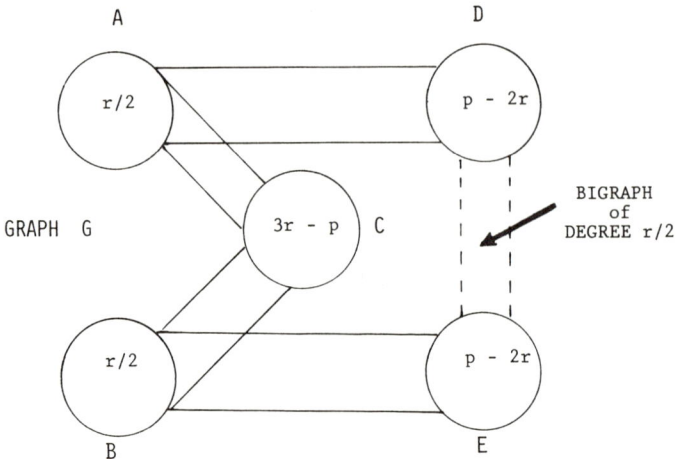

Figure 1 - The graph G

Now since each of A,B,C,D, and E has no internal edges, a necessary condition for G to have a triangle is that the three points of the triangle lie in three distinct subgraphs chosen from A,B,C,D, and E. Clearly from Figure 1 this is not possible. Thus G has no triangle. It is also easy to see that G is regular of degree r.

Now suppose $p < 5r/2$ and that H is a regular triangle free graph on p points with degree r. Since p is odd H can not be bipartite, hence H contains a shortest odd cycle C_m, $m \geq 5$. Now every point in C_m is adjacent to exactly $r-2$ points in $H-C_m$. However a point in $H-C_m$ can be adjacent to at most two points in C_m; otherwise a shorter odd cycle is formed. Hence $m(r-2) \leq 2(p-m)$, implying $r \leq \frac{2p}{m} \leq \frac{2p}{5}$, a contradiction.

The following corollary is immediate.

COROLLARY. For p odd, if $p \equiv a \mod 5$ then the maximum number of lines in any triangle free regular graph on p points is $p(p-a)/5$.

To complete this section we now show how to use these results to generate realizations of regular e.p.d. lists.

THEOREM 2. The list $(r)^q$ is realizable as an e.p.d. list if either:

(a) there exists an even integer p and an integer s such that:
  (1) $1 \leq s \leq p/2$
  (2) $r = 2s - 2$
  (3) $q = ps/2$, or
(b) there exists an odd integer p and an even positive integer s such

that:

(1) $p \geq \dfrac{5s}{2}$

(2) $r = 2s-2$

(3) $q = ps/2$.

3. Realizability of Regular Edge-to-Point Degree Lists by Triangle Free Graphs

As noted in the introduction if $x = \{u,v\}$ is an edge of a triangle free graph G then $dl_x(G) = dp_x(G) = \deg u + \deg v - 2$, which is the degree of x in the line graph $L(G)$. Now it is known that for $L(G)$ to be regular, G must be either regular or bipartite biregular i.e., a bigraph in which all points in a given part have the same degree. This is summarized by the following theorem from [2].

THEOREM 3. There is a connected regular line graph of degree k on p points if and only if one of the following holds:

(1) $p = k+1$

(2) for some integer $c \geq 2$, $k = 2c-2$, $2p \geq (c+1)c$, and $2p \equiv 0 \mod c$

(3) for some integers $c,b \geq 2$, $k=c+b-2$, $p \geq cb$, and $p \equiv 0 \mod lcm(c,b)$.

Thus to find all e.p.d. lists $(r)^q$ which can be realized by triangle free graphs it suffices to examine only bipartite biregular graphs as well as regular triangle free graphs. We have already characterized in Theorem 2 those lists $(r)^q$ which can be realized by connected regular triangle free graphs. The conditions under which a list $(r)^q$ can be realized by a connected bipartite biregular graph are easily extracted from conditions (1) and (3) of Theorem 3. Note that condition (2) of Theorem 3 does not apply, the reason being that the regular graphs characterized by that condition may contain triangles.

THEOREM 4. The e.p.d. list $(r)^q$ is realized by a connected bipartite biregular graph if and only if either:

(1) $q = r+1$

(2) for some integers, $c,b \geq 2$, $r=c+b-2$, $q \geq cb$ and $q \equiv 0 \mod lcm(c.b)$.

We complete this section by noting that both Theorem 2 and Theorem 4 are necessary to characterize all regular e.p.d. lists which can be realized by connected triangle free graphs. To see this, note first that the list $(18)^{125}$ can be realized via Theorem 2 by a regular graph of degree 10 on 25 points, while it can not be realized by any bigraph. However the list $(18)^{96}$ can not be realized by a triangle free regular graph but is realized by $K_{8,12}$.

4. On the Realizability of Lists as Regular Edge-to-Point Degree Lists

In this section we investigate regular lists of the form $(r)^q$ which can be realized as e.p.d. lists of connected graphs. Note that if an edge of a graph has e.p.d. r then G must have at least $r+1$ edges. Hence $q \leq r$ is not possible.

We begin by employing results of the preceeding section.

LEMMA 1. If $(r)^q$ is realized by a triangle free connected graph G then either $q=r+1$, $q=2r$ or $q \geq 3r-3$.

PROOF. We know by Theorem 3 that G must be either regular or bipartite biregular. First suppose G is regular of degree s. Then by Theorem 1 we know that for p either odd or even, $p \geq 2s$. Since $r=2s-2$ we have $q \geq ps/2 \geq (r+2)^2/2 \geq 3r-3$. Now if G is bipartite biregular, by Theorem 4 G can only be either $K_{1,r}$, $K_{2,r}$ or $K_{3,r-1}$ having $(r)^{r+1}$, $(r)^{2r}$ and $(r)^{3r-3}$ as their respective e.p.d. lists.

We now suppose that the list $(r)^q$ is realized as the e.p.d. list of a connected graph G having an edge $e=\{u,v\}$ belonging to a triangle. Let B be the set of points adjacent to both u and v. Let A be those points adjacent to u but not to v and let C be those points adjacent to v but not to u. Then G has a subgraph G' of the form depicted in Figure 2, where $|A| = a$, $|B| = b$ and $|C| = c$.

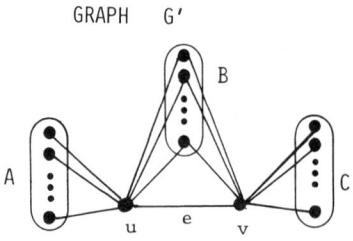

Figure 2 - The subgraph G'

Note that deg $u=a+b+1$, deg $v=b+c+1$ and $dp_e(G)=a+b+c$. Now if G has p points and G' is a spanning subgraph of G then $r=p-2$. If a graph is e.p.d. regular of degree $r=p-2$ we will call it an *e.p.d.r.2* graph. Such graphs are closely related to diameter two graphs. In fact one can easily show that G is an e.p.d.r.2 graph if and only if $G=K_p$ or G has diameter two and no induced five cycle. We will make use of e.p.d.r.2 graphs in the following lemmas.

Denote by $<A>$ the subgraph of G induced by the points of A, and similarly for $<B>$ and $<C>$.

LEMMA 2. Suppose G is an e.p.d.r.2 graph having an edge $e=\{u,v\}$ belonging to a triangle. Let B be the set of points adjacent to u and v. Let A be those points adjacent to u but not to v, and let C be those points adjacent to v but not to u. Then

(1) $<A>=aK_1$ and $<C>=cK_1$

(2) $<B>=bK_1$ or $<B>$ is an e.p.d.r.2 induced subgraph of G of degree $r'=b-2$

PROOF. Suppose $e'=\{y,z\}$ is an edge of $<A>$. Since the e.p.d. of e' is $p-2$, either y or z must be adjacent to v. However if, for example y were adjacent to v, y would be a point of B and not a point of A. Thus $<A>=aK_1$, and similarly $<C>=cK_1$. Now suppose $<B>$ contains an edge $e''=\{r,s\}$. Since the e.p.d. of e'' is $p-2$, e'' must be adjacent to the remaining points of B. This argument applies to each edge in $<B>$, proving (2).

LEMMA 3. Suppose G realizes the e.p.d. list $(r)^q$ and has a subgraph G' of the type described in Lemma 2 with $a=c=0$. Then G is an e.p.d.r.2 graph.

PROOF. First note that $r=b$. Now suppose a point w in $V(G)-V(G')$ is adjacent to a point z in B. Then edge $\{v,z\}$ has an e.p.d. greater than b, a contradiction. Thus $V(G)=V(G')$ and G is an e.p.d.r.2 graph.

LEMMA 3a. If G is as in Lemma 3 then either $q=2r+1$ or $q\geq 3r$.

PROOF. If $<B>=bK_1$, then clearly $q=2r+1$. Otherwise, by Lemma 2, $<B>$ is an e.p.d.r.2 induced subgraph of degree $r-2$ and hence must have at least $r-1$ edges. Therefore G must have at least $3r$ edges.

LEMMA 4. Suppose G realizes the e.p.d. list $(r)^q$ and has a subgraph G' of the type described in Lemma 2 with $a=0$, $c\neq 0$. Then G is an e.p.d.r.2 graph.

PROOF. Clearly $r=b+c$. If a point w in $V(G)-V(G')$ is adjacent to a point z in B then the e.p.d. of edge $\{v,z\}$ is greater than $b+c$, a contradiction. We obtain a similar contradiction by assuming w is adjacent to a point y in C.

LEMMA 4a. If G is as in Lemma 4 then the points of B and C form a complete bigraph.

PROOF. The edges joining point v to the points of B must each have e.p.d. $= b+c$. Since G is an e.p.d.r.2 graph it must be the case that B and C form a complete bigraph (note that the edges joining points in C now also have e.p.d. $= b+c$).

LEMMA 4b. *If G is as in Lemma 4 then either $q=2r+1$ or $q\geq 3r-3$.*

PROOF. Recall that $r=b+c$. By Lemmas 4 and 4a, G must have exactly $bc+2b+c+1$ edges plus those edges which are in $<B>$. If $b=1$ there are no edges in $<B>$ and G has exactly $2c+3$ edges. Thus $r=c+1$ and $q=2r+1$. Now if $b\geq 2$ it suffices to see that $bc+2b+c+1\geq 3(b+c)-3=3r-3$.

LEMMA 5. *Suppose G realizes the e.p.d. list $(r)^q$, where $r\geq 6$, and has a subgraph G' of the type described in Lemma 2 with $a,c\neq 0$. Then $q\geq 3r-3$.*

PROOF. Without loss of generality assue $a\leq c$. Let w be a point of B. Since the e.p.d. of edge $\{u,w\}$ is $a+b$ in G', and since $r=a+b+c$, w must be adjacent to c additional poins in $G-A-B$. Similarly a point z of A must be adjacent to c additional points in $G-A-B$. Thus we have $q\geq ac+bc+2b+a+c+1$. Hence it suffices to show that $ac+bc+2b+a+c+1\geq 3r-3$, or that $bc+ac\geq 2a+2c+b-4$ if $a\leq c$ and $a+b+c\geq 6$. This follows by first examining the case $a,b,c\geq 2$ and then noting that the result holds when $a=c=1$ and when $a=b=1$.

We have proved the following theorem.

THEOREM 5. *The list $(r)^q$ with $r\geq 6$ is realizable as the e.p.d. list of a connected graph for $q=r+1$, $2r$, $2r+1$ and $3r-3$. It has no realization as the e.p.d. list of a connected graph if $r+2\leq q\leq 2r-1$ and $2r+2\leq q\leq 3r-4$.*

5. Final Remarks

The general question of determining when an e.p.d. list is realizable appears to be difficult. However the possibility of finding all lists $(r)^q$ which can be realized by regular e.p.d. graphs may not be out of reach. To do this it will be necessary to study e.p.d. regular graphs with triangles which are not e.p.d.r.2 graphs. Two examples of such graphs are given in Figure 3.

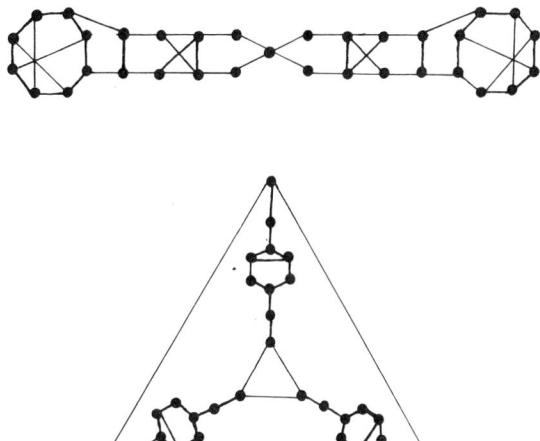

Figure 3 - Two non e.p.d.r.2 graphs

References

[1] B. Andrásfai, P. Erdös and V. T. Sós, On the connection between chromatic number, maximal clique and minimal degree of a graph, *Discrete Mathematics.* 8(1974) 205-218.

[2] D. Bauer, On regular line graphs. *Topics in Graph Theory-Annals of New York Academy of Sciences 328* (F. Harary), ed., New York (1979) 30-31.

[3] D. Bauer, Line-graphical degree sequences, *J. Graph Theory 4 (1980) 219-232.*

[4] G. S. Bloom, J. W. Kennedy, and L. V. Quintas, A characterization of graphs with diameter 2. (Private communication).

[5] F. T. Boesch and F. Harary, Unicycle realizability of degree lists. *Networks 8* (1978) 93-96.

[6] B. Bollobás, *A problem of the theory of communication networks. Theory of Graphs* (G. Katona and P. Erdös, eds.), Akad, Kiado, Budapest (1968) 29-36.

[7] L. Caccetta and R. Häggkvist, On diameter critical graphs. *Discrete Math. 28* (1979) 223-229.

[8] P. Erdös and T. Gallai, Graphs with prescribed degrees of vertices. (Hungarian) *Mat. Lapol 11* (1960) 264-274.

[9] S. Hakimi, On the realizability of a set of integers as degrees of the vertices of a graph. *J. SIAM Appl. Math. 10* (1962) 496-506.

[10] S. L. Hakimi and E. F. Schmeichel, Graphs and their degree sequences: a survey. *Proc. Western Michigan Conf. on Graph Theory and Applications*, Springer-Verlag, New York (1976) 225-233.

[11] F. Harary, *Graph Theory*, Addison-Wesley, Reading (1969).

[12] V. Havel, A remark on the existance of finite graphs (Hungarian) *Časopis Pěst. Mat.* 80 (1955) 477-480.

[13] A. N. Patrinos and S. L Hakimi, Relations between graphs and integer-pair sequences. *Discrete Math.* 15 (1976) 347-358.

[14] S. B. Rao, Characterization of forcibly line-graphic degree sequences, *Utilitas Mathematica* 11 (1977) 357-366.

[15] J. Sheehan, Non-bipartite graphs of girth 4, *Journal of Discsrete Mathematics* 8 (1974), 383-402.

[16] J. Sheehan, An extremal problem in graph theory, *Colloquia Mathematica Societatis János Bolyai*, (1973) 1235-1239.

[17] P. Turán, Eine Extremalaufgabe aus der Graphentheorie. *Mat. Fiz. Lapok* 48 (1941) 436-452.

A Property of k-Optimal Path-Partitions

C. Berge

ABSTRACT

Several attempts have been made to extend the Greene-Kleitman theorem. This note gives a proof, for certain classes of directed graphs, of a property which also extends the Greene-Kleitman theorem, as well as a generalized version of the Gallai-Milgram theorem and the Gallai-Roy theorem. We do not know if this property holds for all directed graphs.

1. k-Optimal Partitions

A path μ, in a directed graph G, is a sequence of vertices, but for convenience, we shall also denote by μ the set of the vertices encountered by this path and by $|\mu|$ the number of these vertices. Let k be a positive integer less than or equal to the maximum cardinality of a path in G. A *partial k-coloring* of G is a family of k disjoint stable sets $S_1, S_2, \ldots, S_k$. If a vertex x belongs to S_j, we say that x is *colored* with j; some of the vertices may bear no color.

Clearly, for a partial k-coloring $(S_1, S_2, \ldots, S_k)$, the number of different colors encountered by a path μ is $\leq \min\{k, |\mu|\}$. We shall say that this coloring is *strong* for a path μ if this path meets exactly $\min\{k, |\mu|\}$ different colors. Now, consider a partition M of the vertex-set into paths $\mu_1, \mu_2, \ldots$. Let x_1 be a vertex of G; let μ_1 be the path of M which contains x_1. The *level* of x_1 (with respect to M) is the cardinality of the portion of μ_1 which starts from x_1 and goes up to the terminal end.

We denote by $L_p(M)$ the set of all the vertices at level p, and for $k \geq 1$, $k \leq \max|\mu|$, we put:

$$\beta_k(M) = |L_1(M)| + |L_2(M)| + \cdots + |L_k(M)|$$
$$= \sum_{\mu \in M} \min\{k, |\mu|\}$$

The partition M is k-*optimal* if it minimizes $\beta_k(M)$. So, for $k = 1$, a k-optimal partition is a partition of X into a minimum number of paths. For $k = \max|\mu|$, $\beta_k(M)$ is always equal to the number of vertices in G, and consequently every path-partition is k-optimal. If G has a hamilton path, a k-optimal partition consists in a single path. We conjectured in [1] that for every graph G and every k, $1 \leq k \leq \max|\mu|$, we have:

Property (A) : *For every k-optimal partition M of G, there exists a partial k-coloring of G which is strong for every path of M.* The Greene-Kleitman theorem can be rephrased as follows :

THEOREM 1. (Green-Kleitman) *Let G be a transitive graph :* $(x,y) \in U$ *and* $(y,z) \in U$ *implies* $(x,z) \in U$. *Then the property* (A) *holds true for all k.*

PROOF. In [7], Greene and Kleitman proved that for the graph of a partially ordered set if we denote by α_k the maximum number of vertices which can be colored in a partial k-coloring, then $\min_M \beta_k(M) = \alpha_k$.

This result extends Dilworth's theorem (case $k = 1$). A shorter proof has been given by M. Saks [10], and an extension has also been given by A. Frank [3].

Now, consider a partition $M = \{\mu_1, \mu_2, \ldots\}$ which minimizes $\beta_k(M)$, and an optimal partial k-coloring $(S_1, S_2, \ldots, S_k)$. Each $\mu_i \in M$ induces a clique, and therefore meets at most once each color. So

$$\alpha_k = |\bigcup_{j=1}^{k} S_j| = \sum_i |\bigcup S_j| \leq$$
$$\leq \sum_i \min\{k, |\mu_i|\} = \beta_k(M) = \alpha_k$$

(by the theorem of Greene-Kleitman).

Hence the number of colored vertices encountered by μ_i is exactly $\min\{k, |\mu_i|\}$, and these vertices have different colors.

Q. E. D.

We conjecture that Property (A) holds for all graphs (1)

---

(1) Recently, this conjecture has been proved by Ch. Payan for symmetric graphs, and by K.B. Cameron for acyclic digraphs. (Note that the symmetric case

2. The Gallai-Milgram Theorem

Another way to state the Gallai-Milgram theorem is as follows:

THEOREM 2. *The property (A) holds true for $k = 1$.*

Furthermore, if there exists a partition M_o with $L_1(M_o) = L_1$, then for every partition M with $L_1(M) \subset L_1$ and $|M|$ minimum relatively to this condition, there exists a stable set S which meets each path of M exactly once.

Gallai and Milgram [5] proved that if the maximum size of a stable set is α, it is always possible to partition the vertex-set into α paths; by the same argument, one can prove a stronger version of this which is equivalent to Theorem 2 (Linial, [8]).

COROLLARY. *Let G be a bipartite graph defined by two vertex-classes X and X'. Then the property (A) holds true for all k.*

Consider first the case : $k = 2p$ even. Let $M = \{\mu_1, \mu_2, \ldots\}$ be any partition. We define a partial $(2p)$-coloring by assigning successively a color to the vertex of μ_i at level 1, then to the vertex of μ_i at level 2, etc... ; the color assigned to $a \in \mu_i$ is the smallest integer (not yet used for μ_i) in $\{1, 2, \ldots, p\}$ if $a \in X$, or in $\{1', 2', \ldots, p'\}$ if $a \in X'$. The vertex a is left uncolored if all of the colors $1, 2, \ldots, p, 1', 2', \ldots, p'$ have been used for μ_i. Clearly, μ_i will be strongly colored, and by processing separately each path μ_i of M, a partial $(2p)$-coloring of G is obtained.

Now consider the case: $k = 2p + 1$ odd. Let M be a partition which minimizes $\beta_k(M)$. Let $\overline{G}$ be the subgraph of G induced by $(X \cup X') - L_1(M) - L_2(M) - \cdots - L_{k-1}(M)$. Let $\overline{M}$ be the trace of M on $\overline{G}$. Then $\overline{M}$ is a partition of $\overline{G}$ with $L_1(\overline{M}) \subset L_k(M)$ which minimizes $|\overline{M}|$: otherwise, we can obtain another partition M' of C with $L_i(M') = L_i(M)$ for $i < k$ and $|L_k(M')| < |L_k(M)|$, which contradicts the minimality of $|L_1(M)| + |L_2(M)| + \cdots + |L_k(M)|$.

By Theorem 2, there exists a stable set $\overline{S}$ of $\overline{G}$ which meets every path of $\overline{M}$. Assign the color 0 to all the vertices in $\overline{S}$. Then color successively the paths of M with the colors $1, 2, \ldots, p, 1', 2', \ldots, p'$ according to the rules defined above (for the case $k = 2p$). Clearly, the colors $0, 1, 2, \ldots, p, 1', 2', \ldots, p'$ define a partial k-coloring of G, and the coloring is strong for each path of M.

Q. E. D.

---

follows immediately from the acyclic case).

3. The Gallai - Roy Theorem

Gallai [6] and Roy [11] have proved independently that all the vertices can be colored with only $\max|\mu|$ colors. A slightly stronger result can be proved by the same argument:

THEOREM 3. *The property (A) is true for $k = \max|\mu|$: furthermore, every partition is k-optimal.*

PROOF. Let G be a graph of order n with $\max|\mu| = k$. Since $\beta_k(M)$ is equal to n for all M, every partition is k-optimal.

We shall define a partial graph H of G by adding successively, to the arcs which belong already to the paths of M, some arcs of G, provided they do not create circuits. When no more arcs can be added, we obtain a partial graph H which is acyclic.

For $x \in X$, put $t(x) = \max\{|\mu| / \mu$ is a path of H starting from $x\}$. If (x,y) is an arc of H, then $t(x)>t(y)$; if (x,y) is an arc of $G-H$, then $t(x)<t(y)$ because $H + (x,y)$ contains a circuit and so H contains a path from x to y.

Thus the function $t(x)$ is a k-coloring of G (i.e: $(x,y) \in U$ implies $t(x) n \neq t(y)$). Clearly, this k-coloring is strong for M.

Q. E. D.

More applications will be found in [1].

References

[1] C. Berge - k-optimal partitions of a directed graph, European J. of Combinatorics, 1982, 3, 7-17.

[2] K .B. Cameron - Polyhedral and Algorithmic Ramifications of anti- chains, Ph. D. Thesis, University of Waterloo, Ontario, 1982.

[3] A. Frank - On chain and antichain families of a partially ordered set J.C.T. B 29, 1980, 176-184.

[4] T. Gallai - On directed Paths and Circuits, Theory of Graphs, (Erdös-Katona), Academic Press, New York, 1968, 115-118.

[5] T. Gallai, A. N. Milgram - Verallgemeinerung eines Graphentheoretishcen Satzes von Rédei, Acta Sc. Math. 21, 1960, 181-186.

[6] C. Greene - Some Partitions associated with a Partially ordered set, J. C. T. A 20, 1976, 69-79.

[7] C. Greene, D.J. Kleitman - The Structure of Sperner k-families, J. C. T. A 20, 1976, 41-68.

[8] N. Linial - Covering Digraphs by Paths, Discr. Math. 23, 1978, 257-272.

[9] B. Roy - Nombre chromatique et plus longs chemins, Rev. Fr. Automat. Informat. 1, 1967, 127-132.

[10] M. Saks - A short proof of the k-saturated partitions, Advances in Math. 33, 1979, 207-211.

Set-Labelled Graphs

I.Z. Bouwer and B. Monson

ABSTRACT

This paper gives a combinatorial description of certain graphs that arise in geometrical contexts, and contains new proofs of some familiar results on symplectic geometries and their groups.

1. Introduction

In this paper we define a family of graphs G whose vertices are all subsets of $X(n) = \{1, 2, \ldots, n\}$, of given cardinalities, with subsets A, B adjacent just when $|AB| \equiv |A||B|$ (mod 2). Our definition was suggested by coordinate choices for the positive vectors in certain root systems, the orthographs of which are defined and described (as G graphs) in 2.2 and 2.21.

By associating to each A its characteristic function we may represent the vertex set of G as a subset of the symplectic space $V = V(n, 2)$, with adjacency corresponding to orthogonality of vectors. (Note that n need not be even.) The symplectic and orthogonal graphs are then easily described in our notation. As particular instances we exhibit in an elementary way the well known isomorphisms of the orthographs for the positive root systems of E_7 and E_8 with the graphs $SP(6,2)$ and $N^+(8,2)$ respectively (2.31 (b),(c)).

Certain symplectic transvections for V correspond naturally to automorphisms ρ_E of G, for certain $E \subset X(n)$: see 3.1 (a). These sets E are regarded as the vertices of a root graph R (cf. 3.2 (a)). In Proposition 3.3 we investigate transitivity properties of the group W generated by the ρ_E's as it acts on R. Associated with W is a Coxeter diagram, copies of which occur as induced subgraphs of R. By

---

During this work the authors were supported by NSERC Grants #A7332 and #A4818 respectively.

enumerating these copies, we easily compute $|W|$ in various cases. This essentially combinatorial approach gives a straightforward and (we believe) new way of investigating certain root systems and finite geometries.

We thank the referees for suggesting several improvements, including example 2.31 (c).

2. The Graphs $G(I;n)$

For any positive integer n let $X(n) = \{1,2,\ldots,n\}$ and let $P(n)$ be the power set of $X(n)$. If $A,B \in P(n)$, then A' will denote the complement of A, $|A|$ the cardinality of A, AB the intersection of A and B and $A + B$ their symmetric difference (or Boolean sum).

DEFINITION 2.1. For any set $I = \{i,j,k,\ldots\} \subset \{0,1,\ldots,n\}$ (of cardinalities) we define a graph $G = G(I;n) = G(i,j,k,\ldots;n)$ whose vertices are all $A \in P(n)$ with $|A| \in I$, and where A is adjacent to B if and only if $|AB| \equiv |A||B|$ (mod 2).

Unless otherwise stated all congruences below are modulo 2. Our definition is motivated by the following geometric considerations.

2.2 ROOT SYSTEMS. Let Π be a polytope in Euclidean n-space $\mathbf{R}^n$ with centroid 0. Further let L be the set of all lines l through 0 and orthogonal to the hyperplanes of symmetry of Π. It is possible to choose a set R^+ of vectors spanning the lines in L so that R^+ is the set of positive roots in a root system $R = R^+ \cup (-R^+)$ [2, pp. 311-314]. Definition 2.1 was suggested by the observation that simple *choices* for coordinates give the vectors in R^+ (cf. the proof 2.21 below). The Weyl group W of R is generated by reflections in the hyperplanes.

For instance, when R is the orthogonal direct sum of root systems of type A_n, D_n, or E_n, the angle between distinct lines in L is either $90°$ or $60°$, so that L (or equally R^+) is represented by an *orthograph* $OG(R^+)$ with vertex set L, where two lines are adjacent when perpendicular [cf. 2, p. 306].

2.21 EXAMPLES. $OG(R^+)$ is isomorphic to

(a) $G(2;n+1)$, when $R = A_n$, $n \geq 1$.
(b) $G(2,n-2;n)$, when $R = D_n$, $n \geq 5$.
(c) $G(2,3,6;6)$, when $R = E_6$.
(d) $G(2,4,6;7)$, when $R = E_7$.
(e) $G(1,2,5,6;8)$, when $R = E_8$.

PROOF. For our choices of roots, see for instance [2, p. 311] for (a) and (b), and [3, p. 419] for (c), (d), and (e).

CASES (a) AND (b): Let $B = \{e_1, \ldots, e_{n+1}\}$ be any orthonormal basis for $\mathbf{R}^{n+1}$. The root system A_n equals the set $B - B$. The positive roots are the vectors $r_{ij} = e_i - e_j$ where $i < j$, so that $r_{ij} \perp r_{kl}$ if and only if $\{i,j\} \cap \{k,l\} = \emptyset$. The positive roots of D_n ($n \geq 5$) are realized by all vectors r_{ij} and $s_{ij} = e_i + e_j$, $1 \leq i < j \leq n$. Assign to r_{ij} the 2-set $\{i,j\}$, and to s_{ij} the $(n-2)$-set $\{1,2,\ldots,n\} \setminus \{i,j\}$.

CASES (c), (d) AND (e): As a typical case, we prove (d). Let $\{e_1, \ldots, e_8\}$ be an orthonormal basis for 8-dimensional Minkowski space satisfying $e_i e_i = 1 = -e_8 e_8$, $1 \leq i \leq 7$. The positive roots of E_7 may be represented by the following vectors $(x_1, \ldots, x_8)$, to each of which corresponds the indicated subset of $X(7)$:

(i) $x_8 = 0$; there are exactly two non-zero coordinates: $x_i = 1 = -x_j$, $1 \leq i < j \leq 7$; subset $= \{i,j\}$.

(ii) $x_8 = 1$; there are three other coordinates equal to 1, and the rest are $x_i = x_j = x_k = x_l = 0$; subset $= \{i,j,k,l\}$.

(iii) $x_8 = 2$; one coordinate $x_i = 0$, and the remaining six are equal to 1; subset $= X(7) \setminus \{i\}$.

Since any two of these vectors are orthogonal if and only if the associated subsets of $X(7)$ have even intersection, the isomorphism follows. □

2.3 SYMPLECTIC GEOMETRIES. Let V be the vector space of n-tuples over $GF(2)$, equipped with the standard inner product "$\cdot$". Let $f \colon P(n) \to V$ map each set A to its characteristic function, which we view as the vector

$$a = f(A) = (a_1, \ldots, a_n), \text{ with } a_i = 1, 0 \text{ for } i \in A, i \notin A,$$

respectively. Thus A is adjacent to B in the above graph G if and only if $a \cdot b = (a \cdot j)(b \cdot j)$, where $j = (1,1,\ldots,1)$ is the characteristic vector for $X(n)$; that is, when $a \perp b$ under the symplectic pairing

$$(a,b) = \#(a \cdot b) - (a \cdot j)(b \cdot j).$$

(This pairing is non-singular if and only if n is even.)

2.31 EXAMPLES.

(a) For $n = 2m$, the *symplectic graph* $SP(2m,2)$ has as its vertices the vectors of $V \setminus \{0\}$, adjacent when orthogonal under the symplectic pairing $(\cdot,\cdot)$ [6, §9], [7, §2]. Thus,

$$SP(2m,2) \simeq G(1,2,\ldots,2m;2m).$$

(b) $G(2,4,6,\ldots,2m;2m+1) \simeq G(1,2,\ldots,2m;2m)$. Indeed, the map taking $A \to A \cap X(2m)$ is an incidence preserving bijection. In particular, in view of Example 2.21 (d), this implies the known isomorphism $OG(E_7^+) \simeq SP(6,2)$: cf. the corresponding statement for automorphism groups in [5, p. 128].

(c) For $n = 2m$, the *orthogonal graph* $O^\epsilon(2m,2)$ is the subgraph of $SP(2m,2)$ induced on the vertex set $\{a \in V\setminus\{0\}: Q^\epsilon(a) = 0\}$, where $\epsilon = +,-$ and Q^ϵ is the quadratic form described in [6, p. 498]. Now Q^ϵ is equivalent to the form $P(a) = \sum_{1 \leq i \leq j \leq n} a_i a_j$, with $\epsilon = +$ (resp. $-$) for $m \equiv 3$ or 4 (resp. 1 or 2) (mod 4). Since $P(a)$ vanishes for $|A| \equiv 3$ or 4 (mod 4) we conclude (for the above choice of ϵ) that

$$O^\epsilon(2m,2) \simeq G(3,4,\ldots,4i-1,4i,\ldots;2m).$$

It follows that the orthograph on the *non-singular* vectors for the form Q^ϵ is

$$N^\epsilon(2m,2) = G(1,2,\ldots,4i+1,4i+2,\ldots;2m).$$

In particular, in view of Example 2.21 (e), this implies the isomorphism $OG(E_8^+) \simeq N^+(8,2)$: cf. the analogous statement for automorphism groups: [5, p. 129].

3. Automorphisms of $G(I;n)$

For any $e \in V$ (with the symplectic pairing $(\cdot,\cdot)$) the transvection $\rho_e: a \to a + (a,e)e$, $a \in V$, is an involutory isometry; and for any isometry τ, $\tau \rho_e \tau^{-1} = \rho_{\tau(e)}$ [1, pp. 138-140]. In the set-theoretic notation of §2 we deduce:

3.1(a) For any $E \in P(n)$, the map

$$\rho_E: P(n) \to P(n)$$

defined by $A \to A$, if $|AE| = |A||E|$; $A \to A \oplus E$, if $|AE| \neq |A||E|$ is an involutory automorphism of the graph $G(I;n)$ *provided* that it preserves the vertex set of $G(I;n)$.

(b) $\rho_F \rho_E \rho_F = \rho_{\rho_F(E)} = \rho_{\rho_E(F)} = \rho_E \rho_F \rho_E$.

(c) $\rho_E \rho_F$ has period 2 or 3 according as $|EF|$ and $|E||F|$ do or do not have the same parity.

Let J be the set of cardinalities of all non-empty $E \in P(n)$ for which ρ_E is in fact an automorphism of $G(I;n)$. Thus the group $W(I;n) = W$ generated by $\{\rho_E: |E| \in J\}$ is a subgroup of $Aut(G(I;n))$.

3.2 REMARKS. (a) Usually, $J \neq I$ and the group W is more conveniently studied using a new *root graph* $R = G(J;n)$. Thus for

$\rho_E, \rho_F \in W$ there are vertices E, F of R which are adjacent if and only if $|EF| \not\equiv |E||F|$ (mod 2). By 3.1 (b), the ρ_E's actually define automorphisms of R, and $W \subset Aut(R)$.

(b) $2 \in J$, since for any $E = \{p,q\}$, ρ_E has the same action on $G(I;n)$ as the transposition (pq).

NOTATION: Let $E_i = \{i, i+1\}$ and $\tau_i = \rho_{E_i}$, $1 \leq i \leq n-1$. Thus $\{\tau_1, \ldots, \tau_{n-1}\}$ generates the symmetric group $S_n \subset W \subset Aut(G)$.

(c) Thus W is transitive on $G(I;n)$-and the graph is regular - if for any distinct i, $k \in I$ there exists some $\rho_E \in W$ taking some i-set to some k-set.

We now examine the transitivity of W on the root graph R. For $1 \leq k \leq n-1$, the subgraph of R induced by any ordered k-tuple $[A_1, \ldots, A_k]$ of vertices is called a *k-chain* if it is isomorphic to the subgraph induced by $[E_1, \ldots, E_k]$ (Figure 1).

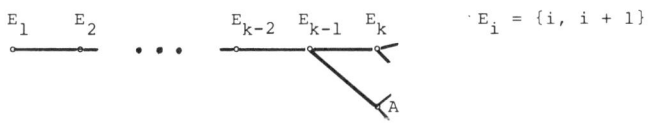

Figure 1.

If for some vertex B of R, $[A_1, \ldots, A_k, B]$ is a $(k+1)$-chain then we say the k-chain $[A_1, \ldots, A_k]$ can be *extended*.

PROPOSITION 3.3. For $1 \leq k \leq n-1$, let $[E_1, E_2, \ldots, E_{k-1}, A]$ be any k-chain in R (Figure 1). Then either

(a) there exists $\sigma \in W$ such that
$$\sigma(E_i) = E_i, 1 \leq i \leq k-1, \text{ and } \sigma(A) = E_k;$$

(b) $A = \{1, \ldots, k-1\}$ and there exists $\sigma \in W$ satisfying (a) if and only if k is even; or

(c) $A = \{k, \ldots, n\}$ and possibly no $\sigma \in W$ satisfies (a).

PROOF. Let $M = \{k+1, \ldots, n\}$. Then for some $L \subseteq M$ we have the following choices for A:

(1) $A = \{1, 2, \ldots, k-1\} \cup L$:

(i) $\overline{k+1 \in L}$. Let $E = E_k \oplus A$; then $\sigma = \rho_E$ satisfies (a).

(ii) $\overline{k+1 \notin L}$. If $L = \emptyset$ we enter case (b). Otherwise, for some $x \in L$, let $F = \{x, k+1\}$ and apply ρ_F to obtain case 1(i).

(2) $A = \{k\} \cup L$:

(i) $\overline{k+1 \notin L}$. For $E = \{k+1\} \cup L$, $\sigma = \rho_E$ satisfies (a).

(ii) $\overline{k+1 \in L}$. If possible choose $x \in ML'$, let $F = \{x, k+1\}$ and apply ρ_F to obtain case 2(i). Otherwise $M = L$ and we obtain case (c).

To conclude case (b) we take $A = \{1, 2, \ldots, k-1\}$. For k odd, $1 \leq k \leq n-2$, the chain $[E_1, \ldots, E_k]$ can be extended while the chain $[E_1, \ldots, E_{k-1}, A]$ cannot; when $k = n-1 = (n+1) - 2$ consider $X(n+1)$ and argue similarly. Thus no σ satisfies (a). For k even, let $F_i = \{1, \ldots, k+1\} \setminus \{i, i+1\}$, $\sigma_i = \rho_{F_i}$, $1 \leq i \leq n-1$. Then $\sigma = \tau_1 \sigma_1 \tau_3 \sigma_3 \cdots \tau_{k-1} \sigma_{k-1}$ satisfies (a) (cf. 3.2(b)). □

COROLLARY 3.4. *W is transitive on those vertices A of R satisfying $|A| < n$. If $A = X(n)$ is a vertex of R, then it is equivalent to the other vertices only when n is even and there exists an odd $j \in J$.*

PROOF. Take $k = 1$ in the Proposition. Case (b) or (c) occurs only for $A = \emptyset$ (excluded by assumption) or $A = X(n)$. Note that $\rho_E(X(n)) = X(n)$ unless n is even and $j = |E|$ is odd. □

We conclude with a few typical examples; for notational convenience below, when $E = \{i, j, \ldots\}$ we denote ρ_E by $\rho_{ij\ldots}$.

3.5 The *Schläfli graph* $G(1, 4, 5; 6)$ has $27 = \binom{6}{1} + \binom{6}{4} + \binom{6}{5}$ vertices which correspond to the 27 lines embedded in a general non-singular cubic surface, where two vertices are adjacent if and only if the corresponding lines intersect [4, p. 78 : pair line a_i with $\{i\}$, b_i with $\{i\}'$, c_{ij} with $\{i,j\}'$]. One easily checks that ρ_E preserves the vertex set of G if and only if $|E| \in J = \{2, 3, 6\}$. Thus the root graph $R = G(2, 3, 6; 6)$ has 36 vertices and is the complement of the orthograph $OG(E_6^+)$ [cf. 2.21 (c)]. Since $\sigma = \rho_{123}$: $\{4\} \to \{1, 2, 3, 4\}$, $\{3, 4, 5, 6\} \to \{1, 2, 4, 5, 6\}$ we see that W is transitive on $G(1, 4, 5; 6)$. By 3.1 (b), $\rho_{456} \rho_{123} \rho_{456} = \rho_{123456} = \hat{\rho}$ (say), so that W is generated by $\{\sigma, \tau_1, \tau_2, \tau_3, \tau_4, \tau_5\}$. We may associate to this set of generators the Coxeter diagram Γ, whose vertices $E, F, \ldots$ correspond to the generators $\rho_E, \rho_F \cdots$ of W, where $\rho_E \rho_F$ has period 3 (or 2) when the corresponding vertices are adjacent (non-adjacent). (See Figure 2).

SET-LABELLED GRAPHS

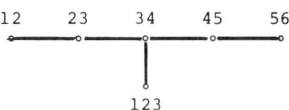

Figure 2.

By 3.1 (c), the Coxeter diagram is actually an induced subgraph of the root graph R!

For $1 \leq k \leq 5$, let N_k be the number of k-chains in R which can be extended to some 5-chain (and let $N_0 = 1$). By 3.3 all such k-chains are equivalent under W (in particular W is transitive on R by 3.4). Hence $\dfrac{N_k}{N_{k-1}}$ is the number of choices for A in Figure 1 (i.e. the number of choices for L in the proof of 3.3). For instance,

$$\frac{N_4}{N_3} = \binom{2}{0} + \binom{2}{1} + \binom{2}{0} = 4$$

(and $A = \{1,2,3\}$, $\{4,5\}$, $\{4,6\}$, $\{4,5,6\}$ resp. - note that $\hat{\rho}$ maps $\{4,5,6\}$ to $\{1,2,3\}$ while fixing all 2-sets). Thus $N_5 = \dfrac{N_5}{N_4} \cdot \dfrac{N_4}{N_3} \cdot \dfrac{N_3}{N_2} \cdot \dfrac{N_2}{N_1} \cdot \dfrac{N_1}{N_0} = 1 \cdot 4 \cdot 9 \cdot 20 \cdot 36 = 25920$. Now in Figure 2, given the 5-chain there are two choices - $\{4,5,6\}$ and $\{1,2,3\}$ - for the pendant vertex, equivalent under $\hat{\rho}$. Since the characteristic vectors for the vertices of Γ form a basis of V, we conclude that $\text{Stab}_W(\Gamma) = \{I\}$ and

$$|W| = 2 \cdot 25920 = 51840.$$

Indeed, W is the Weyl group for the E_6 root system [3, p. 414].

3.6 The *Symplectic graph* $SP(2m,2) \simeq G(1,2,\ldots,2m;2m)$ (cf. 2.31(a)). Here $n = 2m$, $J = \{1,\ldots,2m\}$, $R = G$ and the Coxeter diagram Γ which arises for W is shown in Figure 3.

Figure 3.

Using 3.3, we verify that $N_k/N_{k-1} = 2^{n-k+1} - \delta_k$, where $\delta_k = 0(1)$ for k even (odd), $1 \leq k \leq n$. (The pendant vertex in Γ gives $N_n/N_{n-1} = 2$). Thus W, which is the symplectic group $Sp_{2m}(2)$, has order

$$|W| = \prod_{k=1}^{n}(2^{n-k+1} - \delta_k)$$
$$= 2^{m^2}\prod_{j=1}^{m}(2^{2j} - 1)$$

(cf. [1, p. 147]).

References

[1]   E. Artin, Geometric Algebra *(Interscience, New York, 1957)*.

[2]   P.J. Cameron, J.M. Goethals, J.J. Seidel and E.E. Shult, Line graphs, root systems, and elliptic geometry, *J. Algebra, 43 (1976), 305-327*.

[3]   H.S.M. Coxeter, Extreme Forms, *Canad. J. Math., 3 (1951), 391-441*.

[4]   H.S.M. Coxeter, The equianharmonic surface and the Hessian polyhedron, *Ann. Mat. Pura Appl., 98 (1974), 77-92*.

[5]   H.S.M. Coxeter and W.O.J. Moser, Generators and Relations for Discrete Groups, 3rd Ed. *(Springer-Verlag, Berlin, 1972)*.

[6]   J.J. Seidel, A survey of two-graphs, *Proc. Intern. Colloqu. Teorie Combinatorie (Rome, 1973), Accad. Naz. Lincei, Rome 1976, I 481-511*.

[7]   E.E. Shult, Characterizations of certain classes of graphs, *J. Combinatorial Theory (B), 13 (1972), 142-167*.

Stability of Kings in Tournaments

Michael F. Bridgland

K. B. Reid

ABSTRACT

A *king* in a tournament is a vertex that can reach every other vertex via a 1-path or 2-path. An *all-kings tournament* is a tournament in which every vertex is a king. An all-kings tournament is *k-stable* if the reversal of any k arcs results in an all-kings tournament. It is shown that a 1-stable n-tournament exists if and only if $n \geq 7$ and $n \neq 8$, that almost all tournaments are k-stable (constructions are also given for all $k \geq 1$), that a k-stable tournament of smallest possible order, $4k+3$, exists if and only if a skew-Hadamard $(4k+4) \times (4k+4)$ matrix exists, and that there is a tournament T which can be extended to a tournament W such that exactly t kings of T are not kings of W (i.e. t kings are overthrown) and exactly c kings of W are not vertices of T (i.e. c kings are created) if and only if $(t,c) \neq (t,0)$ with $t > 0$. Several open problems are formulated.

1. Introduction.

A *king* in a tournament is a vertex x which can reach every other vertex in 1 or 2 steps, i.e. such that for every vertex $y \neq x$, either x dominates y or there is some vertex z such that x dominates z and z dominates y. The concept of a king was implicitly introduced by the mathematical sociologist Landau in 1953 [5], who noted that every vertex of maximum score is a king. The idea occurred again in Moon's solution [7] to a problem posed by Silverman [13] and again in Harary et al. [4]. And a Corollary of a result due to Moon [9] (see also Moon and Moser [8]) is that almost all tournaments are all-kings tournaments, i.e. every vertex is a king. However, the term "king" was only introduced by Maurer [6] in 1980. Maurer's delightful introduction to the study of kings in tournaments, written as a lesson in modelling possible patterns of dominance of pecking orders in flocks of chickens, established that there exists an n-tournament with exactly k kings for all integers $n \geq k \geq 1$, save for $k = 2$ with n arbitrary and $n = k = 4$. Maurer's article [6] concluded with several open problems. One such problem was to characterize those 4-tuples (n, k, s, k_s) so that there exists an n-tournament with exactly k kings and exactly s serfs, k_s of which are also kings (a serf is a king in the converse of the tournament); that problem was solved by Reid [11]. As later noted by Maurer, every tournament with no transmitter is the subtournament of kings of some tournament (answering another problem in [6]); related extremal questions were treated by Reid [12]. The results in the present paper were motivated by yet another of Maurer's problems which asks to analyze what single arc reversals do to the set of kings. The results presented here are concerned with "toppling" kings either by single arc reversals (section 2) or by the introduction of new vertices (section 3). That is, these results concern the weakening of strong individuals via reversals of single dominance encounters or by the insertion of outsiders.

For general references on tournaments, see Moon's monograph [9] or Beineke's and Reid's survey [2]. Here $O_T(x)$ (respectively $I_T(x)$) denotes the out-set (respectively, in-set) of a vertex x in a tournament.

2. Arc reversals.

DEFINITION. To *topple a king* x in a tournament means to reverse some arc of T so that x is not a king in the resulting tournament.

EXAMPLE. Many kings might be toppled by a single arc reversal. For, consider the n-tournament T, $n \geq 3$, $n \neq 4, 6$, obtained from any all-kings $(n-2)$-tournament W by adjoining two new vertices y and z so that y dominates exactly z and z dominates exactly every vertex in W. T is an all-kings n-tournament, but the reversal of arc (y, z) results in a tournament in which z is the only king (as z becomes a transmitter). Thus, $(n-1)$ kings have been toppled by a single arc reversal. Moreover, one can easily show that if T is an n-tournament in which $n-1$ kings can be toppled by a single arc reversal, then T must have the structure described above. For $n = 4, 6$, it is easy to see that $(n-2)$ kings can be toppled.

DEFINITION An all-kings tournament in which no king can be toppled is called a *stable tournament*.

LEMMA 1. Suppose that T is a stable n-tournament. Then

    (i)   for each vertex x in T, the subtournament induced by $O(x)$ (resp., $I(x)$) is not trivial and has no transmitter (resp., receiver),

(ii) $n \geq 7$, and

(iii) $n \neq 8$.

PROOF. Part (i) follows from the definition and implies that $d^+(x) \geq 3$ and $d^-(x) \geq 3$ for every vertex x in T. Thus, (ii) follows.

Suppose that there exists a stable 8-tournament. Then, as above, T contains a vertex v such that $d^-(v)+1 = d^+(v) = 4$ and $O(v)$ induces either the strong 4-tournament, denoted A, or the 4-tournament which has a receiver and exactly one 3-cycle, denoted B. Suppose that the arcs of A are given by $(a,b), (a,c), (b,c), (c,d), (d,a), (d,b)$. As T is stable, there exist at least two 2-paths from b to a; those 2-paths must pass through $I(v)$. But then vertex a dominates at most one vertex in $I(v)$, so that there is at most one 2-path from a to v. That is, T is not stable, a contradiction. So $O(v)$ induces B. Suppose that d denotes the receiver in B and c is any other vertex in B. There exist at least two 2-paths from d to c in T; those 2-paths must pass through $I(v)$. But then vertex d dominates at most one vertex in $I(v)$, so that there is at most one 2-path from c to v; again a contradiction. Consequently, no 8-tournament is stable.

An important class of vertex-homogeneous (i.e. vertex-transitive) tournaments are the rotational tournaments, the definition of which is recalled next.

DEFINITION [2]. A regular tournament is called a *rotational tournament* if its vertices can be labelled $0,1,2,\ldots,2m$ in such a way that for some m-set S of $1,2,\ldots,2m$, vertex i dominates $i+j$ (modulo $2m+1$) if and only if $j \in S$. A set S of m positive integers gives rise to such a tournament if and only if, for all j and k in S, $j, k \leq 2m$ and $j + k \neq 2m+1$. For such a set S, the resulting tournament is denoted $R(S)$, and S is called the *symbol* of the tournament.

THEOREM 2. There exists a stable n-tournament if and only if $n \geq 7$, $n \neq 8$.

PROOF. Necessity of the conditions on n follows from the lemma.

Sufficiency is established by construction. Rotational tournaments $R(1,2,4)$ and $R(1,2,4,6)$ are stable, and for $n = 2m+1$, $m \geq 5$, $R(1,2m-1,3,4,\ldots,m)$ is stable. This is seen by checking in each case that 0 can reach every s in the symbol S via a 2-path and that 0 can reach every t in $\{1,2,\ldots,n-1\}-S$ via two 2-paths. That is, each $s \in S$ is the sum of two elements in S, and each $t \notin S$ is the sum of two elements of S in two ways.

The construction for $n = 2m$, $m \geq 6$, utilizes the rotational tournaments described above. In each case arc $(0,1)$ in the rotational tournament is replaced by a copy of $R(1,2,4)$ as follows: Replace 0 by $\{3',5',6'\}$ in such a way that any vertex that dominated (was dominated by) 0 now dominates (respectively, is dominated by) each of $3',5',6'$. Similarly, replace 1 by $\{1',2',4'\}$. Then add a new vertex $0'$ which dominates exactly the m vertices that were dominated by 1 together with $1', 2'$, and $4'$. Complete the construction by adding new arcs so that $\{0',1',2',3',4',5',6'\}$ induces $R(1',2',4')$. For example, when this construction is used on $R(1,2,4)$, a 12-tournament results. Then it can be checked that the resulting tournament of even order is stable. Also, the 10-tournament in Figure 1 is stable.

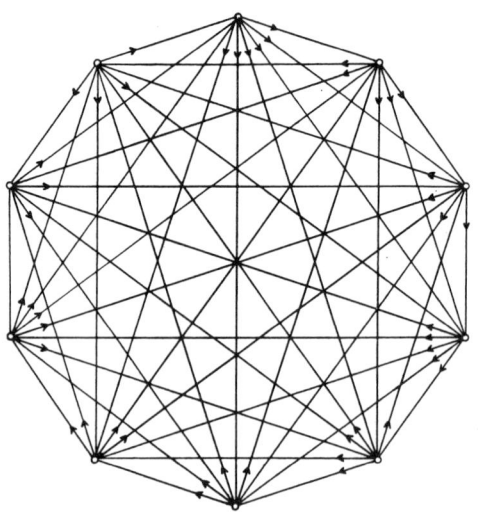

Figure 1

Many all-kings tournaments require several arc reversals in order to produce a non-all-kings tournament.

DEFINITION. An all-kings tournament is *k-stable*, $k \geq 1$, if the reversal of any k arcs results in an all-kings tournament. (Then 1-stable means stable.)

REMARK. If T is a k-stable n-tournament, then $n \geq 4k+3$. For, suppose that $d^-_{O(u)}(v) < k$ for some v in some $O(u)$. Then the reversal of the $d^-_{O(u)}(v)$ arcs from $O(u) \cap I(v)$ to v together with the arc from u

to v results in a tournament in which u is not a king (because of v), a contradiction. Thus for any u, $\binom{d^+(u)}{2} = \sum_{v \in O(u)} d^-_{O(u)}(v) \geq d^+(u) k$ or $d^+(u) \geq 2k+1$. Then $\binom{n}{2} = \sum_u d^+(u) \geq n(2k+1)$ and $n \geq 4k+3$.

DEFINITION.

A k-stable $(4k+3)$-tournament is called an *extremal k-stable tournament*.

In order to state some characterizations of extremal k-stable tournaments recall the following definitions.

DEFINITION.

[2] A tournament is *doubly-regular* if all pairs of vertices jointly dominate the same number of vertices. A tournament is *homogeneous* if every arc lies on the same number of 3-cycles. A *skew-Hadamard* matrix is an $m \times m$ matrix H of $+1$'s and -1's such that

$$HH^t = mI \text{ and } H + H^t = 2I,$$

where H^t is the transpose of H and I denotes the $m \times m$ identity matrix.

THEOREM 3. The following four statements are equivalent for a $(4k+3)$-tournament T:

(i) T is an extremal k-stable tournament.

(ii) T is a doubly regular tournament.

(iii) T is a homogeneous tournament.

(iv) Every $(4k+1)$-subtournament of T has the same score sequence.

Moreover, there exists a $(4k+3)$-tournament satisfying any one (i)-(iv) (and hence all of (i)-(iv)) if and only if there exists a $(4k+4) \times (4k+4)$ skew-Hadamard matrix.

PROOF. The equivalence of (ii) and (iii) and the second part of the Theorem are established in [3]. The equivalence of (iii) and (iv) is established in [10] (and the score sequences in question consist of k $(2k+1)$'s, $2k+1$ $(2k)$'s and k $(2k-1)$'s). Statement (i) implies (ii) since equality must hold throughout in the Remark above. Statement (iii) implies that in T each vertex has score $2k+1$, each pair of distinct vertices jointly dominates exactly k vertices, and each arc is on exactly $k+1$ 3-cycles (see the proof of Theorem 1 in [3]). That is, there are at least $k+1$ paths of length one or two from any vertex to any other vertex, i.e., T is k-stable, and (i) follows.

There exist infinite families of skew-Hadamard matrices [14], so there exist infinitely many extremal k-stable tournaments. Moreover, if T is k-stable and $k \geq l$, then T is l-stable, and thus there exist k-stable tournaments for all $k \geq l$. The existence of extremal k-stable tournaments for all k is unknown as the existence of skew-Hadamard matrices for all possible orders is unknown.

PROBLEM 1. Fix $k \geq 1$. For what values of $n \geq 4k+3$ does there exist a k-stable n-tournament? (The case $k=1$ was treated above; also, there are rotational 2-stable n-tournaments for all odd $n \geq 11$.)

An explicit iterative construction that does not depend on the equivalence of the existence of extremal k-stable tournaments and the existence of skew-Hadamard matrices is as follows: Let T_1, T_2, and T_3 be copies of the same k-stable n-tournament T. The copy of vertex x of T in T_i will be denoted x_i, $1 \leq i \leq 3$. Construct a new $3n$-tournament $T^{(3)}$ from T_1, T_2, and T_3 by joining arcs between the copies of T as follows: x_1 dominates y_2 if and only if y_1 dominates x_1 in T_1, x_2 dominates y_3 if and only if y_2 dominates x_2 in T_2, and x_3 dominates y_1 if and only if y_3 dominates x_3 in T_3, x_1 dominates x_2, x_2 dominates x_3, and x_3 dominates x_1, for all vertices x and y in T. Count the number of paths of length one or two from any vertex x_i in $T^{(3)}$ to another vertex y_j or x_j in $T^{(3)}$. Without loss of generality assume that x dominates y in T. Lower bounds on these counts are presented in three cases.

(i) Suppose that $i=j$ (i.e. x_i and y_j are in the same copy of T in $T^{(3)}$). There is no loss of generality in assuming that $i=j=1$. Any vertex z_2 in $I_{T_2}(x_2)$ in dominated by x_1 in $T^{(3)}$, and any vertex w_2 in $O_{T_2}(y_2)$ dominates y_1 in $T^{(3)}$. Thus, $T^{(3)}$ contains at least $|O_T(y) \cap I_T(x)|$ 2-paths from x_1 to y_1 that pass through T_2. Similarly, $T^{(3)}$ contains at least $|O_T(y) \cap I_T(x)|$ 2-paths from x_1 to y_1 that pass through T_3. And $T^{(3)}$ contains at least $k+1$ paths of length 1 or 2 from x_1 to y_1 that stay in T_1 (as T was chosen to be k-stable). Thus, $T^{(3)}$ contains at least $k+1+2|O_T(y) \cap I_T(x)|$ paths of length 1 or 2 from x_1 to y_1. In a similar manner, one can show that $T^{(3)}$ contains at least $k+2+2|I_T(y) \cap O_T(x)|$ 2-paths from y_1 to x_1 (the "extra" path arises since y_1 dominates y_2 and y_2 dominates x_1).

(ii) Suppose that $i \neq j$ and $x \neq y$ (i.e. x_i and y_j are not in the same copy of T in $T^{(3)}$). There is no loss of generality in assuming that $i=1$ and $j=2$. Note that x_1 dominates x_2, and x_2 dominates y_2. Also each vertex in $I_{T_2}(x_2) \cap I_{T_2}(y_2)$ is on a 2-path from x_1 to y_2, each vertex in $O_{T_1}(x_1) \cap O_{T_1}(y_1)$ is on a 2-path from x_1 to y_2, and each vertex in $O_{T_3}(y_3) \cap I_{T_3}(x_3)$ is on a 2-path from x_1 to y_2. Thus, $T^{(3)}$ contains at least $1 + |I_T(x) \cap I_T(y)| + |O_T(x) \cap O_T(y)| + (O_T(y) \cap I_T(x)|$ 2-paths from x_1 and y_2. Similarly, $T^{(3)}$ contains at least $1 + |I_T(x) \cap I_T(y)| + |O_T(x) \cap O_T(y)| + |O_T(x) + I_T(y)|$ paths of length 1 or 2 from y_2 to x_1 (note that y_2 dominates x_1).

(iii) Suppose that $i \neq j$ and $x = y$ (i.e. y_j is a different copy of x than is x_i). There is no loss of generality in assuming that $i = 1$ and $j = 2$. By construction x_1 dominates x_2. Also, each vertex in $I_{T_2}(x_2)$ is on a 2-path from x_1 to x_2 and each vertex in $O_{T_1}(x_1)$ is on a 2-path from x_1 to x_2. Thus, $T^{(3)}$ contains

$$d_T^-(x) + d_T^+ + 1 = n$$

paths of length 1 or 2 from x_1 to x_2. Similarly, $T^{(3)}$ contains n 2-paths from x_2 to x_1 (note that x_2 dominates x_3 and x_3 dominates x_1). In summary, note that since T is k-stable

$$|O_T(x) \cap I_T(y)| \geq k, \text{ and}$$
$$|O_T(y) \cap I_T(x)| \geq k+1;$$

if l denotes min $\{|I_T(x) \cap I_T(y)| + |O_T(x) \cap O_T(y)|; x \text{ and } y \text{ in } T\}$, then lower bounds on the counts obtained in (i)-(iii) are as follows:

(i)' $3k+3$,

(ii)' $1+l+k$, and

(iii)' n.

Consequently, if $s+1 = \min\{3k+3, k+l+1, n\}$, then $T^{(3)}$ is s-stable. It is possible to pick T so that $l > 0$ and $T^{(3)}$ is more stable than is T.

EXAMPLE. If T is $R(1,2,4)$, then $k=1$, $l=2$, $n=7$ and $T^{(3)}$ is a 3-stable 21-tournament.

Asymptotically, most n-tournaments are k-stable (see Moon [9, pp. 32-34]).

THEOREM 4. Fix $k \geq 1$. The probability that a random n-tournament is k-stable approaches one as n approaches infinity.

PROOF. Let $P_k(n)$ denote the probability that a random (labelled) n-tournament is k-stable. Note that a random n-tournament will fail to be k-stable if for some pair of vertices x and y, either x dominates y and there are $k-1$ or fewer 2-paths from x to y, or y dominates x and there are k or fewer 2-paths from x to y. Then the probability that a random n-tournament is not k-stable is given by

$$1 - P_k(n) \le \binom{n}{2}\sum_{j=1}^{k-1}(\tfrac{1}{2})\binom{n-2}{j}(\tfrac{1}{4})^j(\tfrac{3}{4})^{n-2-j}$$

$$+ \sum_{j=0}^{k}(\tfrac{1}{2})\binom{n-2}{j}(\tfrac{1}{4})^j(\tfrac{3}{4})^{n-2-j}$$

$$= (\tfrac{1}{2})\binom{n}{2}(\tfrac{1}{4})^{n-2}[2\cdot\sum_{j=0}^{k-1}\binom{n-2}{j}3^{n-2-j} + \binom{n-2}{k}3^{n-2-k}].$$

For
$k \le (\tfrac{n}{2})$, $1 - P_k(n) \le (\tfrac{1}{2})\binom{n}{2}(\tfrac{1}{4})^{n-2}[2k\binom{n-2}{k}3^{n-2} + \binom{n-2}{k}3^{n-2}].$
That is,
$1 - P_k(n) \le \binom{2k+1}{2}(\tfrac{3}{4})^{n-2}\binom{n-2}{k}\binom{n}{2}$, so $1 - P_k(n) \le \dfrac{f_k(n)}{(\tfrac{4}{3})^{n-2}}$, where
f_k is a polynomial in n of degree $k+2$. Since $\lim_{n\to\infty}\dfrac{f_k(n)}{(\tfrac{4}{3})^{n-2}} = 0$,
$\lim_{n\to\infty} P_k(n) = 1.$

COROLLARY 5. [1, Theorem 8 and 9, p. 34] *In almost every tournament each arc can be "bypassed" by k 2-paths.*

3. Adjoining new vertices.

Another approach to the overthrow of kings is to study the effect of introducing new individuals rather than reversing the outcome of dominance encounters. That approach is pursued in the remainder of the paper. The simplest case is perhaps most interesting because of its negative conclusion. Namely, no finite number of new vertices can be adjoined to a tournament T to form a new tournament W so that the kings of W are a proper subset of the kings of T.

Theorem 6. *If T and W are tournaments such that (i) T is a subtournament of W, and (ii) each king of W is a vertex of T, then each king of T is a king of W.*

PROOF. Suppose that T and W are tournaments satisfying (i) and (ii). Let x be a king of T which is not a king of W. Define N to be the subtournament of W which is induced by the vertices of W which cannot be reached from x via a 1-path or a 2-path. Then N has at least one vertex, and $V(N)$ is included in $I_W(x) - I_T(x)$, since x is a king of T. Moreover, each vertex in N dominates every vertex in $O_W(x)$.

Let z be a king of N. Each vertex in $I_W(x) - V(N)$ is reachable in W by a 2-path from x. Consequently, each vertex in $I_W(x) - V(N)$ is reachable in W via a 2-path from z, since $O_W(x)$ is included in $O_W(z)$. That is, z is a king of W which is not in $V(T)$, a contradiction.

One extension of the above ideas is the problem of determining whether or not it is ever possible that the addition of new vertices results in a new tournament in which a specified number of the new vertices are kings and a specified number of the old kings are no longer kings.

DEFINITION. Let t and c be nonnegative integers. Tournament T is said to have the (t,c)-*property* if there exists a tournament W such that

(i) T is a subtournament of W

(ii) exactly t kings of T are not kings of W, and

(iii) exactly c kings of W are not vertices of T

Note that Theorem 6 implies that there exists no tournament with the $(t,0)$-property, $t > 0$. For all other values of t and c, existence is established below.

THEOREM 8. Let t and c be nonnegative integers. There exists a tournament with the (t,c)-property if and only if $c > 0$ or $t = c = 0$.

PROOF. The necessity of the conditions follows from the previous theorem.

Every tournament T has the $(0,0)$-property since a tournament W satisfying (i)-(iii) of the definition can be obtained by adding one new vertex x to T such that every vertex in T dominates x. Suppose that $c > 0$. For each pair (t,c) a tournament with the (t,c)-property is constructed.

If $t \neq 0,2$ and $c \neq 0,2$, let T be any tournament with exactly t kings. If N denotes any tournament with c kings, let W be the tournament obtained from T and N such that every vertex in N dominates every vertex in T. Then T and W satisfy (i)-(iii) above.

If $t = 0$ and $c \neq 1,3$, let x be any king in any tournament T. Adjoin c new vertices $z_1,\ldots,z_c$ to T to form a tournament W in which $x, z_1,\ldots,z_c$ induce an all-kings $(c+1)$-tournament in W and z_i dominates (respectively, is dominated by) every vertex in $O_T(x)$ (respectively, $I_T(x)$), $1 \leq i \leq c$. The condition $c \neq 1,3$ ensures the existence of an all-kings $(c+1)$-tournament [6, Theorem 6]. Then T and W satisfy (i)-(iii) above.

The case $(t,c) = (0,3)$ is treated as in the previous case, save that 4 new vertices together with x induce a 5-tournament with exactly 4 kings, one of which is x (see [6, p. 72] for such a 5-tournament). If $(t,c) = (0,1)$, let T be an all-kings regular tournament such that each subtournament induced by the in-set of any vertex has no transmitter (e.g. use the rotational tournament $R(1,3,4,\ldots,l,2l-1)$, $l \geq 3$); to obtain W from T adjoin one new vertex z which dominates exactly $\{x\} \cup I_T(x)$. As each vertex in $I_T(x)$ dominates some vertex in $O_T(x)$, each vertex in $I_T(x)$ is a king in W, and T and W satisfy (i)-(iii) of the definition.

In case $(t,c) = (2,2)$, let T be any tournament with at least 3 kings two of which are x and y such that x dominates y. W is obtained from T by adjoining two new vertices x' and y' such that x' dominates exactly $\{x,y,y'\} \cup O_T(x)$ and y' dominates exactly $\{x,y\} \cup O_T(y)$. Then x (respectively, y) cannot reach x' (respectively, y') via a 1-path or 2-path; however; both x' and y' are kings in W. So T and W satisfy (i)-(iii).

If $t = 2$ and $c > 3$, but $c \neq 5$, let T be a tournament with at least 3 kings. Let W_1 be the tournament constructed from T in case $(t,c) = (2,2)$ above. Since $c-2 \neq 1,3$, the construction in case $(0,c-2)$ above applied to W_1 yields a tournament W such that T and W satisfy (i)-(iii). If $(t,c) = (2,3)$, apply case $(0,1)$ followed by case $(2,2)$; if $(t,c) = (2,5)$, apply case $(2,2)$ followed by case $(0,3)$. If $(t,c) = (2,1)$, let T be any tournament with exactly 4 kings, say x_1, x_2, x_3, x_4. The subtournament of T induced by x_1, x_2, x_3, x_4 has no transmitter (as each king is dominated by some other king) so has score sequence $(1,1,2,2)$ or $(0,2,2,2)$; in either case, adjust notation so that x_1 dominates x_2 and x_2 dominates x_3 and x_4. Form W by adjoining one new vertex z to T such that z is dominated only by x_2; the kings of W are x_1, x_2, and z. This completes the treatment for all pairs $(2,c)$.

If $c = 2$ and $t \neq 2$, let T be obtained from any tournament R with exactly t kings by adding two new vertices a and b such that a dominates exactly every vertex in R and b dominates exactly a. Note that T has $t+2$ kings. Form W from T by adjoining 3 new vertices z_1, z_2, z_3 to T so that

$$O_W(z_1) = \{z_2\} \cup V(R),$$

$$O_W(z_2) = \{z_3, b\} \cup V(R), \text{ and}$$

$$O_W(z_3) = \{z_1, a\} \cup V(R).$$

Each king of T which is in R is not a king of W because of z_2; z_1 is not a king of W because of a. The kings of W are exactly a, b, z_2, and z_3. Thus, T and W satisfy the conditions (i)-(iii) as required. The case $(t,c) = (2,2)$ was treated above.

Consequently, for each pair (t,c), with $c > 0$ or $t = c = 0$, there exists a tournament with the (t,c)-property.

In conclusion, some questions for further study are included below.

PROBLEM 2. Characterize those triples (t,c,d) for which there exists a tournament T with the (t,c)-property such that exactly d nonkings of T are kings of W.

PROBLEM 3. Clearly, the reversal of all arcs interchanges the set of kings and the set of serfs. (Recall that a *serf* is a vertex reachable from every other vertex in one or two steps.) What can be said about the least number of arcs that need be reversed in order to interchange the set of kings and serfs in a non all-kings tournament?

PROBLEM 4. What is the least number $F(n)$ of arcs possible in an oriented graph with n vertices, each of which is a king? G. Katona and E. Szemerédi (Studia Sci. Math. Hung. 2 (1976), 23-28) proved that $\frac{n}{2}\log\frac{n}{2} \leq F(n) \leq n\{\log n\}$.

PROBLEM 5. Suppose that G is an undirected graph with n vertices and at least as many edges as in the answer to Problem 4. Can G be oriented so that all vertices are kings?

PROBLEM 6. Suppose that D is a digraph with king-set A and serf-set B. What can be said about the number of pairs of vertices $x \neq y$ of D which must be considered (by someone with no knowledge of D) in order to determine A and B?

PROBLEM 7. What kind of results can be obtained by assigning weights to vertices as follows: for a vertex x, let
$$w(x) = \alpha d^+(x) + \beta t^+(x),$$
where α and β are nonnegative real numbers $\alpha \geq \beta$, $d^+(x)$ denotes the score of x, and $t^+(x)$ denotes the number of vertices reachable from x via 2-paths?

Acknowledgement.

This paper was written while the second author was Visiting Professor in the Department of Mathematical Sciences at The Johns Hopkins University the first half of 1982. He wishes to acknowledge the hospitality of the members and staff of that Department.

References

[1] B. Alspach, K. B. Reid, and D. P. Roselle, Bypasses in asymmetric digraphs, J. Combinatorial Theory (B) 17 (1974), 11-18.

[2] Lowell W. Beineke and K. B. Reid, Tournaments, Chapter 7 in *Selected Topics in Graph Theory* (edited by Lowell W. Beineke and Robin J. Wilson), Academic Press, London, 1978, 169-204.

[3] E. Brown and K. B. Reid, Doubly regular tournaments are equivalent to skew-Hadamard matrices, J. Combinatorial Theory (A) 12 (1972), 332-338.

[4] F. Harary, R. Z. Norman and D. Cartwright, *Structural Models: An Introduction to the Theory of Directed Graphs*, John Wiley & Sons, New York, 1965, 293-294.

[5] H. G. Landau, On dominance relations and the structure of animal societies, III: The condition for a score structure, Bull. Math. Biophys. 15 (1953), 143-148.

[6] S. B. Maurer, The king chicken theorems, Math. Mag. 53 (1980), 67-80.

[7] J. W. Moon, Solution to problem 463, Math. Mag. 35 (1962), 189.

[8] J. W. Moon and Leo Moser, Almost all $(0,1)$ matrices are primitive, Stud. Sci. Math. Hung. 1 (1966), 153-156.

[9] J. W. Moon, *Topics on Tournaments*, Holt, Rinehart and Winston, New York, 1968.

[10] V. Müller and J. Pelant, On strongly homogeneous tournaments, Czechoslovak Math. J. 24 (1974), 378-391.

[11] K. B. Reid, Tournaments with prescribed numbers of kings and serfs, Congressus Numerantium 29, Proceedings of the Eleventh Southeastern Conference on Combinatorics, Graph Theory and Computing (Utilitas Mathematica, Winnipeg, 1980), 809-826.

[12] K. B. Reid, Every vertex a king, Discrete Math. 38 (1982), 93-98.

[13] D. L. Silverman, Problem 463, Math. Mag. 35 (1962), 189.

[14] G. Szekeres, Tournaments and Hadamard matrices, Enseignement Math. T. XV (1969), 269-278.

The Interchange Graph of Tournaments with the Same Score Vector.

Richard A. Brualdi[1]

Li Qiao[2]

ABSTRACT

We investigate the class of tournaments $T(R)$ having monotone score vector R. A graph whose vertices are the tournaments in $T(R)$ is defined and some of its properties are determined.

1. Introduction

We consider n players $p_1, \ldots, p_n$ who participate in a *round robin tournament*. Thus each player engages every other player in one game, and the result is a win for one player and a loss for the other (the possibility of ties is not entertained). There are two convenient models for such a tournament, one in terms of directed graphs and the other in terms of matrices of $0's$ and $1's$. Consider complete undirected graph with n vertices which have been identified with the players $p_1, \ldots, p_n$. We assign an orientation to the edges by directing the edge joining p_i and p_j *from p_i to p_j* whenever p_i beats p_j. The result is a directed graph T in which every pair of vertices is joined by exactly one arc. Since this directed graph is a model of the tournament, we refer to it as a *tournament*. Our second model arises by identifying $p_1, \ldots, p_n$ with rows $1, \ldots, n$ and columns $1, \ldots, n$ of an $n \times n$ matrix $T = [t_{ij}]$. We set $t_{ij} = 1$ whenever player p_i beats p_j ($i \neq j$) and $t_{ij} = 0$ whenever player p_i is beaten by p_j ($i \neq j$). We also set $t_{ii} = 0$ for $i = 1, \ldots, n$. The resulting $n \times n$ matrix of $0's$ and $1's$ is a combinatorially skew symmetric matrix satisfying $t_{ij} + t_{ji} = 1$ whenever $i \neq j$. We also refer to this model as a *tournament*. Throughout we use both models interchangeably the choice depending

---

1. Research partially supported by a grant from the National Science Foundation.
2. Dedicated by this author to his mother on her 70th birthday.

on which most readily displays the idea being discussed.

Suppose player p_i wins r_i games and thus loses $s_i = n-1-r_i$ games. In terms of directed graphs this means that in the tournament T vertex p_i has *outdegree* r_i and *indegree* $s_i (i = 1, \ldots, n)$. Thus $R = (r_1, \ldots, r_n)$ and $S = (s_1, \ldots, s_n)$ are, respectively, the *outdegree sequence* and *indegree sequence* of the tournament T. In terms of matrices, the tournament T has *row sum vector* $R = (r_1, \ldots, r_n)$ and *column sum vector* $S = (s_1, \ldots, s_n)$. We refer to R also as the *score vector* of T. Since the ordering of the players is arbitrary, we can assume it has been chosen so that $0 \le r_1 \le \cdots \le r_n$ and thus $s_1 \ge \cdots \ge s_n \ge 0$. When such an ordering has been chose, we call T a *monotone tournament* and R a *monotone score vector*. Our tournaments and score vectors are always assumed monotone.

Now let $R = (r_1, \ldots, r_n)$ be a sequence of integers satisfying $0 \le r_1 \le \cdots \le r_n$, and let

$$T(R)$$

denote the collection of all tournaments with score vector R. The set $T(R)$ has been studied in papers by Ryser [9] and Fulkerson [5]. Conditions for $T(R)$ to be nonempty were discovered by Landau [6]; other proofs were given by Moon [7], Ryser [9], and Ford and Fulkerson [4].

1.1 THEOREM. (Landau [6]) There exists a tournament in $T(R)$ if and only if for $i = 1, \ldots, n$

$$r_1 + \cdots + r_i \ge \binom{i}{2}, \tag{1.1}$$

with equality for $i = n$.

A directed graph is called *strong* if from each vertex there are directed paths to all other vertices. A tournament is called *strong* if as a directed graph it is strong. In terms of matrices a tournament $T = [t_{ij}]$ is strong provided there is not a simultaneous reordering of its rows and columns which reduces T to the form

$$\begin{bmatrix} T_1 & O \\ X & T_2 \end{bmatrix}, \tag{1.2}$$

where T_1 and T_2 are square matrices.

Since a simultaneous reordering does not alter the skew-symmetric property of T, it follows that (1.2) is a tournament so that X is a matrix of all 1's. As observed by Ryser [9], if T is monotone and strong, T

already has the form in (1.2). The following corollary is now an easy consequence of Theorem 1.1.

1.2 COROLLARY. The following are equivalent:

(1.3) There exists a strong tournament in $T(R)$.

(1.4) Every tournament in $T(R)$ is strong.

(1.5) $r_1 + \cdots + r_i \geq \binom{i}{2} + 1$ for $i = 1,\ldots,n-1$ and

$$r_1 + \cdots + r_n = \binom{n}{2}.$$

When any (and hence all) of the conditions (1.3), (1.4), and (1.5) are satisfied we call R a *strong score vector*.

Let $T = [t_{ij}] \in T(R)$ and consider the 2×2 submatrix $T[i_1,i_2;j_1,j_2]$ of T lying in rows i_1 and i_2 and columns j_1 and j_2. Suppose $T[i_1,i_2;j_1,j_2]$ is one of the two matrices

$$\begin{bmatrix} 0 & 1 \\ 1 & 0 \end{bmatrix}, \begin{bmatrix} 1 & 0 \\ 0 & 1 \end{bmatrix}. \quad (1.6)$$

Then it follows that $T[j_1,j_2;i_1,i_2]$ is the other. Interchanging these two submatrices in T leads to another tournament T' where also $T' \in T(R)$. Ryser calls this transformation a *double interchange* and proves [9, Theorem 4.2] that each of a pair of tournaments in $T(R)$ is transformable into the other by a finite sequence of double interchanges. Suppose the 2×2 submatrix $T[i_1,i_2;j_1,j_2]$ of T meets the main diagonal. Then the effect of the double interchange is to replace one of the following two 3×3 principal submatrices by the other:

$$\begin{pmatrix} 0 & 1 & 0 \\ 0 & 0 & 1 \\ 1 & 0 & 0 \end{pmatrix}, \begin{pmatrix} 0 & 0 & 1 \\ 1 & 0 & 0 \\ 0 & 1 & 0 \end{pmatrix} \quad (1.7)$$

As observed by Fulkerson [4] and Moon [6] every other double interchange can be accomplished by a sequence of two interchanges of this type. In terms of directed graphs this means the following. Define a Δ-*interchange* to be a transformation which reverses the directions of the arcs of a (directed) cycle of length 3, for short a 3-cycle. Then for tournaments T^1 and T^2 in $T(R)$ there is a finite sequence of Δ-interchanges which transforms T^1 into T^2. In the next section we give a direct proof of this fact. We shall also use the notion of a Δ-interchange to investigate graphically the set of tournaments with a given score vector.

Consider a monotone tournament T with score vector

$R = (r_1,\ldots,r_n)$. Since $r_1 \le \cdots \le r_n$, we say that for $1 \le i < j \le n$, player p_j is *superior* to player p_i and that p_i is *inferior* to p_j. In case $r_i = r_j$ whether p_i is superior or inferior to p_j depends solely on the ordering of the players chosen. An *upset* is said to have occurred in a tournament T whenever a player beats a superior player. The total number of upsets in a tournament T is denoted by $\tau(T)$. The matrix model $T = [t_{ij}]$ of a tournament conveniently displays the upsets. Indeed the upsets correspond to the 1's above the main diagonal and hence

$$\tau(T) = \sum_{j>i} t_{ij}.$$

A tournament with no upsets has only 1's below the main diagonal. In terms of directed graphs, this means that $\tau(T) = 0$ if and only if T is a transitive (monotone) tournament. Let $\tilde{\tau}(R)$ denotes the smallest number of upsets for a tournament in $T(R)$, and $\bar{\tau}(R)$ the maximum number of upsets.

Ryser [10, Theorem 5.2] has characterized tournaments in $T(R)$ for which $\tau(T) = \tilde{\tau}(R)$.

Finally let us mention that surveys of general properties of tournaments are given by Moon [7] and Reid and Beineke [9].

2. The Interchange Graph

Consider a nonempty collection $T(R)$ of tournaments with monotone score vector $R = (r_1, \ldots, r_n)$. We then define an (undirected) graph $G(R)$ whose vertices are the tournaments in $T(R)$. There is an edge joining tournaments $T, T' \in T(R)$ provided T' can be obtained from T by a Δ-interchange. We call $G(R)$ the *interchange graph* of $T(R)$ or of R. This graph is similar to one defined in [1] for the class of $m \times n$ matrices of 0's and 1's with a given row and column sum vector, and we prove some similar theorems about it. It is well known [7, Theorem 4, p. 9] that every tournament $T \in T(R)$ has exactly $\binom{n}{2} - \sum_{i=1}^{n} \binom{r_i}{2}$ 3-cycles so that $G(R)$ is a regular graph of degree $\binom{n}{2} - \sum_{i=1}^{n} \binom{r_i}{2}$.

Call a directed graph *even* if for each vertex v, the indegree of v equals the outdegree of v. It is well known and easy to prove that the arcs of an even directed graph can be partitioned into cycles. For tournaments $T, T' \in T(R)$, let $T - T'$ be the directed graph whose vertices are also $p_1, \ldots, p_n$ and whose arcs are those arcs of T which are not arcs of T'.

2.1 LEMMA. The directed graph $T - T'$ is even.

PROOF. Consider a vertex p_i. Since p_i has the same outdegree in T as in T', it follows readily that the number of vertices that lose to p_i in T but not in T' equals the number of vertices that lose to p_i in T' but not in T. This implies that in $T - T'$ the outdegree and indegree of p_i are equal as required.

2.2 COROLLARY. Let $T \in T(R)$. Then there is a one-to-one correspondence between the set $T(R)$ and the set of all even subgraphs of T.

PROOF. Let $D(T)$ be the collection of even subgraphs of T, and defined $f: T(R) \to D(T)$ by $f(T') = T - T'$. By the lemma, $f(T') \in D(T)$. Then clearly f is injective. Now suppose D is an even subgraph of T. Let T'' be the directed graph obtained from T by reversing the direction of all arcs in D. Since the indegree and outdegree in D are equal for each vertex, it follows that $T'' \in T(R)$. Since $T - T'' = D$, $f(T'') = D$. Hence f is surjective.

Let us note that, as a consequence of the above corollary, each tournament in $T(R)$ has the same number of even subgraphs and this common value is the number of tournaments in $T(R)$.

We denote the Δ-interchange which reverses the arcs of a 3-cycle (p_i, p_j, p_k, p_i) of a tournament by $\Delta(i,j,k)$.

2.3 LEMMA. Let $T \in T(R)$ and let $\mu = (p_{i_1}, \ldots, p_{i_k}, p_{i_1})$ be a cycle of T. Then the tournament T' obtained from T by reversing the arcs of the cycle μ can be obtained from T by a sequence of $k-2$ Δ-interchanges.

PROOF. We prove the lemma by induction on k, it being trivially true for $k = 3$. Let $k \geq 4$, and let $\mu = (q_1, q_2, \ldots, q_k, q_1)$ be any k-cycle of T. We may assume, after relabeling the vertices if necessary, that q_1 loses to some vertex q_j where $j < k$. It follows from the induction hypothesis that the arcs of the j-cycle $(q_1, q_2, \ldots, q_j, q_1)$ may be reversed by $j-2$ Δ-interchanges and, then, that the arcs of the $(k + 2 - j)$- cycle $(q_1, q_j, q_{j+1}, \ldots, q_k, q_1)$ may be reversed by $k-j$ Δ-interchanges. This reverses the arcs of μ in a total of $k-2$ Δ-interchanges as required.

We now obtain the result previously mentioned in the introduction.

2.4 THEOREM. Let $T, T' \in T(R)$. Then T' can be obtained from T by a finite sequence of Δ- interchanges, and hence the graph $G(R)$ is connected.

PROOF. By Lemma 2.1 the directed graph $T-T'$ is even and hence its arcs can be partitioned into cycles. The theorem now follows by repeated application of Lemma 2.3.

A more precise result can be obtained in the following way. For an even directed graph D let $q(D)$ be equal to the largest integer k such that the arcs of D can be partitioned into k cycles. We call $q(D)$ the *cycle packing number of* D. Let the number of arcs of D be denoted by $\sigma(D)$. Now let T, $T' \in T(R)$. By Theorem 2.4 the graph $G(R)$ is connected so that there is a finite distance $\rho(T,T')$ between T and T', which is equal to the smallest length of a chain between T and T'. From Lemma 2.3 we now obtain the following.

2.5 COROLLARY. *For* $T, T' \in T(R)$,

$$\rho(T,T') \leq \sigma(T-T') - 2q(T-T').$$

Actually equality holds in Corollary 2.5 so that a distance formula is available but not easily computable (since $q(T-T')$ is an intractable quantity). Before proving this we establish a stronger result than Theorem 2.4. A connected graph G with at least 3 vertices is called *2-connected* if the deletion of any vertex (along with its incident edges) results in a connected graph.

2.6 THEOREM. *If* $R = (r_1, \ldots, r_n)$ *is a strong score vector with* $n > 3$, *then* $G(R)$ *is 2-connected*.

PROOF. It is well known [7, p.6] that a strong tournament contains cycles of each length k with $3 \leq k \leq n$. Thus suppose R is strong with $n > 3$ and $T \in T(R)$. Then it follows from Corollary 2.2 that $T(R)$ has at least 3 tournaments. We need to show that for tournaments T, T' and T^* in $T(R)$ there is a chain joining T and T' which avoids T^*. For this it suffices to show that for T_1, T_2, T_3 a chain of length 2 in $G(R)$, there exists $T_4 \in T(R)$ with $T_4 \neq T_2$ such that T_1, T_4, T_3 is also a chain of length 2 in $G(R)$.

Suppose T_2 is obtained from T_1 by the Δ- interchange $\Delta(i,j,k)$ and T_3 is obtained from T_2 by $\Delta(u,v,w)$. First suppose $\{i,j,k\}$ and $\{u,v,w\}$ have at most one element in common. Then let T_4 be obtained from T_1 by $\Delta(u,v,w)$. Since T_3 can be obtained from T_4 by $\Delta(i,j,k)$, T_1, T_4, T_3, is a chain of length 2 with $T_4 \neq T_2$. Now suppose $\{i,j,k\}$ and $\{u,v,w\}$ have two elements in common, say $u = j$, $v = i$. Consider the subtournament of T_1 based on the four vertices p_i, p_j, p_k, p_w as indicated in Figure 1.

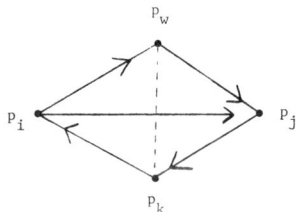

Figure 1

The directed graph $T_3 - T_1$ consists of four arcs which form a 4-cycle $(p_i, p_k, p_u, p_w, p_i)$. First suppose the arc between p_w and p_k is directed from p_w. Then let T_4 be obtained from T_1 by $\Delta(i,w,k)$. Since T_3 can be obtained from T_4 by $\Delta(w,j,k)$, T_1, T_4, T_3 is a chain of length 2 with $T_4 \neq T_2$. A similar argument holds when the arc between p_w and p_k is directed from p_k, and the theorem follows.

We now derive a formula for the distance function of $G(R)$.

2.7 THEOREM. Let $T, T' \in T(R)$. Then

$$\rho(T,T') = \sigma(T-T') - 2q(T-T'). \tag{2.1}$$

PROOF. We prove that (2.1) holds by induction on $\rho(T,T')$. If $\rho(T,T') = 1$, then (2.1) clearly holds. Let $d \geq 1$ and suppose (2.1) holds for each pair of tournaments T, T' with $\rho(T,T') = d$.

Now let $T, T' \in T(R)$ with $\rho(T,T') = d + 1$. Let $T = T^{(0)}, T^{(1)}, \ldots, T^{(d)}, T^{(d+1)} = T'$ be a chain in $G(R)$ of length $d + 1$ joining T and T'. Then it follows that $\rho(T, T^{(d)}) = d$. Hence by the inductive hypothesis the arcs of $T - T^{(d)}$ can be partitioned into $\bar{q}$ cycles $C_1, \ldots, C_{\bar{q}}$ where $\bar{q} = q(T-T^{(d)})$ and $d = \rho(T,T^{(d)}) = \sigma(T-T^{(d)}) - 2\bar{q}$. Let T' be obtained from $T^{(d)}$ by reversing the arcs of the 3-cycle C in $T^{(d)}$. Let l be the number of arcs common to C and $T^{(d)} - T$. Then

$$\sigma(T-T') = \sigma(T-T^{(d)}) + 3 - 2l,$$

and since $T-T'$ has $\bar{q} + 1 - l$ pairwise arc disjoint cycles,

$$q(T-T') \geq \bar{q} + 1 - l.$$

Using the fact that $d = \rho(T,T^{(d)}) - 2\bar{q}$, we now obtain

$$\sigma(T-T') - 2q(T-T') \leq \sigma(T-T^{(d)}) - 2\bar{q} + 1 = d+1$$

Since $\rho(T,T') = d+1$, it follows that $\rho(T,T') \geq \sigma(T-T') - 2q(T-T')$. It now follows from Corollary 2.5 that (2.1) holds. Hence the theorem is true by induction.

2.8 COROLLARY Let $T \in T(R)$ and let T' be obtained from T by reversing the arcs of a k-cycle. Then

$$\rho(T,T') = k-2$$

Let G be a connected graph. The *eccentricity* $e(v)$ of a vertex v is defined to be the largest distance from v to a vertex of G. The *diameter* $\delta(G)$ of G is the largest of the eccentricities of its vertices. Thus $\delta(G)$ is the largest distance between a pair of vertices of G. Theorem 2.7 can be used to give a non-trivial lower bound on the eccentricity of a tournament in $G(R)$ for R strong and hence on the diameter of $G(R)$.

2.9 COROLLARY. Suppose $R = (r_1, \ldots, r_n)$ is a strong score vector. Then for each tournament $T \in T(R)$, its eccentricity $e(T)$ in $G(R)$ satisfies $e(T) \geq n-2$. Hence $\delta(G(R)) \geq n-2$.

PROOF. Suppose R is strong and $T \in T(R)$. Then it is well known that T has a Hamiltonian cycle μ, that is, an n-cycle. Let T' be obtained from T by reversing the arcs of μ. By Corollary 2.8, $\rho(T,T') = n-2$ and the corollary follows.

In the next section we shall discuss a class of tournaments for which the diameter equals $n-2$. We also remark that it is easy to check that $\delta(R) = n-2$ whenever R is a strong score vector and $n = 3$ or 4. It is also easy to check that for R a strong score vector and $n = 5$, then $\delta(R) = 3$, unless $R = (2,2,2,2,2)$ in which case $\delta(R) = 4$. The following is an internal characterization of the eccentricity of a tournament.

2.10 COROLLARY. Let $T \in T(R)$. Then

$$e(T) = \max\{\sigma(D) - 2q(D)\}$$

where the maximum is taken over all even subgraphs D of T.

PROOF. This result is immediate from Theorem 2.7 and Lemma 2.2.

To conclude this section we determine some of the general structure of interchange graphs. The key idea used is that of an upset. Recall that for a monotone score vector R, $\check{\tau}(R)$ and $\bar{\tau}(R)$ denote, respectively, the minimum and maximum number of upsets for a tournament $T \in T(R)$. For k an integer with $\check{\tau}(R) \leq k \leq \bar{\tau}(R)$, let $T_k(R)$ denote the set of all tournaments $T \in T(R)$ with $\tau(T) = k$ upsets. The sets $T_k(R)$, $\check{\tau}(R) \leq k \leq \bar{\tau}(R)$, are pairwise disjoint, and so

$$T(R) = T_{\hat{\tau}}(R) \cup T_{\hat{\tau}+1}(R) \cup \cdots \cup T_{\bar{\tau}}(R)$$

is a partition of $T(R)$, where we have written $\hat{\tau}$ and $\bar{\tau}$ in place of $\hat{\tau}(R)$ and $\bar{\tau}(R)$. Now suppose Then $T \in T_k(R)$ where $k = \tau(T)$. Let T' be obtained from T by an Δ-interchange. Then it follows easily that $\tau(T') = \tau(T) \pm 1$ so that $T' \in T_{k-1}(R)$ or $T' \in T_{k+1}(R)$. In terms of the interchange graph this means the following.

2.11 THEOREM. The vertices of the interchange graph $G(R)$ can be paritioned into $\bar{\tau}(R) - \hat{\tau}(R) + 1$ sets $T_{\hat{\tau}}(R), T_{\hat{\tau}+1}(R), \ldots, T_{\bar{\tau}}(R)$ so that all edges of $G(R)$ join a vertex in $T_k(R)$ with a vertex in $T_{k+1}(R)$ for some $k = \hat{\tau}, \ldots, \bar{\tau}-1$.

In graph theory language, the graph $G(R)$ is a $(\bar{\tau}(R) - \hat{\tau}(R) + 1)$ - partite graph, but one of a very special type. By grouping the tournaments with an odd number of upsets and those with an even number of upsets we obtain the following.

2.12 COROLLARY. The interchange graph $G(R)$ is bipartite.

In spite of Theorem 2.11 the structure of an interchange graph can be very complicated. Let $T \in T(R)$ and suppose $T \in T_k(R)$. If $k = \hat{\tau}(R)$, then T is joined in $G(R)$ only to tournaments in $T_{k+1}(R)$. If $k = \bar{\tau}(R)$, then T is joined only to tournaments in $T_{k-1}(R)$. Now suppose that $\hat{\tau}(R) < k < \bar{\tau}(R)$. Then T is joined only to tournaments in $T_{k-1}(R)$ and $T_{k+1}(R)$. But it is possible that T is joined only to tournaments in $T_{k-1}(R)$ or only to tournaments in $T_{k+1}(R)$. Before giving an example, we make some new definitions and prove a lemma.

Let T be a tournament with a monotone score $R = (r_1, \ldots, r_n)$. The *upset directed graph* T_u of T is the directed graph corresponding to the upsets in T. Thus the vertices of T_u are $p_1, \ldots, p_n$ and there is an arc from p_i to p_j provided $1 \le i < j \le n$ and p_i beats p_j. In terms of matrices, T_u corresponds to the 1's above the main diagonal of T. The *non-upset directed graph* T_{nu} of T is defined in a similar way using the non-upsets of T and corresponds to the 1's below the main diagonal. It follows that the arcs of T_u and T_{nu} partition the arcs of T. We call the arcs of T_u the *upset arcs* of T and the arcs of T_{nu} the *non-upset arcs* of T. A directed graph is called *transitive* provided for any three vertices u, v, and w, whenever there are arcs from u to v and from v to w, there is also an arc from u to w.

2.13 LEMMA. Let $T \in T(R)$. Then no interchange can decrease the number of upsets of T if and only if T_u is transitive. Similarly, no interchange can increase the number of upsets of T if and only if T_{nu} is transitive.

PROOF. Consider vertices p_i, p_j, p_k where $1 \leq i < j < k \leq n$ such that there are arcs in T from p_i to p_j and from p_j to p_k. If the arc between p_i and p_k is from p_k, then (p_i, p_j, p_k, p_i) is a 3-cycle, T_u is not transitive, and $\Delta(i,j,k)$ decreases the number of upsets of T. If the arc between p_i and p_k is from p_i for all such triples p_i, p_j, p_k, then T_u is transitive and no interchange can decrease the number of upsets of T. The lemma now follows.

If $T \in T_{\tilde{\tau}}(R)$, then clearly no interchange can decrease the number of upsets of T. The preceding lemma suggests that this property does not characterize tournaments in $T(R)$ with the minimum number of upsets. Indeed, let $R = (2, 2, 2, 2, 2,)$. Then clearly $\tilde{\tau}(R) \geq 3$. Since

$$\tilde{T} = \begin{pmatrix} 0 & 0 & 0 & 1 & 1 \\ 1 & 0 & 0 & 0 & 1 \\ 1 & 1 & 0 & 0 & 0 \\ 0 & 1 & 1 & 0 & 0 \\ 0 & 0 & 1 & 1 & 0 \end{pmatrix}$$

is a tournament in $T(R)$ with 3 upsets, $\tilde{\tau}(R) = 3$. By considering rows 4 and 5 of tournaments in $T(R)$ one obtains that $\bar{\tau}(R) \leq 7$. But then $\bar{\tau}(R) = 7$, since the tournament $\bar{T}$ which is the transpose of $\tilde{T}$ (in terms of directed graphs, $\bar{T}$ is obtained from $\tilde{T}$ by reversing all arcs) is in $T(R)$ and has 7 upsets. Now consider the tournament $T \in T(R)$ given by

$$T = \begin{pmatrix} 0 & 0 & 1 & 0 & 1 \\ 1 & 0 & 0 & 1 & 0 \\ 0 & 1 & 0 & 0 & 1 \\ 1 & 0 & 1 & 0 & 0 \\ 0 & 1 & 0 & 1 & 0 \end{pmatrix}$$

Then $\tau(T) = 4$, and it is easy to check that T_u is transitive. The tournament T^* obtained from T by transposition has the property that $\tau(T^*) = 6$ and T^*_{nu} is transitive. It follows, using Theorem 2.11, that the graph $G(R)$ is a 5-partite graph whose vertices are partitioned into $T_3(R), \ldots, T_7(R)$ such that T is a vertex of $T_4(R)$ joined only to vertices in $T_5(R)$. Similarly, T^* is in $T_6(R)$ and is joined only to vertices in $T_5(R)$.

3. Special Score Vectors.

In this section we prove several theorems for the class of tournaments $T(R)$ for some special score vectors R. These are

$$\hat{R} = (\hat{r}_1, \ldots, \hat{r}_n) = (1, 1, 2, \ldots, n-2, n-2) \quad (n \geq 3)$$

and

$$\bar{R} = (\bar{r}_1, \ldots, \bar{r}_n) = \begin{cases} (\frac{n-1}{2}, \ldots, \frac{n-1}{2}), & \text{for } n \text{ odd}, n \geq 3 \\ (\frac{n-2}{2}, \ldots, \frac{n-2}{2}, \frac{n}{2}, \ldots, \frac{n}{2}), & \text{for } n \text{ even}, n \geq 4 \end{cases}$$

A tournament $T \in T(\bar{R})$ is called *regular* when n is odd and *near-regular* when n is even. We refer to $\bar{R}$ as a *regular score vector* when n is odd and a *near-regular score vector* when n is even. The vector $\hat{R}$ is the smallest strong score vector and the vector $\bar{R}$ the largest score vector in a sense to be defined now. Let $R = (r_1, \ldots, r_n)$ and $R' = (r_{1'}, \ldots, r_{n'})$ be two monotone score vectors. We write $R \leq R'$ if

$$r_1 + \cdots + r_i \leq r_{1'} + \cdots + r_{i'} \quad (i = 1, \ldots, n). \quad (3.1)$$

Since R and R' are both score vectors, we have equality in (3.1) when $i = n$. This defines a partial order on score vectors, which is the converse order of majorization (see e.g. [10, p. 104]).

3.1 LEMMA. For all strong score vectors $R = (r_1, \ldots, r_n)$,

$$\hat{R} \leq R \leq \bar{R}.$$

PROOF. Using Corollary 1.2 we calculate that

$$1 + 1 + \cdots + (i-1) = 1 + \binom{i}{2} \leq r_1 + \cdots + r_i$$

for $i = 1, \ldots, n$. Hence $\hat{R} \leq R$. Let n be odd and suppose that for some

$$i = 1, \ldots, n-1$$

$$r_1 + \cdots + r_i > i(\frac{n-1}{2}).$$

Then it follows from the monotonicity of R that $r_n \geq \cdots \geq r_i \geq r_{i+1} \geq \frac{n+1}{2}$, so that

$$r_1 + \cdots + r_n > i(\frac{n-1}{2}) + (n-i)(\frac{n+1}{2}) > \binom{n}{2}$$

From this contradiction we conclude $R \leq \bar{R}$.

Now let n be even. Suppose that for some $i \leq \frac{n}{2}$,

$$r_1 + \cdots + r_i > i(\frac{n-2}{2}).$$

Then $r_n \geq \cdots \geq r_{i+1} \geq \frac{n}{2}$ and

$$r_1 + \cdots + r_n > i(\frac{n-2}{2}) + (n-i)\frac{n}{2} > \binom{n}{2}$$

a contradiction. Now suppose that for some i with $\frac{n}{2} < i < n$,

$$r_1 + \cdots + r_i > \frac{n}{2}(\frac{n-2}{2}) + (i-\frac{n}{2})\frac{n}{2}.$$

Then $r_n \geq \cdots \geq r_{i+1} \geq \frac{n}{2}$ and

$$r_1 + \cdots + r_n > \frac{n}{2}(\frac{n-2}{2}) + (i-\frac{n}{2})\frac{n}{2} + (n-i)\frac{n}{2} = \binom{n}{2}$$

a contradiction. It follows that $R \leq \bar{R}$.

With the partial order defined above, it follows from Lemmas 3.1 that $\hat{R}$ is the strong score vector which is furthest from being regular (n odd) or near-regular (n even). We now investigate $T(\hat{R})$ more thoroughly. It follows readily that $\hat{\tau}(\hat{R}) = 1$ (see (3.2) below). To indicate the number of components of $\hat{R}$ we now write $\hat{R}_n$ in place of $\hat{R}$.

Let $T = [t_{ij}]$ and $T' = [t'_{ij}]$ be two tournaments, not necessarily having the same score vector. We define their *join* $T * T'$ by

$$T * T' = \begin{pmatrix} T & 0 & 0 \\ 0 & 0 & 0 \\ 1 & 1 & T' \end{pmatrix},$$

where the center zero block is $|X|$. We define $T_1 * \cdots * T_m$ inductively by $T_1 * \cdots * T_m = T_1 * (T_2 * \cdots * T_m)$. Let k be an integer with $k \geq 3$ and let

$$\hat{T}_{(k)} = \begin{pmatrix} 0 & 0 & \ldots & 0 & 1 \\ 1 & 0 & \ldots & \ldots & 0 \\ \cdot & \cdot & \cdot & \cdot & \cdot \\ \cdot & \cdot & \cdot & \cdot & \cdot \\ \cdot & \cdot & \cdot & \cdot & \cdot \\ 1 & \cdot & \ldots & 0 & 0 \\ 0 & 1 & \ldots & 1 & 0 \end{pmatrix}, \qquad (3.2)$$

the tournament in $T(\hat{R}_k)$ with exactly 1 upset

3.3 THEOREM. Let $T = [t_{ij}] \in T(\hat{R}_n)$ and let $\tau(T) = u$. Then there exist integers $k_1, \ldots, k_u$ such that

$$T = \hat{T}_{(k_1)} * \ldots * \hat{T}_{(k_u)}.$$

PROOF. Since $\hat{r}_1 = 1$, T has exactly one 1 in its first row. Let this 1 be in column k_1. If $k_1 = n$, $T = \hat{T}_n$. Suppose $k_1 < n$. Then it follows that the leading $k_1 \times k_1$ submatrix of T is a $\hat{T}_{(k_1)}$ and that $T = \hat{T}_{(k_1)} * T'$ for some tournament $T' \in T(\hat{R}_{n-k_1+1})$. The theorem now follows by induction.

In terms of directed graphs, Theorem 3.3 implies that for each tournament $T \in T(\hat{R}_n)$, the upset directed graph T_u is just a path from p_1 and to p_n, namely $(p_1, p_{k_1}, p_{k_1+k_2-1}, \ldots, p_n)$. In particular, it follows that $\bar{\tau}(\hat{R}_n) = n-1$.

As a corollary we are able to determine easily the diameter of the interchange graph $G(\hat{R})$.

3.4 COROLLARY. $\delta(G(\hat{R}_n)) = n-2$

PROOF. Let $T, T' \in T(\hat{R}_n)$ with $T \neq T'$. By Theorem 2.4, the distance T and T' in $T(\hat{R}_n)$ is given by

$$\rho(T,T') = \sigma(T-T') - 2q(T-T') \qquad (3.3)$$

where $\sigma(T-T')$ is the number of arcs of the directed graph $T-T'$ and $q(T-T')$ is the cycle packing number of $T-T'$. By Theorem 3.3 there are integers $k_1, \ldots, k_u$ and $l_1, \ldots, l_v$ such that

$$T = \hat{T}_{(k_1)} * \cdots * \hat{T}_{(k_u)} \text{ and } T' = \hat{T}_{(l_1)} * \cdots * \hat{T}_{(l_v)}.$$

From the structure of the tournaments $T_{(k)}$ defined by (3.2) and the definition of join, it follows that $\sigma(T-T') \leq n$. Since $q(T-T') \geq 1$, it follows from (3.3) that

$$\rho(T,T') \leq n-2.$$

Hence $\delta(G(\hat{R}_n)) \leq n-2$. By Corollary 2.9, $\delta(G(\hat{R}_n)) \geq n-2$, and thus we have equality.

We note that the property of the diameter equalling $n-2$ does not characterize $\hat{R}_n (n = 3,4,...)$ among (strong) score vectors. For example, let $R = (1,2,2,2,3) \neq \hat{R}_5$, and let $T \in T(R)$. As a directed graph T has one vertex, namely p_5, of outdegree 3 (and so indegree 1). By Corollary 2.10 the eccentricity of T is given by

$$e(T) = \max\{\sigma(D) - 2q(D)\} \tag{3.4}$$

where the maximum is taken over all even nonempty subgraphs D of T. Any such D can include at most one of the three arcs of T that goes from p_5. Of the two arcs of T from p_5 not in D, at least one goes to a vertex of outdegree 2(either p_1, p_3, or p_4). It follows that

$$\sigma(D) \leq 10 - 2 - 1 = 7.$$

If $\sigma(D) = 6$ or 7, then clearly $q(D) \geq 2$ and $\sigma(D) - 2q(D) \leq 7 - 4 = 3$. If $\sigma(D) \leq 5$, then $q(D) \geq 1$ and $\sigma(D) \leq 5 - 2 = 3$. Thus by (3.4), $e(T) \leq 3$ so that $\delta(G(R)) \leq 3$. Hence $\delta(G) = 3$.

We give a characterization of $\hat{R}_n$ in the next theorem. Let R be a score vector. Then R is said to be *uniquely Hamiltonian* if every $T \in T(R)$ has a unique Hamiltonian cycle. Necessarily, a uniquely Hamiltonian score vector is strong. Douglas [3] has given a characterization of tournaments with n vertices which have exactly one Hamiltonian cycle. His characterization implies the following.

3.5 LEMMA. Let T be a tournament with $n \geq 5$ vertices that has exactly one Hamiltonian cycle μ. Then there is a vertex v_1 of T with outdegree 1 and vertices v_0 and v_2 such that the Hamiltonian cycle has the form $(\ldots, v_0, v_1, v_2, \ldots)$ where there is an arc from v_0 to v_2 in T.

We now show that for every strong score vector $R \neq \hat{R}_n$, there is a $T \in T(R)$ with at least two Hamiltonian cycles.

3.6 THEOREM. Let $R = (r_1, \ldots, r_n)$ be a monotone score with $n \geq 4$. Then R is uniquely Hamiltonian if and only if $R = \hat{R}_n$.

PROOF. We first observe that every $T \in T(\hat{R}_n)$ has exactly one Hamiltonian cycle and that each upset-arc of T is an arc of this cycle. This follows easily from the canonical form for tournaments in $T(\hat{R}_n)$ given in Theorem 3.3, which implies in particular that the removal of any upset arc of a tournament in $T(\hat{R}_n)$ leaves a directed graph which is not strong.

Now suppose R is uniquely Hamiltonian. We prove that $R = \hat{R}_n$ by induction on n. When $n = 4$, the only strong score vector is $\hat{R}_4$, so that the conclusion holds in this case. Now let $n \geq 5$ and let $T \in T(R)$ where T has vertices $p_1, \ldots, p_n$. Then T has a unique Hamiltonian cycle and by Lemma 3.5 it has the form $(\cdots p_j, p_l, p_k \cdots)$ where the outdegree of p_l is 1 (that is, $r_l = 1$) and there is an arc in T from p_j to p_k. We delete the vertex p_l from T (and all incident arcs) to obtain a tournament T' with score $R' = (r_1', \ldots, r_{l-1}', r_{l+1}', \ldots, r_n')$ where

$$r_i' = \begin{cases} r_i \text{ if } i = k, \\ r_i - 1 \text{ if } i \neq k. \end{cases}$$

Since there is an arc from p_j to p_k in T, T' has a Hamiltonian cycle. Thus R' is a strong score vector, although not necessarily monotone. We claim that R' (the monotone rearrangement thereof) is uniquely Hamiltonian. Suppose not. Then there is a tournament T^* with score vector R' which has two different Hamiltonian cycles μ_1 and μ_2. By adding the vertex p_l to T^* and arcs from p_l to p_k and p_i to p_l for $i = 1, \ldots, n$ but $i \neq k, l$, we obtain a tournament in $T(R)$ and easily construct from μ_1 and μ_2 two distinct Hamiltonian cycles. This contradiction proves that R' is uniquely Hamiltonian. Hence by the induction hypothesis, the monotone rearrangement is $(1, 1, 2, \ldots, n-3, n-3)$. If we can show that $r_k' = 1$, then it will follow that $R = \hat{R}_n$. Thus suppose $r_k' = d$ where $1 < d < n-2$. Then

$$R = (1, 2, 2, 3, \ldots, d, d, d+2, \ldots, n-2).$$

Then the tournament T_0 whose only upset arcs are arcs from p_1 to p_{d+2} (p_{d+2} has outdegree d) and from p_2 to p_n has score vector R and has the two Hamiltonian cycles

$$(p_1, p_{d+2}, p_{d+1}, \ldots, p_2, p_n \cdots p_{d+3}, p_1),$$

and

$$(p_1, p_{d+2}, p_2, p_n, \ldots, p_{d+3}, p_{d+1}, \ldots, p_3, p_1)$$

This contradicts the assumption that R is uniquely Hamiltonian. Hence it follows by induction that $R = \hat{R}_n$.

A problem of considerable interest and considerable difficulty is that of determining the number of tournaments having a particular score vector R, that is, the cardinality of the set $T(R)$. Thus we are asking for the number of *labelled* tournaments with score vector R and not the number of nonisomorphic tournaments with score vector R. The latter have been listed by Moon [7, Appendix, pp. 91-96] for $n \leq 6$. The total number of nonisomorphic tournaments with n vertices (score vector not specified) has been determined by Davis (see Moon [7, pp. 84-89].)

It is possible to count the number of tournaments with score vector $\hat{R}_n$, due to its specific nature. For general R, such a count seems very difficult.

3.7 THEOREM.
$$|T(\hat{R}_n)| = 2^{n-2}.$$

PROOF. Consider the special tournament $\hat{T}_{(n)} \in T(\hat{R}_n)$ for which the only upset arc is an arc from p_1 to p_n. It follows from Corollary 2.2 that the number of tournaments in $T(\hat{R}_n)$ equals the number of even subgraphs of $\hat{T}_{(n)}$. Since the directed graph obtained from $\hat{T}_{(n)}$ by removing the upset arc has no cycles, it follows that every even subgraph of $\hat{T}_{(n)}$ with at least one arc is a cycle one of whose arcs is the arc from p_1 to p_n. The number of such k-cycles is clearly $\binom{n-2}{k-2}$ for $k = 3, \ldots, n$. Hence

$$|T(\hat{R}_n)| = 1 + \sum_{k=3}^{n} \binom{n-2}{k-2} = \sum_{i=0}^{n-2} \binom{n-2}{i} = 2^{n-2}$$

It seems likely that $\hat{R}_n$ is the strong score vector with n coordinates with the fewest number of tournaments, and we *conjecture* that

$$|T(R)| > 2^{n-2}$$

for all strong score vectors $R = (r_1, \ldots, r_n) \neq \hat{R}_n$. To support this conjecture, we have this inequality for strong score vectors R satisfying $\tilde{\tau}(R) = 2$ or 3, but the proofs are not given here.

We conclude this section with some observations concerning the class of regular and near-regular tournaments. Consider $T(\bar{R}_n)$ where $\bar{R}_n = (\frac{n-1}{2}, \ldots, \frac{n-1}{2})$, n odd. The tournaments in $T(\bar{R}_n)$ are even directed graphs and hence their arcs can be partitioned into cycles. By [2]

$$\hat{\tau}(\bar{R}_n) = \binom{\frac{n+1}{2}}{2}, \quad \bar{\tau}(\bar{R}_n) = \binom{n}{2} - \binom{\frac{n+1}{2}}{2}$$

It follows from Theorem 2.11 that the diameter of the interchange graph $G(\bar{R}_n)$ satisfies

$$\delta(\bar{R}_n) \geq \bar{\tau}(\bar{R}_n) - \hat{\tau}(\bar{R}_n) = \binom{n}{2} - 2\binom{\frac{n+1}{2}}{2} = \frac{(n-1)^2}{4}.$$

We show that there is a tournament in $T(\bar{R}_n)$ with $\hat{\tau}(\bar{R}_n)$ upsets and a tournament in $T(\bar{R}_n)$ with $\bar{\tau}(\bar{R}_n)$ upsets whose distance is exactly $\bar{\tau}(\bar{R}_n) - \hat{\tau}(\bar{R}_n)$. First we prove the following theorem. Let $\bar{T}_n$ be the rotational tournament [9] whose first row is $(0, 1, \ldots, 1, 0, \ldots, 0)$ where there are $\frac{n-1}{2}$ 1's. Thus

$$\bar{T}_n = P + P^2 + \cdots + P^{\frac{n-1}{2}}$$

where P is the full cycle permutation matrix

$$\begin{pmatrix} 0 & 1 & 0 & \cdots & 0 \\ 0 & 0 & 1 & \cdots & 0 \\ \cdot & \cdot & \cdot & & \cdot \\ 0 & 0 & 0 & \cdots & 1 \\ 1 & 0 & 0 & \cdots & 0 \end{pmatrix}$$

Note that $\tau(\bar{T}_n) = \hat{\tau}(\bar{R}_n)$. Let $\hat{T}_n$ be the tournament obtained from $\bar{T}_n$ by matrix transposition (reversing all arcs). Thus $\hat{T}_n$ is the rotational tournament whose first row is $(0, 0 \ldots, 0, 1, \ldots 1)$ where again there are $\frac{n-1}{2}$ 1's. Moreover,

$$\hat{T}_n = P^{\frac{n+1}{2}} + \cdots + P^{n-1}$$

Note that $\tau(\hat{T}_n) = \bar{\tau}(\bar{R}_n)$.

3.8 THEOREM. When n is odd, the arcs of $\bar{T}_n$ can be partitioned into $\binom{\frac{n-1}{2}}{2}$ 4-cycles and $\frac{n-1}{2}$ 3-cycles.

PROOF. Rather than give a complete description of the cycles in the partition, we illustrate, in terms of a matrix, the partition when $n = 9$.

$$(3,5)\bar{T}_9 = \begin{bmatrix} 0 & a & b & c & x_1 & 0 & 0 & 0 & 0 \\ 0 & 0 & d & e & x_2 & a & 0 & 0 & 0 \\ 0 & 0 & 0 & f & x_3 & b & d & 0 & 0 \\ 0 & 0 & 0 & 0 & x_4 & c & e & f & 0 \\ 0 & 0 & 0 & 0 & 0 & x_1 & x_2 & x_3 & x_4 \\ x_1 & 0 & 0 & 0 & 0 & 0 & a & b & c \\ a & x_2 & 0 & 0 & 0 & 0 & 0 & d & e \\ b & d & x_3 & 0 & 0 & 0 & 0 & 0 & f \\ c & e & f & x_4 & 0 & 0 & 0 & 0 & 0 \end{bmatrix}$$

In the above matrix, each of the letters $a, b, \ldots, f$ corresponds to a 4-cycle, while each of the letters x_1, x_2, x_3, x_4 corresponds to a 3-cycle. Thus, for instance, the letter d corresponds to the cycle $(p_2, p_3, p_7, p_8, p_2)$ while the letter x_3 corresponds to the cycle (p_3, p_5, p_8, p_3). The procedure is quite general, and the theorem follows.

3.9 COROLLARY. The distance between $\hat{T}_n$ and $\bar{T}_n$ in $G(\bar{R}_n)$ satisfies

$$\rho(\bar{T}_n, \hat{T}_n) = \bar{\tau}(\bar{R}_n) - \hat{\tau}(\bar{R}_n) = \frac{(n-1)^2}{4}$$

PROOF. Since $\tau(\bar{T}_n) = \bar{\tau}(\bar{R}_n)$ and $\tau(\hat{T}_n) = \hat{\tau}(\bar{R}_n)$
it follows that

$$\rho(\bar{T}_n, \hat{T}_n) \geq \bar{\tau}(\bar{R}_n) - \hat{\tau}(\bar{R}_n).$$

On the other hand, by Theorem 2.7,

$$\rho(\bar{T}_n, \hat{T}_n) = \sigma(\bar{T}_n - \hat{T}_n) - 2q(\bar{T}_n - \hat{T}_n)$$
$$= \sigma(\bar{T}_n) - 2q(\bar{T}_n)$$
$$= \binom{n}{2} - 2q(\bar{T}_n).$$

It follows from Theorem 3.8 that

$$q(\bar{T}_n) \geq \binom{n-1}{2} + \frac{n-1}{2} = \binom{n+1}{2}$$

Hence

$$\rho(\overline{T}_n,\hat{T}_n) \le \binom{n}{2} - 2\binom{\binom{n+1}{2}}{2} = \overline{\tau}(\overline{R}_n) - \hat{\tau}(\overline{R}_n)$$

The corollary now follows.

From the above proof we also obtain the following.

3.10 COROLLARY.
$$q(\overline{T}_n) = \binom{\binom{n+1}{2}}{2}$$

As a final corollary we estimate the number of regular tournaments with n vertices. The determination of the number of regular tournaments with n vertices was posed as an open problem by Palmer [8, p. 414]. Spencer [11] has determined the asymptotic number of regular tournaments.

3.11 COROLLARY.
$$|T(\overline{R}_n)| \ge 2^{\frac{n^2-1}{8}} \quad \text{for } n \text{ odd}$$

PROOF. It follows from Corollary 2.2 that the number of tournaments in $T(\overline{R}_n)$ equals the number of even subgraphs of the rotational tournament $\overline{T}_n$. Since the arcs of $\overline{T}_n$ can be partitioned into $\binom{\binom{n+1}{2}}{2} = \frac{n^2-1}{8}$ cycles, $\overline{T}_n$ has at least $2^{\frac{n^2-1}{8}}$ even subgraphs and the result follows.

The estimate in Corollary 3.11 can be improved by noticing that for n odd, there is also a partition of the arcs of $\overline{T}_n$ into $\frac{n-1}{2}$ Hamiltonian cycles which give rise to different even subgraphs of $\overline{T}_n$. This partition, which is well known, is illustrated in terms of a matrix for $n = 9$ in (3.6).

$$\bar{T}_9 = \begin{bmatrix} 0 & y_1 & y_2 & y_3 & y_4 & 0 & 0 & 0 & 0 \\ 0 & 0 & y_1 & y_2 & y_3 & y_4 & 0 & 0 & 0 \\ 0 & 0 & 0 & y_1 & y_2 & y_3 & y_4 & 0 & 0 \\ 0 & 0 & 0 & 0 & y_1 & y_2 & y_3 & y_4 & 0 \\ 0 & 0 & 0 & 0 & 0 & y_1 & y_2 & y_3 & y_4 \\ y_4 & 0 & 0 & 0 & 0 & 0 & y_1 & y_2 & y_3 \\ y_3 & y_4 & 0 & 0 & 0 & 0 & 0 & y_1 & y_2 \\ y_2 & y_3 & y_4 & 0 & 0 & 0 & 0 & 0 & y_1 \\ y_1 & y_2 & y_3 & y_4 & 0 & 0 & 0 & 0 & 0 \end{bmatrix} \qquad (3.6)$$

Each y_i corresponds to a Hamiltonian cycle C_i, and we obtain $2^{\frac{n-1}{2}}$ even subgraphs of $\bar{T}_n$. Suppose a nonempty union Ω of (the arcs of) $C_1, \ldots, C_{\frac{n-1}{2}}$ equalled the union Ω' of (the arcs of) 4-cycles and 3-cycles of Theorem 3.8 (illustrated for $n = 9$ in (3.5)). Since each C_i has an arc in common with one of the 3-cycles, Ω contains an arc of one of the 3-cycles. Since $C_{\frac{n-1}{2}}$ has an arc in common with each 3-cycle, it follows that $C_{\frac{n-1}{2}}$ is included in Ω and each 3-cycle is included in Ω'. But now it follows that each C_i is included in Ω. We conclude that only the empty union and the union of all the C_i's can equal a union of 4-cycles and 3-cycles of Theorem 3.8. Hence we obtain the following.

THEOREM 3.12. For n odd,
$$|T(\bar{R}_n)| \geq 2^{\frac{n^2-1}{8}} + 2^{\frac{n-1}{2}} - 2.$$

The estimate for the number of tournaments in $T(\bar{R}_n)$ should be contrasted with the *total* number of (labelled) tournaments with n vertices. The later number is $2^{\frac{n^2-n}{2}}$ since each of the $\binom{n}{2} = \frac{n^2-n}{2}$ edges of a (labelled) complete graph with n vertices can be independently oriented to obtain a tournament.

Now consider $T(\bar{R}_n)$ where n is even so that $\bar{R}_n = (\frac{n-2}{2}, \ldots, \frac{n-2}{2}, \frac{n}{2}, \ldots, \frac{n}{2})$. The tournaments in $T(\bar{R}_n)$ are not regular, so their arcs cannot be partitioned into cycles. By

Theorem 2.11 and [2],

$$\delta(\overline{R}_n) \geq \overline{\tau}(\overline{R}_n) - \tilde{\tau}(\overline{R}_n) = \binom{n}{2} - \binom{\frac{n+2}{2}}{2} - \binom{\frac{n}{2}}{2}$$

$$= \frac{n(n-2)}{4}$$

Let $\overline{T}_n$ be the tournament in $T(\overline{R}_n)$ defined by

$$\overline{T}_n = \begin{bmatrix} \underline{X} & \underline{X}' \\ \underline{Y} & \underline{X} \end{bmatrix}$$

where $\underline{X}$ is the $\frac{n}{2} \times \frac{n}{2}$ matrix with o's on and below the main diagonal and 1's elsewhere, and where $\underline{Y}$ is obtained from $\underline{X}$ by replacing the o's on the main diagonal with 1's.

Then by [2],

$$\tau(\overline{T}_n) = \overline{\tau}(\overline{R}_n) = \binom{n}{2} - \binom{\frac{n+2}{2}}{2}$$

THEOREM 3.13. When n is even, the tournament $\overline{T}_n$ contains $\binom{n/2}{2}$ are disjoint 4-cycles.

PROOF. We illustrate the proof for $n = 8$:

$$\overline{T}_8 = \begin{bmatrix} 0 & a & b & c & 0 & 0 & 0 & 0 \\ 0 & 0 & d & e & a & 0 & 0 & 0 \\ 0 & 0 & 0 & f & b & d & 0 & 0 \\ 0 & 0 & 0 & 0 & c & e & f & 0 \\ 1 & 0 & 0 & 0 & 0 & a & b & c \\ a & 1 & 0 & 0 & 0 & 0 & d & e \\ b & d & 1 & 0 & 0 & 0 & 0 & f \\ c & e & f & 1 & 0 & 0 & 0 & 0 \end{bmatrix}$$

Now consider the tournament $\tilde{T}_n \in T(\overline{R}_n)$ obtained from $\overline{T}_n$ by reversing all arcs except those from p_{n-i} to $p_{\frac{n}{2}-i}$ for $i = 0, 1, \ldots, \frac{n}{2} - 1$. Then it follows from [2] that

$$\tau(\tilde{T}_n) = \hat{\tau}(\bar{R}_n) = \binom{n/2}{2}. \text{ Hence}$$

$$\rho(\bar{T}_n, \tilde{T}_n) \geq \bar{\tau}(\bar{R}_n) - \hat{\tau}(\bar{R}_n) = \frac{n(n-2)}{4}$$

COROLLARY 3.14 $\rho(\bar{T}_n, \tilde{T}_n) = \bar{\tau}(\bar{R}_n) - \hat{\tau}(\bar{R}_n) = \frac{n(n-2)}{4}$

PROOF. By Theorem 2.7 and 3.13

$$\rho(\bar{T}_n, \tilde{T}_n) = \sigma(\bar{T}_n, \tilde{T}_n) - 2q(\bar{T}_n, \tilde{T}_n)$$

$$= \binom{n}{2} - \frac{n}{2} - 2q(\bar{T}_n, \tilde{T}_n)$$

$$\leq \binom{n}{2} - \frac{n}{2} - 2\binom{n/2}{2} = \frac{n(n-2)}{4}.$$

The corollary now follows.

COROLLARY 3.15. $|T(\bar{R}_n)| \geq 2^{\frac{n^2-2n}{8}}$ for n even.

In Corollary 2.9 we obtained a lower bound for the diameter $\delta(G(R))$ of the interchange graph $G(R)$ when R is a strong score vector. It would be of interest to have a non-trivial upper bound for $\delta(G(R))$. We have noted that for the regular score vector $\bar{R}_n$ (n odd), $\delta(G(\bar{R}_n)) \geq (n-1)^2/4$, and for the near-regular score vector $\bar{R}_n$ (n even), $\delta(\overline{G(R_n)}) \geq \frac{n(n-2)}{4}$.

CONJECTURE 3.16. For a score vector $R = (r_1, \ldots, r_n)$,

$$\delta(G(R)) \leq \begin{cases} \frac{(n-1)^2}{4}, & \text{for } n \text{ odd} \\ \frac{n(n-2)}{4}, & \text{for } n \text{ even} \end{cases}$$

The validity of the conjecture would imply that the above inequalities for the diameter $\delta(G(\bar{R}_n))$ are equalities.

In Theorem 3.8 we showed that the special rotational tournament $\bar{T}_n$ in the regular tournament class $T(\bar{R}_n)$, $\bar{R}_n = (\frac{n-1}{2}, \ldots, \frac{n-1}{2})$ where n is odd, has the property that its arcs can be partitioned into $\binom{\frac{n-1}{2}}{2}$ 4-cycles and $\frac{n-1}{2}$ 3-cycles. Whether this property holds for all rotational tournaments (see [9, p. 172] for the general definition of a rotational tournament) or even for all regular tournaments is unknown to us.

PROBLEM 3.17. Can the arcs of every tournament in $T(\bar{R}_n)$ be partitioned into $\binom{\frac{n-1}{2}}{2}$ 4-cycles and $\frac{n-1}{2}$ 3-cycles? If not, does this hold for the rotational tournaments?

We think it is evident from the development given in this paper that the class $T(R)$ has a rich and fascinating combinatorial structure and that much remains to be determined.

References

[1] R.A. Brualdi, Matrices of zeros and ones with fixed row and column sum vectors, *Linear Alg. and its Applics.*, 33(1980), 159-231.

[2] R.A. Brualdi and Li Qiao, Upsets in round robin tournaments, *J. Combin. Theory, B*, 35 (1983), 62-77.

[3] R.J. Douglas, Tournaments that admit exactly one Hamiltonian circuit, Proc. London Math. Soc. (3) 3(1970), 716-730.

[4] L.R. Ford, Jr. and D.R. Fulkerson, *Flows in Networks* (Princeton Univ. Press), 1962.

[5] D.R. Fulkerson, Upsets in Round Robin Tournaments, *Canad. J. Math.*, 17(1965), 957-969.

[6] H.G. Landau, On dominance relations and the structure of animal societies, III: The condition for a score structure, *Bull. Math. Biophys.* 15(1953), 143-148.

[7] J.N. Moon, *Topics on Tournaments* (Holt, Rinehart, and Winston, New York), 1968.

[8] E.M. Palmer, The enumeration of graphs, Ch. 14 of *Selected Topics in Graph Theory*, ed. by L.W. Beineke and R.J. Wilson (Academic Press, New York), 1978, 385-415.

[9] K.B. Reid and L.W. Beineke, Tournaments, op. cit., Ch. 7., 169-204.

[10] H.J. Ryser, Matrices of zeros and ones in combinatorial mathematics, *Recent advances in matrix theory*, ed. H. Schneider (University of Wisconsin Press, Madison), 1964, 103-124.

[11] J. Spencer, Random regular tournaments, *Per. Math. Hung.* 5(1974), 105-120.

Linear Algorithms for Convex Drawings of Planar Graphs

Norishige Chiba
Tadashi Yamanouchi
Takao Nishizeki

ABSTRACT

In a convex drawing of a planar graph, all the edges are drawn by straight line segments in such a way that every face boundary is a convex polygon. In this paper, we present two linear algorithms for the convex drawing problem of planar graphs: drawing and testing algorithms. The former draws a given planar graph G convex if possible: it extends a given convex polygon of an outer facial cycle of G into a convex drawing of G. The latter tests the possibility: it determines whether G has an outer facial cycle extendable to a convex drawing of G. Moreover it finds all these facial cycles. These two algorithms together can efficiently yield various convex drawings of a planar graph.

1. Introduction

The problem of drawing a planar graph often arises in Design Automation for electrical networks, particularly VLSI circuits. However, we do not here concentrate on a particular practical application. We are interested in drawing a planar graph so that one can easily and rapidly recognize its structure such as the adjacency of vertices and the connectivity-property. One of the feasible methods for the purpose is a convex drawing in which all the edges are drawn by straight line segments without any crossing so that all the face boundaries are convex polygons.

Clearly not every planar graph has a convex drawing. Tutte [5] proved that every 3-connected planar graph has a convex drawing, and established a necessary and sufficient condition for a planar graph to have a convex drawing with prescribed outer polygon. Furthermore he gave a "barycentric mapping" method for finding a convex drawing, which solves a system of linear equations and so usually requires $O(n^3)$ computation time [6]. Throughout this paper we denote by n the number of vertices in a graph.

In this paper we present two linear algorithms for the convex drawing problem of planar graphs: drawing and testing algorithms. The former draws a given planar graph G convex if possible: it extends a given convex polygonal drawing of an outer facial cycle of G into a convex drawing of G. The latter tests the possibility. That is, it determines whether a given planar graph has a convex drawing or not, and moreover finds all the outer facial cycles extendable to a convex drawing of the graph. These two algorithms together yield various convex drawings of a given planar graphs. Both the time and space of these algorithms are linear, so optimal to within a constant factor. Our linear drawing algorithm compares favorable with the known $O(n^3)$ one. The algorithms have been implemented in PASCAL. In Section 5 we give a computational example: a convex drawing of an input planar graph and all its extendable facial cycles.

Recently Thomassen gave a short proof to a strong version of the Tutte's Result [4]. Our drawing algorithm is based on it. On the other hand we modify their results into a form suitable for the convex testing, which is represented in terms of 3-connected components. Using the form, we show that the convex testing of a graph G can be reduced to the planarity testing of a certain graph obtained from G. Our testing algorithm employs this fact together with two linear algorithms of Hopcroft and Tarjan [2] [3]: one for testing the planarity of a graph; and the other for dividing a graph into 3-connected components.

2. Preliminaries

In this section, we first define some terms, then give illustrative examples, and finally present a known result on the convex drawing.

Let $G = (V, E)$ be a *graph* with vertex set V and edge set E. The vertex set of a graph G is often denoted by $V(G)$. We consider only a drawing of a *2-connected simple graph*, that is, a graph with no cut vertices, multiple edges or loops. A graph is *planar* if it is embeddable in the plane without edge crossing. A *plane* graph G is a planar graph which is embedded in the plane. A plane graph divides the plane into connected regions called *faces*. The unbounded face is called the *outer face* of G. A cycle C of a planar graph G is *(outer) facial* if C bounds an (outer) face of a plane graph G. A path joining vertices x and y is called an *x-y path*. A *convex drawing* of a planar graph is a representation of the graph on the plane such that all edges are drawn by straight line segments without any crossing and that all the face boundaries are convex polygons. Since all the edges are drawn by straight lines, a convex drawing of a plane graph is uniquely determined only by the positions of the vertices.

Clearly not every 2-connected planar graph has a convex drawing. For example, the 2-connected planar graph depicted in Figure 1(a) cannot be drawn convex even if any facial cycle is chosen as an outer cycle. Next consider the graph G in Figure 1(b). In this case it depends on the facial cycle chosen as an outer cycle whether G can be drawn convex or not. If the outer facial cycle S is 1-2-3-4-1, G cannot be drawn convex for any polygonal drawing S^* of S, as shown in Figure 1(c). On the other hand, G can be drawn convex if the facial cycle $S = 1-2-3-4-5-6-1$ is chosen as an outer cycle and moreover S is drawn as a convex polygon S^* such that vertices 1, 2, 3, 4, and 6 are the apices (i.e. geometric vertices) of S^*, as shown in Figure 1(d). However, if S^* is a convex polygon with the apices 1, 2, 3, and 4, then G cannot be drawn convex as shown in Figure 1(e). Thus we may define: a convex polygonal drawing (for short, a convex polygon) S^* of a facial cycle S of a graph G is *extendable* if there exists a convex drawing of G having S^* as the outer polygon; a facial cycle S is *extendable* if S has an extendable convex polygon S^*.

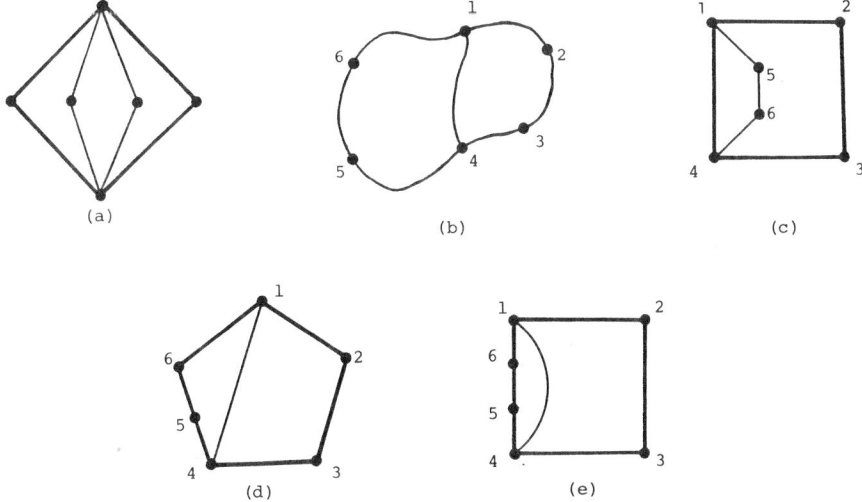

Figure 1. Examples of drawings.

Tutte established a necessary and sufficient condition for a convex polygon to be extendable [5]. The following lemma is a strong version of his result obtained by Thomassen [4].

LEMMA 1. (Thomassen [4]) Let G be a 2-connected plane graph with the outer facial cycle S, and let S^* be a convex polygon of S. Let $P_1, P_2, \ldots, P_k$ be the paths in S, each corresponding to a side of S^*. (Thus S^* is a k-gon. It should be noted that not every vertex of the cycle S is an apex of the polygon S^*.) Then S^* is extendable if and only if Condition I below holds.

CONDITION I

(a) for each vertex v of $G - V(S)$ having degree at least three in G, there exist three paths disjoint except v, each joining v and a vertex of S;

(b) $G - V(S)$ has no connected component C such that all the vertices on S adjacent to vertices in C lie on a single path P_i; and no two vertices in each P_i are joined by an edge not in S; and

(c) any cycle of G which has no edge in common with S has at least three vertices of degree ≥ 3 in G.

3. Convex Drawing Algorithm

In this section we give a linear convex drawing algorithm of planar graphs. Suppose that a 2-connected plane graph G is given together with an extendable convex polygon S^* of the outer facial cycle S. The algorithm extends S^* into a convex drawing of G in linear time.

Our drawing algorithm is based on Thomassen's short proof of Lemma 1. The outline of the drawing algorithm is as follows. We reduce the convex drawing of G to those of several subgraphs of G as follows: delete from G an arbitrary apex v of S^* together with the edges incident to v; divide the resulting graph $G' = G - v$ into the blocks $B_1, B_2, \ldots, B_p$, $p \geq 1$ (see Figure 2); determine a convex polygon S_i^* of the outer facial cycle S_i of each B_i so that B_i with S_i^* satisfies Condition I; and recursively apply the algorithm to each B_i with S_i^* to determine the positions of vertices not in S_i. The detail of our algorithm is as follows.

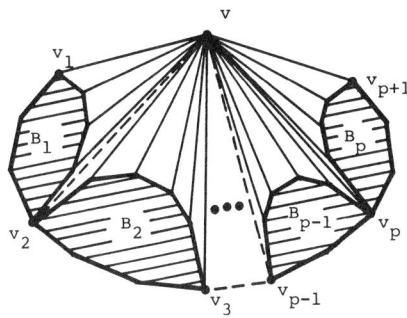

Figure 2. Reduction of the convex drawing of G into subproblems.

Algorithm Convex-Drawing. Let G be a given 2-connected plane graph with the outer facial cycle S, and let S^* be an extendable convex polygon of S. For simplicity we reduce the drawing of the graph G into that of a graph G' which has no vertex of degree two not on S.

STEP 1. For each vertex v of degree two not on S, replace v together with the two edges incident to v by a single edge joining the vertices adjacent to v. Let G' be the resulting graph.

STEP 2. Call the procedure DRAW (G', S, S^*) below to extend S^* into a convex drawing of G'.

STEP 3. For each deleted vertex of degree 2 determine its position on the straight line segment joining the two vertices adjacent to the vertex. *Stop.*

DRAW (G, S, S^*). *Comment:* This procedure extends a convex polygon S^* of the outer facial cycle S of a plane graph G into a convex drawing of G, where G has no vertex of degree 2 not on S.

STEP 1. If G has at most three vertices, a convex drawing of G has been obtained, so *return*. Thus assume that G has at least four vertices. Select an arbitrary apex v of S^*, and let $G' := G - v$. Divide the plane graph G' into the blocks B_i ($1 \le i \le p$). Let v_1 and v_{p+1} be the two vertices on S adjacent to v, and let v_i, $2 \le i \le p$, be the cut vertices of G' such that $v_1 \in V(B_1)$, $v_{p+1} \in V(B_p)$ and $v_i = V(B_{i-1}) \cap V(B_i)$. (See Figure 2.) Every v_i, $1 \le i \le p+1$, is necessarily on S since the extendable S^* with G satisfies Condition I and G has no vertex of degree two not on S.

STEP 2. Draw each block B_i convex by applying the following procedures.

STEP 2.1. Determine a convex polygon S_i^* of the outer facial cycle S_i of B_i as follows. Since the positions of the vertices in $V(S_i) \cap V(S)$ have been already determined on S^*, one should determine the positions of the vertices in $V(S_i) - V(S)$. Locate the vertices in $V(S_i) - V(S)$ in the interior of the triangle $v \cdot v_i \cdot v_{i+1}$ in such a way that the vertices adjacent to v are apices of a convex polygon S_i^* and the others are on the straight line segments of S_i^*.

STEP 2.2. Recursively call the procedure DRAW(B_i, S_i, S_i^*) to extend S_i^* to a convex drawing of B_i. (Note that the convex polygon S_i^* with B_i satisfies Condition I and B_i has no vertex of degree two not on S_i.)

STEP 3. *Return.*

We have the following result on the algorithm.

THEOREM 1. Let G be a 2-connected plane graph with the outer facial cycle S, and let S^* be an extendable convex polygon of S. Then the algorithm CONVEX-DRAWING extends S^* into a convex drawing of G, and uses linear time and space.

PROOF. Clearly the boundaries of all the inner faces containing the apex v are convex polygons. Therefore, in order to prove inductively the correctness of the algorithm, one should show that every block B_i with S_i^* satisfies Condition I. We omit the proof since it is similar to that of Theorem 5.1 in [4]. Thus we shall establish the claims on time and space.

As a data structure to represent a plane graph $G = (V, E)$, we use doubly linked adjacency lists, in each of which the edges adjacent to a vertex are stored in the order of the plane embedding, clockwisely around the vertex. The two copies of each edge (v, w) in the adjacency lists of v and w are linked each other so that one can be accessed directly from the other. Given an edge e, one can directly access to the edge clockwisely next to e around an end vertex of e. Clearly this data structure requires linear space.

Clearly Steps 1 and 3 of CONVEX-DRAWING can be executed in $O(n)$ time. Therefore we shall prove that Step 2, and so DRAW, spends linear time. Let P be the $v_1 - v_{p+1}$ path in the outer facial cycle S' of $G - v$ which newly appears on S'. While traversing P, one can easily (1) find the cut vertices v_i, $2 \le i \le p$, which are also on S, (2) obtain the outer facial cycle S_i of B_i as the union of the traversed

v_i-v_{i+1} path and the $v_{i+1}-v_i$ path on S, and (3) decides the positions of the vertices of S_i as specified in Step 2.1. Thus, we can implement the algorithm so that the time required by the procedure DRAW, exclusive of recursive calls to itself, is proportional to the number of the traversed edges in P, that is, the edges newly appeared on the boundaries of the outer facial cycles. Since every edge appears on a boundary of an outer facial cycle at most once, the number of edges traversed during an execution of DRAW is at most $|E|$ in total. Thus DRAW runs in linear time.

REMARK: In Theorem 1 above we assume that an arithmetic operation of infinite decimal requires one unit time. Tutte's proof of Theorem II in [5] yields another convex drawing algorithm of polynomial time complexity, but possibly of higher order.

4. Testing Algorithm

In this section, we present a convex testing algorithm which determines whether a given 2-connected planar graph has a convex drawing and moreover finds all the extendable facial cycles, both in linear time.

One can construct a linear algorithm which only determines whether a given convex polygon of a particular outer facial cycle is extendable, that is, satisfies Tutte's condition [5] or Condition I. However a planar graph G may have an exponential number of facial cycles so it is impractical to test all the facial cycles of a graph one by one through the algorithm. Note that the plane embedding of G is not always unique unless G is 3-connected [1]. Thus we shall modify Condition I in Lemma 1 into a form suitable for our purpose. One may easily notice that the existence of a convex drawing of a graph G heavily depends on the structure of 3-connected components of G.

This section is organized as follows: Section 4.1 gives definitions of 3-connected components and separation pairs. In Section 4.2 we represent Condition I in terms of 3-connected components. Section 4.3 gives a linear convex testing algorithm. In Section 4.4 we show how to find all the extendable facial cycles of a graph.

4.1. Definitions

We first borrow the definition of some terms from [2]. In Section 4.1 "graph" is used instead of "multigraph" since only multigraphs are concerned. A pair $\{x,y\}$ of vertices of a 2-connected graph $G=(V,E)$ is a *separation pair* if there exist two subgraphs $G'_1=(V_1,E'_1)$ and $G'_2=(V_2,E'_2)$ satisfying the following conditions (a) and (b):

(a) $V=V_1\cup V_2$, $V_1\cap V_2=\{x,y\}$;

(b) $E = E'_1 \cup E'_2$, $E'_1 \cap E'_2 = \emptyset$, $|E'_1| \geq 2$, $|E'_2| \geq 2$.

A 2-connected graph G is said to be *3-connected* if G has no separation pair. For a separation pair $\{x,y\}$, $G_1 = (V_1, E'_1 + (x,y))$ and $G_2 = (V_2, E'_2 + (x,y))$ are called *split graphs* of G. The new edges (x,y) added to G_1 and G_2 are called *virtual edges*. Dividing a graph G into two split graphs G_1 and G_2 is called *splitting*. Reassembling the two split graphs G_1 and G_2 into G is called *merging*. Merging is the inverse of splitting. Suppose a graph G is split, the split graphs are split, and so on, until no more splits are possible (each remaining graph is 3-connected). The graphs constructed in this way are called the *split components* of G. The split components of a graph G are of three types: *triple bonds* (i.e. a set of three multiple edges), *triangles* (i.e. a cycle consisting of three edges), and 3-connected graphs. The *3-connected components* of G are obtained from the split components of G by merging triple bonds into a bond and triangles into a ring, as far as possible. Here a *bond* is a set of multiple edges, and a *ring* is a cycle (we use "ring" instead of "polygon" used in [2] in order to avoid the confusion). The split components of a graph G are not necessarily unique, but the 3-connected components of G are unique.

We illustrate the decompositions of a 2-connected graph in Figure 3. The graph G depicted in Figure 3(a) has six separation pairs $\{1,2\}$, $\{1,3\}$, $\{2,3\}$, $\{2,7\}$, $\{3,6\}$, and $\{4,5\}$. The graph G is decomposed into nine split components as shown in Figure 3(b), and into seven 3-connected components $F_1, F_2, \ldots, F_7$ as shown in Figure 3(c). The components F_1, F_2, and F_6 are 3-connected graphs; F_3, F_5, and F_7 are rings; and F_4 is a bond.

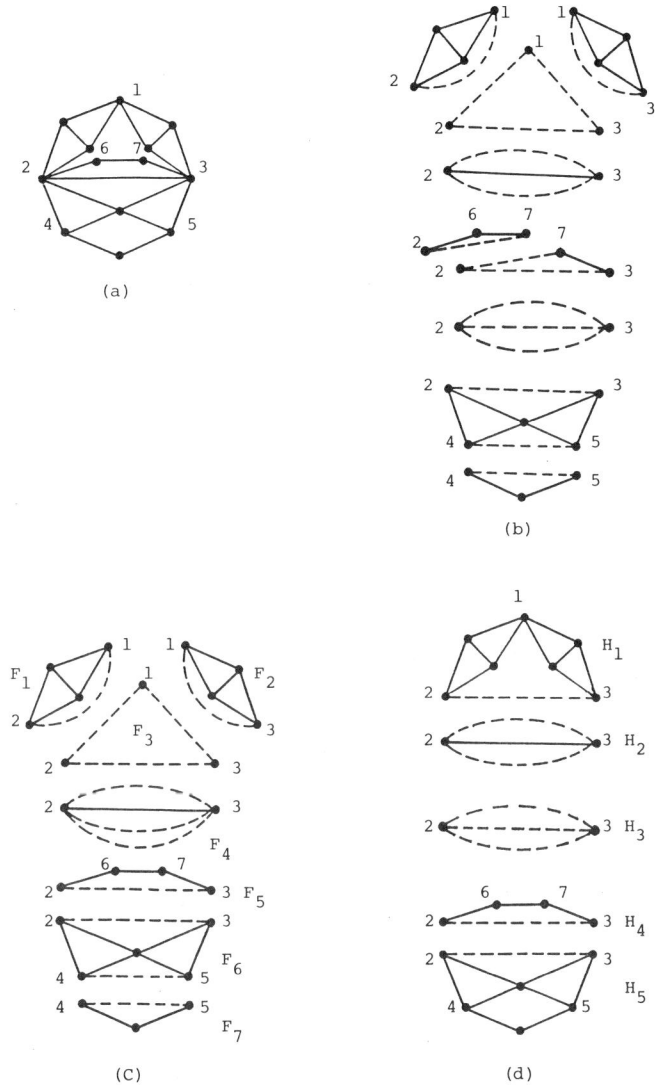

Figure 3. Decompositions of a graph, where virtual edges are written by dashed lines:
(a) a 2-connected graph G;
(b) split components of G;
(c) 3-connected components of G; and
(d) $\{2,3\}$-split components of G with one exception H_3.

Next we introduce new terms. Suppose that $\{x,y\}$ is a separation pair of a graph G and that G is split at $\{x,y\}$, the split graphs are split, and so on, until no more split are possible at $\{x,y\}$ (the remaining graphs are not necessarily 3-connected). A graph constructed in this way is called an $\{x,y\}$-*split component* of G if it has at least one real (i.e. non-virtual) edge. In Figure 3(d), the components H_1, H_2, H_4 and H_5 are the $\{2,3\}$-split components. A separation pair $\{x,y\}$ is *prime* if x and y are the end vertices of a virtual edge contained in a 3-connected component. In other words, a separation pair $\{x,y\}$ is prime if there exist two subgraphs G'_1 and G'_2 such that G'_1 and G'_2 satisfy the conditions (a) and (b) aforementioned and either G'_1 or G'_2 is 2-connected or is a subdivision of an edge joining two vertices of degree three or more. As known from Fig. 3(c), separation pairs $\{1,2\}$, $\{1,3\}$, $\{2,3\}$, and $\{4,5\}$ are prime, but $\{2,7\}$ and $\{3,6\}$ are not.

In some cases it can be easily known from the number of $\{x,y\}$-split components or the structures that a graph can never be drawn convex. A *forbidden separation pair* $\{x,y\}$ is a prime separation pair which has either (i) at least four $\{x,y\}$-split components or (ii) three $\{x,y\}$-split components none of which is either a ring or bond. Note that an $\{x,y\}$-split component corresponds to an edge (x,y) of G if it is a bond, and corresponds to a subdivision of an edge (x,y) if it is a ring. The graph in Figure 3(a) has exactly one forbidden separation pair $\{2,3\}$. It will be shown later in Section 4.2 that if a planar graph G has a forbidden separation pair then G can never be drawn convex: G has no extendable facial cycle.

On the other hand the converse of the fact above is not true. In order to say more precisely, we need one more term. A *critical separation* pair $\{x,y\}$ is a prime separation pair which has either (i) three $\{x,y\}$-split components including a ring or a bond, or (ii) two $\{x,y\}$-split components neither of which is a ring. In the graph of Figure 3(a), prime separation pairs $\{1,2\}$ and $\{1,3\}$ are critical, but $\{4,5\}$ is neither forbidden nor critical. When G has no forbidden separation pair, there happen two cases: if G has no critical separation pair either, then G is a subdivision of a 3-connected graph, and so every facial cycle of G is extendable; otherwise, that is, if G has critical separation pairs, a facial cycle S of G may or may not be extendable, depending on the interaction of S and critical separation pairs. (The detailed criterion, called Condition II, will be given in Section 4.2.)

4.2. Condition II

We now give a condition suitable for the testing algorithm, which is equivalent to Tutte's condition [5] or to Condition I under a restriction that S^* is *strict*, that is, every vertex of S is an apex of S^*.

LEMMA 2. Let $G=(V,E)$ be a 2-connected plane graph with the outer facial cycles S, and let S^* be a strict convex polygon of S. Then S^* is extendable if and only if G and S satisfies the following Condition II.

CONDITION II
(a) G has no forbidden separation pair;
(b) For each critical separation pair $\{x,y\}$ of G there exists at most one $\{x,y\}$-split component having no edge of S. Moreover, such an $\{x,y\}$-split component is either a bond if $(x,y) \in E$ or a ring otherwise.

PROOF. We shall show that Condition II is equivalent to Condition I under the restriction that every vertex of S is an apex of S^*.

CONDITION I IMPLIES CONDITION II: Let $\{x,y\}$ be a prime separation pair of G, and let $H_1, H_2, \ldots, H_m$ be the $\{x,y\}$-split components having no edges of S. Then we can show that $m=0$ or 1 and that if $m=1$ then H_1 is neither a ring or a bond, as follows. First suppose that one of the $\{x,y\}$-split components, say H_1, is neither a ring nor a bond, then H_1 has a vertex v ($\neq x, y$) of degree three or more. Clearly there exist no three paths disjoint except v, each joining v and a vertex of S, since such a path must contain either x or y. This contradicts Condition I(a). Thus every H_i must be either a ring or a bond. Next suppose that $m \geq 2$, then H_1 together with H_2 forms a cycle in G which has no edge in common with S. The cycle has exactly two vertices x and y of degree ≥ 3 in G, contrary to Condition I(c).

Since at most two $\{x,y\}$-split components contain edges of S, there are at most three $\{x,y\}$-split components and moreover one of them is a bond or a ring if there are three. Thus $\{x,y\}$ is not forbidden.

Let $\{x,y\}$ be critical and $m=1$. If $(x,y) \notin E$, then clearly H_1 is not a bond, so H_1 must be a ring. Thus we may assume that $(x,y) \in E$. Suppose that H_1 is a ring. Then edge (x,y) is in S, and so the $x-y$ path in H_1, which is a connected component of $G - V(S)$, is adjacent only with the vertices x and y in S, contradicting Condition I(b). Note that edge (x,y) is a side of S^*. Thus H_1 must be a bond.

CONDITION II IMPLIES CONDITION I: First suppose that G has a vertex v of degree three or more, not satisfying Condition I(a). Then, using Menger's theorem [1, p.47] one can easily show that there exists a prime separation pair $\{x,y\}$ such that one of the $\{x,y\}$-split components H_i contains v and has no edge of S. Since H_i contains a vertex v of degree three or more, H_i is neither a ring nor a bond, contradicting Condition II.

Next suppose that Condition I(b) is not satisfied. Then there exists a connected component C of $G-V(S)$ such that only the vertices x and y of an edge (x,y) on S are adjacent with vertices in C, because G has no multiple edges and S^* is strict. Therefore $\{x,y\}$ is a prime separation pair, and clearly the $\{x,y\}$-split component containing C has no edge in common with S and is not a bond since G is simple. This contradicts Condition II.

Finally suppose that there exists a cycle Z in G violating Condition I(c). Since G is 2-connected, Z has exactly two vertices x and y of degree ≥ 3. Of course $\{x,y\}$ is a prime separation pair. If $(x,y) \in E$, then an $\{x,y\}$-split component having no edges of S is a ring. Otherwise, there are two $\{x,y\}$-split components having no edges of S. Either case contradicts Condition II.

It should be noted that Condition II does not depend on the drawing S^* of S at all. One may suspect that the restriction on S^* loses the generality: there would be an extendable convex polygon of S even if there is no extendable strict S^*. However the following lemma dispels this suspicion.

LEMMA 3. Assume that G is a 2-connected plane graph with the outer facial cycle S, and that S has an extendable convex polygon. Then every strict convex polygon of S is extendable.

PROOF. Immediately follows from either Theorems I and II of [5] or Condition I.

Lemmas 2 and 3 immediately lead to the following theorem.

THEOREM 2. A facial cycles S of a 2-connected *planar* graph G is extendable if and only if S and G satisfies Condition II.

REMARK: Thomassen claims without proof that under the same restriction as ours Condition I is equivalent to condition (d) in [4, p. 256] which is similar to Condition II. However one can easily find a counter example to his claim.

4.3 Testing Algorithm.

Condition II is more suitable for testing than Condition I. In this section, we show that the convex testing, i.e. checking Condition II, can be reduced to the planarity testing of a certain graph.

Theorem 2 immediately yields the following corollaries.

COROLLARY 1. A 2-connected planar graph G has no convex drawing if G has a forbidden separation pair.

COROLLARY 2. A 2-connected planar graph G has a convex drawing for any facial cycle of G if G has no forbidden or critical separation pairs.

COROLLARY 3. A 3-connected planar graph G has a convex drawing for any facial cycle of G.

COROLLARY 4. If a facial cycle S of a 2-connected planar graph G satisfies Condition II, then S contains every vertex of critical separation pairs of G.

PROOF. Assume that $\{x,y\}$ is a critical separation pair of G and that S does not contain a vertex x of the pair $\{x,y\}$. Then exactly one $\{x,y\}$-split component contains all the edges of S. On the other hand, Condition II implies that there exists exactly one $\{x,y\}$-split component not containing edges of S, and it must be a ring or a bond. Thus there are exactly two $\{x,y\}$-split components, one of which is a ring or a bond. Then $\{x,y\}$ could not be critical, contrary to the assumption.

We will show in Theorem 3 that the converse of the claim of Corollary 4 is also true in a certain sense. Before presenting Theorem 3 we need the following two lemmas.

LEMMA 4. Suppose that a 2-connected planar graph G has no forbidden separation pair and has exactly one critical separation pair. Then G has a convex drawing.

PROOF. Let $\{x,y\}$ be the critical separation pair. By the definition of a critical separation pair, one can easily observe that G is one of the seven types in Figure 4, in which a shaded part corresponds to an $\{x,y\}$-split component which is neither a ring nor a bond. In each case, one can easily verify that the cycle indicated by a bold line satisfies Condition II (b). Q.E.D.

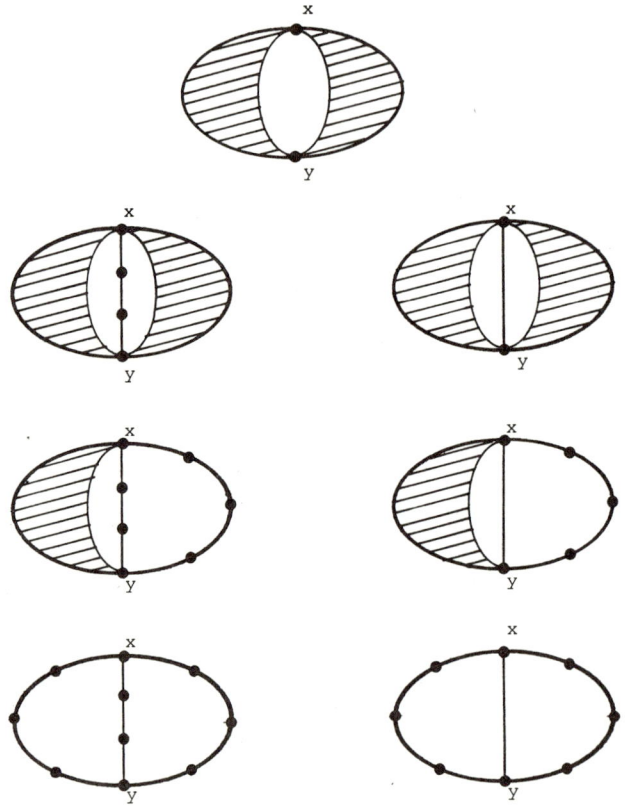

Figure 4. Seven types of a plane graph having exactly one critical separation pair $\{x,y\}$.

Thus we concentrate on a graph having two or more critical separation pairs.

LEMMA 5. Let G be a 2-connected planar graph, and let $\{x,y\}$ be a prime separation pair of G. If a facial cycles S of G contains the vertices x and y, then exactly two $\{x,y\}$-split components contain edges of S.

PROOF. Since S is a cycle, at most two $\{x,y\}$-split components contains edges of S. Furthermore, since S is a facial cycle, not all the edges of S are contained in a single $\{x,y\}$-split component. Thus exactly two $\{x,y\}$-split components contain edges of S.

We are now ready to present Theorem 3 which plays a crucial role in our testing algorithm.

THEOREM 3. Suppose that a 2-connected planar graph G has no forbidden separation pair and has two or more critical separation pairs. Apply the following operation to G for every critical separation pair $\{x,y\}$ of G: if $(x,y) \in E$, then delete edge (x,y) from G; otherwise and if exactly one $\{x,y\}$-split component is a ring, then delete the $x-y$ path in the component from G. Let G_1 be the resulting graph. (Graphs G, and G_1 are illustrated in Figs. 5(a), and (b), respectively.) Then S is an extendable facial cycle of G if and only if S is a facial cycle of G_1 which contains all the vertices of critical separation pairs of G.

PROOF. *Necessity:* Assume that S is the extendable outer cycle of a *plane* graph G. Then, since S satisfies Condition II, all the deleted edges or paths are not on S, and hence S remains to be the outer cycle of the plane subgraph G_1 of G. Moreover S contains all the vertices of the critical separation pairs by Corollary 4.

Sufficiency: Assume that S is a facial cycle of G_1 which contains all the vertices of critical separation pairs of G. Clearly S is also a facial cycle of G. Let $\{x,y\}$ be a critical separation pair of G. Then, by Lemma 5, exactly two $\{x,y\}$-split components, say H_1 and H_2, contain edges of S. Therefore G has at most one $\{x,y\}$-split component containing no edges of S. Suppose that there exists such a component H_3 and that H_3 is neither a ring nor a bond. Then H_3 contains no vertex of critical separation pairs except x and y: if H_3 contains a vertex $z(\neq x,y)$ of a critical separation pair, H_3 would contains an edge of S since S contains all the vertices of critical separation pairs, contrary to the supposition. Therefore there is a critical separation pair $\{u,v\}$, different from $\{x,y\}$, such that vertex u or v is not contained in H_3. Note that G has at least two critical separation pairs. Therefore H_1 or H_2, say H_1, is neither a ring nor a bond, and the other H_2 is a ring or a bond. Then edge (x,y) or the $x-y$ path in H_2 should have been deleted in G_1, so H_2 could not contain edges of S, contrary to the assumption. Hence H_3 must be either a ring or a bond. Thus we have shown that S with G satisfies Condition II (b), so S is extendable.

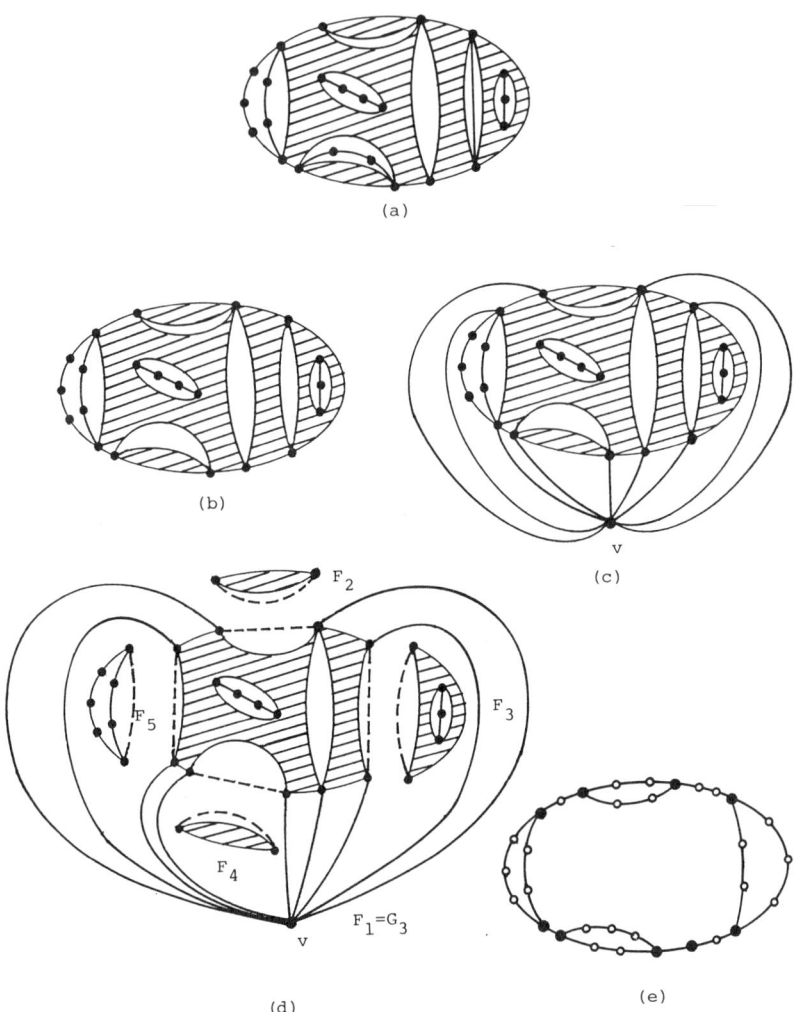

Figure 5. Illustrations of graphs:
(a) G; (b) G_1; (c) G_2;
(d) $F_1 = G_3$; F_2, F_3, and F_4
are in group (b); and F_5 are
in group (c); and
(e) the union of all the extendable facial cycles of G.

We immediately have the following corollaries from Theorem 3. We define here one more term. Let v be a vertex of a 2-connected plane graph G_2 and let $G_1 = G_2 - v$ be 2-connected. Then the *v-cycle* of G_2 is the cycle of the plane subgraph G_1 of G_2 which bounds the face of G_1 in which v lay.

COROLLARY 5. Suppose that a 2-connected planar graph G has no forbidden separation pair and has two or more critical separation pairs. Let G_1 be the graph defined in Theorem 3. Let G_2 be the graph obtained from G_1 by adding a new vertex v and joining v to all the vertices of critical separation pairs of G. (Graph G_2 is illustrated in Figure 5(c).) Then S is an extendable facial cycle of G if and only if

(a) G_2 is planar; and

(b) S is the v-cycle of a plane embedding of G_2.

COROLLARY 6. Suppose that a 2-connected planar graph G has no forbidden separation pair and has two or more critical separation pairs. Let G_2 be the graph defined in Corollary 5. Then G has a convex drawing if and only if G_2 is planar.

Combining Corollaries 1, 2, 5, 6, and Lemma 4, we immediately have the following linear testing algorithm.

Algorithm CONVEX-TESTING

STEP 1. Finds all the separation pairs of the given 2-connected planar graph G by the linear algorithm of Hopcroft and Tarjan [2] for finding 3-connected components. Determine three sets of separation pairs: the sets PSP of prime separation pairs; the set CSP of critical separation pairs; and the set FSP of forbidden separation pairs. If FSP≠∅ then stop with a message "G has no convex drawing" (see Corollary 1). If FSP=CSP=∅ then print a message "All facial cycles are extendable" (see Corollary 2), and go to Step 3. If $|$CSP$|$ =1 then let S be an extendable outer facial cycle depicted in Figure 4, and go to Step 3 (see Lemma 4).

STEP 2. Obtain a graph G_1 from G by applying the operation in Theorem 3 to every critical separation pair of G. Construct a graph G_2 from G_1 by adding a new vertex v and joining v to all the vertices of CSP. Test the planarity of G_2 by the linear algorithm of Hopcroft and Tarjan [3] (see Corollary 6). If G is nonplanar, then stop with a message "G has no convex drawing (There is no facial cycles containing every vertex of CSP)." Otherwise let S be the v-cycle of a plane graph G_2. Necessarily S satisfies Condition II (see Corollary 5).

STEP 3. Stop with a message "The given graph G has a convex drawing."

4.4 Finding all Extendable Facial Cycles.

Algorithm CONVEX-TESTING in Section 4.3 finds an extendable facial cycle S if any. In this section we show how to generate all the extendable facial cycles of G. If the given graph G has exactly one critical separation pair, then one can easily enumerate all the extendable facial cycles, since G has at most four such facial cycles, as verified by checking the seven types in Figure 4. Thus we may assume that a graph G has two or more critical separation pairs.

Suppose that the outer facial cycle S of a 2-connected plane graph G is extendable. Let G_2 be the graph defined in Corollary 5. Consider how to generate all the v-cycles of plane embeddings of G_2. The 3-connected components of G_2 are classified into five groups:

(1) the 3-connected graph containing the vertex v;
(2) 3-connected graphs containing the vertices of exactly one critical separation pair of G (Only the graph in (1) can contain vertices of two or more critical separation pairs);
(3) rings containing the vertices of noncritical separation pair G (each of these corresponds to a subdivision of an edge);
(4) rings containing the vertices of a critical separation pair $\{x,y\}$ of G (if there exists exactly one such ring then an $x-y$ path corresponding to the ring should have been deleted in G_1, so there must exist exactly two such rings); and
(5) triple bonds containing the vertices of a critical separation pair.

We next merge some of these 3-connected components into graphs which are necessarily subdivisions of 3-connected graphs. The resulting graphs are classified into three groups, depending on the types of 3-connected components involved in the merge:

(a) the graph G_3 obtained from the 3-connected graph of (1) and 3-connected components of type (3);
(b) graphs obtained from 3-connected components of type (2) or (3); and
(c) graphs obtained from 3-connected components of type (4) or (5).

The resulting graphs are illustrated in Figure 5(d), in which F_1 is G_3, F_2, F_3, and F_4 are in group (b), and F_5 is in group (c).

We now consider the embeddings of G_2. Since a 3-connected planar graph and its subdivision graph are uniquely embeddable [1], the graph G_3 has the unique plane embedding. Now consider a planar graph G'_3 obtained by merging G_3 and one of the graphs in group (b) or (c), say F. Since every graph in group (b) or (c) does not contain a critical separation pair, that is, is a subdivision of a 3-connected graph, G'_3 has exactly two distinct plane embeddings: one is obtained by merging the plane graph G_3 and a plane graph F; and the other is obtained from it by replacing F with its mirror image. Thus, if there exist k graphs in groups (b) and (c), the planar graph G_2 has 2^k distinct plane embeddings, from each of which a distinct v-cycle of G_2 arises. Not that all the graphs in groups (b) and (c) are, of course, disjoint.

Thus we can find all the extendable facial cycles. However, a graph may have an exponential number of these facial cycles, so we do not enumerate these facial cycles. We simply represent these cycles by the union of them, from which one can easily pick up whichever one likes. As an example, we illustrate in Figure 5(e) the representation of all the extendable facial cycles of G in Figure 5(a), in which every cycle containing all the "black" vertices of the critical separation pairs of G is extendable. Replace Step 3 of the algorithm CONVEX-TESTING by the following Steps 3 and 4, then the revised algorithm constructs the representation of all the extendable facial cycles of a given graph G.

STEP 3. If $|CSP|=1$ then enumerate all the extendable facial cycles, and go to Step 4. Otherwise, construct the union of all the extendable facial cycles of G as follows. Divide G_2 into the 3-connected components and merge these 3-connected components into graphs of the three groups (a), (b) and (c) above. For each graph F in group (b) or (c) apply the following procedure: let (x,y) be the virtual edge in F, and let P_{xy} be the $x-y$ path on S joining x and y in F; let F' be the plane graph obtained from a plane graph F by deleting the virtual edge (x,y); find another $x-y$ path Q_{xy} in F such that $P_{xy} \cup Q_{xy}$ is a facial cycle of F'; and add Q_{xy} to S by joining them at the vertices x and y. (Clearly the cycle S' obtained from S by replacing P_{xy} by Q_{xy} is one of the v-cycles of G_2, and so S' is extendable.)

STEP 4. Stop with a message "All facial cycles have been found."

We have the following result on the above algorithm.

THEOREM 4. The revised CONVEX-TESTING determines whether a 2-connected planar graph has a convex drawing, and moreover finds all the extendable facial cycles. It uses linear time and space.

Finally we remark that one can easily find all extendable convex polygons of an extendable facial cycles (left to the reader).

5. Example.

The algorithms CONVEX-TESTING and CONVEX-DRAWING have been implemented in PASCAL and run on a small computer FACOM 230/38s. Experiments indicate that the algorithm can find all the extendable facial cycles of a graph having 150 edges in less than 6 seconds. We illustrate a computational example in Figure 6. Fig. 6 (a) depicts an input graph G having 50 vertices and 83 edges. Algorithm CONVEX-TESTING finds all the extendable facial cycles of G, the union of which is represented as shown in Figure 6 (b). A pair of paths, either of which can be a part of an extendable facial cycle, are in parentheses. Since there are two pairs of such paths, Figure 6(b) represents the four extendable facial cycles:

$C_1 = 15-16-1-2-3-4-5-6-7-8-9-10-11-12-13-14$

$C_2 = 15-16-1-2-3-4-5-6-7-8-9-10-11-49-48-47$

$C_3 = 15-16-1-2-3-31-32-7-8-9-10-11-12-13-14$

$C_4 = 15-16-1-2-3-31-32-7-8-9-10-11-49-48-47.$

The cycle C_1 is indicated by bold lines in Figure 6(a). When both a plane graph G with the outer cycle C_1 and a convex polygon (regular 16-gon) of C_1 are given, algorithm CONVEX-DRAWING obtains a convex drawing of G, which is depicted in Figure 6(c).

(a)

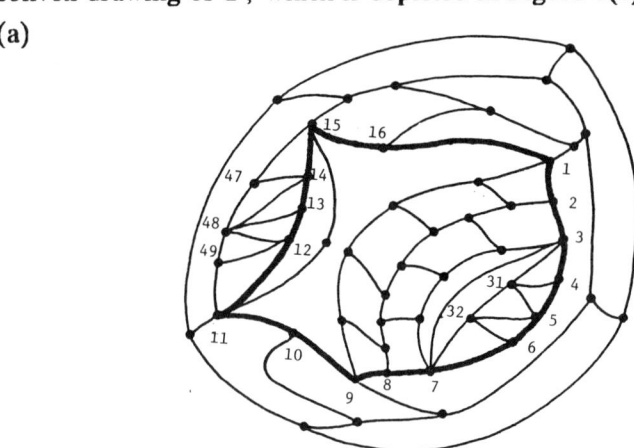

(b) THE EXTENDABLE FACIAL CYCLES OF GIVEN GRAPH
15 16 1 2 3 (4 5 6)(31 32) 7 8 9 10 11 (12 13 14)(49 48 47)

(c)

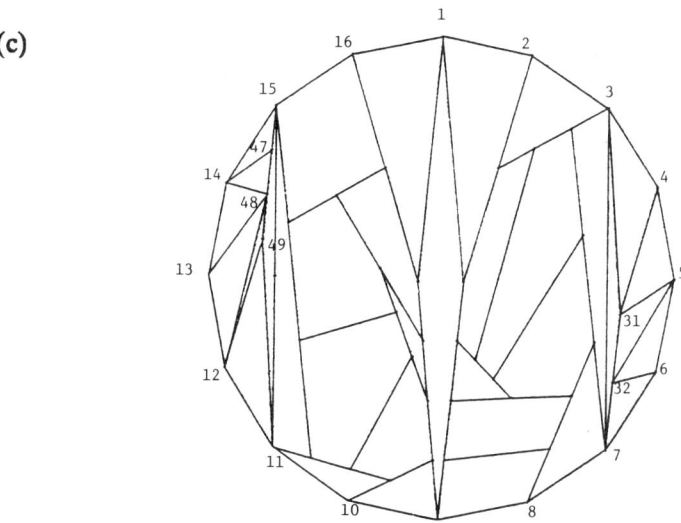

Figure 6. A computational example: (a) input graph G (some vertex numbers are not shown for simplicity); (b) output of CONVEX-TESTING; and (c) output of CONVEX-DRAWING.

References

[1] F. Harary, Graph Theory, Addison-Wesley, Reading, Mass., 1972 (revised).

[2] J. E. Hopcroft and R. E. Tarjan, Dividing a graph into triconnected components. SIAM J. Comput. 2, 3, 1973.

[3] J. E. Hopcroft and R. E. Tarjan, Efficient planarity testing, J. of ACM, 21, 549-568, 1974.

[4] C. Thomassen, Planarity and duality of finite and infinite graphs, J. of Combinatorial Theory, Series B 29, 244-271, 1980.

[5] W. T. Tutte, Convex representations of graphs, Proc. London Math, Soc. (3) 10, 304-320, 1960.

[6] W. T. Tutte, How to draw a graph, Proc. London math. Soc 13, 743-768, 1963.

Triangle-Free Graphs with Restricted Bandwidth

F. R. K. Chung

W. T. Trotter, Jr.*

ABSTRACT

The bandwidth of a graph G is the smallest integer m for which the vertices in G can be labeled $v_1, v_2, \ldots, v_n$ so that $|i-j| \leq m$ whenever $\{v_i, v_j\}$ is an edge of G. Essentially, it gives the smallest edge length achievable when the vertices of G are arranged at distinct integer-valued points on the line. In this paper we study a Turán-type extremal graph problem: What is the maximum number $t(n,m)$ of edges in a triangle-free graph having n vertices and bandwidth at most m? We show that

$$.586\, nm = (2-\sqrt{2})\, nm \leq t(n,m) \leq \frac{5+\sqrt{3}}{11}\, nm = .612\, nm.$$

I. Introduction

Let $G = (V, E)$ be a graph having n vertices. The *bandwidth* of G is the smallest integer m for which the vertices of G can be labelled $v_1, v_2, \ldots, v_n$ so that $|i-j| \leq m$ whenever $\{v_i, v_j\}$ is an edge of G. In this paper, we investigate the problem of determining the maximum number $t(n,m)$ of edges in a triangle-free graph having n vertices and bandwidth at most m. When m is small, it may be possible to determine $t(n,m)$ precisely. For example, it is easy to see that $t(n,1) = n-1$ and $t(n,2) = \lfloor 3n/2 - 5/2 \rfloor$. On the other hand, the well-known theorem of Turán gives $t(n,m) = \lfloor n^2/4 \rfloor$ when $m \geq n-1$. However, it appears to be a difficult problem to determine $t(n,m)$ for all values of the parameters so we will be concerned primarily with providing inequalities for $t(n,m)$. We will prove

---

* Research supported in part by NSF grants ISP-8011451 and MCS-8202172.

$$.586\ nm = (2-\sqrt{2})nm \le t(n,m) \le \frac{5+\sqrt{3}}{11} nm = .612\ nm.$$

II. Inequalities for $t(n,m)$

We begin with the following elementary result.

THEOREM 1: $\frac{1}{2} nm \le t(n,m) \le (\frac{3}{4} + o(1))\ nm$

PROOF: Consider the graph H containing all edges of the form $\{i,j\}$ where $|i-j| \le m$ and i and j have opposite parity. H is triangle-free and has $\frac{1}{2} nm$ edges, so $\frac{1}{2} nm \le t(n,m)$.

On the other hand, let G be any triangle-free graph having $t(n,m)$ edges. Then let d_i denote the number of edges $\{i,j\}$ where $i < j \le i + m$. Then $t(n,m) = \sum_{i=1}^{n} d_i$. For each $i = 1, 2, \ldots, n-m$, let G_i denote the restriction of G to the vertices $\{i, i+1, \ldots, i+m\}$. Since G_i is triangle-free, it has at most $m^2/4$ edges. However, G_i has at least $\sum_{k=0}^{m-1} d_{i+k} - k = (\sum_{k=0}^{m-1} d_{i+k}) - m^2/2$ edges. Therefore,

$$m\ t(n,m) \le \sum_{i=1}^{n-m+1} \sum_{k=0}^{m-1} d_{i+k} + m^3 \le \frac{3}{4} nm^2 + \frac{1}{4} m^3$$

so $t(n,m) \le (\frac{3}{4} + o(1))nm.$ □

The remainder of the paper is devoted to improving these elementary inequalities. We begin with a construction for improving the lower bound.

THEOREM 2: $(2-\sqrt{2})nm = .586nm \le t(n,m).$

PROOF: Let c be a real number with $\frac{1}{2} < c < 1$. Partition the set $\{1, 2, \ldots, n\}$ into $s = n/mc$ blocks $B_1, B_2, \ldots, B_s$ each consisting of mc consecutive integers.[1] Then let G_c be the graph containing all edges of the form $\{i,j\}$ where $|i-j| \le m$ and i and j come from distinct consecutive blocks. Since G_c is bipartite, it is triangle-free. A simple computation shows that G_c has $(2 - c - \frac{1}{2c})\ nm$ edges, and that this function is maximized when $c = \sqrt{2}/2$. For this value, G_c has

---

[1]. We remark that although n/mc or mc may not be integers, such statements are always made with the implicit understanding that the graphs (and quantities) involved may have to be adjusted slightly by adding or deleting (asymptotically) trivial subgraphs (and amounts) so as to make the stated inequalities true.

$(2-\sqrt{2})nm$ edges. □

It will take a bit more work to improve the elementary upper bound $t(n,m) \leq (\frac{3}{4} + 0(1))nm$.

THEOREM 3: $t(n,m) \leq \dfrac{5+\sqrt{3}}{11} nm = .612nm$.

PROOF: Let G be a triangle-free graph with $t(n,m)$ edges. Let A be the adjacency matrix of G, i.e., A is the $n \times m$ matrix where $a_{ij} = 1$ when $\{i,j\}$ is an edge, and $a_{ij} = 0$ otherwise. For convenience, we let $\bar{A}$ denote the adjacency matrix of the complement of G, i.e., $\bar{a}_{ij} = 1 - a_{ij}$. For each $i=1,2,\ldots,n$, we let R_i denote the vector of length n formed by the entries in the ith row of A. We then let $R_i \circ R_j$ denote the inner product of the vectors. Note that $R_i \circ R_j$ counts the number of paths of the following form:

Figure 1

Next, let $S = \sum_{i=1}^{n}\sum_{\substack{j=1 \\ i \neq j}}^{n} R_i \circ R_j$. In view of our previous comment on counting paths, it follows that $S = \sum_{k=1}^{n} r_k^2 - r_k$ where r_k is the degree of vertex k. Note that r_k is also the sum of the entries in R_k so that $t(n,m) = \dfrac{1}{2}\sum_{k=1}^{n} r_k$. We now proceed to obtain an upper bound on S.

First we note that $R_i \circ R_j = 0$ unless $|i-j| \leq 2m$ and $a_{ij}=0$. We write $S = S_1 + S_2$ where

$$S_1 = \sum_{i=1}^{n}\left(\sum_{j=i-2m}^{i-m-1} R_i \circ R_j + \sum_{j=i+m+1}^{i+2m} R_i \circ R_j\right) \text{ and } S_2 = \sum_{i=1}^{n}\sum_{\substack{j=i-m \\ j \neq i}}^{i+m} R_i \circ R_j.$$

We observe that S_1 denotes the number of paths of the type shown in Figure 1 where $m < |i-j| \leq 2m$. Therefore, we can provide an alternate expression for S_1.

$S_1 = 2\sum_{k=1}^{n} w_k$ where w_k denotes the number of pairs (i,j) with $k - m \leq i < k < j \leq k + m$, $i + m < j$, and $a_{ki} = a_{kj} = 1$. Hereafter, we refer to the positions a_{ki} and a_{kj} as being *separated* in R_k when $k - m \leq i < k$, $k < j \leq k + m$ and $i + m < j$. We find it helpful to view w_k as counting the number of "separated ones" in the vector R_k. We then let w_k' count the number of separated zeroes in R_k.

Now let i be fixed. If $i - m \leq j < i$, then

$$R_i \circ R_j = \sum_{k=1}^{n} a_{ik} a_{jk}$$

$$= \bar{a}_{ij} \sum_{k=j-m}^{j+m} a_{ik} a_{jk}$$

$$\leq \bar{a}_{ij} \sum_{k=j-m}^{j+m} a_{ik}$$

$$= \bar{a}_{ij} \sum_{k=j-m}^{i+m} a_{ik} - \bar{a}_{ij} \sum_{k=j+m+1}^{i+m} a_{ik}$$

$$= \bar{a}_{ij} r_i - \bar{a}_{ij} \sum_{k=j+m+1}^{i+m} (1 - \bar{a}_{ik})$$

$$= \bar{a}_{ij} r_i - \bar{a}_{ij} |i-j| + \sum_{k=j+m+1}^{i+m} \bar{a}_{ij} \bar{a}_{ik}$$

Similarly, if $i < j \leq i + m$, then

$$R_i \circ R_j \leq \bar{a}_{ij} r_i - \bar{a}_{ij} |i-j| + \sum_{k=i-m}^{j-m-1} \bar{a}_{ij} \bar{a}_{ik}.$$

Therefore,

$$\sum_{\substack{j=i-m \\ j \neq i}}^{i+m} R_i \circ R_j \leq (2m - r_i) r_i - \sum_{j=i-m}^{i+m} \bar{a}_{ij} |i-j|$$

$$+ \sum_{j=i-m}^{i-1} \sum_{k=j+m+1}^{i+m} \bar{a}_{ij} \bar{a}_{ik}$$

$$+ \sum_{j=i+1}^{i+m} \sum_{k=i-m}^{j-m-1} \bar{a}_{ij} \bar{a}_{ik}.$$

However, $\sum_{j=i-m}^{i-1} \sum_{k=j+m+1}^{i+m} \bar{a}_{ij} \bar{a}_{ik} = \sum_{j=i+1}^{i+m} \sum_{k=i-m}^{j-m-1} \bar{a}_{ij} \bar{a}_{ik} = w_i'$.

Thus we have

$$S_2 = \sum_{i=1}^{n}(2m-r_i)r_i - \sum_{i=1}^{n}\sum_{j=i-m}^{i+m} \bar{a}_{ij}|i-j| + 2\sum_{i=1}^{n} w_i'.$$

It follows that

$$\begin{aligned}S &= S_1 + S_2 \\ &\leq \sum_{i=1}^{n}(2m-r_i)r_i - \sum_{i=1}^{n}\sum_{j=i-m}^{i+m} \bar{a}_{ij}|i-j| + 2\sum_{i=1}^{n}(w_i+w_i') \\ &= \sum_{i=1}^{n} 2mr_i - r_i^2 + E_i \text{ where } E_i = 2(w_i+w_i') - \sum_{j=i-m}^{i+m} \bar{a}_{ij}|i-j|\end{aligned}$$

In order to complete our argument we need to establish an upper bound on the expression $E_i = 2(w_i+w_i') - \sum_{j=i-m}^{i+m} \bar{a}_{ij}|i-j|$ for all vectors R_i with fixed row sum r_i. The proof of the following Lemma will be given later.

LEMMA

$$E_i \leq \begin{cases} -\dfrac{r_i^2}{4} & \text{if } r_i \leq \left(\dfrac{2\sqrt{23}+30}{101}\right) m \\ \dfrac{39r_i^2 - 30mr_i + 4m^2}{46} & \text{if } \left(\dfrac{2\sqrt{23}+30}{101}\right) m \leq r_i \leq 8m/7 \\ 3mr_i - 2m^2 - 3r_i^2/4 & \text{if } r_i \geq 8m/7 \end{cases}$$

From the previous theorem, we know that $t = t(n,m) = \frac{1}{2}\sum_{i=1}^{n} r_i \geq .58nm$ so that the average value of r_i is at least $1.16m$. It follows that we should expect to use the upper bound $E_i \leq 3mr_i - 2m^2 - 3r_i^2/4$ for most values of i. However, this inequality does not hold when r_i is small so we will have to correct for these terms. To accomplish this, we define $A = \{i : r_i > 8m/7\}$, $B = \{j : (2\sqrt{23}+30)m/101 < r_j \leq 8m/7\}$ and $C = \{k : r_k \leq (2\sqrt{23}+30)m/101\}$. We then let $a = |A|$, $b = |B|$, and $c = |C|$. Note that $a + b + c = n$.

Now we return to the inequality $S \leq \sum_{i=1}^{n} 2mr_i - r_i^2 + E_i$. We substitute $S = \sum_{i=1}^{n} r_i^2 - r_i$, replace the expression E_i by the upper bounds. The resulting inequality is

$$\sum_{i=1}^{n} r_i^2 - r_i \le \sum_{i=1}^{n} (2mr_i - r_i^2 + 3mr_i - 2m^2 - 3r_i^2/4)$$
$$+ \sum_{j \in B} ((39r_j^2 - 30mr_j + 4m^2)/46 - (3mr_j - 2m^2 - 3r_j^2/4))$$
$$+ \sum_{k \in C} ((-r_k^2/4) - (3mr_k - 2m^2 - 3r_k^2/4)).$$

Simplifying, we obtain

$$\frac{11}{4}\sum_{i=1}^{n} r_i^2 - \sum_{j \in B}(147r_j^2 - 336mr_j + 192m^2)/92$$
$$- \sum_{k \in C}\left[\frac{1}{2}r_k^2 - 3mr_k + 2m^2\right] \le \sum_{i=1}^{n}((5m+1)r_i - 2m^2).$$

At this point, we need to establish the following inequality for real nonnegative values r_i satisfying $\sum_{i=1}^{n} r_i = $ constant.

$$\frac{11}{4n}\left[\sum_{i=1}^{n} r_i\right]^2 \le \frac{11}{4}\sum_{i=1}^{n} r_i^2 - \sum_{j \in B}(147r_j^2 - 336mr_j + 192m^2)/92$$
$$- \sum_{k \in C}\left[\frac{1}{2}r_k^2 - 3mr_k + 2m^2\right] \qquad (*)$$

The standard method for proving inequalities of this type is to use a "local exchange" to uniformize the r_i's. In order to establish (*), we first show that it suffices to establish the inequality when $r_j = 8m/7$ for all $j \in B$ and $r_k = (2\sqrt{23}+30)m/101$ for all $k \in C$. To see that this statement is valid, let F denote the expression on the right hand side of (*). Then suppose that $r_k < (2\sqrt{23}+30)m/101$ for some $k \in C$. Next choose $i \in A$ with r_i as large as possible. We know that $r_i \ge 1.16m$. Then let F' denote the value of the expression on the right hand side of (*) when we decrease r_j by ϵ and increase r_k by ϵ where ϵ is a positive value. A simple calculation shows

$$F - F' > \epsilon\left[\frac{11}{2}r_i - \frac{9}{2}r_k - 3m\right] > 0$$

Thus if (*) holds for F', it also holds for F. This exchange can be applied as many times as required to increase each r_k to $(2\sqrt{23}+30)m/101$ where $k \in C$.

Next, we show that it suffices to prove that (*) is valid when C is empty. For suppose C is nonempty but that $r_k = (2\sqrt{23}+30)m/101$ for every $k \in C$. We then let F denote the value of the expression on the right hand side of (*). We then choose $i \in A$ with r_i as large as possible and let F' be the value of the right hand side of (*) when we decrease r_i by ϵ and increase r_k by ϵ. Note that this moves k from C to B. It can be shown that $F - F' = \left[\frac{11}{2}r_i - \frac{53}{23}r_k - \frac{84}{23}m\right]\epsilon$ which is positive. We may then apply this exchange as many times as is required to remove all elements of C.

Under the assumption that $C = \phi$, we may then show that it suffices to show that (*) is valid when $r_j = 8m/7$ for every $j \in B$. To see that this is true we observe that decreasing r_j by ϵ when $i \in A$ and increasing r_j by ϵ when $j \in B$ and $r_j < 8m/7$ produces a net change of $\left[\frac{11}{2}r_i - \frac{53}{23}r_k - \frac{84}{23}m\right]\epsilon$ which is positive.

So we have reduced the problem of establishing (*) in the general case to the problem of proving (*) is valid when $C = \phi$ and $r_j = 8m/7$ for every $j \in B$. But when $r_j = 8m/7$, the term $2r_j^2 - 4mr_j + 2m^2$ is zero, so in this case, the inequality reduces to

$$\frac{11}{4n}\left[\sum_{i=1}^{n}r_i\right]^2 \leq \frac{11}{4}\sum_{i=1}^{n}r_i^2.$$

This is now the well-known Cauchy-Schwarz inequality. With this observation, our proof that the inequality (*) is valid is complete. We may conclude that

$$\frac{11}{4n}\left[\sum_{i=1}^{n}r_i\right]^2 \leq \sum_{i=1}^{n}(5m+1)r_i - 2m^2$$

We can then solve this quadratic inequality to obtain $\sum_{i=1}^{n}r_i \leq \frac{20 + 4\sqrt{3}}{22}nm$ and thus

$$t(n,m) = \frac{1}{2}\sum_{i=1}^{n}r_i \leq \frac{5+\sqrt{3}}{22}nm = .612nm.$$

It remains to prove the Lemma.

LEMMA

$$E(R_i) = E_i \leq \begin{cases} -\dfrac{r_i^2}{4} & \text{if } r_i \leq \left(\dfrac{2\sqrt{23}+30}{101}\right)m \\ \dfrac{39r_i^2 - 30mr_i + 4m^2}{46} & \text{if } \left(\dfrac{2\sqrt{23}+30}{101}\right)m \leq r_i \leq 8m/7 \\ 3mr_i - 2m^2 - 3r_i^2/4 & \text{if } r_i \geq 8m/7 \end{cases}$$

for a row vector R_i with row sum r_i.

PROOF: The proof here involves some calculation which is done by the symbolic computation system VAXIMA. The details of manipulations are sometimes omitted.

Let R_i denote a vector with r_i 1's such that (a) E_i is maximized; (b) $\sum_{j=i-m}^{i+m} \bar{a}_{ij}|i-j|$ is as small as possible among all R_i satisfying (a).

We will prove a sequence of facts on R_i from which the Lemma will then follow.

CLAIM 1: If $a_{ij} = 1$ and $a_{ik} = 0$ for $i < j < k \leq i+m$, then we have $3(k-j) < 4 \sum_{l=j-m}^{k-m-1} \bar{a}_{il}$

PROOF: Let R_i' be the vector obtained from R_i by exchanging the values of the entries a_{ij} and a_{ik} in R_i. Since $\sum_{j=i-m}^{i+m} \bar{a}_{ij}|i-j| = (k-j) + \sum_{j=i-m}^{i+m} \bar{a}'_{ij}|i-j| > \sum_{j=i-m}^{i+m} \bar{a}'_{ij}|i-j|$, we have $E(R_i) > E(R_i')$, where a_{ij}' denotes the jth entry of R_i'. This implies

$$E(R_i) > E(R_i') = E(R_i) + (k-j) + 2 \sum_{l=j-m}^{k-m-1} (a_{il} - \bar{a}_{il})$$
$$= E(R_i) + 3(k-j) - 4 \sum_{l=j-m}^{k-m-1} \bar{a}_{il}$$

Therefore

$$3(k-j) < 4 \sum_{l=j-m}^{k-m-1} \bar{a}_{il}$$

CLAIM 2: If $a_{ij} = 0$ and $a_{ik} = 1$ for $i < j < k \leq i+m$, then we have $3(k-j) \geq 4 \sum_{l=j-m}^{k-m-1} \bar{a}_{il}$.

PROOF: Let R_i' be the vector obtained from R_i by exchanging the values of the entries a_{ij} and a_{ik} in R_i.

Since $\sum_{j=i-m}^{i+m} \bar{a}_{ij}|i-j| < \sum_{j=i-m}^{i+m} \bar{a}_{ij}'|i-j|$, we have $E(R_i) \geq E(R_i')$, which implies then

$$3(k-j) \geq 4 \sum_{l=j-m}^{k-m-1} \bar{a}_{ik}.$$

Claims 3 and 4 can be proved in a similar way.

CLAIM 3: If $a_{ij} = 0$ and $a_{ik} = 1$ for $i - m \leq j < k < i$, then we have $3(k-j) < 4 \sum_{l=j+m+1}^{k+m} \bar{a}_{il}$.

CLAIM 4: If $a_{ij} = 1$ and $a_{ik} = 0$ for $i - m \leq j < k < i$, then we have $3(k-j) \geq 4 \sum_{l=j+m+1}^{k+m} \bar{a}_{il}$.

Now we partition R_i into "blocks" $B_1, B_2, \ldots, B_s$ where a block is a set of consecutive entries a_{il}, $j \leq l < k$, all of value 0 or all of value 1. R_i is chosen so that the number of (maximal) blocks in $\{i-m, \ldots, i+m\} - \{i\}$ is minimized among all R_i satisfying (a) and (b) (see p. 182).

CLAIM 5: Suppose B is a block consisting of entries a_{il}, $i \leq l < k$ for fixed $i < j < k \leq i+m+l$ and all a_{il} are 0's. Then we have

$$0 = a_{i,j-m-1} = a_{i,j-m} = \cdots = a_{i,j-m+s-1} \text{ for some } s, \quad (i)$$

and $1 = a_{i,j-m+s} = \cdots = a_{i,k-m-1};$

if $i+1 < j < k \leq i+m$ then $s = \dfrac{3(k-j)}{4}$. \quad (ii)

PROOF: (i) is an immediate consequence of Claim 1. We only have to prove (ii).

Since $a_{i,j-1} = 1$ and $a_{i,k-1} = 0$ by Claim 1 we have $3(k-j) < 4 \sum_{l=j-m-1}^{k-m-1} \bar{a}_{il}$. Since $a_{i,j} = 0$ and $a_{i,k} = 1$, by Claim 2 we have

$$3(k-j) \geq 4 \sum_{l=j-m}^{k-m-1} \bar{a}_{il}.$$

This implies $\bar{a}_{i,j-m-1} = 0$ and $\bar{a}_{i,j-m-1} = 1$ and

$$3(k-j) = 4 \sum_{l=j-m}^{k-m-2} \bar{a}_{il}.$$

Suppose $a_{i,j'} = 1$ and $a_{i,k'} = 0$ for $k-m-1 \leq j' < k' < j-m-1$. Then by Claim 3 we have

$$3(k'-j') \le 4 \sum_{l=j'+m+1}^{k'+m} \bar{a}_{il} = 0$$

which is impossible. Thus we have

$$0 = a_{i,j-m-1} = a_{i,j-m} = \ldots a_{i,j+m+s-1} \text{ where } s = \frac{3}{4}(k-j)$$

$$1 = a_{i,j+m+s} = a_{i,j+m+s+1} = \cdots a_{i,j-m-1}$$

Similarly we can prove

CLAIM 6: Suppose B is a block consisting of entries a_{il}, $i < j \le l < k$, for fixed $j < k$. If all a_{il} are 1's, then for some s we have

$$1 = a_{i,j-m-1} = a_{i,j-m} = \ldots = a_{i,j-m+s-1},$$

$$0 = a_{i,j-m+s-1} = a_{i,k-m-1}.$$

If $i+1 < j < k \le i+m$, then $s = \frac{k-j}{4}$.

CLAIM 7: Suppose B is a block consisting of entries a_{il}, $j < l \le k$, for fixed $j < k \le i$. If all a_{il} are 0's, then for some k we have

$$1 = a_{i,j+m+1} = \ldots = a_{i,j+m+s}$$

$$0 = a_{i,j+m+s+1} = \ldots = a_{i,k+m+1}$$

If $i-m \le j < k < i$, then we have $s = \frac{(k-j)}{4}$.

CLAIM 8: Suppose B is a block consisting of entries a_{il}, $j < l \le k$, for fixed $j < k \le i$. If all a_{il} are 1's then for some s we have

$$0 = a_{i,j+m+1} = \ldots = a_{i,j+m+s+1},$$

$$1 = a_{i,j+m+s+1+} = \cdots = a_{i,k+m+1}.$$

If $i-m \le j < k < i$, then $s = \frac{3(k-j)}{4}$.

Now suppose the entries $\{a_{il} : i < l \le i+m\}$ are partitioned into blocks $B_1, \ldots, B_r$ and entries $\{a_{il} : i-m \le l < i\}$ are partitioned into blocks $B_1', \ldots, B_{r'}'$. Let ϵ_i be 0 if B_i consists only of 0's and be 1 otherwise. ϵ_i' is the corresponding value for B_i'. Then the sequence $(\epsilon_1', \epsilon_2', \ldots, \epsilon_{r'}'; \epsilon_1, \epsilon_2, \ldots, \epsilon_r)$ is called the block pattern of R_i. $B_1, B_r, B_1', B_{r'}'$ are called boundary blocks and the rest are interior blocks.

CLAIM 9: All interior 0-blocks are of the same size t. All interior 1-

blocks are of the size $3t$.

PROOF: Suppose B_u is an interior 1-block consisting of entries $\{a_{il} : j \le l < k\}$. The entries $\bar{B}_u = \{a_{i,l-m} : a_{il} \in B_u\}$ are in a 1-block B_w' and a 0-block B'_{w+1}. Then $3|B_w' \cap \bar{B}_u| = |B'_{w+1} \cap \bar{B}_u|$. If B_u is an interior 0-block and the entries in $\bar{B}_u$ can be partitioned into a 0-block B_w' and a 1-block B'_{w+1}, then

$$|B_w' \cap \bar{B}_u| = 3|B'_{w+1} \cap \bar{B}_u|.$$

Now suppose B_u is an interior 1-block with $\bar{B}_u \subseteq B_w' \cup B'_{w+1}$. If B'_{w+1} is an interior block, then we have

$$|B'_{w+1} \cap \bar{B}_u| = \frac{3}{4}|B_u|$$

$$|B'_{w+1} \cap \bar{B}_{u+1}| = \frac{3}{4}|B_{w+1}|$$

Since B'_{w+1} is also an interior block, we have

$$|B'_{w+1} \cap \bar{B}_u| = \frac{1}{4}|B'_{w+1}|$$

$$|B'_{w+1} \cap \bar{B}_{u+1}| = \frac{3}{4}|B'_{w+1}|$$

Therefore $3|B_u| = |B_{u+1}| = |B'_{w+1}|$ and Claim 9 will follow.

CLAIM 10: Suppose B_u is interior in $\{i, \ldots, i+m\}$. If $\bar{B}_u \subseteq B_w' \cup B'_{w+1}$, then B_w' is not interior.

PROOF: Suppose $B_u = \{a_{il} : j \le l < k\}$ and $B_w' = \{a_{i,l} : j' \le l < k'\}$ are interior. We now consider R_i' obtained from R_i by exchanging the entries $a_{i,j-t}$ with $a_{i,k-t}$ and the entries $a_{i,j'-t}$ with $a_{i,t}$ for $1 \le t \le p$ where B'_{w-1} consists of $\{a_{i,l} : j'' - p \le l < j''\}$. It is easy to verify that $E(R_i') \ge E(R_i)$ and the number of blocks in R_i' is fewer than that of R_i, which is a contradiction.

It follows from Claim 10 and the symmetric version of Claim 10 that

CLAIM 11: $r \le 3$ and $r' \le 3$. If $r=3$ and $r'=3$, then there is one interior 0-block and one interior 1-block.

CLAIM 12: One of r and r' is less than 3.

PROOF: If $r = r' = 3$, we may assume without loss of generality that the block pattern is $(0,1,0:1,0,1)$ and $\bar{B}_2 \subset B_1' \cup B_2'$. Now we consider R_i' with entries a'_{il} such that

$a'_{i,l} = a_{i,l-1}$ if $l \neq i+1$ and $l \neq i-m$; $a'_{i,i+1} = a_{i,i+m}$; and $a'_{i,i-m} = a_{i,i-1}$. It can be easily checked that $E(R_i') > E(R_i)$, which is impossible.

From now on, we may assume $r \leq 2$. From Claims 5, 6 and 11, we conclude that the block pattern for R_i is one of the following

(i) $(0,1:1,0)$
(ii) $(1,0,1:1,0)$
(iii) $(0,1,0:0,1)$
(iv) $(1,0:0,1)$

CLAIM 13: If the block pattern in $(0,1:1,0)$, then $E(R_i) \leq -\dfrac{r_i^2}{4}$.

PROOF: If $||B_1| - |B_2'|| \geq 2$, we can adjust the size of B_1, B_2' and obtain an R_i' with larger $E(R_i')$ value, which is impossible. We may then assume $||B_1| - |B_2'|| \leq 1$. By straightforward calculation we have $E(R_i) \leq -\dfrac{r_i^2}{4}$.

CLAIM 14: If the block pattern is $(1,0:0,1)$, then $E(R_i) \leq 3mr_i - 2m^2 - 3r_i^2/4$.

PROOF: It can be similarly shown that $||B_i'| - |B_2|| \leq 1$. Thus we can calculate $E(R_i)$ and obtain

$$E(R_i) \leq 3mr_i - 2m^2 - 3r_i^2/4.$$

The remaining two cases are more complicated.

CLAIM 15: If the block pattern is $(1,0,1:1,0)$, then $r_i \leq m/2$ and

$$E(R_i) \leq \dfrac{39r_i^2 - 30mr_i + 7m^2}{98} \quad \text{for} \quad r_i \geq \dfrac{31}{79}m \quad \text{and}$$

$$E(R_i) \leq \dfrac{17r_i^2 - 10mr + 2m^2}{18} \quad \text{otherwise.}$$

PROOF: In this case the pattern is $(1,0,1:1,0)$. Let b_i denote $|B_i'|$. Then we have,

$$2b_1 + b_3 + \dfrac{b_2}{4} = r_i \text{ (by Claim 7)}$$

$$b_1 + b_2 + b_3 = m$$

Therefore $b_1 = \dfrac{3}{4}b_2 + r_i - m \geq 0$ and $b_3 = 2m - r_i - \dfrac{7}{4}b_2 \geq 0$. This implies $\dfrac{4(m-r_i)}{3} \leq b_2 \leq \dfrac{4}{7}(2m-r_i)$. Now by straightforward calculation $E(R_i) = -\dfrac{16m^2 - 32b_2b_3 - 48b_1^2 - 24b_1b_2 - 15b_2^2}{32}$. Substituting for b_1 and b_3 and setting $b = b_2$, we obtain

$$E(R_i) = f(b) = \frac{12r_i^2 + (16b - 24m)r_i + 8m^2 - 8bm + b^2}{8}$$

Since $\dfrac{d^2f}{db^2} = \dfrac{1}{4} > 0$, there is no interior maximum. Using Claim 1, we have $b_1 \leq \dfrac{1}{4}|B_1| = \dfrac{1}{4}(b_1 + \dfrac{b}{4})$ and $b_3 \leq \dfrac{1}{4}|B_2| = \dfrac{1}{4}(\dfrac{3}{4}b + b_3)$. Thus $b_1 \leq \dfrac{1}{12}b$ and $b_3 \leq \dfrac{b}{4}$. This implies $\dfrac{2m - r_i}{2} \leq b \leq \dfrac{(3m - r_i)}{2}$ (It can be easily shown that in this case $r_i \leq \dfrac{m}{2}$).

We then have

$$E(R_i) \leq \begin{cases} f\left[\dfrac{(3m-r_i)}{2}\right] = \dfrac{39r_i^2 - 30mr_i + 7m^2}{48} & \text{if } r_i \geq \dfrac{31}{79}m \\ f\left[\dfrac{4}{3}(m-r_i)\right] = \dfrac{-17r_i^2 + 10mr_2 - 2m^2}{18} & \text{otherwise} \end{cases}$$

CLAIM 16: If the block pattern is $(0,1,0:0,1)$, then $r_i \leq \dfrac{5m}{4}$,

$$E(R_i) \leq \frac{39r_i^2 - 30mr_i + 4m^2}{46} \text{ if } \frac{8}{7}m \geq r_i \text{ and}$$

$$E(R_i) \leq \frac{75r_i^2 + 174mr_i - 100m^2}{2} \text{ if } r_i < \frac{8}{7}m.$$

PROOF: Let b_i denote $|B_i'|$ for $i=1,2,3$. Then we have

$$b_2 + \frac{b_2}{4} + b_3 = r_i$$

$$b_1 + b_2 + b_3 = m$$

Therefore

$$b_3 = r_i - \frac{5}{4}b_2 \geq 0$$

$$b_1 = m - r_i + \frac{b_2}{4} \geq 0$$

This implies $4(r_i - m) \leq b = b_2 \leq \dfrac{4}{5}r_i$ and $r_i \leq \dfrac{5}{4}m$. Also

$$b_3 \le \frac{3}{4}|B_2| = \frac{3}{4}\left[\frac{1}{4}b+b_3\right]$$

and

$$b_1 \le \frac{3}{4}\left[b_1+\frac{3}{4}b\right]$$

i.e.

$$b_3 \le \frac{3}{4}b, \quad b_1 \le \frac{9}{4}b$$

$$r_i - \frac{5}{4}b \le \frac{3}{4}b, \quad m - r_i + \frac{b}{4} \le \frac{9}{4}b$$

Therefore $b \ge \frac{r_i}{2}$, $b \ge \frac{m-r_i}{2}$.

By straightforward computation we have

$$E(R_i) \le \frac{4r_i^2 + 16br_i - 8mr_i + 8bm - 23b^2}{8} = f(b)$$

Since $\frac{d^2f}{db^2} < 0$, the maximum value of $f(b)$ is achieved at $\frac{df}{db}(b) = 0$,

i.e., $b = \frac{8r + 4m}{23}$. We have

$$E(R_i) \le f\left[\frac{8r+4m}{23}\right] = \frac{39r_i^2 - 30mr_i + 4m^2}{46} \quad \text{if } \frac{8}{7}m \ge r_i \text{ and}$$

$$E(R_i) \le f(4(r_i-m)) = \frac{75r_i^2 + 174mr_i - 100m^2}{2} \quad \text{if } \frac{5}{4}m \ge r_i > \frac{8}{7}$$

Now we can summarize the following

$$E(R_i) \le \begin{cases} -\frac{r_i^2}{4} & \text{if } r_i \le \left(\frac{2\sqrt{23}+30}{101}\right)m \\ \frac{39r_i^2 - 30mr_i + 4m^2}{46} & \text{if } \left(\frac{2\sqrt{23}+30}{101}\right)m \le r_i \le 8m/7 \\ 3mr_i - 2m^2 - 3r_i^2/4 & \text{if } r_i \ge 8m/7 \end{cases}$$

This completes the proof of our theorem.

III. Related problems

The function $t(n,m)$ we studied in this paper came up in connection with the following problem. For an integer $n \geq 2$, let $T(n)$ denote the directed graph whose vertex set is $\{1,2,\ldots,n\}$ and whose edge set is $\{(i,j): 1 \leq i < j \leq n\}$. Recall that a directed graph is said to be *unipathic* when it contains at most one directed path between any two vertices. We then let $u(n,m)$ denote the maximum number of edges in a subgraph of $T(n)$ so that the restriction to any m consecutive vertices is unipathic. Since a unipathic directed graph is triangle-free, it follows that $u(n,m) \leq t(n,m-1) + \binom{n-m+1}{2}$. Unlike the situation with $t(n,m)$, it is possible to provide an explicit formula for $u(n,m)$.

This formula was provided by Maurer, Rabinovitch, and Trotter [2] who showed that the problem of determining $u(n,m)$ was equivalent to determining the rank of a certain class of partially ordered sets.

THEOREM 4: Let n and m be integers with $n \geq m \geq 2$. Then $u(n,m) = \binom{q}{2}(m-1)^2 + q(m-1)r + \lfloor \frac{r^2}{4} \rfloor$ where $n = q(m-1) + r$ and $\lceil \frac{1}{2}(m-1) \rceil \leq r < \lceil \frac{3}{2}(m-1) \rceil$. □

We refer the reader to [2] for the argument for this theorem and to [1] for a discussion of the combinatorial theory of rank for partially ordered sets.

Here we studied a Turán type problem with restrictions imposed on the bandwidth. There are numerous variations that can be formulated in the following framework.

Let **H** denote a family of graphs and **F** denote another family of graphs. Define $t(n,\mathbf{H},\mathbf{F})$ to be the maximum number of edges in a graph G in **F** on n vertices which does not contain any graph in **H**. If we take **H** to contain only one graph K_s and **F** to be the family of all graphs, then $t(n,\mathbf{H},\mathbf{F})$ is just the Turán number. The problem of determining $t(n,\mathbf{H},\mathbf{F})$ is of interest for various families **H** and **F**. Here we mention a few:

(i) $\mathbf{H} = \{K_s\}$ and $\mathbf{F} = \{\text{all graphs with bandwidth} \leq m\}$;

(ii) $\mathbf{H} = \{\text{all graphs with chromatic number } k\}$ and $\mathbf{F} = \{\text{all graphs with bandwidth } m\}$;

(iii) $\mathbf{H} = \{\text{all graphs with chromatic number } k\}$ and $\mathbf{F} = \{\text{all graphs with chromatic number } l\}$,

(iv) $\mathbf{H} = \{\text{cycles of length} \leq k\}$ and $\mathbf{F} = \{\text{all graphs with chromatic number } k\}$.

In this paper we studied the case that **H** consists of one graph K_3

and $\mathbf{F} = B_m$ is the set of all graphs with bandwidth $\leq m$. We proved that

$$(2-\sqrt{2})nm \leq t(n,\{K_3\},B_m) \leq \frac{5+\sqrt{3}}{11}nm$$

Probably the lower bound is closer to the "truth" here. However, the likely candidate for an extremal graph (see Theorem 2) does not have constant degree and cannot achieve the upper bound obtained here by using averaging arguments. Some new idea is need to close up the gap.

References

1. S. B. Maurer, I. Rabinovitch, and W. T. Trotter, Jr., "Large minimal realizers of partial order, II", Discrete Math. 31 (1980) 297-313.
2. S. B. Maurer, I. Rabinovitch, and W. T. Trotter, Jr., "A Generalization of Turán's Theorem to Directed Graphs", Discrete Math. 32 (1980) 167-189.

On Clique Partition Numbers of Regular Graphs

Peter Eades

Norman J. Pullman

Peter J. Robinson

ABSTRACT

This paper is concerned with partitioning edges of a regular graph with a minimum number of cliques. In particular, those pairs of integers (n,x) for which there is a k-regular graph on n vertices with a minimum clique partition of size x are determined for $k=3,4$.

1. Introduction

For our purposes, graphs have no self adjacent vertices or multiple edges. A *clique* is a complete subgraph of a graph, and a clique with k vertices is a *k-clique*. A *clique partition* of a graph G is a family C of cliques of G such that every edge of G is in exactly one member of C. A clique partition C of G is *minimum* if $|C| \leq |C'|$ for every clique partition C' of G. The cardinality of a minimum clique partition of G is called the *clique partition number* of G and is denoted by $cp(G)$. Other notation and terminology is consistent with [1].

Bounds for $cp(G)$ for connected k-regular graphs G on n vertices were given in [2].

(1.1) THEOREM ([2], Theorem 3.2). If $3 \leq k+1 < n$, then

$$8n(k+2)/[2(k+2)^2+(-1)^k-1] \leq cp(G) \leq kn/2.$$

The upper bound $kn/2$ is known to be achievable for all even $n \geq 2k$ and all odd $n \geq 5k/2$ (see [2] and [3]). The lower bound given by (1.1) is achievable when it is an integer (P.J. Robinson, private communication). The best (that is, the achievable) lower bounds for $cp(G)$ were determined in [2] for $k=3,4$. When $k=3$ and $n>4$, the best lower bound is $\lceil 5n/6 \rceil$ as in (1.1), or $1+\lceil 5n/6 \rceil$ depending on whether $n \equiv 0$

(mod 6) or not. When $k=4$ and $n \geq 14$, the best lower bound is $\lceil 2n/3 \rceil$ as in (1.1), or $1 + \lceil 2n/3 \rceil$ depending on whether $n \equiv 0$ (mod 3) or not.

In this paper we characterize all pairs of integers (n,x) such that a k-regular connected graph with n vertices and clique partition number x exists, for $k=3$ and $k=4$.

The following notation is useful. For each pair (k,n), denote by $S_k(n)$ the set of integers x for which a k-regular connected graph with n vertices and clique partition number x exists. In section 2 we compute $S_3(n)$; in section 3 we compute $S_4(n)$. The main results are contained in Theorem 2.1 and 3.1. They are:

THEOREM (2.1c) If $m>3$, then $S_3(2m)$ is the set of all integers in the interval $[5m/3, 3m]$ of the same parity as m.

THEOREM (3.1b,c) If $n>13$, then $S_4(n)$ is a set of all but three integers in the interval $[2n/3, 2n]$.

2. Cubic Graphs

A 3-regular graph is said to be *cubic*. The following theorem gives $S_3(n)$ for all even $n \geq 4$. (When k is odd, n must be even because a k-regular graph on n vertices has $kn/2$ edges.)

(2.1) THEOREM

(a) $S_3(4) = \{1\}$.

(b) $S_3(6) = \{5,9\}$.

(c) If $n>6$ is even then

$$S_3(n) = \{x : 5n/6 \leq x \leq 3n/2 \text{ and } x \equiv \frac{1}{2}n \ (\text{mod } 2)\}.$$

PROOF: Parts (a), and (b) are easily proved; we consider only part (c).

First suppose that G is a cubic graph with n vertices and with $cp(G)=x$. From (1.1) we can assume that $5n/6 \leq x \leq 3n/2$. If c_i, $i=2,3$, denotes the number of i-cliques in a minimum clique partition of G, then, since G contains no 4-cliques, $c_2+c_3=x$. Also, counting edges we obtain $c_2+3c_3=3n/2$. It follows that $x \equiv \frac{1}{2}n$ (mod 2).

Now suppose that x is an integer such that $x \equiv \frac{1}{2}n$ (mod 2) and $5n/6 \leq x \leq 3n/2$. We explicitly construct a cubic connected graph with n vertices and clique partition number x.

Suppose that $\frac{1}{2}(3n/2-x)$ is even; let $s = \frac{1}{4}(3n/2-x)$. Consider the graphs $G_{s,n}$ in figure 2.1.

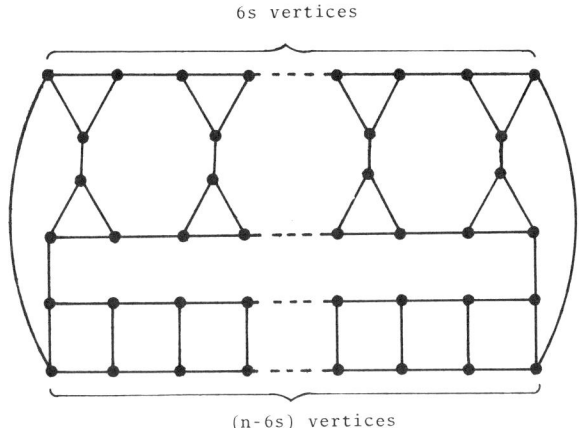

Figure 2.1. $G_{s,n}$

The subgraph of $G_{s,n}$ induced by the $6s$ vertices at the top contains $2s$ 3-cliques. Since $n>6$, there is only one way that other 3-cliques could occur, that is, if $s=1$ and $n=8$. $G_{1,8}$ is illustrated in figure 2.2.

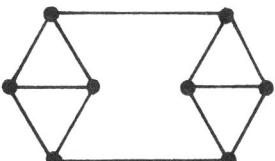

Figure 2.2. $G_{1,8}$

It is clear that $cp(G_{1,8}) = 8 = x$. If $s=0$ or $s>1$, then a minimum clique partition of $G_{s,n}$ consists of the $2s$ 3-cliques at the top, and the remaining $(3n/2 - 6s)$ edges; hence $cp(G) = x$.

Suppose that $\frac{1}{2}(3n/2 - x)$ is odd; let $s = \frac{1}{4}(3n/2 - x) - \frac{1}{2}$. Now let $G'_{s,n}$ be $G_{s,n}$ with triangle τ contracted to a point. Comparing $G'_{s,n}$ to $G_{s,n}$, it is clear that n decreases by 2, c_3 by 1 and c_2 by 0, hence x decreases by 1. Thus $G'_{s,n}$ handles the case where $(1/2)(3n/2-x)$ is odd.

3. 4-Regular Graphs

In this section we calculate $S_4(n)$ for each $n>4$. The results are summarized in the following Theorem.

(3.1) THEOREM.

(a) $S_4(5)=\{1\}$, $S_4(6)=\{4\}$, $S_4(7)=\{8,10\}$, $S_4(8)=\{6,8,16\}$,

$S_4(9)=\{6,8,9,10,12,14\}$, $S_4(10)=\{9,10,11,12,14,16,18,20\}$,

$S_4(11)=\{x : 9\leq x\leq 18\} \cup \{20,22\}$,

$S_4(12)=\{8,9\} \cup \{x : 11\leq x\leq 20\} \cup \{22,24\}$

$S_4(13)=\{x : 11\leq x\leq 22\} \cup \{24,26\}$

(b) if $14\leq n\equiv 0 \pmod{3}$, then

$S_4(n)=\{2n/3\} \cup \{x : (2n+6)/3\leq x\leq 2n-4\} \cup \{2n-2, 2n\}$

(c) If $14\leq n\not\equiv 0 \pmod{3}$, then

$S_4(n)=\{x : \lceil 2n/3\rceil + 1\leq x\leq 2n-4\} \cup \{2n-2, 2n\}$

The proof has two parts. The first part (Lemma (3.3), (3.7), (3.8), (3.9)) establishes necessary conditions for $x \in S_4(n)$. The second part shows how to construct the graphs with clique partition number x which satisfy the necessary conditions. This is done inductively. Transformations are performed on graphs in $S_4(n-4)$ increasing the number of vertices by 4, while increasing their clique partition numbers by 3 or 8. Together with standard constructions for the smallest and largest members of $S_4(n)$, they enable us to build all the members of $S_4(n)$.

First we establish some negative results, that is, necessary conditions for $x \in S_4(n)$. Suppose that G is a 4-regular connected graph with $n>5$ vertices and C is a minimum clique partition with $|C| = x$. If c_i, $i=2,3,4$, is the number of i-cliques in C, then

$$c_2 + c_3 + c_4 = x, \tag{3.2}$$

and

$$c_2 + 3c_3 + 6c_4 = 2n.$$

These equations do not have nonnegative integral solutions for $x=2n-1$ or $x=2n-3$, and the next Lemma can be deduced.

(3.3) LEMMA $(2n-1) \notin S_4(n)$, $(2n-3) \notin S_4(n)$. □

Now suppose that c_4 is positive, that is, C contains a 4-clique A. Each vertex of A is incident with exactly one edge of G not in A, and so each of these edges must be a 2-clique in C. Hence we obtain

(3.4) if $c_4 > 0$ then $c_2 \geq 4$.

Further, each of the 2-cliques of C is incident with at most 2 4-cliques of C, and so

(3.5) $c_2 \geq 2c_4$.

Note that if $c_2 = 2c_4$ then every 2-clique of C is incident with a 4-clique of C and every vertex in the subgraph induced by the 4-cliques of C has degree 4. Hence

(3.6) if $c_2 = 2c_4$, then $n = 4c_4$ and $c_3 = 0$.

We can use these observations in the proof of the next Lemma.

(3.7) LEMMA. If $t > 4$ then $2t+1 \notin S_4(3t)$.

PROOF. Using the notation above, suppose that $n = 3t$ and $x = 2t+1$. The equations (3.2) ensure that c_4 is odd, and $2c_2 = 3c_4 + 3$. From (3.5) we deduce that $c_4 = 1$ or 3. If $c_4 = 3$, then by (3.6) we have $n = 12$, contradicting $t > 4$. If $c_4 = 1$, then $c_2 = 3$, contradicting (3.4). □

The following bounds are proved in [2].

(3.8) LEMMA. Suppose that $x \in S_4(n)$.
  (a) If $n = 7$ then $x = 8$ or $x = 10$.
  (b) If $n = 8$ then $6 \leq x \leq 16$.
  (c) If $n = 9$ then $6 \leq x \leq 14$.
  (d) If $n = 10$ then $9 \leq x \leq 20$.
  (e) If $n = 13$ then $11 \leq x \leq 26$.
  (f) If $n > 9$ and $n \equiv 0 \pmod{3}$ then $2n/3 \leq x \leq 2n$.
  (g) If $n > 10$, $n \neq 13$, and $n \not\equiv 0 \pmod{3}$ then $[2n/3] + 1 \leq x \leq 2n$.

Our final negative result concerns small graphs.

(3.9) LEMMA
(a) If $x \in S_4(8)$ then $x \neq 7, 9, 10, 11, 12, 14$.
(b) If $x \in S_4(9)$ then $x \neq 7, 11, 13$.
(c) If $x \in S_4(10)$ then $x \neq 13, 15$.
(d) $10 \notin S_4(12)$.

PROOF: We prove only (d). The proofs of (a), (b) and (c) are similar. Suppose that G is a 4-regular connected graph with 12 vertices and a minimum clique partition C with $|C|=10$. Equations (3.2) imply that either C contains three 2-cliques and seven 3-cliques, or else six 2-cliques, two 3-cliques and two 4-cliques. In the first case, consider the graph $\hat{G}$ obtained by deleting the edges of each of the seven 3-cliques from G. Every vertex in $\hat{G}$ has even degree, and $\hat{G}$ contains 3 edges. Hence $\hat{G}$ contains a 3-clique, contradicting the minimality of C. If C contains two 4-cliques and two 3-cliques, then the two 4-cliques cannot share a vertex with each other or with the 3-cliques, and the two 3-cliques can have at most one vertex in common. Hence G has at least 13 vertices, contradicting hypothesis. □

The next 2 Lemmas are the elementary induction tools for the proof of Theorem (3.1).

(3.10) LEMMA. If there is a 4-regular connected graph G with a minimum clique partition which includes 2 nonadjacent 2-cliques, then there is a 4-regular connected graph H such that $|V(H)| = |V(G)| + 4$, $cp(H) = cp(G) + 3$, and there is a minimum clique partition of H which contains 4 mutually nonadjacent 2-cliques.

PROOF: Replace the specified subgraph of G by a particular subgraph with 4 more vertices and 8 more edges (as indicated in Figure 3.1) to form H. Then the specified minimum clique partition of G can be modified to form the desired minimum clique partition of H. □

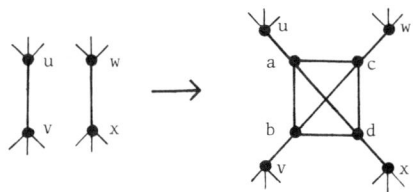

Figure 3.1

(3.11) LEMMA. Suppose that there is a 4-regular connected graph G with a minimum clique partition which includes four 2-cliques with the property that each of these 2-cliques is adjacent to at most one of the others. Then there is a 4-regular connected graph H such that $|V(H)| = |V(G)| + 4$, $cp(H) = cp(G) + 8$, and there is a minimum clique partition of H which contains 4 mutually nonadjacent 2-cliques.

PROOF: Same as that of Lemma 3.10 using the transformation indicated in Figure 3.2. □

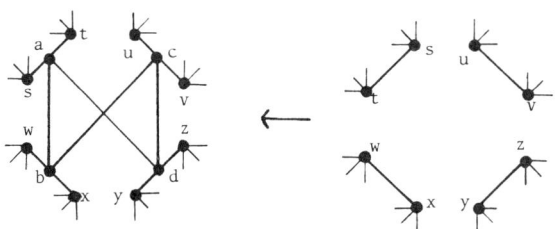

Figure 3.2

(Note: It is possible that $t=u$ and/or $x=y$; but all other vertices are distinct)

The following 2 Lemmas assist with the construction of a starting point for the inductive proof of Theorem (3.1).

(3.12) LEMMA

If there is a 4-regular connected graph G with a minimum clique partition which contains 3 mutually nonadjacent 2-cliques, then there is a 4-regular connected graph H such that

$$|V(H)| = |V(G)| + 3, \ cp(H) = cp(G) + 4,$$

and there is a minimum clique partition of H which contains 3 mutually nonadjacent 2-cliques.

PROOF: Similar to that of Lemma 3.10 using the transformation indicated in Figure 3.3.

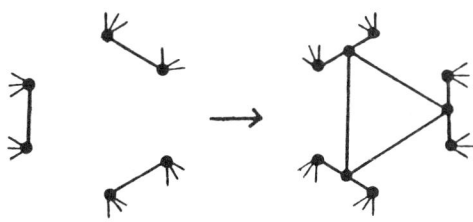

Figure 3.3

(3.13) LEMMA. Suppose that G and H are k-regular connected graphs each of which has a minimum clique partition which contains a 2-clique. Then there is a k-regular connected graph L with

$|V(L)| = |V(G)| + |V(H)|$, and $cp(L) = cp(G) + cp(H)$.

PROOF: Suppose that uv and wx are 2-cliques in minimum clique partitions of G and H respectively. Form K by deleting edges uv and wx, and adding edges uw and vx to the disjoint union of G and H. □

The following two Lemmas show how the constructive Lemmas (3.10) and (3.11) can be applied.

(3.14) LEMMA. If G is a 4-regular connected graph with n vertices and $cp(G) > 2n/3$, then every minimum clique partition of G contains at least 4 2-cliques; at least 2 of these are not adjacent.

PROOF: Suppose that C is a minimum clique partition of G. Suppose that C contains a 4-clique A. Each of the 4 vertices of A is incident with an edge of G not in A; these edges must be 2-cliques in C. If these 4 edges were all adjacent then $G = K_5$, contradicting $cp(G) > 2n/3$.

Next suppose that C does not contain a 4-clique. Denote by $\hat{G}$ the graph obtained from G by deleting the edges of all the 3-cliques of C. If $\hat{G}$ has no edges then C consists entirely of 3-cliques and so $cp(G) = 2n/3$, contrary to hypothesis. Because G is 4-regular, vertices have degrees 0, 2 or 4 in $\hat{G}$. If $\hat{G}$ has no vertices of degree 4 then it consists of disjoint cycles, and by the minimality of C, $\hat{G}$ does not contain a 3-clique, and so $\hat{G}$ contains a cycle of length at least 4 with a pair of nonadjacent vertices. If $\hat{G}$ has a vertex u of degree 4, then consider two edges ua and ub incident with u. The vertex a must be incident with another edge ac of $\hat{G}$, and since $\hat{G}$ does not contain a 3-clique, $b \neq c$, and the 2-cliques ub and ac are nonadjacent.

(3.15) LEMMA. If G is a 4-regular connected graph with n vertices and $cp(G) > (2n + 10)/3$, then every minimum clique partition of G contains four 2-cliques such that each one of these four is adjacent to at most one of the others.

PROOF Using the equations (3.2) we can show that, since $cp(G) > (2n + 10)/3$, each clique partition C of G contains at least six 2-cliques. Consider the graph $\hat{G}$ obtained from G by deleting all the edges of 4-cliques and 3-cliques of C. Now $\hat{G}$ has at least 6 edges, contains no 3-cliques, and each vertex has degree 0, 1, 2, or 4. Let P be a longest path in $\hat{G}$. If P has 5 edges then we are done — see figure 3.4(a). If P has 4-edges, then there are at least 2 edges of $\hat{G}$ not in P. If there are 2 edges disjoint from P we are done; otherwise, since $\hat{G}$ has no vertex of degree 3, one of figure 3.4 (b) or (c) must be a subgraph of $\hat{G}$. In either case there is a suitable set of 2-cliques. We can deal with the cases where P has fewer than 4 edges in a similar fashion. □

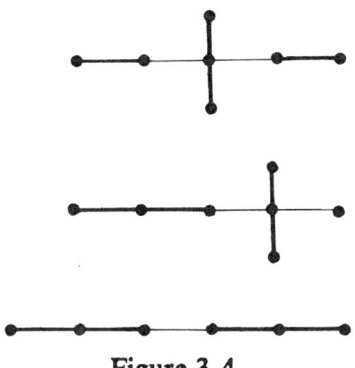

Figure 3.4

The next result follows from (3.10), (3.11), (3.14) and (3.15).

(3.16) COROLLARY
(a) If $x \in S_4(n)$ and $x > 2n/3$ then $(x+3) \in S_4(n+4)$.
(b) If $x \in S_4(n)$ and $x > (2n+10)/3$ then $(x+8) \in S_4(n+4)$. □

To start the inductive argument, it is necessary to construct a few graphs explicitly. Unfortunately, lack of space prevents us from displaying the graphs used to prove the following Lemma.

(3.17) LEMMA
(a) $8 \in S_4(8)$.
(b) $\{8,9,10,12\} \subseteq S_4(9)$.
(c) $\{10,11,12,14,16,18\} \subseteq S_4(10)$.
(d) $\{10,11,12,13,14,15,16,17,18,20\} \subseteq S_4(11)$.
(e) $\{15,20,22\} \subseteq S_4(12)$.
(f) $\{19,21\} \subseteq S_4(13)$. □

PROOF OF THEOREM 3.1. Lemma (3.17), the graphs constructed in [2], and some tedious applications of Lemmas (3.12), (3.13), and Corollary (3.16), can be used to prove Theorem 3.1 for all $n \leq 20$. Suppose that $N > 20$ and that Theorem (3.1) holds for $n < N$.

We consider the cases $N = 3t$, $N = 3t+1$, $N = 3t+2$, separately.

Suppose $N = 3t$; since $N - 4 > 13$ and $N - 4 \not\equiv 0 \pmod{3}$, induction guarantees that

$$S_4(N-4) = \{x: 2t-1 \leq x \leq 6t-12\} \cup \{6t-10, 6t-8\}$$

Using Corollary (3.16) we obtain

$$\{x: 2t+2 \leq x \leq 6t-9\} \subseteq S_4(N)$$

and

$$\{x: 2t+9 \leq x \leq 6t-4\} \cup \{6t-2, 6t\} \subseteq S_4(N).$$

Now $t > 4$, and so $6t-9 > 2t+9$. Hence

$$\{x: 2t+2 \leq x \leq 6t-4\} \cup \{6t-2, 6t\} \subseteq S_4(N)$$

Further, from [2, Theorem 5.2], $2t \in S_4(N)$. Hence

$$\{2t\} \cup \{x: 2t+2 \leq x \leq 6t-4\} \cup \{6t-2, 6t\} \subseteq S_4(N)$$

By Lemmas (3.3) and (3.7), it follows that

$$S_4(N) = \{2N/3\} \cup \{x: 2N/3+2 \leq x \leq 2N-4\} \cup \{2N-2, 2N\}.$$

Next suppose that $N = 3t+1$. In this case $N-4 \equiv 0 \pmod 3$, and so

$$S_4(N-4) = \{2t-2\} \cup \{x: 2t \leq x \leq 6t-10\} \cup \{6t-8, 6t-6\}$$

From Corollary (3.16) we can deduce that

$$\{x: 2t+3 \leq x \leq 6t-7\} \subseteq S_4(N)$$

and

$$\{x: 2t+10 \leq x \leq 6t-2\} \cup \{6t, 6t+2\} \subseteq S_4(N).$$

Now $t > 4$ so $2t+10 < 6t-7$, and from [2, Theorem 5.2], $2N/3 + 1 \in S_4(N)$. Hence, using Lemma (3.3), $S_4(N) = \{x: \lceil 2N/3 \rceil +1 \leq x \leq 2N-4\} \cup \{2N-2, 2N\}$.

A similar argument can be applied to the case $N = 3t+2$. □

Acknowledgments

We would like to thank the referee for his suggestions.

This work was supported in part, by the Natural Science and Engineering Research Council of Canada under grants A4041 and T1821.

References

[1] J. A. Bondy and U. S. R. Murty, *Graph Theory with applications*, MacMillan, London and Basingstoke, 1977

[2] Norman J. Pullman and Dom de Caen, "Clique coverings of graphs I - clique partitions of regular graphs", *Utilitas Math.* 19 (1981), 177-205.

[3] N. J. Pullman and N. C. Wormald, "Regular Graphs with Prescribed Odd Girth" (to appear), *Utilitas Math.*

Cube-Supersaturated Graphs and Related Problems

P. Erdős

M. Simonovits

ABSTRACT

In this paper we shall consider ordinary graphs, that is, graphs without loops and multiple edges. Given a graph L, $ex(n,L)$ will denote the maximum number of edges a graph G^n of order n can have without containing any L. Determining $ex(n,L)$, or at least finding good bounds on it will be called

TURÁN TYPE EXTREMAL PROBLEM. Assume that a graph G^n has $E > ex(n,L)$ edges. Then it must contain some copies of L. Such a graph will be called supersaturated, or *L-supersaturated*. L-supersaturated graphs mostly contain not only one, but very many copies of L. The problem discussed here (and called "the problem of L-supersaturated graphs") is

Determine the minimum number of copies of L a graph G^n with $E > ex(n,L)$ edges must contain.

The main results of this paper are two "recursion theorems" motivated by the case when L is the graph determined by the vertices and edges of a cube.

Notation

Below we shall consider ordinary graphs, that is, graphs without loops and multiple edges. $G, H, \ldots, S$ and $G^n, H^n, \ldots, S^n$ will denote graphs and the upper indices will always denote the number of vertices. Also, we shall use $v(G)$, $e(G)$ and $\chi(G)$ to indicate the number of vertices, edges, and the chromatic number, respectively. K_p is the complete graph on p vertices, C_p is the cycle of length p, $K_{p,q}$ denotes the complete bipartite graph with p and q vertices in its color-classes. $G(A,B)$ denotes the bipartite subgraph of G induced by A and B, $(A \cap B = \emptyset)$.

Below $c_1, \ldots, c_j, \ldots$ will always denote positive constants, *independent of n*, but they may depend on the graphs in consideration and they may be different in different parts of the paper.

1. Introduction

In this paper we shall consider problems on supersaturated graphs. These type of problems are strongly related to extremal graph problems of Turán type.

Turán type extremal graph problems. Let $\mathbb{L}$ be a given family of (so called forbidden) graphs. Determine the maximum number of edges a graph G^n can have without containing any $L \in \mathbb{L}$ as a subgraph.

The maximum will be denoted by $ex(n, \mathbb{L})$, the family of graphs attaining this maximum by $EX(n, \mathbb{L})$. These graphs will be called *extremal*. If $p+1$ denotes the minimum chromatic number in G^n, then, by a theorem of Erdős and Simonovits [3],

$$ex(n, \mathbb{L}) = \left(1 - \frac{1}{p}\right)\binom{n}{2} + o(n^2) \qquad (1)$$

An extremal graph problem will be called *degenerate* if $p=1$, that is, *if there is a bipartite graph among the forbidden ones*. By (1), the problem is degenerate iff $ex(n, \mathbb{L}) = o(n^2)$. (By [8], if $L \in \mathbb{L}$ is bipartite and $c = 2/v(L)$, then $ex(n, \mathbb{L}) = O(n^{2-c})$.) In this paper we restrict our consideration to this degenerate case.

Let G^n have more than $ex(n, \mathbb{L})$ edges. Then it will be called *supersaturated* and obviously, it contains at least *one* forbidden subgraph. It is surprising that G^n contains in such cases not only one but very many forbidden subgraphs.

THE PROBLEM OF SUPERSATURATED GRAPHS. Given a graph G^n with $E > ex(n, \mathbb{L})$ edges, at least how many subgraphs of G^n are in $\mathbb{L}$? This minimum will be denoted by $f(n, \mathbb{L}, E)$.

First we mention a general theorem on supersaturated graphs.

THEOREM A. (Erdős-Simonovits, [5].) Given a family $\mathbb{L}$ of forbidden graphs, each of which has v vertices. For every $c > 0$ there exists a $c' > 0$ such that if

$$e(G^n) > ex(n, \mathbb{L}) + cn^2, \qquad (2)$$

then G^n contains at least $c'n^v$ forbidden $L \in \mathbb{L}$.

Obviously, this result is sharp, since G^n has only $O(n^v)$ subgraphs of order v. The following problem seems much more difficult:

PROBLEM OF WEAKLY SUPERSATURATED GRAPHS. Let $\mathbb{L}$ be a family of forbidden graphs of order v. What is the minimum number of forbidden subgraphs in a graph G^n with

$$e(G^n) = ex(n, L) + k$$

edges if $k = o(n^2)$?

REMARK 1. The above definitions, Theorem A and problems carry over to the case of h-uniform hypergraphs without any difficulty. There, a problem is degenerate if $ex(n, il) = o(n^h)$.

As we mentioned, the problem of *weakly* supersaturated graphs is much more difficult than the case $k = cn^2$. The problem of "*very weakly supersaturated*" graphs seems even more intractable. Thus, for example, we can easily handle the problem of C_4-supersaturated graphs, but see no hope to prove the following

CONJECTURE 1. If $e(G^n) = ex(n, C_4) + 1$, then G^n contains at least $\sqrt{n} + o(\sqrt{n})$ copies of C_4.

(If true, then this conjecture is sharp.)

Below we shall consider only the case, when $\mathbb{L} = \{L\}$, (and use the simpler notation $ex(n, L)$ and $f(n, L, E)$). In extremal graph problems and in many other similar situations we find that the extremal configurations tend to behave either in a very regular pattern, or in a very chaotic way: they have almost random structure. We feel that in case of bipartite L *the graphs G^n with a given number E of edges and having roughly the minimum number of copies of L tend to look like random graphs*. This is the background of the conjectures formulated below, and motivates the results of this paper and of some other ones [5], [6]. To formulate the main conjecture, let us count first the expected number of L's in a random graph G^n, where the edges are chosen independently, at random, and with probability $E / \binom{n}{2}$, so that the expected number of edges be E!).

Let $e = e(L)$, $v = v(L)$. The complete graph K_n contains $a_L \cdot \binom{n}{v}$ copies of L and each occurs in our random graph with probability $\left(E / \binom{n}{2} \right)^e$. Hence the expected number of occurrences of L in G^n is

$$a_L \cdot \binom{n}{v} \cdot \left[E / \binom{n}{2} \right]^e \approx c_L \cdot \frac{E^e}{n^{2e-v}}. \tag{3}$$

CONJECTURE 2. Given a bipartite graph L with $v=v(L)$ and $e=e(L)$ and a $c>0$, then there exists a $c'=c'(c)>0$ such that if

$$E=e(G^n) \geq (1+c) \cdot ex(n,L), \tag{4}$$

then G^n contains at least $c' \cdot \dfrac{E^e}{n^{2e-v}}$ copies of L.

One can easily show that if we use another random graph model, where the graphs G^n of E edges are chosen with the same probability, then (3) still holds. This shows that Conjecture 2 is sharp if true. The main goal of this paper is to prove Conjecture 2 in various cases. One of the main difficulties is that in most degenerate extremal graph problems, even if we have satisfactory upper bounds on $ex(n,L)$, *we are unable to prove good lower bounds.* Therefore we formulate a weaker form of the conjecture.

CONJECTURE 2^*. Assume that for a given forbidden L $ex(n,L)=O(n^{2-\alpha})$. Then there exist an $\tilde\alpha \leq \alpha$ and two constant c and $c'>0$ such that if $E=e(G^n)>cn^{2-\tilde\alpha}$, then G^n contains (for $e=e(L)$ and $v=v(L)$) at least $c' \cdot \dfrac{E^e}{n^{2e-v}}$ copies of L.

We cannot prove the assertion of Conjecture 2^* even under the stronger assumption that $E=e(G^n)>\dfrac{n^2}{\log n}$.

2. Main Results

PROPOSITION 1. If T is a tree, then Conjecture 2 holds with $L=T$.

In case, when L is an even cycle, Conjecture 2^* with $\tilde\alpha=1-\dfrac{1}{k}$ was proved by Simonovits, [12]. More precisely, a theorem of Erdős, (unpublished, see [1] or [2]) asserts that

$$ex(n,C_{2k})=O(n^{1+1/k}). \tag{5}$$

In [12] it is proved that there exist a $c_k>0$ and a $c_k'>0$ such that if

$$E=e(G^n)>c_k n^{1+1/k}, \tag{6}$$

then G^n contains at least $c_k' \cdot \left(\dfrac{E}{n} \right)^k$ copies of C_{2k}. This is sharp by

Remark 2, and proves Conjecture 2^*. (For $k=2,3$, and 5 we know that (5) is sharp. Hence (6) proves Conjecture 2 for C_4, C_6 and C_{10}.)

If we wish to prove general degenerate extremal theorems, or corresponding results on supersaturated extremal graphs, one way to do this is to look for *recursion theorems*. A recursion theorem is a result asserting that

(a) if we have an upper bound on $ex(n,L)$ and construct a new graph L' from L in some given way, then we obtain an upper bound on $ex(n,L')$ in a simple way, in terms of $ex(n,L)$ and some other parameters of L';

(b) or, correspondingly, we can obtain lower bounds on $f(n,L',E)$ in terms of $f(n,L,E)$.

In [10] we have proved the following recursion theorem:

DEFINITION 1. Given a bipartite graph L and a fixed two-coloring of it by blue and red, L_t denotes the graph obtained by fixing a $K_{t,t}$ (which is vertex-disjoint from L and is colored by blue and red) and then joining all the red vertices of $K_{t,t}$ to all the blue vertices of L and all the blue ones of $K_{t,t}$ to all the red ones of L, (see Figure 1).

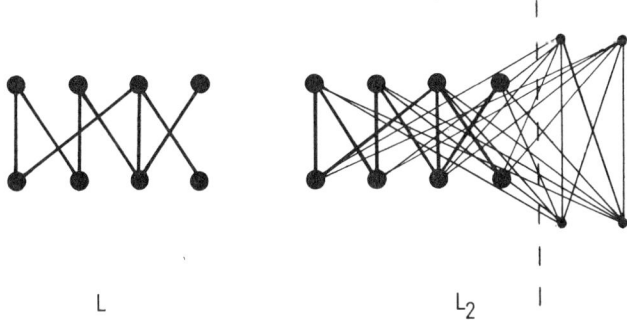

$L \qquad\qquad L_2$

Figure 1.

THEOREM B. [4]. Let L be a bipartite graph with a fixed 2-coloring. Let $ex(n,L) = O(n^{2-\alpha})$ and define $\beta = \beta_t$ by

$$\frac{1}{\beta} - \frac{1}{\alpha} = t. \qquad (7)$$

Then

$$ex(n,L_t) = O(n^{2-\beta}). \tag{8}$$

This theorem was needed primarily to prove the Cube theorem: Turán asked that if L denotes one of the five graphs corresponding to the five regular polyhedra, can one determine $ex(n,L)$? For tetrahedron (K_4) the answer is given by Turán's theorem, [13,14], for octahedron by us, for dodecahedron and icosahedron by Simonovits, see the survey [11], or [10]. The cube problem seems to be the most difficult. We [4] have proved:

THEOREM C. If Q denotes the graph determined by the vertices and edges of the cube and Q^* is the graph obtained by joining two opposite vertices of the cube by an edge, (see Figure 2), then

$$ex(n,Q) < ex(n,Q^*) = O(n^{8/5}). \tag{9}$$

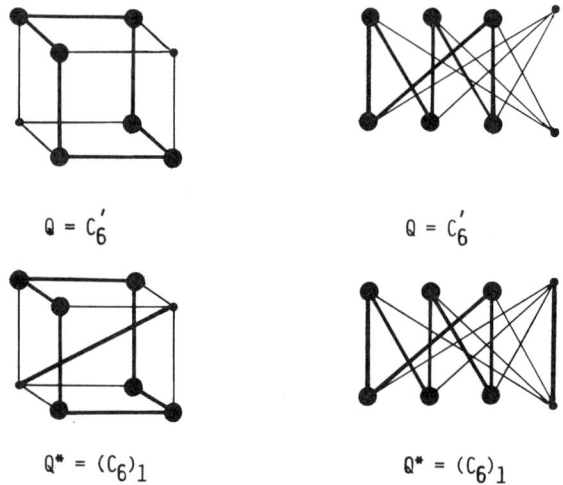

Figure 2.

(One can see in Figure 2 that if $L = C_6$, then $L_1 = Q^*$. Hence the cube theorem immediately follows from $ex(n, C_6) = O(n^{4/3})$ and Theorem B.)

Here we shall prove

THEOREM 1. Let L be a bipartite graph with a fixed 2-coloring and $ex(n,L) = O(n^{2-\alpha})$ for some $\alpha \in (0,1)$. If β is defined by (7), and

Conjecture 2* holds for L, then it also holds for L_t: there exists a constant $C>0$ such that if $e(G^n)=E>Cn^{2-\beta}$, then G^n contains at least $C_{L,t} \cdot \dfrac{E^{e'}}{n^{2e'-v'}}$ copies of L_t, where $e'=e(L_t)$, $v'=v(L_t)$.

THEOREM 2. Let L be a bipartite graph with a fixed 2-coloring in red and blue and L^* be the graph obtained from L by taking two new vertices, x and y and joining x to all the blue vertices of L and y to all the red ones (but not joining x to y). If $ex(n,L)=0(n^{2-\alpha})$ with some $\alpha \in (0,1)$ and β is defined by (7), and Conjecture 2* holds for L, then it also holds for L^* in the sense that there exists a constant $C>0$ such that if $E=e(G^n)>Cn^{2-\beta}$ then G^n contains at least $C_L^* \cdot \dfrac{E^{e'}}{n^{2e'-v'}}$ copies of L^*, where $e'=e(L^*)=e(L)+v(L)$ and $v'=v(L^*)=v(L)+2$.

These theorems immediately imply the following

THEOREM 3. There exist three constant $C_Q, c, c^* >0$ such that if $E=e(G^n)>C_Q \cdot n^{8/5}$, then G^n contains at least cE^{12}/n^{16} copies of Q and at least $c^* \cdot E^{13}/n^{18}$ copies of Q^*.

REMARK. None of the theorems above imply the other, since e' is different in them. Hence Theorem 1 yields (generally) more copies of a smaller sample graph.

Using Proposition 1 and the above theorems we obtain

THEOREM 4. If L is a bipartite graph and x,y are two vertices of L such that $L-\{x,y\}$ is a tree, then $(ex(n,L)=O(n^{3/2})$, and there exist two constants C and c^* such that if $E=e(G^n)>Cn^{3/2}$, then G^n contains at least $c^* \cdot \dfrac{E^e}{n^{2e-v}}$ copies of L, where $e=e(L)$ and $v=v(L)$.

Theorem 4 has many applications. Thus e.g. if Q' is the graph obtained from the cube graph Q by deleting *one* edge, then Q' can be obtained from a path P^6 by the operation described in Theorem 2. Thus Theorem 2 can be applied with $\alpha=1$, $\beta=\dfrac{1}{2}$, yielding that if $e(G^n)>c_{Q'} \cdot n^{3/2}$, then G^n contains at least $c' \cdot \dfrac{E^7}{n^6}$ copies of Q'. There are many other cases, where Theorem 4 immediately yields the sharp result.

3. Proofs

The main tool of the proof of the Cube Theorem (Theorem C) is counting C_4's in G^n. Here we need a result sharper than that of [4]. We shall count C_4's in bipartite graphs where the sizes of the color-classes are m and $h=n-m$. Since our method works for $K_{p,q}$ as well, first we shall prove

LEMMA 1. Given p and q, there exist two constant c and $c'>0$ such that if $G^n \subseteq K_{m,h}$, and $E=e(G^n)>c \cdot m \cdot h^{1-1/p}$, then G^n contains at least

$$c' \cdot \frac{E^{pq}}{m^{(q-1)p} \cdot h^{(p-1)q}} \tag{10}$$

copies of $K_{p,q}$.

(Lemma 1 implies that Conjecture 2* holds for $K_{p,q}$ with $\tilde{\alpha} = \frac{1}{p}$, since for any L it is sufficient to prove Conjecture 2* for the case of bipartite G^n.)

PROOF. We shall use a convexity argument for which we extend $\binom{n}{k}$ to all the reals by

$$\binom{x}{k} = \begin{cases} \frac{x(x-1)\ldots(x-k+1)}{k!} & \text{if } x \geq k-1 \\ 0 & \text{if } x < k-1 \end{cases}$$

One can easily show that $\binom{x}{k}$ is convex in $(-\infty, +\infty)$, and in $(0, +\infty)$

$$\frac{x^k}{k!} \geq \binom{x}{k} = \frac{x^k}{k!} + O(x^{k-1}) \geq \frac{1}{2} \cdot \frac{x^k}{k!} + O(1). \tag{11}$$

Let M and H be the color-classes of $K_{m,h} \supseteq G^n$. We shall call $(x, \{y_1, \ldots, y_q\})$ a *cap* if each y_j is joined to $x \in M$. Let us count them. If S is their number and $d_1, \ldots, d_m$ are the degrees in M, then

$$S = \sum_{i \leq m} \binom{d_i}{q} \geq m \cdot \binom{2E/m}{q} \geq c_1 \cdot \frac{E^q}{m^{q-1}} - O(m) \geq c_2 \cdot \frac{E^q}{m^{q-1}}, \tag{12}$$

Here we have used Jensen's Inequality and that $E/m \to \infty$. (The case $h = O(1)$ is trivial!)

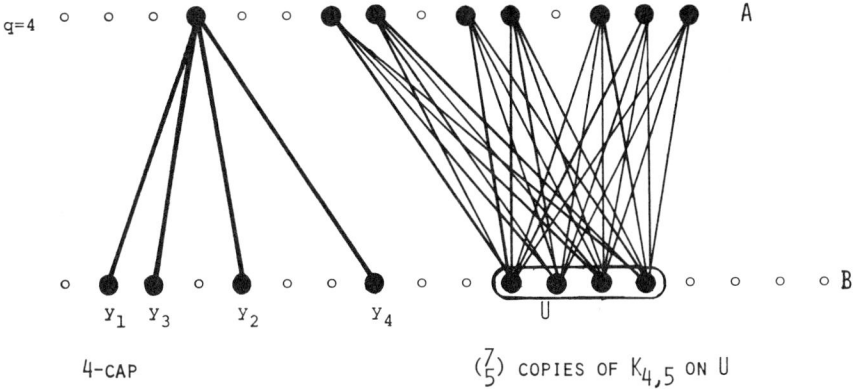

Figure 3

Let D_U denote the number of caps on a fixed $U=\{y_1,\ldots,y_q\}$. Then $\sum D_U = S$ and the number of $K_{p,q} \subseteq G^n$ is exactly

$$\sum \binom{D_U}{p} \geq \binom{h}{q} \left(\frac{\sum D_U}{\binom{h}{q}} \atop p \right) \geq c_3 \cdot \frac{S^p}{h^{qp-p}} - O(h^q)$$

$$\geq c_4 \cdot \frac{E^{pq}}{m^{(q-1)p} \cdot h^{(p-1)q}}.$$

(Here again we replaced $\binom{h}{q}$ by $c_5 \cdot h^q$, which is permitted unless $h \leq q-1$. However, $h = O(1)$ have been excluded.)

LEMMA 2. Let $G(M,H)$ be a bipartite graph with the colour-classes M and H, where $|M| = m \geq |H| = h$. Assume that L is a bipartite graph which satisfies Conjecture 2^* with a given $\tilde{\alpha} > 0$. If $E = e(G(M,H)) \geq c_6 \cdot m \cdot h^{1-\tilde{\alpha}}$ then $G(M,H)$ contains at least $\tilde{c} \cdot \frac{E^e}{h^{e-v+1} \cdot m^{e-1}}$ copies of L, where $e = e(L)$, $v = v(L)$.

PROOF. We partition M into the subsets $U_1, \ldots, U_{\lceil m/h \rceil}$, each of which is of size $\approx h$. Let F_i denote the number of edges joining U_i and H. On average, $G(U_i, H)$ has $\geq c_6 \cdot h^{2-\tilde{\alpha}}$ edges, hence at least half of the edges belong to $G(U_i, H)$'s with $F_i \geq c_6 \cdot \frac{1}{2} \cdot h^{2-\tilde{\alpha}}$: we may apply the assumption to them, they contain at least $c_7 \cdot (F_i)^e / h^{2e-v}$ copies of L. Thus we get at least

$$c_7 \cdot \sum \frac{F_i^e}{h^{2e-v}} = c_8 \cdot \left(\frac{Eh}{m}\right)^e \cdot \frac{m}{h} \cdot \frac{1}{h^{2e-v}} = c_8 \cdot \frac{E^e}{h^{e-v+1} \cdot m^{e-1}}$$

copies of L.

REMARK. This estimate is asymmetric. A symmetric weakening is that, under the assumptions of the lemma, $G(M, H)$ contains at least

$$c_8 \cdot \frac{E^e}{(hm)^{e-v+1} \cdot (h+m)^{v-2}} \tag{13}$$

copies of L.

PROOF OF THEOREM 1. (A) The operation $\mathbf{A}_t : L \to L_t$ is such that

$$\mathbf{A}_1(\mathbf{A}_{t-1}(L)) = \mathbf{A}_t(L).$$

A similar identity holds for the exponents: denote the β obtained from (7) by $\beta_t(\alpha)$. Then

$$\beta_1(\beta_{t-1}(\alpha)) = \beta_t(\alpha).$$

Using this, the reader can easily check that if Theorem 1 is proved for $t=1$, then a trivial induction on t proves Theorem 1 for every t. Hence *it is enough to prove Theorem 1 for $t=1$*

(B) Take a graph G^n with E edges and partition its vertex set into two parts M and H so that the bipartite graph $G(M, H)$ — defined by the edges of G^n joining M to H — has the maximum number of edges for all possible partitions. Let $d(x)$ denote the degree in G^n, $d'(x)$ the degree in $G(M, H)$. Then one can easily see that $d'(x) \geq \frac{1}{2} d(x)$ for every vertex x.

In the argument below, the vertices of *high valency are disturbing*, therefore we apply below a *two-step regularization procedure* to $G(M, H)$. We define the constants $r_i = 2^{i-2}/i^2$ for $i = 4, 5, \ldots$ and $r_3 = 0$. Let $|M| = m$ and $|H| = h$ and

$$A_i = \{x \in M: r_{i-1} \cdot \frac{E}{m} \leq d'(x) < r_i \cdot \frac{E}{m}\}, \quad i=4,5,\ldots \tag{14}$$

Clearly, for at least one $i \geq 5$ $|A_i| \geq \frac{m}{2^i}$. We fix an $A^* \subseteq A_i$ with $|A^*| = \lceil \frac{m}{2^i} \rceil$. The bipartite graph $G(A^*, H)$ has at least $\frac{E}{4i^2}$ edges. (This means that the degrees in the first class of $G(A^*, H)$ are already roughly the same and still, $G(A^*, H)$ has almost (?) the original number of edges. Next, we partition H into the subsets

$$B_j = \{y \in H: r_{j-1} \cdot \frac{E}{4i^2 h} \geq d''(y) < r_j \cdot \frac{E}{4i^2 h}\} \quad j=4,5,\ldots \tag{15}$$

where $d''(y)$ denotes the degree of y in $G(A^*, H)$. As in the previous case, we can fix a $j \geq 5$ such that $|B_j| \geq \frac{h}{2^j}$, then a $B^* \subseteq B_j$ of $\lceil \frac{h}{2^j} \rceil$ vertices. Now the graph $G(A^*, B^*)$ has $E^* \geq \frac{E}{16 i^2 j^2}$ edges. (This means that the degrees in B^* are now roughly the same and *at most* $r_j \cdot \frac{E}{4i^2 h}$. At the same time, we have lost the "almost regularity" in A^*. However, we still have that the degrees in A^* are *at most* $r_i \cdot \frac{E}{m}$.)

(C) Now we shall count the number of C_4's in $G^* = G(A^*, B^*)$. Let $N(x,y)$ denote the number of paths P_4 joining an $x \in A^*$ to a $y \in B^*$. If (x,y) is an edge, then $N(x,y)$ is just the number of C_4's containing (x,y). Hence, by Lemma 1, applied with $p=q=2$, we get

$$\sum N(x,y) \geq c_1 \cdot \frac{(E^*)^4}{(a^* \cdot b^*)^2} \tag{16}$$

C_4's, if $|A^*| = a^*$ and $|B^*| = b^*$. Of course, to get (16) we need that if e.g. $a^* \geq b^*$, then $e(G^*) \geq c_2 \cdot a^* \cdot \sqrt{b^*}$. Observe that $\beta \leq \frac{1}{2}$ in all our applications, hence $E \geq c_3 n^{3/2}$ can be assumed with a "very large" c_3, which (since $a^* = \lceil \frac{m}{2^i} \rceil$, $b^* = \lceil \frac{h}{2^j} \rceil$ and $E^* \geq \frac{E}{16 i^2 j^2}$) implies that $e(G^*) \geq c_2 \cdot a^* \cdot \sqrt{b^*}$. (The reader is reminded at this point that the constants c_i are not necessarily the same in different proofs.)

Let now $d^*(x)$ denote the degree of x in G^*. The edge (x,y) where $x \in A^*, y \in B^*$, will be called of *type A* if $d^*(x) \geq d^*(y)$. Otherwise, it will be called of *type B*. By symmetry, we may assume that (16) holds also for the edges of type A. Assume first (which is not always true) that we can apply Lemma 2 to *each* of these (at most E^*) edges. Each (x,y) defines a subgraph $G_{x,y} \subseteq G^*$ spanned by all the neighbours of x in G^* but y, and by all the neighbours of y but x. (See

Figure 4.) If $G_{x,y}$ contains an L, then G^* contains a corresponding L_1. Thus we obtain at least

$$c_4 \cdot \frac{N(x,y)^e}{d^*(x)^{e-1} \cdot d^*(y)^{e-v+1}} \geq c_5 \cdot \frac{N(x,y)^e}{\left(2^i \cdot \frac{E}{m}\right)^{e-1} \cdot \left(2^j \cdot \frac{E}{h}\right)^{e-v+1}} \quad (17)$$

copies of L_1 containing (x,y). (Here, we used only the weaker bounds $d^*(x) \leq 2^i \cdot \frac{E}{m}$ and $d^*(y) \leq 2^j \cdot \frac{E}{h}$.)

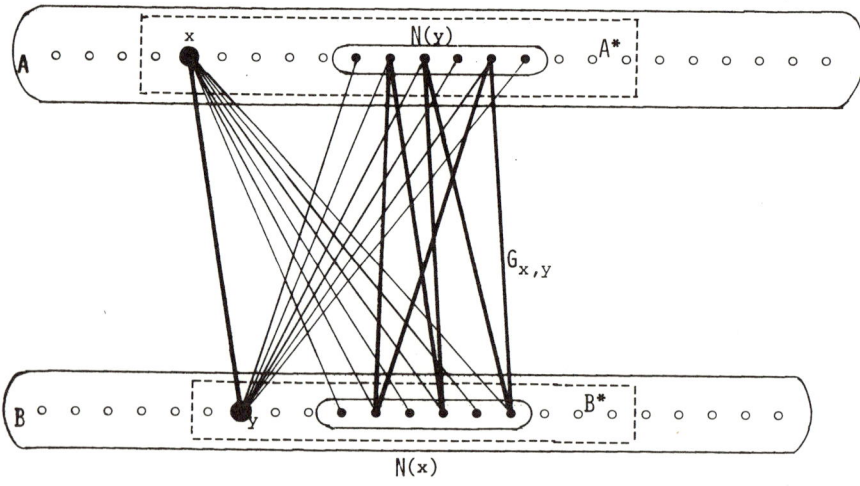

Figure 4

Since, by Jensen's Inequality,

$$\sum N(x,y)^e \geq c_6 \cdot \left[\frac{(E^*)^3}{(a^* b^*)^2}\right]^e \cdot E^* = c_6 \frac{(E^*)^{3e+1}}{(a^* b^*)^{2e}}, \quad (18)$$

and since $a^* = \lceil \frac{m}{2^i} \rceil$, $b^* = \lceil \frac{h}{2^j} \rceil$, we obtain — by adding up the contribution (17) for each edge of type A — at least

$$\sum c_4 \cdot \frac{N(x,y)^e}{d^*(x)^{e-1} \cdot d^*(y)^{e-v+1}} \tag{19}$$

$$\geq c_7 \cdot \frac{(E^*)^{3e+1}}{m^{2e} \cdot h^{2e}} \cdot \frac{1}{\left(\frac{E}{m}\right)^{e-1} \cdot \left(\frac{E}{h}\right)^{e-v+1}} \cdot \frac{2^{ie} \cdot 2^{je}}{2^{i(e-1)} \cdot 2^{j(e-v+1)}}$$

copies of $L_1 \subsetneq G^*$. Since $E^* \geq \frac{E}{16 i^2 j^2}$, (19) yields at least

$$c_8 \cdot \frac{E^{e+v+1}}{m^{e+1} \cdot h^{e+v-1}} \cdot \frac{2^{i+j}}{P(i,j)} \tag{20}$$

copies of L_1, where $P(i,j)$ is some polynomial in i and j. Clearly, $\frac{2^{i+j}}{P(i,j)} \to \infty$ if either $i \to \infty$ or $j \to \infty$. Hence we may assume that i and j are bounded, which means that the last factor may be ignored. Since $e' = e+v+1$ and $v' = v+2$, and since $m^{e+1} \cdot h^{e-v+1}$ attains its maximum (for $m+h=n$) if $m = c_8 n$, (20) yields at least

$$c_9 \cdot \frac{E^{e'}}{n^{2e'-v'}}$$

copies of L_1, each obtained $O(1)$ times. This *would* complete our proof. Unfortunately, we do not know if Lemma 2 is applicable to each $G_{x,y}$.

(D) One can easily check that (16) guarantees that *"on average"* Lemma 2 is applicable to $G_{x,y}$. More precisely, the average of $N(x,y)$ is at least $c_1 \cdot \frac{(E^*)^3}{(a^* \cdot b^*)^2}$. We may fix the constant C of Theorem 1 so large that if $E = e(G^*) > C n^{2-\beta}$, then either $d^*(x) \geq d^*(y)$ and

$$\frac{(E^*)^3}{(a^* \cdot b^*)^2} \geq C^* \cdot d^*(x) \cdot d^*(y)^{1-\alpha}, \tag{21}$$

or $d^*(x) < d^*(y)$ and

$$\frac{(E^*)}{(a^* \cdot b^*)^2} \geq C^* \cdot d^*(x)^{1-\alpha} \cdot d^*(y). \tag{21*}$$

Indeed, $E^* \geq \frac{E}{16 i^2 j^2}$ and $d^*(x) \leq 2^i \cdot \frac{E}{m}$, $d^*(y) \leq 2^i \cdot \frac{E}{h}$, from which a short calculation (using also (7)!) yields (21). Now we take only those edges of G^* which satisfy

$$N(x,y) \geq \frac{1}{2E^*} \sum N(x,y).$$

Let us call them "good" edges. Clearly (16) holds for these edges too, if c_1 is replaced by $\frac{1}{2}c_1$. Hence we may assume that if we take only the "good" edges of type A, (16) still holds, with $\frac{1}{4}c_1$ instead of c_1. Let E^{**} denote the number of these edges. Trivially, (18) and (19) remain valid if we use (as above) E^* — and not E^{**} — in our formulas, except that the summation is taken only for the E^{**} good edges of type A. Indeed,

$$\sum N(x,y)^e \geq c_6^* \left(\frac{(E^*)^4}{(a^* \cdot b^*)^2} \cdot \frac{1}{E^{**}} \right)^e \cdot E^{**} \geq c_6 \cdot \frac{(E^*)^{3e+1}}{(a^*b^*)^{2e}}. \quad (18^*)$$

In (19) we used only (18), thus it remains valid and the last part of the proof carries over without any change. Thus Theorem 1 is proved.

PROOF OF THEOREM 2. This proof is very similar to that of Theorem 1. We start exactly as above: choose a bipartite $G(M,H)$ in G^n and then apply the "two-step regularization", that is, find $G^* = G(A^*, B^*)$, as above. $N(x,y)$ again denotes the number of paths P_4 joining an $x \in A^*$ to a $y \in B^*$. However, here we shall count the P_4's in G^* instead of counting C_4's and will use the summation for all the pairs, not only for the joined ones, to establish a formula corresponding to (16). First we assert that

$$\sum N(x,y) \geq c_1 \cdot \frac{(E^*)^3}{a^* \cdot b^*}. \quad (16')$$

Indeed, $\sum N(x,y)$ counts the number of P_4's in G^*. If $E^* \geq 4a^*$, then the number of P_3's ($=K_{1,2}$'s) with the middle vertex in A^* is at least $c_2 \cdot \frac{(E^*)^2}{a^*}$. (Here we use Lemma 1 in a sharper form: we use also that we can guarantee that the *first class of $K_{p,q}$ is in the second class of $K_{m,h}$*; as a matter of fact, we have proved this! Of course, counting the $K_{1,q}$'s in G^* is just "counting the caps" in the proof of lemma 1.)

Anyway, we have at least $c_2 \cdot \frac{(E^*)^2}{a^*}$ P_3's in G^* with the middle vertex in A^*. The degrees in B^* are between $r_{j-1} \cdot \frac{E}{i^2 h}$ and $r_j \cdot \frac{E}{i^2 h}$, and therefore they are bounded from below by $c_3 \cdot \frac{E}{b^*}$ (= the average degree in G^* in B^*). Hence we may extend each P_3 in at least $2c_3 \cdot \frac{E^*}{b^*}$

ways. Thus we have at least $c_4 \cdot \frac{(E^*)^3}{a^* b^*}$ 4-walks in G^*, and least $c_5 \cdot \frac{(E^*)^3}{a^* b^*}$ of them are paths P_4 if E^*/b^* is sufficiently large (which can be assumed here).

From here on, the proof is roughly the same as in the previous case. Instead of (18) we get (summing for all the $a^* b^*$ pairs (x,y))

$$\sum N(x,y)^e \geq c_6 \cdot \left(\frac{(E^*)^3}{(a^* \cdot b^*)^2}\right)^e \cdot a^* \cdot b^* = c_6 \cdot \frac{(E^*)^{3e}}{(a^* b^*)^{2e-1}}. \tag{18'}$$

Thus the number of L^*'s in G^* is at least

$$\sum c_7 \cdot \frac{N(x,y)^e}{d^*(x)^{e-1} \cdot d^*(y)^{e-v+1}} \tag{19'}$$

$$\geq c_8 \cdot \frac{(E^*)^{3e}}{(mh)^{2e-1}} \cdot \frac{1}{\left(\frac{E}{m}\right)^{e-1} \cdot \left(\frac{E}{h}\right)^{e-v+1}} \cdot \frac{2^{ie} \cdot 2^{je}}{2^{i(e-1)} \cdot 2^{j(e-v+1)}}.$$

Since $E^* \geq \frac{E}{16 i^2 j^2}$, (19') gives at least

$$c_9 \cdot \frac{E^{e+v}}{m^e \cdot h^{e+v-2}} \cdot \frac{2^{i+j}}{P(i,j)} \tag{20'}$$

copies of L^* in G^*, where $P(i,j)$ is again a polynomial. Again, the last term in (20') tends to ∞, hence it can be deleted: G^* contains at least

$$c_{10} \cdot \frac{E^{e+v}}{m^e h^{e+v-2}} = c_{10} \cdot \frac{E^{e'}}{m^{e'-v'+2} \cdot h^{e'-2}}$$

copies of L^* (since $v' = v + 2$, $e' = e + v$). Here the minimum is achieved (under the condition that $m + h = n$) if $m = c_{11} \cdot n$, $h = c_{12} \cdot n$, with $c_{11}, c_{12} > 0$. This would complete our proof.

As in the proof of Theorem 1, we still have a technical problem: "Can we apply Lemma 2 to the graphs $G_{x,y}$? Do the graphs $G_{x,y}$ have sufficiently many edges?" However, from (16') we get that $G_{x,y}$ has, on average, $c_1 \cdot \frac{(E^*)^3}{(a^* \cdot b^*)^2}$ edges. This is exactly the same as in the previous proof. Therefore the last part of that proof, namely (D), can be repeated without any significant changes. (The insignificant changes are that the summation is taken for $a^* b^*$ terms instead of E^* terms.) This completes the proof.

REMARK. The fact that the average number of edges in $G_{x,y}$ is the same in both proofs is surprising at first sight but becomes evident (?) if we remember that most parts of our proofs aim at proving that our graph behaves (from certain point of view) as a random graph. In a random graph $e(G_{x,y})$ is "independent" of whether (x,y) is an edge of G^* or not.

We can prove Conjecture 2* in several other special cases, but shall not discuss them here.

The proof of Theorem 2 (or of Theorem 1) yields a proof of the Cube theorem as well. However, this is not radically different from the original one. Perhaps one interesting difference is that here we used a slightly weaker regularization method.

On the other hand, there are some instances where, in proving Conjecture 2*, we obtain new (and often simpler) proofs of existing extremal graph theorems.

References

[1] J. A. Bondy and M. Simonovits: Cycles of even length in graphs, Journal of Combin. Theory, 16 (1974) 97-105.

[2] B. Bollobás: Extremal Graph Theory, London Math. Soc. Monographs, N11, Academic Pres,, 1978.

[3] P. Erdős and M. Simonovits: A limit theorem in graph theory, Studia Sci. Math. Hungar. 1 (1966) 51-57.

[4] P. Erdős and M. Simonovits: Some extremal problems in graph theory, Coll. Math. Soc. J. Bolyai 4 (1969) 377-390.

[5] P. Erdős and M. Simonovits: Supersaturated graphs and hypergraphs, Combinatorica, 2(3) (1982) 275-288.

[6] P. Erdős and M. Simonovits: Compactness results in extremal graph theory, Combinatorica, 3 (2) (1983).

[7] R. J. Faudree and M. Simonovits: On a class of degenerate extremal problems, Combinatorica, 3 (1) (1983), 97-107.

[8] T. Kővári, V. T. Sós and P. Turán: On a problem of Zarankiewicz, Coll. Math., 3 (1954), 50-57.

[9] M. Simonovits: Extremal graph problems with symmetrical extremal graphs, additional chromatic conditions, Discrete Math. 7 (1974), 279-319.

[10] M. Simonovits: The extremal problem of the icosahedron, J. Combin. Theory, B 17 (1) (1974), 69-79.

[11] M. Simonovits: Extremal graph theory, a survey, to appear in " Selected Topics in Graph Theory, II", Ed. Beineke and Wilson, Academic Press (1983), 161-200.

[12] M. Simonovits: Cycle-supersaturated graphs, to be published

[13] P. Turán: On the theory of graphs, Coll. Math. 3 (1954), 19-30.

[14] P. Turán: On an extremal problem in graph theory, (in Hungarian) Mat. Fiz. Lapok 48 (1941) 436-452.

On the Complexity of Finding Sets of Edges in Graphs and Digraphs

T.I. Fenner

A.M. Frieze

ABSTRACT

Suppose we are given a graph (V,E) and a function $c:E \to Z$ where $c(e)$ is said to be the colour of edge e of E. A subset $S \subseteq E$ is said to be polychromatic if $|c(S)| = |S|$. In this paper we study the complexity of recognizing whether such an edge coloured graph has certain types of polychromatic subsets, such as spanning trees, paths, cutsets, matchings and Hamiltonian cycles.

1. Introduction

Erdős, Simonovits and Sós [6] studied the following problem: given a graph H, what is the maximum number of colours m that can be used to colour the edges of a complete graph K_n on n vertices so that no subgraph isomorphic to H has all its edges coloured differently? They considered, in particular, the cases when H is a complete graph, a path or a cycle. Hahn [9] [10] considered a similar problem when H is a star. In his doctoral thesis, Bate [1] [2] investigates, *inter alia*, related problems on the existence of polychromatic paths and cycles. (See also Galvin [7].)

In this paper we study the problem of the complexity of determining whether such a subgraph exists.

An edge colouring of a graph $G = (V,E)$ is a function $c:E \to Z$ for some set Z. A subset $X \subseteq E$ is said to be *polychromatic* if $|c(S)| = |S|$, i.e. each edge of S is coloured differently. A similar definition is assumed for digraphs.

Possible applications of this model could be to problems where each colour represents an indivisible resource which can only be used once.

In the paper we study the problem of determining whether an edge coloured graph or digraph has various types of polychromatic subset, including spanning trees, spanning arborescences, $s-t$ paths, cycles, cutsets, matchings and Hamiltonian paths and cycles. Most cases turn out to be NP-Complete and in particular it is interesting to note that the case of spanning arborescences, which is a special case of a matroid-greedoid intersection problem, turns out to be NP-Complete, thus showing that this problem is unlikely to be polynomially solvable.

2. Polychromatic Spanning Trees and Arborescences

Spanning Trees

We first consider the problem of determining whether or not an edge coloured graph G contains a polychromatic spanning tree. Fortunately this is a matroid intersection problem: indeed a polychromatic spanning tree is a set of $n-1$ edges independent in both

(1) the cycle matroid of G, and

(2) the partition matroid defined by polychromatic sets of edges.

Thus the existence of a polychromatic spanning tree can be tested in polynomial time using a matroid intersection algorithm, e.g. Lawler [13] pp. 313-314. In fact, this algorithm could easily be implemented to run in time $O(|V||E|)$.

Furthermore, it follows from Edmonds' matroid intersection duality theorem [5] that G contains a polychromatic spanning tree if and only if

$$(1.1)\ r_1(S) + r_2(E-S) \geq n-1 \qquad \text{for all } S \subseteq E$$

where

(a) $n - r_1(S) =$ *the number of components in the graph* (V,S)

(b) $r_2(T) =$ *the number of distinct colours used in* T.

We can deduce from (1.1) the following simple but interesting result:

THEOREM 1.1. *Let G be a complete graph on n vertices and let the edges of G be coloured with $n-1$ colours such that each colour is used at least once and no more than $\lceil n/2 \rceil + 1$ times. Then G contains a polychromatic spanning tree.*

PROOF. If G does not have a polychromatic spanning tree then (1.1) fails for some S. Suppose (V,S) has p components containing $n_1, n_2, \ldots, n_p$ vertices respectively and that $r_2(E-S) = t$. We can

quickly eliminate the following values for p:
(i) $p = 1$: here $r_1(S) = n-1$;
(ii) $p = 2$: here $|E-S| \geq n-1$ implies $t \geq 1$;
(iii) $p = n$: here $S = \emptyset$ implies $t = n - 1$;
(iv) $p = n-1$: here $|S| = 1$ implies $r_1(S) = 1$ and $t \geq n-2$.

We can thus assume that $3 \leq p \leq n-2$ and $n \geq 5$. Since (1.1) fails we have $t \leq p-2$ and so some colour occurs at least $|E-S|/(p-2)$ times. Now

$$|E-S| \geq \binom{n}{2} - \sum_{i=1}^{p}\binom{n_i}{2}$$

$$\geq \binom{n}{2} - \binom{n-p+1}{2} = \binom{p-1}{2} + (p-1)(n-p+1).$$

Thus some colour is used at least $m = \lceil (p-1)/2 + (p-1)(n-p+1)/(p-2) \rceil$ times. It is easy to show, given our range of values for p, that $m \geq \lceil n/2 \rceil + 2$ for $n \geq 4$. The result follows.

Spanning Arborescences

We consider next the problem (PSA) of determining whether an arc coloured digraph D contains a polychromatic spanning arborescence rooted at a given vertex r. Here an arborescence is a spanning tree in which each arc is directed away from the root and so each vertex other than r has indegree 1. The uncoloured version is solvable in time $O(|V| + |E|)$, and it appears plausible that PSA could be polynomially solvable as it involves the intersection of a greedoid, i.e. the arborescence (see Korte and Lovász [12]), and the partition matroid defined by the colouring. We show, however, that PSA is NP-Complete:

THEOREM 1.2. PSA is NP-Complete.

PROOF. We prove this by showing that SATISFIABILITY $\propto$ PSA. Suppose we are given a set $X = \{x_1, x_2, \ldots, x_n\}$ of boolean variables and a set $C = \{C_1, C_2, \ldots, C_m\}$ of clauses over X. We define an instance of PSA as follows: let $L = \{x_1, \bar{x}_1, \ldots, x_n, \bar{x}_n\}$ be the set of literals, where we can assume each literal appears in at least one clause, otherwise the complementary literal can be taken as true. We start with a bipartite digraph D_0 with vertices $L \cup C$ and arcs of the form (σ, C_j) whenever literal σ occurs in clause C_j. Each arc of D_0 is differently coloured. We now add a new vertex r and join it by an arc to each member of L (see Figure 1(a)). Using n new colours

$a_1, \ldots, a_n$, we colour these arcs, so that (r, x_i) and $(r, \bar{x}_i)$ are both coloured a_i for $i = 1, 2, \ldots, n$.

We next add a new vertex s and an arc (r, s) using a new colour. We then join s to each literal σ by a path P_σ (from s to σ) where the sets of interior vertices of the P_σ are disjoint. Suppose that literal σ occurs in k_σ clauses; then P_σ contains k_σ arcs. The arcs of P_σ are coloured with exactly the same k_σ colours used to colour those arcs of D_0 with initial vertex σ. Finally, each σ is joined to each interior vertex of P_σ by an arc, using a new colour each time. This completes the description of the arc coloured digraph D (see Figure 1(b)).

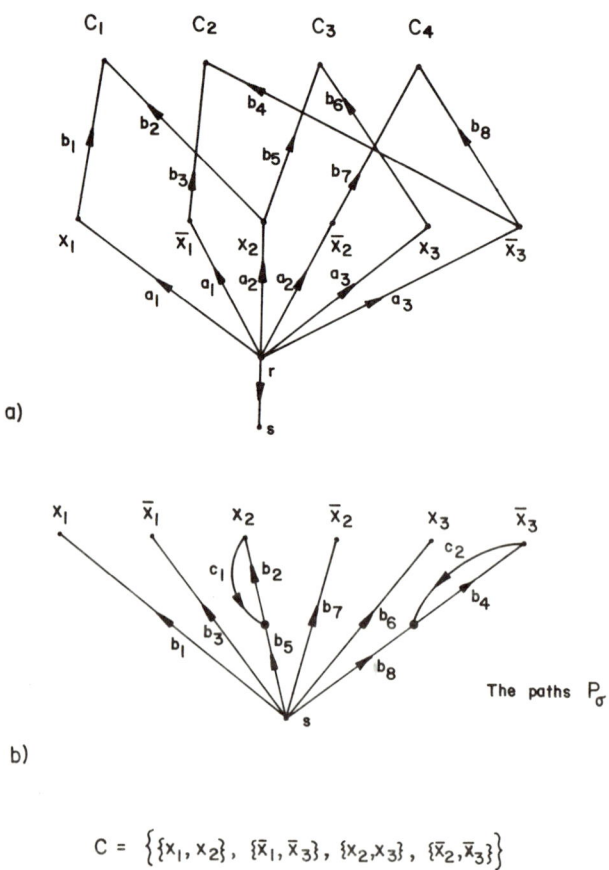

$C = \{\{x_1, x_2\}, \{\bar{x}_1, \bar{x}_3\}, \{x_2, x_3\}, \{\bar{x}_2, \bar{x}_3\}\}$

Figure 1

Suppose first that C is satisfied by some assignment of truth values to X. Then there exists a polychromatic spanning arborescence rooted at r with the following set of arcs:

(1) the arc (r,σ) for each true literal σ;

(2) for each $C_i, i = 1, 2, \ldots, m$, an arc (σ, C_i) where σ is some true literal occurring in C_i (the existence of some such σ is guaranteed since C is satisfied);

(3) the arc (r,s);

(4) the path P_σ for each false literal σ;

(5) all arcs of the form (σ, v) where σ is a true literal and v is an interior vertex of P_σ.

It is easy to see that (1) to (5) define a polychromatic spanning arborescence.

Now suppose, conversely, that D contains a polychromatic spanning arborescence rooted at r with arc set A. Let the true literals be those σ for which $(r,\sigma) \in A$. Note that, since (r,x_j) and $(r,\bar{x}_j)$ have the same colour, this defines a proper assignment of truth values. Any variable x_j not assigned a truth value in this way will be assigned the value *true*. We now show that this assignment of truth values satisfies C.

We note first that for each literal σ there are precisely 2 arcs in D that enter σ, i.e. (r,σ) and the terminal arc u_σ of P_σ. Now it can easily be shown, by tracing P_σ in reverse, that to avoid cycles we must have either $u_\sigma \notin A$ or $P_\sigma \subseteq A$; but if $P_\sigma \subseteq A$ then $(\sigma, C_i) \notin A$ because P_σ uses the colours of all such arcs. Thus $(\sigma, C_i) \in A$ implies that $(r,\sigma) \in A$ and hence that σ is true in our assignment. Since A spans the vertices of D this assignment satisfies C. □

We note that the problem is still hard if we do not specify the root since in the above proof r is the root of any spanning arborescence.

We further note that PSA remains NP-Complete when we restrict attention to those cases in which the number of colours used is $n-1$, where n is the number of vertices. This is easily seen by "splitting" r as in Figure 2.

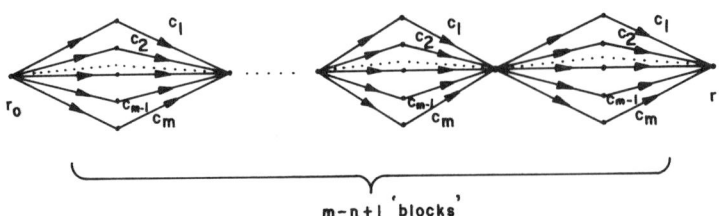

(We assume the original colours of D were $c_1, c_2, \ldots, c_m$, where $m > n - 1$. Arcs without a specified colour are given distinct new colours. No. of colours = no. of vertices - 1.)

Figure 2

The problem remains NP-Complete if we further restrict attention to complete digraphs; we simply add a new vertex O and a new colour, red say, and complete the digraph with red arcs.

However, we observe that if the given digraph has no directed cycles then the problem is solvable in polynomial time. Indeed, suppose that the vertices of D are numbered $1, 2, \ldots, n$ such that $i < j$ if (i,j) is an arc of D. We can obtain an arborescence of D by selecting, for each $j > 1$, an arbitrary arc entering j. Thus PSA reduces to a bipartite matching problem in which the bipartite graph G has one vertex for each $j = 2, 3, \ldots, n$ and another vertex for each colour. The arcs are of the form (j, c) whenever c is the colour of some arc entering j in the digraph D. Each matching of G of cardinality $n-1$ corresponds to a polychromatic spanning arborescence of D and vice versa.

We finally note that, unless NP = Co-NP, there cannot be any *good* characterization of the set of coloured digraphs which contain a polychromatic spanning arborescence. This is in clear contradistinction to the undirected case where we have the characterization given by (1.1). These remarks will also apply to other classes of polychromatic subgraphs for which the corresponding recognition problems are NP-Complete.

3. Polychromatic Paths, Cycles and Cutsets

Paths and Cycles

If an edge coloured graph has a polychromatic spanning tree then every pair of vertices is connected by a polychromatic path. If, however, we specify a pair of vertices s and t of a graph and ask whether it contains a polychromatic path from s to t then, surprisingly perhaps, this problem is NP-Complete. We prove this by showing that 3-DIMENSIONAL MATCHING (3DM) is polynomially transformable to this problem which we call POLYCHROMATIC PATH (PP).

An instance of 3DM is described by
(a) disjoint sets X,Y,Z where $|X| = |Y| = |Z| = m$;
(b) a subset $T \subseteq X \times Y \times Z$.

The problem is to determine whether or not there exists $M \subseteq T$ such that (a) $|M| = m$, (b) each element of $X \cup Y \cup Z$ occurs in exactly one member of M. A set M satisfying (a) and (b) is called a 3-dimensional matching.

THEOREM 2.1. PP is NP-Complete.

PROOF. Given an instance of 3DM we construct the edge coloured graph G indicated in Figure 3.

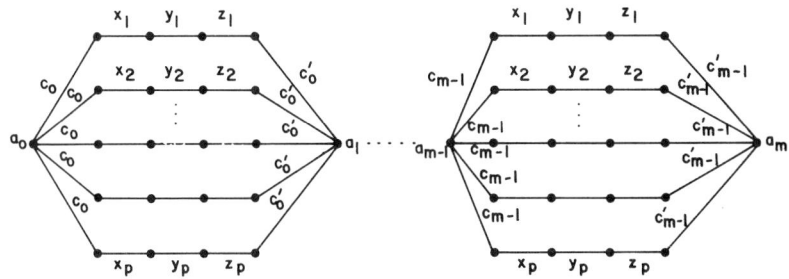

Figure 3

This can be described informally as follows. Starting with a path $a_0, a_1, \ldots, a_m$, we duplicate each edge (a_i, a_{i+1}) p times, where $p = |T|$. Each of the p edges joining a_i to a_{i+1}, $i = 0, 1, \ldots, m-1$ is then replaced by a distinct path of length 5. This gives us a graph with

$m + 1 + 4mp$ vertices and $5mp$ edges as in Figure 3. Let
$$T = \{(x_i, y_i, z_i) \mid i = 1, 2, \ldots, p\}.$$
For $i = 0, 1, \ldots, m-1$ we order the p paths joining a_i to a_{i+1}. The 5 edges of the kth path joining a_i to a_{i+1} are coloured c_i, x_k, y_k, z_k, c_i', respectively, where c_i and c_i', $i = 0, 1, \ldots, m-1$ are $2m$ new colours.

It is now easy to see that T contains a 3-dimensional matching if and only if G contains a polychromatic path from a_0 to a_m. □

We note that the graph G constructed in Theorem 2.1 is bipartite, planar and edge series-parallel [14].

By adding a large monochromatic clique to the graph we can make the ratio of vertices to colours as large as we like; alternatively, by adding a large polychromatic clique we can make this ratio as small as we like.

By applying the transformation indicated in Figure 4 to each of the sets of edges coloured $c_0, c_0', c_1, c_1', \ldots, c_{m-1}'$, we can obtain a graph in which no vertex has degree exceeding 3.

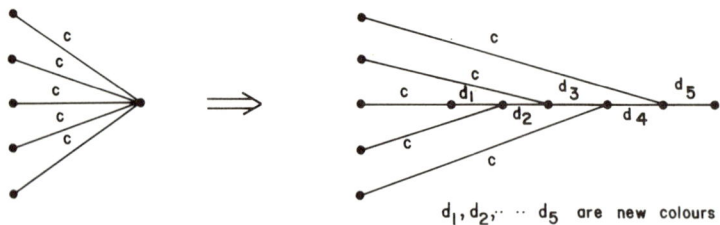

Figure 4 (for $p = 5$)

The graph is still planar, edge series-parallel and can easily be made bipartite. Thus we have

COROLLARY 2.1. PP is NP-Complete for bipartite, planar, edge series-parallel graphs with maximum degree 3. □

If we add the edge (a_o, a_m) to the graph constructed in Theorem 2.1 using a new colour (red say), we see that the problem (PC) of determining whether or not an edge coloured graph has a polychromatic cycle is also NP-Complete. Moreover, we have

COROLLARY 2.2. PC is NP-Complete for bipartite, planar, edge series-parallel graphs with maximum degree 3.

It is easy to show (Bate [1]) that an edge coloured *complete* graph contains a polychromatic cycle if and only if it contains a polychromatic triangle (if cycle C is polychromatic, any chord splits C into 2 smaller cycles one of which must be polychromatic). Clearly, checking for polychromatic triangles can be done in polynomial time.

All of the above results, except the last remark on complete graphs, carry over to digraphs without difficulty. We note that the uncoloured versions of these problems are solvable in time $O(|V| + |E|)$.

Cutsets

We now consider the problem (PCS) of whether or not an edge coloured graph has a polychromatic cutset.

THEOREM 2.2. Planar PCS is NP-Complete.

PROOF. Starting with the graph G constructed in Theorem 2.1 (see Figure 3) together with the red edge (a_0, a_m), we add "new" red edges as shown in Figure 5 to create an edge coloured plane graph H in which the common boundary of any 2 neighbouring faces is a single edge.

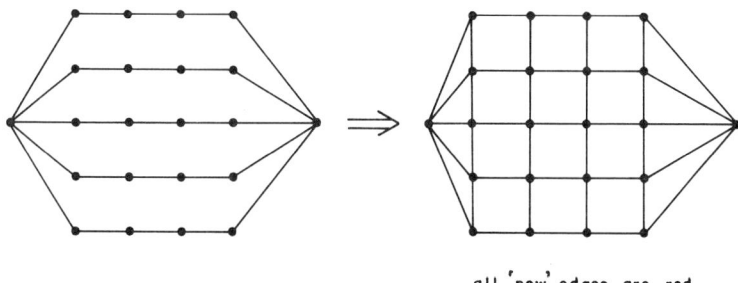

all 'new' edges are red

Figure 5 (for $p = 5$)

It is straightforward to show that none of these "new" red edges can occur in any polychromatic cycle of H. Now we construct the dual graph H^*, the vertices of which correspond to the faces of H. For an edge e of H, let $h(e)$ denote the edge of H^* joining the 2 faces containing e; by construction, h is a bijection. We colour $h(e)$ the same as e. Now it is well known [3] that C is a cycle of H if and only if $h(C)$ is a minimal cutset of H^*. Since C and $h(C)$ are coloured in the same way, C is polychromatic if and only if $h(C)$ is. It follows that 3DM $\propto$ Planar PCS.

For complete graphs PCS remains NP-Complete: given an edge coloured graph G, we can complete the graph, colouring each new edge with a different new colour, to create an edge coloured complete graph G' which has a polychromatic cutset if and only if G has.

4. Polychromatic Matchings

It is easy to show that the problem (PPM) of determining whether an edge coloured graph has a polychromatic perfect matching is NP-Complete, even for bipartite graphs. This contrasts with the uncoloured matching problem which is solvable in time $O(|V|^{2.5})$.

THEOREM 3.1. PPM is NP-Complete, even when the number of colours is one half of the number of vertices.

PROOF. Given an instance of 3DM, construct a bipartite graph G with vertex sets X and Y. For each $(x,y,z) \in T$ we include the edge (x,y) and colour it z. Now T contains a 3-dimensional matching if and only if G contains a polychromatic perfect matching. Note that the number of colours used is m which is half the number of vertices and hence is minimal. The above construction may produce parallel edges. We can remove parallel edges by replacing each edge by a path of length 3 as shown in Figure 6. The resulting graph is still bipartite. □

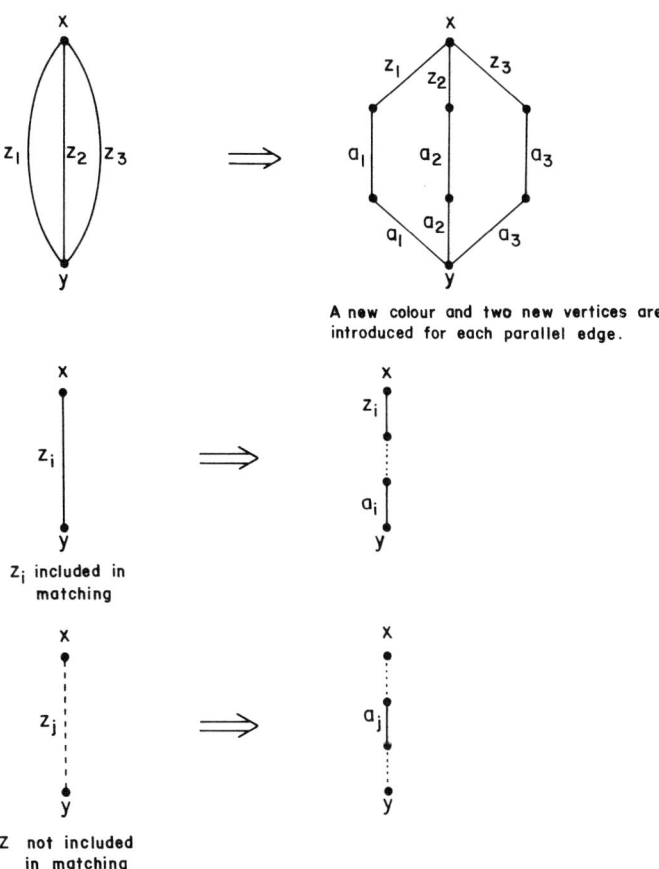

Figure 6

We show next that PPM remains NP-Complete when the graph G is a complete bipartite graph or a complete graph. Thus, even when the existence of many perfect matchings is assured, the problem remains difficult.

COROLLARY 3.2. PPM is NP-Complete for complete bipartite graphs and complete graphs.

PROOF. Consider an arbitrary edge coloured bipartite graph. Add 2 new vertices a and b, putting them into different parts of the vertex partition. Complete the graph by adding new edges all of the same new colour, red say. Any polychromatic perfect matching must contain the edge (a,b) and after deleting the red edges we are left with the previous problem. The same argument is valid whether we construct a complete graph or complete bipartite graph. □

5. Polychromatic Hamiltonian Cycles and Paths

The general problem (PHC) of whether an edge coloured graph contains a polychromatic Hamiltonian cycle is NP-Complete: we can colour each edge differently and then every Hamiltonian cycle is polychromatic. It is not obvious that the problem remains hard when we restrict it to the case (CPHC) where we have a complete graph with n vertices and only n colours are used. We give an outline proof of this result.

THEOREM 4.1. CPHC is NP-Complete.

PROOF. Suppose that we are given an arbitrary edge coloured graph G. If $m > n$ colours are used, we split vertex 1 to obtain G_1 as indicated in Figure 7, where each "block" of m parallel edges uses each colour. Then G_1 contains a polychromatic Hamiltonian cycle if and only if G does. (For the case $m=n+1$, we must add a dummy colour which is not used in G so that we have $m=n+2$.)

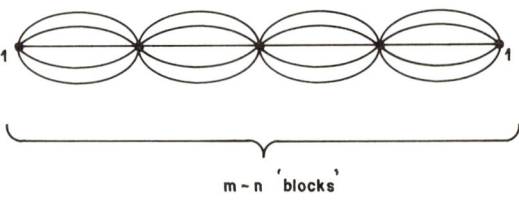

Figure 7

To remove the parallel edges, we replace each "block" of m parallel edges as indicated in Figure 8, to obtain a simple graph G_2. It is not difficult to show that G_2 has a polychromatic Hamiltonian cycle if and only if G_1 does.

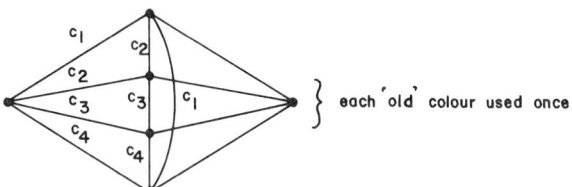

(*m* new colours and *m* new vertices are added for each of the $m-n$ "blocks".)

Figure 8 (for $m = 4$)

In order to obtain a complete graph we split one vertex as indicated in Figure 9. The resulting graph has the same number of colours as vertices. □

(r, b, and g are new colours; the graph is completed by colour b.)
Figure 9

Using similar constructions it can be shown that, given a complete graph G on n vertices, edge coloured using $n-1$ colours, the problem of determining whether G has a polychromatic Hamiltonian path is NP-Complete. This problem remains NP-Complete when either one or both of the endpoints of the path is specified.

The corresponding Hamiltonian path and cycle problems for directed graphs can also be shown to be NP-Complete.

References

[1] N.G. Bate, "Circuits of Edge-Coloured Complete Graphs", Ph. D. Thesis, University of Keele, England, 1981.

[2] N.G. Bate, "Complete Graphs Without Polychromatic Circuits" to appear in Discrete Mathematics.

[3] J.A. Bondy and U.S.R. Murty, "Graph Theory with Applications", Macmillan Press, London, 1976.

[4] S. A. Cook, "The Complexity of Theorem Proving Procedures", in Proceedings 3rd Annual ACM Conference on Theory of Computing, ACM, New York (1971) 151-158.

[5] J. Edmonds, "Submodular Functions, Matroids and Certain Polyhedra" in Proceedings International Conference on Combinatorics at Calgary, Gordon and Breach, New York (1970) 69-87.

[6] P. Erdős, M. Simonovits and V.T. Sós, "Anti-Ramsey Theorems", Infinite and Finite Sets (Colloq. Keszthely, 1973) Vol. 2, Colloq. Math. Soc. János Bolyai 10, North Holland, Amsterdam (1975) 633-643.

[7] F. Galvin, Problem 6034, Amer. Math. Monthly 1975, Solution: Amer. Math. Monthly 1977, p 224.

[8] M.R. Garey and D.S. Johnson, "Computers and Intractability: a Guide to the Theory of NP-Completeness", W.H. Freeman and Co., San Francisco, 1979.

[9] G. Hahn, "Some Star Anti-Ramsey Numbers" Proceedings 8th S.E. Conference on Combinatorics, Graph Theory and Computing, Louisiana State University, Baton Rouge (1977) 303-310.

[10] G. Hahn, "More Star Sub-Ramsey Numbers", Discrete Mathematics 34 (1981) 131-139.

[11] R.M. Karp, "Reducibility among Combinatorial Problems" in Complexity of Computer Computations (eds. R.E. Miller and J.W. Thatcher), Plenum Press, New York (1972) 85-103.

[12] B. Korte and L. Lovász, "Mathematical Structures Underlying Greedy Algorithms", Fundamentals of Computation Theory, Lecture Notes in Computer Science 117, Springer-Verlag, Berlin (1981) 205-209.

[13] E.L. Lawler, "Combinatorial Optimization: Networks and Matroids", Holt, Rinehart and Winston, New York, 1976.

[14] J. Valdes, R.E. Tarjan and E.L. Lawler, "The Recognition of Series Parallel Digraphs", SIAM Journal on Computing 11 (1981) 298-313.

Cycle Decompositions, 2-Coverings, Removable Cycles, And The Four-Color-Disease

Herbert Fleischner

ABSTRACT

Sabidussi's conjecture (GSC) asserts that given an Euler trail T in a connected Eulerian graph G without 2-valent vertices, there exists a cycle decomposition S of G such that consecutive edges of T belong to different cycles of S. For planar graphs, the GSC has been settled in a more general form which establishes a nontrivial necessary condition for the validity of the Four-Color-Theorem. Moreover, this generalization of the GSC is connected with the existence of cycle decompositions into even cycles in planar Eulerian graphs and with the existence of so-called removable cycles in such graphs.

Let G denote a connected Eulerian (pseudo) graph with $\delta(G) \geq 4$ unless explicitly stated otherwise. For every $v \in V(G)$, let $X(v)$ denote a partition of the set of edges incident to v into classes of two elements. Each class is called a (forbidden) transition, and we call $X = \bigcup_{v \in V(G)} X(v)$ a (complete) system of transitions. Both an Euler trail T and a cycle decomposition S of G induce in a natural way systems of transitions denoted by X_T, X_S respectively (a transition of X_T corresponds to edges consecutive in T, and a transition of X_S is a pair of adjacent edges in a cycle of S).

Two systems of transitions X_1, X_2 of G are called compatible (or - as some suggested - orthogonal), if $X_1 \cap X_2 = \phi$. Correspondingly, for given X and T (X and S), we say that X and T (X and S) are compatible if $X \cap X_T = \phi$ ($X \cap X_S = \phi$). In particular, if $X_S \cap X_T = \phi$, then S and T are said to be compatible.

In 1975, Sabidussi stated the following conjecture.

Sabidussi's conjecture (GSC).

Given an Euler trail T of G, there exists a cycle decomposition S of G compatible with T.

While the converse of the GSC (i.e., with S and T switching positions in the GSC), is easily proved (Sabidussi has done that with the help of a result of Kotzig; see also [1, Satz 1]), a proof or counterexample to the GSC has yet to be developed. However, for planar graphs even a more general statement than the GSC is true. For this purpose, we call X (for a given G) nonseparating if G - t is connected for every t ∈ X. (Incidentally, if G has a loop vv, then $\{vv,vv\}$ is considered as a (separating) transition of G although this set consists of only one element; however such problems can be avoided by allowing vertices of degree 2 and defining X on $V(G) = \{v/deg\ v > 2\}$).

THEOREM A. ([1, Satz 2]). If G is planar, and if X is nonseparating in G, then G contains S compatible with X.

As has been shown in [1] the planarity condition cannot be deleted; the ultimate reason lies in the fact that the Petersen graph is not 3-edge-colorable.

The hardest part of the proof of Theorem A is the case where G is 2-connected, 4-regular and has (in some embedding in the plane) no intersecting transitions (see Fig. 1).

a. $X(v)=\{\{x_1x,x_2x\},\{x_3x,x_4x\}\}$ *b.* $X(v)=\{\{x_2x,x_3x\},\{x_1x,x_4x\}\}$

Figure 1.
The non-intersecting transitions for a vertex of degree 4.

In this case G is related in a natural way to a 2-connected, 3-regular planar graph G_3 (see Figure 2).

Figure 2.
G_3 obtained from G.

Here, the edges $x'x''$ form a 1-factor of G_3. If we use the 4CT (or the validity of the 4CC-depending on your degree of scepticism), then a 3-edge-coloring of G_3 yields a compatible cycle decomposition S of G with $\chi(I(S)) \leq 3$. This relation between the 3-edge colorability of G_3 and such cycle decomposition S is, in fact, reversible. (I(S) denotes the intersection graph of S with every $C \in S$ considered as a subgraph of G, and $\chi(H)$ denotes the chromatic number of graph H).

Theorem A yields the following statement on 3-regular graphs which is a necessary condition for the 4CT.

COROLLARY B. (special case of [1, Satz 3]). Let G_3 be a 2-connected, 3-regular, planar graph, and let $\{L,Q\}$ be a decomposition of $E(G_3)$ into a 1-factor L and a 2-factor Q. Then G_3 has a cycle covering S_3 using every L-edge exactly twice and every Q-edge exactly once. - That is, the cycles of S_3 are alternating with respect to L and Q.

$S_3 \cup Q$ is a cycle covering of G_3 using every edge exactly twice. A more trivial cycle covering of G_3 using every edge exactly twice, is the set of face boundaries of an embedding of G_3 in the plane. In fact, the following conjecture is due to P.D. Seymour.

Seymour's Conjecture (PSC).

Every nontrivial connected bridgeless graph admits a cycle covering using each edge exactly twice (shortly 2-covering). It is easy to see that in proving the PSC one can restrict oneself to 2-connected 3-regular graphs.

Another conjecture due to G. Haggard states that every 2-connected (3-regular) graph admits an embedding in some orientable or nonorientable surface such that every face boundary is a cycle of the graph. If true, this conjecture implies, of course, the validity of the PSC. However, we want to try to relate the GSC to the PSC.

LEMMA 1. In proving the GSC it suffices to consider such G which have 4- and 6-valent vertices only.

PROOF. Suppose the GSC is true for Eulerian graphs having 4- and 6-valent vertices only. We show by induction on $S_G := \sum_{v \in V(G)} (deg(v)-4) = 2q-4p$ that the validity of the GSC holds in general. Let G be a graph having a vertex v with deg $v \geq 8$, and let $t_1, t_2 \in X(v)$ be chosen arbitrarily. Form a new graph G_1 by "splitting away" t_1 and t_2 from v; i.e., introduce a new vertex v_1 with $xv_1 \in E(G)$ iff $xv \in t_i, i=1,2, yv \in E(G_1)$ iff $yv \in E(G) - t_1 \cup t_2$; all other incidences remain unchanged. Clearly, T corresponds to an Euler trail T_1 of G_1 (since t_1, t_2 are induced by T), and $S_{G_1} < S_G$. Applying the validity of the GSC to G_1 we obtain a corresponding S_1; and any cycle of S_1 passing through v_1 and v (if such cycle exists), corresponds in G to two cycles each of which contains no element of X_T. Hence, any S_1 in G_1 compatible with T_1 is easily transformed into S of G compatible with T.

Graphs as described in Lemma 1 can be related to 2-connected, 3-regular graphs as well: For a vertex of degree 4 we proceed as indicated by Figures 1 and 2, while for a vertex of degree 6 we perform the operation outlined in Figure 3.

Figure 3.
The transformation of a vertex of degree 6.

The graph G_3 obtained from G by applying these two (local) transformations has the following property: It contains a cycle C (corresponding to T) such that the vertices not on C form an independent set of G. We follow other authors in calling such C a *Tutte cycle*. Hence, the validity of the GSC is equivalent to the statement that every 2-connected 3-regular graph G_3 having a Tutte cycle C admits a 2-

covering S_3 with $C \in S_3$. This shows the immediate relation between the GSC and the PSC.

The following conjecture was stated by the author several years ago.

Conjecture (FTC).

Every 3-regular, cyclically 4-edge-connected graph is either 3-edge-colorable, or it contains a Tutte cycle.

Incidentally, C. Thomassen conjectures that every 4-connected line graph is Hamiltonian. We note in passing that the validity of this conjecture would imply the validity of the FTC. Moreover, C. Thomassen believes (private communication) that 4-connected line-graphs are 2-tough; thus Chvatal's 2-toughness conjecture would imply Thomassen's conjecture. In fact, P. Ash and B. Jackson believe that one can drop the condition of 3-edge colorability in the FTC. This stronger form of the FTC would imply a conjecture put forward at this conference by F. Jaeger: Every cyclically 4-edge-connected 3-regular graph has a cycle C such that G - C is a forest. As Jaeger remarks, Seymour's 6-flow-theorem follows easily from the validity of this conjecture.

LEMMA 2. The validity of the GSC and FTC implies the validity of PSC. The proof of Lemma 2 can easily be derived from the above and the fact that a smallest counterexample to PSC must be cyclically 4-edge-connected. In fact, P.D. Seymour put forward the even stronger conjecture that any prescribed cycle can be assumed to belong to a 2-covering. Obviously, this conjecture would imply GSC.

More recently, P.D. Seymour proved the following result, [3].

THEOREM C. A planar graph has a cycle decomposition into even cycles iff it is Eulerian and every block has an even number of edges.

In proving Theorem C, P.D. Seymour proceeds indirectly. His strategy aims at narrowing the scope of possible counterexamples G to graphs with the following properties:

a) G is 2-connected
b) $\Delta(G) = 4$
c) every vertex of degree four belongs to a triangle containing a vertex of degree two; and any two such triangles are edge-disjoint. - For such graphs one then replaces each of these triangles with an edge (see Fig. 4) to obtain a 2-connected, planar graph H with an even number of edges and - by construction - having vertices of degree two and three only.

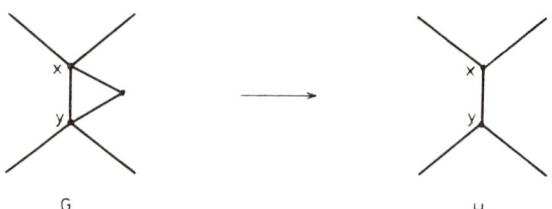

Figure 4.
The construction of H from G.

For such H, P.D. Seymour applies the following Lemma (see [3, 2.4 and 2.5]).

LEMMA D. Let H be a 2-connected planar graph with $\Delta(H)=3$, and for each edge e let $p(e) \in \{1,2\}$. Suppose that for every $v \in V(H)$, $\sum_{e \text{ inc. } v} p(e) \in \{2,4\}$. Then the following is true:

a) There is a set S_3 of cycles using each edge $p(e)$ times.

b) If, in addition, $|E(H)|$ is even, then for every such S_3 and $F = \{e/p(e) = 2\}$ there is a function $t: F \to S_3$ such that

(1) $f \in t(f)$ for every $f \in F$

(2) $|\{f \in F/t(f) = C\}| + |E(C)|$ is even for every $C \in S_3$.

While part a) of Lemma D is nothing but a modification of Corollary B, part b) of that Lemma tells us how to extend S_3 to a cycle decomposition of G into even cycles (namely, precisely those f counted in (2) are replaced with the two edges of G which are not in H; see Figure 4). In fact, Theorem C and Theorem A are more closely related than by the above application of Corollary B.

LEMMA 3. Let G' be plane and 4-regular, and let X' be a system of nonintersecting, nonseparating transitions of G'. Assume the validity of Theorem C. Then G' contains a cycle decomposition S' compatible with X'.

PROOF. Construct from G' the 3-regular graph H as indicated by Figures 1 and 2. Form G from H by reversing the operation described in Figure 4. Let $G_0 = G$ if $|E(G)|$ is even; otherwise, let G_0 be obtained from G by subdividing an edge corresponding to an edge of G'. By Theorem C, G_0 has a cycle decomposition S_0 into even cycles. Hence, none of the triangles of G_0 lies in S_0; this is true, in particular, for those triangles which share no edge with G'. Therefore, every cycle $C_0 \in S_0$ corresponds to a cycle C' of G', and no element of X' is a subset of C'. The set S' of these C' is as required.

In other words, not only can Theorem C be proved with the help of Theorem A, but, in turn, the application of Theorem C condenses what we have called above "the hardest part of the proof of Theorem A", into a short statement. One is, therefore, tempted to ask for a proof of Theorem C which does not use Lemma D nor the 4CT. Incidentally, the planarity condition of Theorem C cannot be dropped. The reason is the same as for the planarity condition of Theorem A: The Petersen graph is not 3-edge-colorable.

Let us return to the GSC. Are there any nontrivial classes of graphs in which the GSC holds? As noted in [1], if $deg(v) \equiv 0 \pmod 4$, for every $v \in V(G)$, then an alternating blue-red-coloration of the edges following T yields a decomposition of G into two Eulerian graphs G_1, G_2. If S_i is any cycle decomposition of G_i, $i=1,2$, then $S_1 \cup S_2$ is a cycle decomposition of G compatible with T. On the other hand, by Lemma 1 we may restrict ourselves to G having vertices of degree 4 and 6 only. We shall use the following Theorem of C.A.B. Smith to prove the GSC in a stronger form if G has at most one vertex of degree 6.

THEOREM E. *Every edge of a 3-regular graph G_3 belongs to an even number of Hamiltonian cycles.*

We reinterpret Theorem E. Suppose G_3 is Hamiltonian, and let H and H' be two Hamiltonian cycles passing through a certain edge e. Denote $Q = H$ and $L = E(G_3) - E(H)$. It is easy to see that the edges of Q not in H' and the edges of L in H' form a disjoint set of Q,L-alternating cycles. By interpreting $Q = H$ as obtained from an Euler trail T in the 4-regular graph G (with G and G_3 related as indicated by Figures 1 and 2) we are led to the next result.

COROLLARY 4. *Let T be an Euler trail in the 4-regular graph G. Then, for every edge e, there is a set of totally disjoint cycles $C_1, \ldots, C_r, r \geq 1$, compatible with T and not containing e such that the set $X_{T'}$, of transitions t of X_T with $t \subset G' = G - \bigcup_{i=1}^{r} E(C_i)$ corresponds to an Euler trail T' of G'.*

We say that $T'(X_{T'})$ is induced by $T(X_T)$ if $t \in X_T$ and $t \subseteq E(G')$ implies $t \in X_{T'}$.

THEOREM 5. *Let T be an Euler trail of G whose vertices have degree 4 with the exception of at most one vertex v of degree 6. Then* there exists a set $\{C_1,\ldots,C_r;r\geq 1\}$ of totally disjoint cycles compatible with T such that T induces an Euler trail T' in $G' = G - \bigcup_{i=1}^{r} E(C_i)$.

PROOF. Corollary 4 implies the theorem for 4-regular G. Hence we assume G to have precisely one vertex x of degree 6.

For $t = \{vx,xw\} \in X_T$ we form $\overline{G} = (G-t) \cup \{vw\}$ and $X_{\overline{T}}$ where $\overline{T} = (T-t) \cup \{vw\}$ is the Euler trail in $\overline{G}$ corresponding to T. $\overline{G}$ is 4-regular; hence Corollary 4 applies to $\overline{G}$ with $e = vw$. The corresponding cycles $C_1,\ldots,C_r$ are compatible with X_T as well since they avoid e and, therefore, are cycles in G as well. From the graph $\overline{G}'$ (see Corollary 4) we obtain $G' = (\overline{G}' - \{vw\}) \cup t$ with $G' \subset G$ and $\Delta(G') = 4$ or 6 depending on whether or not some C_i contains edges incident to x, $1 \leq i \leq r$. In the second case we have an Eulerian trail T' in G' defined by $X_{T'} = X_{\overline{T}'} \cup \{t\}$ which is, in fact, induced by T. In the first case we define $X_{T'} = X_{\overline{T}'} \cup \{t,\{g,h\}\}$ where g,h are the edges incident to x in $\overline{G}'$. Clearly, also $X_{T'}$ is induced by X_T and T' is an Euler trail of G' induced by T (note that g and h belong to different elements of X_T).

If one repeats the procedure expressed by Theorem 5 and its proof, then one obtains step by step sets of cycles $S_j, j=1,\ldots,m$ for a certain m, such that S_j is compatible with X_T, and $S = \bigcup_{j=1}^{m} S_j$ is a cycle decomposition of G compatible with X_T.

Unfortunately, this procedure can no longer be deployed if G has more than one vertex of degree 6. This is demonstrated by the following construction.

Consider in the Petersen graph the cycle (written as a vertex sequence) $C = 1,2,a,3',1',b,2',3,c,1$ (Fig. 5a). Following the construction indicated by the Figures 1 - 3 one obtains the graph G_P of Figure 5.b. with the Euler trail T_P corresponding to C indicated by the transitions. It is easy to check that there is a unique cycle decomposition compatible with T_P, namely $S_P = \{C_1,C_2,C_3,C_4\}$ (see Fig. 5b).

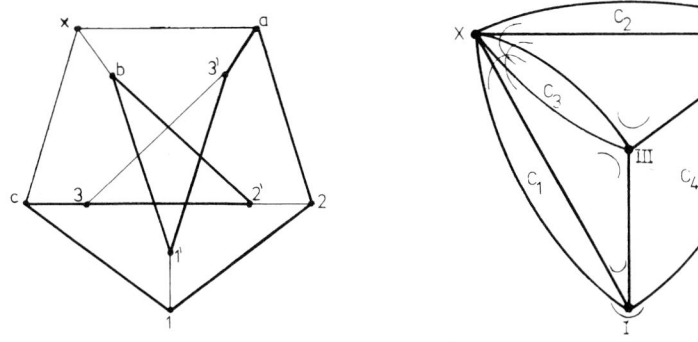

Figure 5

We now take a copy of G_P and a mirror image of G_P and form the graph G exhibited in Figure 6.

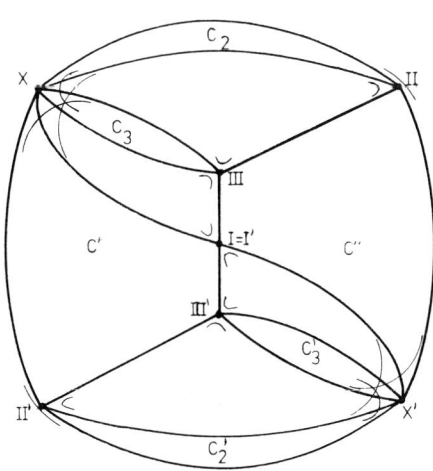

Figure 6.
The graph G with Euler trail T.

It follows from the construction of G that every cycle decomposition S of G compatible with T induces such in G_P with regard to T_P and vice versa. Hence, the only adequate cycle decomposition in G is $S = \{C', C'', C_2, C_3, C'_2, C'_3\}$. Moreover, it is easy to check that for G and T, the conclusion of Theorem 5 no longer holds. The reason for this lies in the fact, that C_4 is the only cycle satisfying Theorem 5 with regard to G_P and T_P.

Since the Petersen graph is not 3-edge-colorable we must have $\chi(l(S_P)) > 3$, $\chi(l(S)) > 3$; in fact, $\chi(l(S_P)) = \chi(l(S)) = 4$ (see Figures 5 and 6). This leads us to the following questions:

For G with vertices of degree 4 and 6 only and an Euler trail T in G, and having S compatible with T, how large can $n(G,T,S) := \max_T \min_S \chi(S))$ be (the minimum taken over all S compatible with T, the maximum taken over all Euler trails T in G)? If, in addition, G is planar, is it true that $n(G,T,S) \leq 4$ for every such G? In other words: Does, in general, $\max_G n(G,T,S)$ exist, and is $\max_{G \text{ planar}} n(G,T,S) = 4$?

If we construct from such G a 3-regular graph G_3 (see Figures 1 - 3), then it is clear what these questions mean with respect to edge-colorings of G_3. Theorem 5 yields a nice little result for 3-regular graphs which seems to be new.

COROLLARY 6. If $G_3 - v$ is Hamiltonian for some vertex v of the 3-regular graph G_3, then $G_3 - v$ has at least two Hamiltonian cycles, or else G is Hamiltonian.

Let us look at G_3 of Fig. 7 (corresponding to G of Figure 6).

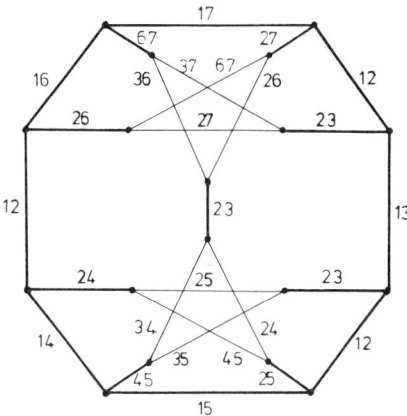

Figure 7

While it is true that a 3-edge-colorable 3-regular graph has a covering with even cycles where every edge is used twice (just take the 2-factors induced by the 3-edge-coloring), the converse is not true as P.D. Seymour, [3], showed by a graph similar to our G_3 and thus proved a claim of Szekeres to be false. Following Seymour's notation, the even cycle covering of G_3 is indicated by labelling the edges with 2 digits, each digit representing a cycle (Seymour's example has as many vertices; however, it is cyclically 4-edge-connected, hypohamiltonian and only six cycles are needed in the covering). The Petersen graph, however, has no such covering with even cycles. That is, one cannot add to the PSC the condition that the cycles are even. We note, however, that M. Preissmann showed that there is an infinite family of so-called "flower-snarks" having a 2-covering by even cycles and not being 3-edge-colorable. Finally, as F. Jaeger pointed out to the author, there are still other 2-covering conjectures which are stronger than the PSC and which are related to Haggard's conjecture.

We now turn to the concept of removable cycles which was introduced by A.M. Hobbs who asked the following question: Given a 2-connected Eulerian graph with no vertices of degree 2, does there exist a cycle C such that $G - E(C)$ is 2-connected? Such C is called a removable cycle. It soon became clear that the admission of multiple edges implies a negative answer to this question: Just take a 1-factor L in the Petersen graph, subdivide the edges of L and "double" the edges incident to the vertices of degree 2. The reason for the nonexistence of a removable cycle in the graph thus obtained is the same as to why the planarity condition of Theorems A and C cannot be dropped: The Petersen graph has no 1-factorization.

However, if one restricts oneself to simple 2-connected graphs, then one can show that any such graph G (be it Eulerian or not) has a removable cycle provided $\delta(G) \geq 4$. This was achieved by B. Jackson (in fact, he even showed that one can find such cycle of length $\geq \delta(G) - 1$). It should be noted, however, that similar results were obtained by Mader, and Thomassen and Toft.

For planar graphs one can allow multiple edges. This has been shown in a forthcoming paper by B. Jackson and the author, [2].

THEOREM F. If G is a planar, 2-connected Eulerian graph with $\delta(G) \geq 4$, then G has a removable cycle avoiding a prescribed edge e.

Again, the "hardest" part of the proof of Theorem A follows easily from Theorem F. That is, replacing Theorem C with Theorem F in the statement of Lemma 3 yields a true statement. This can be seen from the construction outlined in Figure 8.

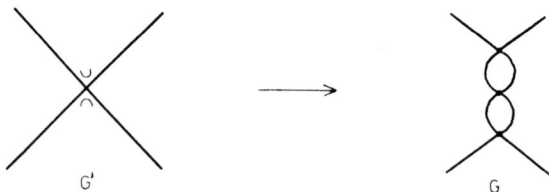

Figure 8.
Transforming G' into G without transitions

Clearly, since G' has only nonintersecting and nonseparating transitions, therefore, G is planar, 2-connected and 4-regular. Moreover, no digon of G is removable. Assuming the validity of Theorem F, G has a removable cycle C_1 which corresponds to a cycle of G' compatible with X. Suppressing in $G_1 = G - E(C_1)$ the vertices of degree 2, we can apply Theorem F again; the corresponding cycle of G' is compatible with X. This way, S compatible with X can be constructed step by step.

However, Hobbs and the author knew of this relation between Theorem A and Theorem F, but we didn't anticipate that Theorem A was a useful tool to prove Theorem F. This was B. Jackson's idea. Namely: One proceeds indirectly by first showing that a possible smallest counterexample H must be 3-connected. Then one defines a system of transitions X for H such that $E(Z) \in X$ for every digon Z of H (the other transitions are defined arbitrarily). In a cycle decomposition S of H compatible with X, $H - E(C)$ is not two-connected for every $C \in S$. One then choses $C_0 \in S$ such that $H_1 = H - E(C_0)$ has an endblock B_0 (that is, a 2-connected component or a block containing precisely one cutpoint of H_1), with as few vertices as possible and $e \in \ell(B_0)$. B_0 contains a cycle $C_1 \in S$. One then concludes that H has a 2-vertex-cut if $H - E(C_1)$ is not 2-connected, contradicting the first step of the proof.

The conclusion of Theorem F remains true if one drops the condition "Eulerian", [2]. In this case, however, one cannot apply Theorem A. In fact, one transforms H into a 3-regular, planar, 2-connected graph H_3 in such a way that the application of the 4CT to H_3 yields a cycle-path-decomposition S of H with no digon belonging to S unless it is a removable cycle. Moreover, the rest of the proof is much more involved than the proof of Theorem F.

To avoid the application of the 4CT in this generalization of Theorem F, we are presently working on a generalization of Theorem A for arbitrary planar graphs H. In this case, one aims at obtaining a cycle-path-decomposition S compatible with X, with X defined for a fixed plane embedding of H in such a way that $t \in X$ implies that t is part of the edge set of some face boundary; and every path-element of S has its endvertices among the odd vertices of H.

Finally, we would like to point out that there is no general directed version of the GSC. This is demonstrated by Figure 9.

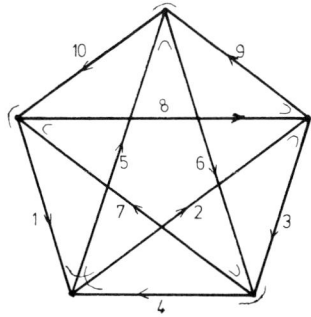

Figure 9.
T=1,...,10 has no compatible oriented cycle decomposition.

References

[1] H. Fleischner, Eulersche Linien and Kreisüberdeckungen, die vorgegebene Durchgänge in den Kanten vermeiden. JCT(B) 29, no. 2 (1980), 145-167.

[2] H. Fleischner and B. Jackson, Removable cycles in planar graphs (to appear).

[3] P.D. Seymour, Even circuits in planar graphs. JCT(B) 31, no.3 (1981), 327-338.

Strong Edge-Coloring
of Cubic Planar Graphs

J. L. Fouquet
J. L. Jolivet

ABSTRACT

In this paper, we study some different ways of coloring the edges of cubic planar graphs, particularly the multi k-gons. We give algorithms to obtain such colorings for some classes of multi-k-gons. These colorings are similar to Heawood colorings for multi-3-gons.

1. Introduction and Preliminary Results

All graphs considered in this paper will be finite simple, cubic, planar and bridgeless. Each graph will always be imbedded in the plane. The set of such graphs will be denoted by **M**.

$E(G)$ and $V(G)$ will denote respectively the edge-set and the vertex-set of the graph G. If $F \subseteq E(G)$, $<F>$ will denote the subgraph of G induced by the set of ends of edges in F. For $k \in \{2,3,4,5\}$, $\mathbf{M}_k$ will denote the class of graphs such that each face has a length congruent to 0 mod k. Such graphs are known as multi-k-gons [10].

Our purpose is the study of particular matchings and colorings in these different classes of multi-k-gons.

DEFINITION 1: A subset F of $E(G)$ is a strong matching if and only if $<F>$ has no edge in $E(G)\backslash F$.

DEFINITION 2: A strong edge coloring of G is a partition of $E(G)$, each part of it being a strong matching.

The dual of a graph G of **M** is 4-colorable [1]. Let **C** = $\{A,B,C,D\}$ be the set of colors used; if $I \neq J$ are two different colors, an (I,J)-edge is an edge bounding a face colored I and a face colored J. Let us denote by $E_{I,J}$ the set of (I,J)-edges, and let $E_1 = E_{A,B} \cup E_{C,D}$, $E_2 = E_{A,C} \cup E_{B,D}$, $E_3 = E_{A,D} \cup E_{B,C}$. It is easy to verify that the partition $\{E_1, E_2, E_3\}$ of $E(G)$ is a 3-edge coloring of G.

Let $\phi : E(G) \to \{1,2,3\}$ be defined by $\phi(e) = i$ if $e \in E_i$.

The Heawood valuation ϕ^* of $V(G)$ associated with ϕ is a function from $V(G)$ into the set $\{-1, +1\}$ derived from ϕ by considering the sign of the rotation defined by the colors 1,2,3 around each vertex x (Figure 1).

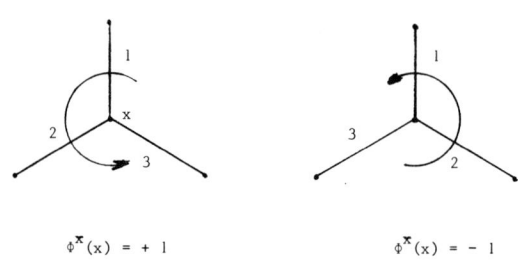

$\phi^x(x) = +1$ \qquad\qquad $\phi^x(x) = -1$

Figure 1

The total sum of ϕ is $T(\phi) = \sum_{x \in V(G)} \phi^*(x)$ and $\epsilon(\phi)$ will denote the set of edges $[x,y]$ such that $\phi^*(x) = \phi^*(y)$.

2. The Class M_3 of Multi-3-Gons

It is well-known that we always can color the faces of a multi-3-gon with four colors in such a way that each of them appears around each edge. Figure 2 shows an example of this.

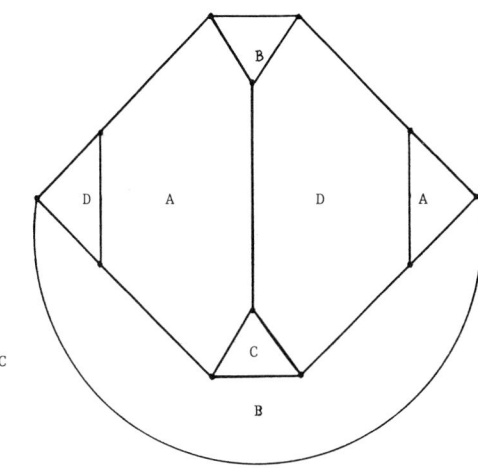

Figure 2

It can easily be verified that this face-coloring induces six strong matchings $E_{I,J}$ of cardinality $|E(G)|/6$, and that each E_i is a perfect matching.

From the preceding remarks, several results previously known can be deduced:

1- The vertex set $V(G)$ has a cardinality $|V(G)| \equiv 0 \ [\mathrm{mod}\ 4]$

2- THEOREM (Vigneron [17] and Kotzig [12]). For each 3-edge coloring ϕ of G in $\mathbf{M}_3$, one has : $T(\phi) \equiv 0 \ [\mathrm{mod}\ 4]$

3- THEOREM (Fleischner [5]). For each 3-edge coloring ϕ of G in $\mathbf{M}_3$, one has : $|\epsilon(\phi)| \equiv 0 \ [\mathrm{mod}\ 2]$

Another application of these ideas is the following study of graphs of $\mathbf{M}_3$.

Let us consider two strong matchings $E_{I,J}$ and $E_{K,L}$, with $\{I,J\} \cap \{K,L\} = \emptyset$ defined above, $E_{A,B}$ and $E_{C,D}$ for example. A face $\mathbf{F}$ of G certainly has one edge (or several) in one of the sets $E_{A,B}$, $E_{C,B}$, but not in both: suppose that $\mathbf{F}$ is colored in A, then the neighbour faces of $\mathbf{F}$, (in a cyclic order around $\mathbf{F}$) are colored $B,C,D,B,C,D,\ldots$ or $B,D,C,B,D,C,\ldots$ so the edges of $\mathbf{F}$ are never in $E_{C,D}$. Consequently the faces of G can be partitioned into two sets: those which possess edges in $E_{A,B}$, and the others (which possess edges in $E_{C,D}$). Let $\mathbf{P}_1$ and $\mathbf{P}_2$, be these two sets. The vertex set $\mathbf{P}^*$ of the dual G^* of G is similarly partitioned into $\mathbf{P}_1^*$ and $\mathbf{P}_2^*$.

Let $\hat{G}$ be the partial subgraph of G^*, the edges of which are the edges between two vertices of $\mathbf{P}_1^*$ or two vertices of $\mathbf{P}_2$; let $\hat{\epsilon}_1$ and $\hat{\epsilon}_2$ be these sets of edges.

The Figures 3, 4 show these sets for the preceding graph G.

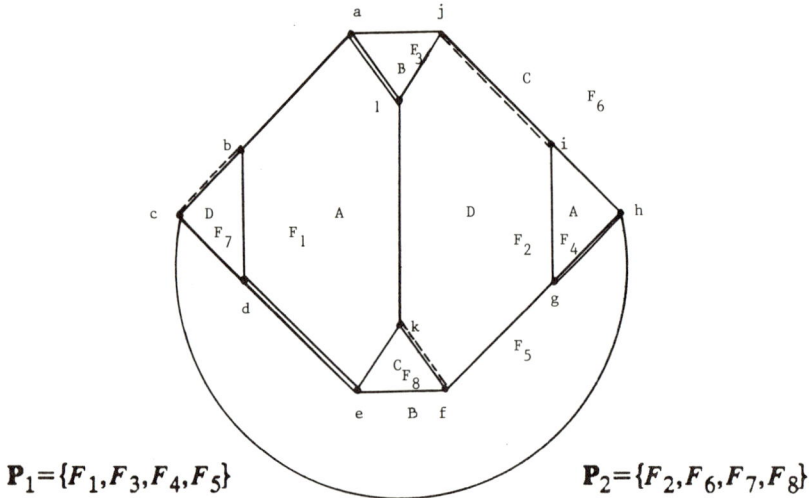

Figure 3

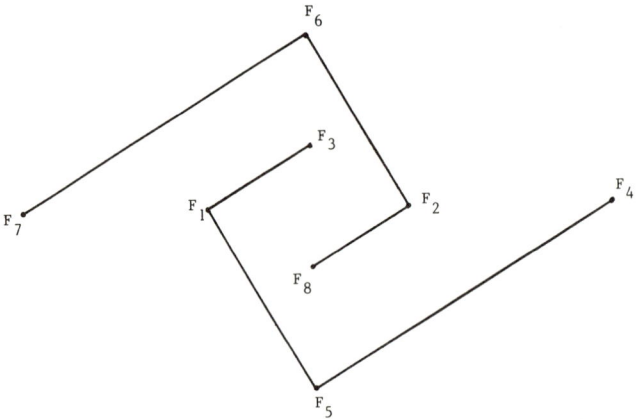

Figure 4

The previous remarks show that a vertex v of $\hat{G}$, corresponding to a face F of length $3k$ has degree k in $\hat{G}$.

Let us define the face-vector of a graph $G \in M_3$ as $(p_3, p_6, \ldots, p_{3k})$ where p_{3i} is the number of faces having a length $3i$, for $i = 1\ldots k$. An immediate corollary of Euler's formula for planar graphs gives the following:

$$\sum_{i=1}^{k}(6-3i)p_{3i}=12$$

If a graph $G \in M_3$ only possesses faces of length 3 and 6, the preceding formula implies $p_3 = 4$, so the graph $\hat{G}$ has exactly four vertices of degree 1, all the other having degree 2. We so obtain the results of Grünbaum and Motzkin [11] who exhibited the structure of these graphs, subsequently used by Goodey [9] for proving that they are hamiltonian.

If a graph $G \in M_3$ only possesses faces of length 3, 6, and just one of length 9, then $\hat{G}$ has five vertices of degree one, one of degree 3, and the others of degree 2. In the same way the structure of those graphs, in particular hamiltonian properties [7], can be studied.

Both cases are pictured in Figure 5:

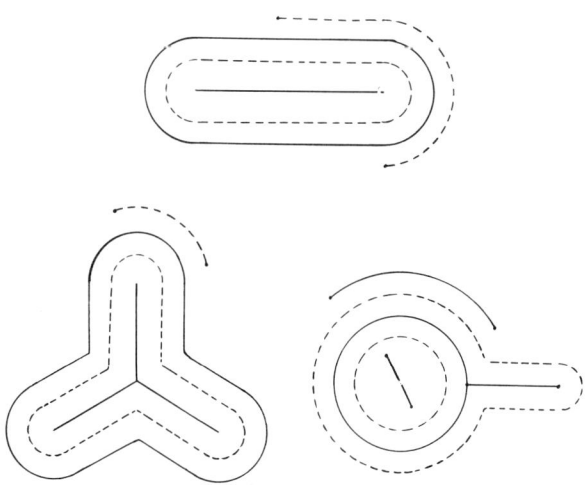

Figure 5

3. The Strong-Chromatic Index of a Graph G:

Every graph of M_3 has a strong-coloring with six strong matchings, and for general cubic graphs it can be easily seen that 12 strong matchings are always sufficient by applying Brook's theorem [4] to the square of the line-graph of G. It is natural to define the strong chromatic index of a graph $X_s(G)$ as the minimum number of strong matchings necessary to realise a partition of $E(G)$. A graph G of M_3 satisfies $X_s(G) = 6$ (the equality is proved by considering a triangle of G and the three adjacent edges), while a general cubic graph satisfies : $5 \leq X_s(G) \leq 12$. We look now at the situation of graphs in class M_5.

4. The Class M_5

THEOREM. Let G be a graph of M_5, then $X_s(G) = 5$; we can find a five strong-coloring of $E(G)$ such that the five colors appear around each face in a circular permutation, and can be obtained by the following algorithm, called CF_5.

STEP 1: We choose an arbitrary face F and we color its edges using $\sigma = (1,2,3,4,5)$ in a circular permutation.

STEP 2: We color edges incident to F with the only color suitable for each of them (Figure 6).

STEP 3: We choose a face F' adjacent to F and we color it, according to the constraint on colors which must be strong-matchings (Figure 7).

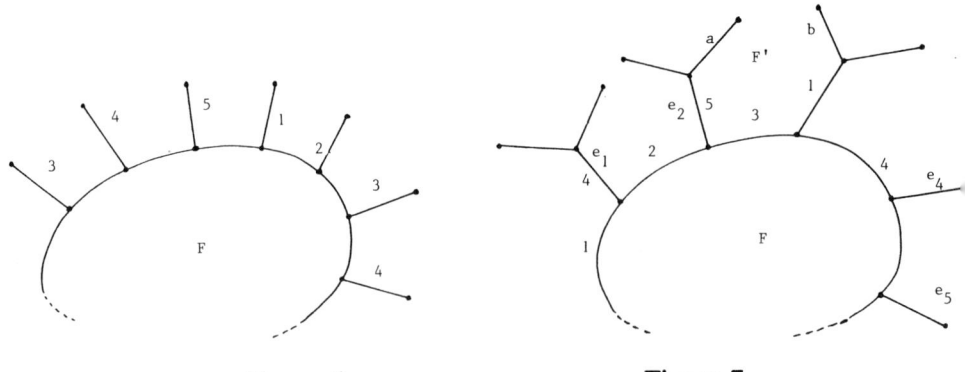

Figure 6   Figure 7

The edge a can not get any of the colors 2,3,5 ; the face F' has three edges colored 1,3,5 namely e_3, e, e_2 and so the following edge around F' cannot be colored with 1. Hence the color of a is 4; for the same reason the color of b is 2, so the permutation used for coloring F' will be $\sigma' = (1,3,5,4,2)$.

STEP 4: We repeat this for other faces adjacent to F' and so on. □

We can get a proof of the theorem by using results of T. Gallaï [8] it was noticed by F. Jaeger. A second way is to use the three following theorems A,B,C, first proved by Grünbaum [10] for theorem A and B, and Grünbaum-Motzkin [11] for theorem C in M_3, (see J. Malkevitch [14] for new proofs and extensions, useful in the proof below).

THEOREM A. There is no cubic planar graph having all faces except one with length congruent to 0 modulo k.

THEOREM B. There is no cubic planar graph having all faces except two adjacent ones with length congruent to 0 modulo k.

THEOREM C. All the subgraphs obtained in a cubic planar graph of M_k by alternatively choosing left and right edges (as shown in Figure 8) are elementary cycles, called L,R-cycles. □

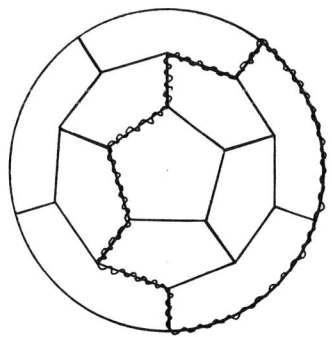

Figure 8. A Left-Right path in the dodecahedron D_{20}.

PROOF OF THE THEOREM. Let G be a graph of M_5. It is easy to verify that it has at least 20 vertices, and that the only such graph with 20 vertices is the dodecahedron D_{20}. The following Figure 9 shows D_{20} with a strong coloring satisfying the conclusions of the theorem, so we suppose the property established for all graphs with $n' < n$ vertices in M_5 and we show it for a graph G with n vertices.

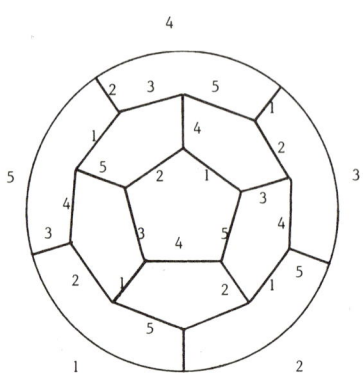

Figure 9

a) Let C be a L,R-cycle of G, then : $l(C) \equiv 0$ [10], where $l(C)$ denote the length of C. For it is clear that $l(C) \equiv 0$ [2] because of the construction of C; let us suppose that $l(C)$ is not congruent to 0 modulo 5. Consider the cycle C of G and replace all the edges and vertices situated inside C as follows:

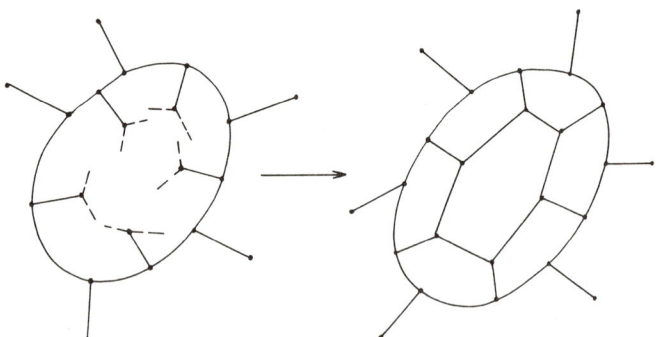

Figure 10

We create a new graph G' which is identical to G outside the cycle C, but inside C, G' possesses faces of length congruent to 0 modulo 5, except one which has length $l(C)/2$; this contradicts the theorem A, so our assertion is proved.

b) If there is exists an L,R-cycle which is not bounded (inside or outside) by pentagonal faces, the above transformation on the inside (resp. outside) of G gives us a graph G_1 (resp. G_2) which belongs to M_5 and has fewer vertices than G; if not, it is possible to reduce the problem to several configurations which lead us to the subcase c), the proof being rather technical. Our induction hypothesis applies and the algorithm CF_5 furnishes a 5-strong-coloring of G_1 (and G_2). It is easy to verify that these colorings may be used to color all the graph G (changing the names of colors).

c) If every L,R-cycle is bounded (inside or outside) by pentagonal faces.
- If both inside or outside are bounded by pentagonal faces, for some L,R-cycle, the graph is a generalised dodecahedron D_n with $n \equiv 0$ [20] which is easily colorable by CF_5. (Such a graph has two non-adjacent faces of length $n/4$ while the others are pentagons).
- If every L,R-cycle C has its inside or its outside (but not both) bounded by pentagon faces, we proceed as follows.

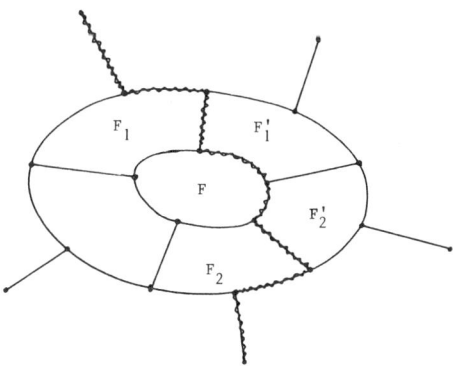

Figure 11

Consider the situation described in Figure 11. Suppose that F is a pentagon, then either F_1 and F_2 are pentagons, for F'_1 and F'_2 are pentagons. In both cases using iterations of this argument, it can be quickly proved that G is isomorphic to a graph D_n, $n \equiv 0 \mod 20$. □

The algorithm CF_5 seems to be a powerful tool in the study of

Class M_5. Here are several combinatorial properties of multi 5-gons obtained from the theorem.

PROPOSITION 1: The subgraph of edges colored with three arbitrary fixed colors among $\{1,2,3,4,5\}$ is formed of stars $K_{1,3}$ and elementary cycles of length congruent to 0 modulo 6. □

PROPOSITION 2: Let α,β,γ be three colors among $\{1,2,3,4,5\}$. Then a vertex x and a vertex y are separated by one of the cycles described in proposition 1, if the colors incident on them are respectively (α,β,γ) and (α,γ,β) in an orientation of the plane. For the dodecahedron D_{20} one can notice that two such vertices are antipodal. □

PROPOSITION 3: The number of vertices of G is divisible by 20. One can show that $V(G)$ is partitioned into 20 equicardinal classes by the 20 possible triples among the color-set of cardinality 5.

PROPOSITION 4: For any graph G in M_5, one can find a 4-coloring of the faces (from which a 3-coloring of the edges can easily be deduced).

PROOF: Let us consider the following partition of the faces: a part A consists of all faces having a permutation τ such that $\tau(1) = 2$. Parts B,C,D are formed analogously, the value of $\tau(1)$ being 3,4 or 5 respectively. Two faces in the same class are never adjacent because each color is a strong coloring.

The preceding assertion leads us to conjecture:

CONJECTURE 1: Every planar cubic bridgeless graph is obtained by a-reductions on a graph G in M_5 (see Figure 12)

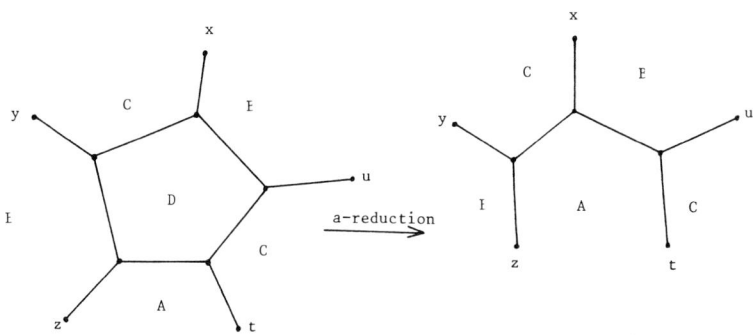

Figure 12

The conjecture trivially implies the four-color theorem and is not a great help for proving it; nevertheless it gives an analogous of Headwood's work on multi-3-gons [15]. In connection with our conjecture there is the following conjecture of J. Malkevitch [15].

CONJECTURE 2: Every cubic bridgeless graph gives a graph of M_5 by a suitable sequence of p-extensions (see Figure 13).

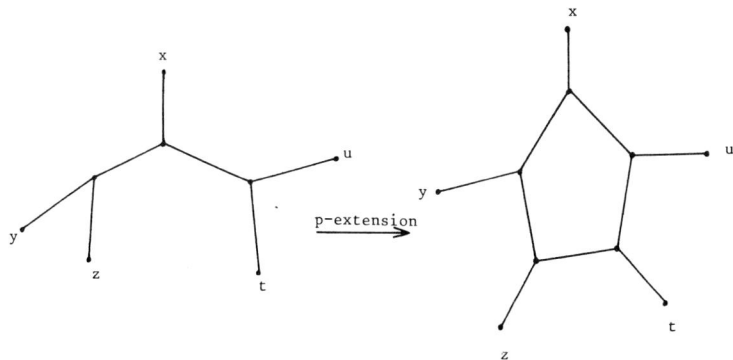

Figure 13

Notice that no consideration on a face coloring is done here, so the converse transformation of a p-extension is not an a-reduction; so the two conjectures are certainly not equivalent.

REMARK 1: The Petersen graph considered as a mult-5-gon in the projective plane, admits a 5-strong coloring obtained by the same method; but this is not true in general for multi-5-gon in that surface.

REMARK 2: The use of circular permutations of colors around faces of length greater than 5, clearly leads to a unique 5-coloring; but if we remove this condition we do not preserve the unicity (see Figure 14 a).

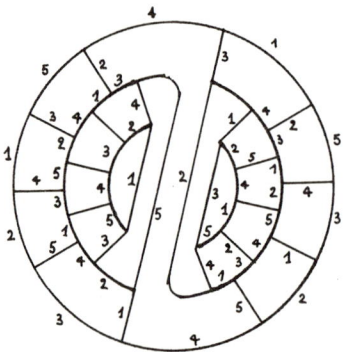

Figure 14 a

A multi 5-gon with 40 vertices having a 5-coloring such that a face F is not colored with a circular permutation of the colors.

REMARK 3: The class C of cubic planar and bridgeless graphs which can be colored with 5 strong matchings is larger than M_5; if we look at the characterisation of members of C which only possess pentagonal or hexagonal faces, we can prove that the only such graph is the "football" below (see Figure 14 b).

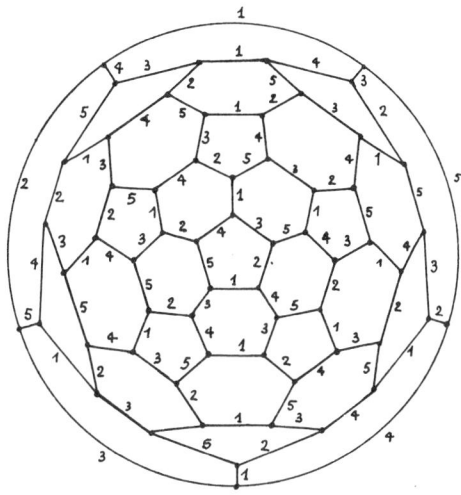

Figure 14 b

5. The Class M_4

The graphs of M_4 have a 6-strong coloring. It can be proved, as before, that the following algorithm CF_4 gives such a coloring for those graphs.

ALGORITHM CF_4 for coloring a graph of M_4 with six colors denoted $\{1,2,3,4,5,6\}$.

STEP 1: Initially, we choose a face F_1, and an orientation of its boundary. We color the edges around this face with four colors chosen among the six possible by using a fixed permutation of them.

STEP 2: We alternately color the edges adjacent to F_1 with the two unemployed colors. See Figure 15.

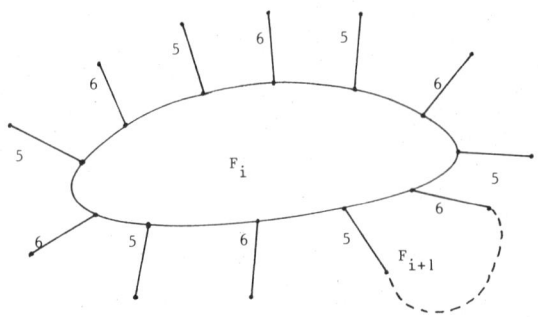

Figure 15

STEP 3: We choose a face $F_{i'}$ having a common edge, say $[a,b]$, with a face, say F_i, the edges of which are colored, and we color the boundary of F_i' with the three colors already used and with the very one not appearing around a or b. We repeat step 3 until all edges are colored. In the example of Figure 14 the permutation used would be $(5,1,6,3)$ for the face F_i.

As with the coloring of multi-5-gons, this coloring gives us several combinatorial properties of multi-4-gons; in particular propositions 1 and 2 are still valid. If the first permutation chosen is $(1,2,3,4)$, the faces of G will be colored by permutations in the three following pairs:

$$P_1 = \{(1,2,3,4), (1,4,3,2)\},$$
$$P_2 = \{(2,5,4,6), (2,6,4,5)\},$$
$$P_3 = \{(1,5,3,6), (1,6,3,5)\},$$

Each set of faces colored with one of the permutations in a pair P_i, or with the other, constitutes a 2-factor of G.

At the end of this paragraph devoted to class M_4, we notice that a coloring of a multi-4-gon with 4 colors (each of them being a matching) can be obtained by using only circular permutations of $\{1,2,3,4\}$ around the faces; as before this coloring gives several combinatorial properties of class M_4.

6. Relations Between These Different Classes:

In this part we look at the following problem: giving a graph G in M_k, does there exist a "natural" way to transform it in a graph G' in $M_{k'}$ (with $k \neq k'$)?

1 - Transformation of $G \in M_3$ into $G' \in M_4$.

Let us consider a perfect matching L of G which is the union of two strong matchings, as discussed in the first paragraph. Let us substitute for each edge of L a 4-cycle as pictured in Figure 16a. Then the graph obtained is a multi 4-gon. Notice that we can transform a cubic, planar, 2-connected graph in a multi 3-gon if we know a 3-edge coloring of it (see O. Ore [16]), but the operation of Figure 16 b done on each edge of a perfect matching gives a graph of M_4. This answers a question of J. Malkevitch [13], [14].

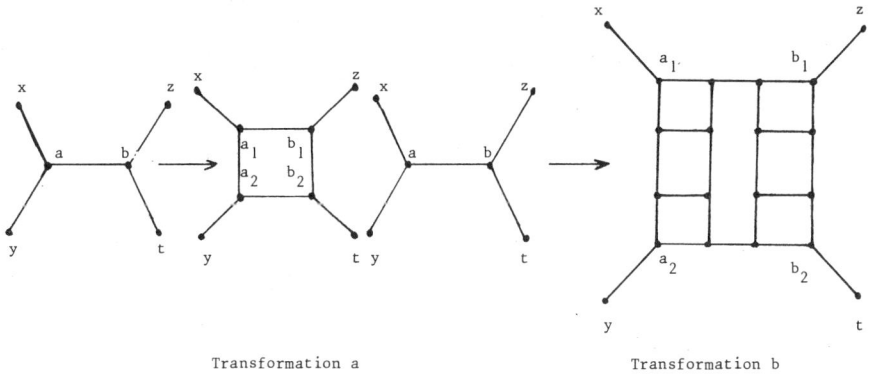

Transformation a   Transformation b

Figure 16

2 - Transformation of $G \in M_4$ into $G' \in M_5$.

We use the following operation:

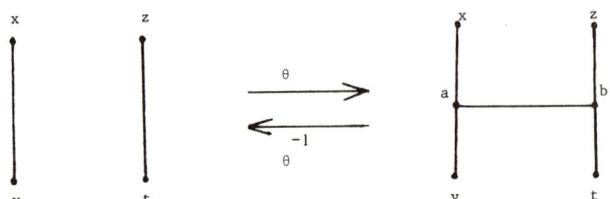

Figure 17

Let G be a graph in M_4, then G is colored with the algorithm CF_4, one can easily see that by doing this operation on the edges colored 1 and 3 in the faces of P_1, on the edges 2 and 4 on the faces of P_2, and on the edges 5 and 6 on the faces of P_3, the graph G' so obtained is a multi-5-gon.

3 - Transformation of $G \in M_5$ into $G' \in M_4$ (*)

It suffices to reduce all the edges of a strong matching obtained by CF_5, with the operation θ^{-1}.

4 - Transformation of $G \in M_4$ into $G' \in M_3$

The operation used here is the converse of the transformation a described in Figure 16. It is more complicated than the preceding other operations and we just picture it in Figure 18.

(*) This operation does not always give a connected graph, so G' is formed with one or several graphs in M_4.

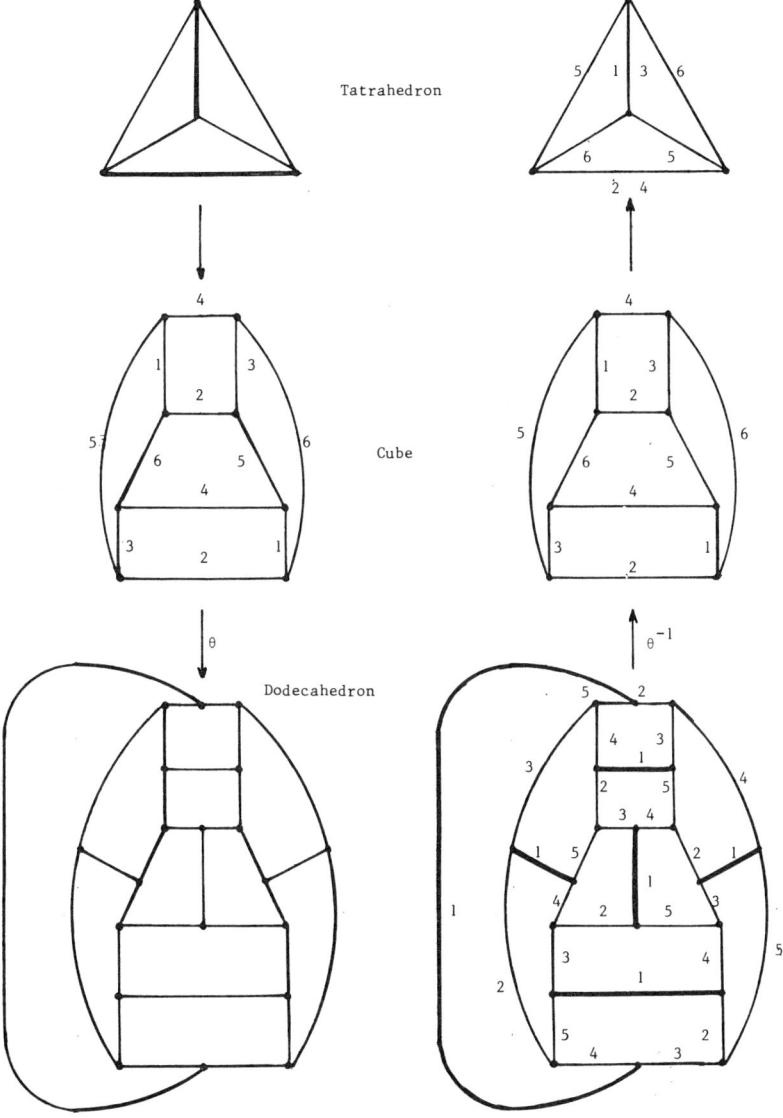

Figure 18

References

[1] K. I. Appel, W. Haken, J. Koch, Every map is four-colorable Part 2: reducibility, Illinois J. Math. 20 (1977) 491-567.

[2] D. Barnette, Conjecture 5, Recent progress in combinatorics, ed. W. T. Tutte, Academic Press, New-York, 1969, 343.

[3] J. C. Bermond, J. L. Fouquet, M. Habib, B. Peroche, On k linear arboricity, to appear.

[4] R. L. Brooks, On coloring the nodes of network, Proc. Cambridge Phil. Soc. 37 (1941) 194-197.

[5] H. Fleischner, A note on line colorings of cubic graphs, Arch. Math. (Brno) 10 (1974) n°4, 195-197.

[6] J. L. Fouquet, These Université de Paris 6, Juin 1981.

[7] J. L. Fouquet, B. Peroche, On covering by hamilton walks in certain class of multi-3-gon. In prepartion.

[8] Gallai, Signierte Zellenzerlengungen I. Acta Math. Acad. Sci. Hungar. 22 (1971), 51-63.

[9] P. Goodey, A class of hamiltonian polytopes. Journal of Graph theory, vol. 1 (1977), 181-185.

[10] B. Grunbaum, Convex polytopes. Interscience Publishers, John Wiley and Sons. London N.Y. Sydney (1967).

[11] B. Grunbaum, T. S. Motzkin, the number of hexagons and the simplicity of geodesics on certain polyhedra. Can J. Math. 15 (1963), 744-751.

[12] A. Kotzig, Coloring of trivalent polyhedra. Canad. J. of Math. 17 1965), 659-664.

[13] J. Malkevitch, Problem 3 "Hamiltonian circuits in polyhedral graphs" in "open problem" p. 568. Second international conference on combinatorial mathematics Editors A. Gewirtz, L. V. Quintas, New-York. Academy of Sciences (1979).

[14] J. Malkevitch, Properties of planar graphs with uniform vertex and face structure, Memoirs of the American Math. Soc. 99.

[15] J. Malkevitch, Private communication.

[16] O. Ore, The four color problem. Academic Press, N.Y. London 1967.

[17] L. Vigneron, Coloration des réseaux cubiques. C.R. Acad. R. Acad. Sc. Paris 223 (1946) 705 and 770.

Galois Theory of Tree Extensions *

Georg Gati

ABSTRACT

Two of the main Galois theoretic results, namely a Fundamental Theorem and a solvability criterion, are derived for trees, a class of structures that is quite different from the usual objects of Galois theory, like fields and rings. This Galois theory of trees arose from investigations into the complexity of algorithmic problems. It points to the possibility of similarly considering more general classes of graphs.

1. Introduction and Notation

Generalizations of algebraic concepts to model theory go back to A. Robinson [8]. For Galois theory one considers any first order theory τ of the predicate calculus, the special case being the elementary theory of fields. A *model extension* for τ is a pair (M_1, M_2) of models of τ such that M_1 is a submodel of M_2. The *Galois group* $\mathrm{Gal}(M_2/M_1)$ of this model extension is the group of all automorphisms of M_2 that fix M_1 elementwise. Engeler [3] has considered the role of these groups when M_1 and M_2 are certain models of theories arising in connection with algorithmic problems. Then in several cases [2, 4] these models have tree structure. This led us to deriving for trees analogues of the Fundamental Theorem and the solvability criterion of the Galois theory of fields. For this the characterization of the automorphism groups of trees by G. Pólya [7] proved useful. In the following we set up the terminology and notation for a more detailed description of our results. Our graph theoretical terminology follows Harary [5].

DEFINITION 1.1. A *tree extension* is a pair (T_1, T_2) of trees where T_1

---

* Supported under Grants No. 2.861.-0.77 and 82.813-0.80 of Swiss National Science Foundation. This work is part of the author's "Habilitationsschrift".

is a subtree of T_2.

DEFINITION 1.2. The *Galois group* $\text{Gal}(T_2/T_1)$ of a tree extension (T_1, T_2) consists of those automorphisms of T_2 that fix T_1 pointwise. We write only Gal for $\text{Gal}(T_2/T_1)$ if it is clear which tree extension (T_1, T_2) is meant.

It T_1 is the empty tree then a tree extension is a tree. If T_1 consists of a single point only then a tree extension is a rooted tree. In both cases the Galois group of a tree extension is the automorphism group of these structures in the usual sense.

We denote by S_n the unique - up to permutation group isomorphism - permutation group on n objects of order $n!$. Let G and H be two finite permutation groups. Then $G \cdot H$ denotes the *direct product* of G and H and $H * G$ the *wreath of H around G* (see [5, p. 169]). Let G_1, G_2 and H be permutation groups acting on the sets A_1, A_2 and B. Then the group $H * (G_1 \cdot G_2)$ is isomorphic to the group $(H * G_1) \cdot (H * G_2)$; the element $((g_1, g_2), (\underline{h}_1, \underline{h}_2))$ is mapped onto the element $((g_1, \underline{h}_1), (g_2, \underline{h}_2))$ under this isomorphism where $g_i \in G_i$, $i = 1, 2$, $\underline{h}_i$ is a mapping from A_i into H, $i = 1, 2$, and $(\underline{h}_1, \underline{h}_2)$ is the mapping from $A_1 \cup A_2$ into H which is equal to $\underline{h}_i$ when restricted to A_i, $i = 1, 2$. Also the group $H * (G_1 * G_2)$ is isomorphic to the group $(H * G_1) * G_2$.

DEFINITION 1.3. The class of *tree groups* is the smallest class of permutation groups that contains all S_n, $n \in \mathbb{N}$, and is closed under the operations of direct product and wreath.

The definition above permits us to recall a result of G. Pólya [7, p. 209]:

LEMMA 1.1. *The automorphism group of a tree is a tree group.*

This is also true for rooted trees:

LEMMA 1.2. *The automorphism group of a rooted tree is a tree group.*

PROOF. We use induction on the height h of the rooted tree T. If $h = 0$ then $T = K_1$, its automorphism group is S_1 and the assertion holds. Assume that it is proved for all rooted trees of height smaller than n and let T be a rooted tree of height n. Denote by $\underline{T}_1, \ldots, \underline{T}_k$ the isomorphism classes of the maximal planted subtrees of T and let $c_i := |\underline{T}_i|$, $1 \le i \le k$. If τ_i is the automorphism group of some element of $\underline{T}_i$, $1 \le i \le k$, then the automorphism group τ of T is isomorphic to

$$\prod_{j=1}^{k}(\tau_i * S_{c_i}) \cdot S_1$$

where the S_1 on the right hand side accounts for the root of T. By the induction hypothesis the τ_i are tree groups. Hence, τ is a tree group.

We now come to the analogous property of tree extensions:

LEMMA 1.3. The Galois group Gal of a tree extension (T_1,T_2) is a tree group.

PROOF. Let V_1 and V_2 be the point sets of T_1 and T_2 and let V_{1e} be the set of endpoints of T_1. Then $W = <V_2 - V_1>$ has no cycles, hence is a forest. Let $T^1, \ldots, T^m$ be the component trees of W and $V_{21}, \ldots, V_{2m}$ their point sets. As T_2 has no cycles each T^j, $1 \le j \le m$, has exactly one nearest endpoint v_i in V_{1e}. For assume otherwise, then there would be two paths s_1 and s_2 from some point in T^j to two different endpoints v_i and v_i' of T_1 that meet exactly one endpoint of T_1 and otherwise completely lie in $V_2 - V_1$. As T_1 is a tree there is a path s_3 in T_1 from v_i to v_i'. Hence the concatenation of the paths s_1, s_3 and s_2 yields a cycle in T_2, a contradiction to the fact that T_2 is a tree. Denote by W_i', $1 \le i \le n$, the point set of the forest consisting of exactly those T^j whose nearest endpoint is v_i and by Γ_i the automorphism group of the rooted tree $(<W_i' \cup \{v_i\}>, v_i)$, $1 \le i \le n$. Then each element of the group $G := \prod_{i=1}^{n}\Gamma_i$ induces an automorphism of T_2 that fixes T_1. Hence, G is isomorphic to a subgroup of $\text{Gal}(T_2/T_1)$. Note that G is the maximal subgroup of the automorphism group of T_2 that stabilizes the trees $W_i := <W_i' \cup \{v_i\}>$ and that $\text{Gal}(T_2/T_1)$ also stabilizes these trees. Hence, G is isomorphic to $\text{Gal}(T_2/T_1)$. But, by Lemma 1.2, G is a tree group, thus $\text{Gal}(T_2/T_1)$, too.

In the preceding proof we also have shown the following.

LEMMA 1.4. Let (T_1,T_2) be a tree extension. Then there exist subtrees $W_1, \ldots, W_n$ of T_2 with the following properties:

(a) Two such subtrees have no point in common.
(b) Each W_i contains exactly one endpoint of T_1.
(c) The union of the W_i and of T_1 is T_2.
(d) Let Γ_i be the automorphism group of the tree W_i rooted at its unique endpoint of T_1, $1 \le i \le n$. Then $\text{Gal}(T_2/T_1) = \prod_{i=1}^{n}\Gamma_i \cdot E_{|V_1 - V_{1e}|}$.

2. The Unwindings of a Tree Group

We introduce a concept that will be needed for most of our further investigations. Let G be a permutation group and d a natural number. Then G^d denotes the direct product of d copies of G and E_d denotes S_1^d. We denote by $S(G)$ the set of subgroups of G. We introduce the symbol S_0 by the following convention: $S_0 \cdot G = G \cdot S_0 = G$ and $S_0 * G = G * S_0 = S_0$.

Let K be a tree group. It follows from the associative and distributive laws for the direct product and the wreath that K satisfies exactly one of the following three characterizations:

(i) K is a symmetric group;
(ii) K is a direct product of two tree groups;
(iii) K is the wreath of a tree group around a symmetric group.

DEFINITION 2.1. Let K be a tree group. We recursively define the set $U(K)$ of *unwindings* of K as follows:

(i)   if $K = S_d$ then $U(K) = \{S_{d-f} \cdot E_f | 0 \leq f \leq d\}$;
(ii) if $K = G \cdot H$ then $U(K) = \{G' \cdot H' | G' \in U(G), H' \in U(H)\}$;
(iii) if $K = G * S_d$ then $U(K) = \{(G * S_{d-f}) \cdot \prod_{i=1}^{f} G_i | 0 \leq f \leq d, G_i \in U(G), 1 \leq i \leq f\}$.

It follows that $U(K) \subset S(K)$. If K is the Galois group of the tree extension (T_1, T_2) then we can characterize the elements of $U(K)$. Let V_2 be the point set of T_2', $v \in V_2$ and $V \subset V_2$. Then $K \cdot v$ denotes the orbit of v under the action of K. Let K_V denote the stabilizer subgroup of V in K which contains exactly those $g \in K$ with $g \cdot v \in V$ for all $v \in V$. We define an equivalence relation $\sim_V$ on the elements of K_V by $g \sim_V h$ iff $g \cdot v = h \cdot v$ for all $v \in V$. We denote by g' the equivalence class of the element g. Then a group structure is defined on $K_V/\sim_V$ by $g'h' := (gh)'$. Note that this definition is independent of the choice of representatives: Let $g_1 \in g'$ and $h_1 \in h'$. Then $g_1 h_1(v) = g_1(h(v)) = g(h(v))$ for all $v \in V$. Hence, $(g_1 h_1)' = (gh)'$. The group $K_V/\sim_V$ operates on V by $g' \cdot v := g \cdot v$ for all $v \in V$. Again this definition is independent of the choice of the representative of g'. We denote by $K_{|V}$ the permutation group consisting of the abstract group $K_V/\sim_{V'}$, the set V and the operation described above.

THEOREM 2.1. Let H be the Galois group $\text{Gal}(T_2/T_1)$ of a tree extension (T_1, T_2) and G a subgroup of H. Then G is an unwinding of H if and only if the following three conditions are satisfied:

(i) G is a tree group;

(ii) if v is a point of T_2 then $G_{|G \cdot v} = H_{|G \cdot v}$;

(iii) if w is a point of T_2 fixed by G and u and v are points adjacent to w with $H \cdot u = H \cdot v$ but $G \cdot u \neq G \cdot v$ then G fixes u or v.

PROOF. As H is a tree group we can prove this theorem by an induction on the structure of a tree group. Let H first be a symmetric group S_d. Then $d = 1$. But then $G \leq H$ implies $G = H$ and the assertion holds vacuously. Let now $H = S_d \cdot S_1$. We use induction on d. The case $d = 1$ is handled as above.

Assume that the assertion is proved for all $d < n$. We consider $S_n \cdot S_1$:

"only if": $G \in U(S_n \cdot S_1)$ implies

$$G \in \{S_n \cdot S_1\} \cup \{G' \cdot S_1 | G' \in U(S_{n-1} \cdot S_1)\}.$$

Hence, G is a tree group and (i) is satisfied. If $G = S_n \cdot S_1$ then $G = H$ and (ii) holds vacuously. If $G = G' \cdot S_1$ with $G' \in U(S_{n-1} \cdot S_1)$ then let G' operate on V' and $V'' := V_2 - V'$. As V'' is a singleton orbit we have $G_{|V''} = S_1 = H_{|V''}$. If $v \in V'$ then $G \cdot v = G' \cdot v$. As $G' \in U(S_{n-1} \cdot S_1)$ the induction hypothesis implies $H_{|G \cdot v} = H_{|G' \cdot v} = (S_{n-1} \cdot S_1)_{|G' \cdot v} = G'_{|G' \cdot v} = G_{|G' \cdot v} = G_{|G \cdot v}$. Thus (ii) is satisfied. If $G = S_n \cdot S_1$ then $G = H$ and (iii) is vacuously satisfied. Using the notation from above let $G = G' \cdot S_1$. If u or v are in V'' then (iii) is trivially satisfied. If u and v are in V' then already G' fixes w, and u and v are in the same orbit of $S_{n-1} \cdot S_1$ but in different orbits of G'. By the induction hypothesis G' fixes u or v. Hence, also $G = G' \cdot S_1$ fixes u or v. Thus (iii) is satisfied.

"if": Assume first that there is no triple (u,v,w) satisfying (iii). Then $G \cdot v = H \cdot v$ for all points v of T_2. Hence, by (ii)

$$H_{|H \cdot v} = H_{|G \cdot v} = G_{|G \cdot v} = G_{|H \cdot v}.$$

It follows that $G = H$. But $H \in U(H)$, which proves the assertion. Let now (u,v,w) be a triple satisfying (iii) and let v be fixed by G. Then $G = G' \cdot S_1$ with $G' \leq S_{n-1} \cdot S_1$. By (i) G' is a tree group. Let G' operate on V'. For all $v \in V'$ we have $G' \cdot v = G \cdot v$ and thus by (ii)

$$G'_{|G' \cdot v} = G_{|G \cdot v} = H_{|G \cdot v} = S_{n-1} \cdot S_{1 |G' \cdot v}.$$

For all $w \in V'$ the group G fixes w iff G' fixes w. For all $u,v \in V'$ we have $v \in H \cdot u$ iff $v \in (S_{n-1} \cdot S_1) \cdot u$ and $v \notin G \cdot u$ iff $v \notin G' \cdot u$. It follows that G' satisfies (i), (ii) and (iii) with respect to $S_{n-1} \cdot S_1$.

Thus by the induction hypothesis $G' \in U(S_{n-1} \cdot S_1)$. Hence, $G \in U(S_n \cdot S_1)$.

Let H be now direct product of two tree groups H_1 and H_2 for which the assertion is proved. We prove the assertion for H.

"only if": As $G \in U(H)$ we have $G = G_1 \cdot G_2$ with $G_i \in U(H_i)$, $I = 1, 2$. As the G_i are tree groups, G is a tree group, too, and (i) follows. Also $G \cdot v = G_i \cdot v$ for $i = 1$ or $i = 2$. Thus

$$G_{|G \cdot v} = G_{i|G_i \cdot v} = H_{i|G_i \cdot v} = H_{|G \cdot v}$$

and (ii) follows. If $v \in H \cdot u$ then $v \in H_i \cdot u$ for $i = 1$ or $i = 2$. Then G fixes w implies G_i fixes w and $v \notin G \cdot u$ implies $v \notin G_i \cdot u$. By the induction hypothesis G_i fixes u or v. Hence, G fixes u or v and (iii) follows.

"if": As $G \leq H$ we have $G = G_1 \cdot G_2$ with $G_i \leq H_i$, $i = 1, 2$. As G satisfies (i), the G_i satisfy (i), too. As $G \cdot v = G_i \cdot v$ for $i = 1$ or $i = 2$ we have

$$H_{i|G_i \cdot v} = H_{|G \cdot v} = G_{|G \cdot v} = G_{i|G_i \cdot v}.$$

Hence, the G_i satisfy (ii). If $v \in H_i \cdot u$ then $v \in H \cdot u$. If G_i fixes w then G fixes w. If $v \notin G_i \cdot u$ then $v \notin G \cdot u$. Thus, if a triple (u,v,w) satisfies the conditions of (iii) for G_i then G fixes u or v. Hence, G_i fixes u or v and thus satisfies (iii). By the induction hypothesis $G_i \in U(H_i)$, $i = 1, 2$. Hence, $G \in U(H)$.

Now let $H = (H' * S_d) \cdot S_1$ where H' is a tree group for which the assertion is proved. We prove it for H.

"only if": $G \in U(H)$ implies

$$G \in \{(H' * S_{d-f}) \cdot S_1 \cdot \prod_{j=1}^{f} H_i' | H_i' \in U(H'), 1 \leq i \leq f, 0 \leq f \leq d\}.$$

As by the induction hypothesis the H_i' and H' satisfy (i), G satisfies (i). If H_i' operates on v then

$$G_{|G \cdot v} = H_i'|_{H_i' \cdot v} = H'|_{H_i' \cdot v} = H_{|G \cdot v}.$$

If $H' * S_{d-f} \cdot S_1$ operates on v then

$$G_{|G \cdot v} = H' * S_{d-f} \cdot S_{1|((H' * S_{d-f}) \cdot S_1) \cdot v}$$

$$= H_{|((H' * S_{d-f}) \cdot S_1) \cdot v} = H_{|G \cdot v}.$$

Hence, G satisfies (ii). To see that G satisfies (iii) consider the induced subtree $T = (V,E)$ of T_2 which consists of all points of T_2 that can be reached from w on paths that contain no point T_1 but the endpoint z of T_1 that is nearest to w. Then the tree extension $(\{z\},T)$ has a Galois group H' and $G_{|V} \in U(H')$. But if G would not satisfy (iii) for the triple (u,v,w) then $G_{|V}$ would not satisfy (iii) with respect to H', contrary to the induction hypothesis.

"if": Let $T = (V,E)$ be one of the subtrees W_i of T_2 described in Lemma 1.4 and z the unique endpoint of T_1 contained in V. Denote by V^1 the points of T adjacent to z. Note that $H_{|V^1} = S_d$. Assume that G has an orbit $U \subset V^1$ with $|U| \geq 2$. As $H_{|U} = S_{|U|}$ it follows from (ii) that $G_{|U} = S_{|U|}$. Denote by U' the set of points of V that can be reached from U on paths that do not conain z. As $G_{|U}$ fixes no point, also $G_{|U'}$ fixes no point. It follows from (iii) that the orbits of $G_{|U'}$ and $H_{U'}$ are equal, and, by (ii), the restrictions of the groups G and H to these orbits are equal. Thus

$$G_{|U'} = H_{|U'} = H' * S_{|U|}.$$

It follows from (iii) that

$$G_{|V^1 - U} = E_{|V^1 - U|}.$$

Let v be a point of $V^1 - U$ and $\{v\}'$ the set of points of V that can be reached from v on paths that do not contain z. Then

$$G_{|\{v\}'} = H''$$

where $H'' \leq H'$. H'' satisfies (i), (ii) and (iii) because otherwise G would not satisfy (i), (ii) and (iii). Thus, by the induction hypothesis $H'' \in U(H')$. It follows that

$$G = (H' * S_{d-f}) \cdot \prod_{i=1}^{f} G_i \cdot S_1$$

where $G_i \in U(H')$, $1 \leq i \leq f$, and $f = |V^1 - U|$, i.e. $0 \leq f \leq d$. Hence, $G \in U(H)$. It remains to consider the case $H = H' * S_2$ which is completely analogous to the case $H = (H' * S_d) \cdot S_1$.

3. The Fundamental Theorem

Let (T_1,T_2) be a tree extension with Galois group Gal. We denote by $I(T_1,T_2)$ the set of subtrees of T_2 that contain T_1. We write $T' \leq T$ if T' is a subtree of T. For $T \in I(T_1,T_2)$ we denote by $g(T)$ the subgroup $\mathrm{Gal}(T_2/T)$ of Gal. Thus, $g(T_1) = \mathrm{Gal}$, $g(T_2)$ is the identity group, and if $T' \in I(T_1,T_2)$ is a subtree of T then $g(T) \leq g(T')$. For $G \in S(\mathrm{Gal})$ we denote by $t(G)$ the maximal subtree of T_2 which is pointwise fixed by all elements of G. Note that $t(G) \in I(T_1,T_2)$, $T_1 \le t(\text{Gal})$, $t(E_{|V_2|}) = T_2$, and $G' \le G$ implies $t(G) \le t(G')$. Thus the mappings $t: S(\text{Gal}) \to I(T_1,T_2)$ and $g: I(T_1,T_2) \to S(\text{Gal})$ form the (canonical) Galois correspondence of the tree extension (T_1,T_2). We have $G \le g(t(G))$ and $t(G) = t(g(t(G)))$ for all $G \in S(\text{Gal})$ and $T \le t(g(T))$ and $g(T) = g(t(g(T)))$ for all $T \in I(T_1,T_2)$. A criterion for $G \in S(\text{Gal})$ and $T \in I(T_1,T_2)$ to satisfy the equations $g(t(G)) = G$ and $t(g(T)) = T$, respectively, is called the Fundamental (or Main) Theorem of the Galois theory (see, e.g., [1, p. 186 and p. 236] for field extensions; the unwindings of the Galois group play a role analogous to Krull's closed subgroups of the Galois group). Let v be a point of a tree T. Then $\text{Adj}(v,T)$ denotes the set of points in T adjacent to v.

LEMMA 3.1. Let (T_1,T_2) be a tree extension with Galois group Gal and Galois correspondence (t,g). Then for all $T \in I(T_1,T_2)$ we have $g(T) \in U(\text{Gal})$.

PROOF. We have to show that $g(t)$ satisfies conditions (i), (ii) and (iii) of Theorem 2.1 for all $T \in I(T_1,T_2)$. By definition $g(T) = \text{Gal}(T_2/T)$. Hence, by Lemma 1.3, (i) is satisfied. Let v be a point of T_2. If v is also a point of T then $g(T) \cdot v = \{v\}$. But $g(T)_{|\{v\}} \simeq S_1 \simeq \text{Gal}_{|\{v\}}$, and (ii) is trivially satisfied. If v does not lie in T then consider the tree extension (T,T_2). Then, by Lemma 1.4(c), v must lie in one of the trees W_i of (T,T_2), say in the tree W_{i_0}. By Lemma 1.4(d) also $g(T) \cdot v$ must completely lie in W_{i_0}. It follows that $g(T)_{|g(T) \cdot v} \simeq S_{d_k} * \cdots * S_{d_1}$. Similarly, $\text{Gal}_{|\text{Gal} \cdot v} \simeq S_{e_k} * \cdots * S_{e_1}$. As $\text{Gal} \cdot v \supset g(T) \cdot v$ it follows that $g(T)_{|g(T) \cdot v} \le \text{Gal}_{|\text{Gal} \cdot v}$. Hence, $\text{Gal}_{|g(T) \cdot v} \simeq S_{d_k} * \cdots * S_{d_1}$, and (ii) is satisfied. Now let w be a point of T_2 fixed by $g(T)$, let W_{i_0} be the subtree of T_2 in which w lies according to Lemma 1.4(c) applied to the tree extension (T_1,T_2) and let z be the unique endpoint of T_1 contained in W_{i_0} according to Lemma 1.4(b). It follows that if two points u and v adjacent to w satisfy $\text{Gal} \cdot v = \text{Gal} \cdot w$ then none of them can lie on the unique path from w to z. By Lemma 1.3, $\text{Gal}_{|\text{Gal} \cdot v} \simeq S_{|\text{Gal} \cdot v|}$. Hence, there is an element $\gamma \in \text{Gal}$ which maps u onto v and fixes all points of T_2 that can be reached from w on paths not containing u or v. If $g(T)$ fixes neither u nor v then γ fixes T pointwise, hence $\gamma \in g(T)$, a contradiction to $u \notin g(T) \cdot v$. Thus (iii) is satisfied.

THEOREM 3.1. (Fundamental Theorem of the Galois theory of tree extensions).

Let (T_1, T_2) be a tree extension with Galois group Gal and Galois correspondence (t, g). Let $G \in S(Gal)$ and $T = (V, E) \in I(T_1, T_2)$.

(i) $g(t(G)) = G$ if and only if $G \in U(Gal)$.

(ii) $t(g(T)) = T$ if and only if T contains no point v such that there exists a line $\{v, w\}$ in T_2 where w satisfies the equation

$$(Adj(v, T_2) \cap Gal \cdot w) - V = \{w\}. \qquad (*)$$

PROOF. It follows from Lemma 1.4 that it suffices to consider the case $T_1 = K_1$. T_2 is then a rooted tree and $I(T_1, T_2)$ is the set of subtrees of T_2 that contain the root. We define a *stem* of a rooted tree as a line incident with the root. Let us now consider (i). Assume $g(t(G)) = G$. As $t(G) \in I(T_1, T_2)$ Lemma 3.1 implies $g(t(G)) \in U(Gal)$, thus $G \in U(Gal)$. For the other direction we assume $G \in U(Gal)$ and use induction on the height $h(T_2)$ of T_2. The case $h(T_2) = 1$ is trivial. Assume that the assertion is proved for all trees of height smaller than m and let $h(T_2) = m$. Let the degree of the root r of T_2 be d and denote by $T^1, \ldots, T^d$ the d component subtrees of the forest $T_2 - r$. Then

$$Gal = \prod_{i=1}^{c} G_i * S_{d_i}) \cdot S_1$$

where the G_i are isomorphic to the automorphism groups of the T^k considered as rooted at their unique point adjacent to r. Now $G \in U(Gal)$ implies

$$G = \prod_{i=1}^{c}((G_i * S_{d_i} - f_i) \cdot \prod_{j=1}^{f_i} G_{ij}) \cdot S_1$$

where $G_{ij} \in U(G_i)$, $1 \le j \le f_i$, $1 \le i \le c$.

Denoting by $V(T)$ the point set V of a tree $T = (V, E)$ we have

$$t(G) = \langle \bigcup_{i=1}^{c} \bigcup_{j=1}^{f_i} V(t(G_{ij})) \cup \{r\} \rangle$$

and

$$g(t(G)) = \prod_{i=1}^{c}((G_i * S_{d_i} - f_i) \cdot \prod_{j=1}^{f_i} g(t(G_{ij}))) \cdot S_1.$$

By induction hypothesis we have $g(t(G_{ij})) = G_{ij}$, $1 \le j \le f_i$, $1 \le i \le c$, and hence $g(t(G)) = G$. We now consider (ii). Let $t(g(T)) = T$. Assume that there exists a point v in $T = (V, E)$ and a line $\{v, w\}$ in T_2 satisfying (*). Then $w \notin V$, which implies

$$g(T) \cdot w \cap V = \emptyset. \qquad (**)$$

As $g(T)$ fixes v and $g(T) \le Gal$ we have $g(T) \cdot w \subset Adj(v, T_2) \cap Gal \cdot w$. It follows with (**) and (*) that

$$\{w\} \subset g(T) \cdot w \subset (Adj(v, T_2) \cap Gal \cdot w) - V \subset \{w\}.$$

Hence $g(T) \cdot w = \{w\}$. Thus $w \in V(t(g(T)))$. As $w \notin V$ we have $t(g(T)) \ne T$, a contradiction to our assumption. Hence, there cannot exist a line $\{v,w\}$ in T_2 satisfying (*). In order to prove the "if" direction assume that there is no adjacent point pair $(v,w) \in V \times V(T_2)$ that satisfies (*). If $t(g(T)) \ne T$ then there is a point $w \in V(t(g(T))) - V$, and we can choose w in such a way that w is adjacent to a point $v \in V$. As $Gal \cdot w \cap Adj(v, T_2) \subset g(T) \cdot w \cup V$ we have

$$\{w\} \subset (Gal \cdot w \cap Adj(v, T_2)) - V \subset g(T) \cdot w \subset \{w\}$$

where the rightmost inclusion follows from $w \in V(t(g(T)))$. Hence, $\{v,w\}$ satisfies (*), a contradiction. Thus $t(g(T)) = T$.

We now want to shed some light on the role of those unwindings of the Galois group that are also normal subgroups.

DEFINITION 3.1. Let G be a tree group. We define the set $NU(G)$ of *normal unwindings* of G as $\{H \in U(G) | H \triangleleft G\}$ where $H \triangleleft G$ denotes that H is a normal subgroup of G.

LEMMA 3.2. The following equations recursively characterize the set of normal unwindings of a tree group:

(i) $NU(S_d) = \{S_d, E_d\}$;

(ii) $NU(G \cdot H) = \{G' \cdot H' | G' \in NU(G), H' \in NU(H)\}$;

(iii) $NU(G * S_d) = \{G'^d | G' \in NU(G)\} \cup \{G * S_d\}$.

PROOF. (i) and (ii) immediately follow from the definitions. Hence, we only consider (iii) and assume that S_d operates on the set A and G operates on the set B and that the assertion holds for the group G. Let $K \in NU(G * S_d)$. Then $K \in U(G * S_d)$ implies

$$K = G * S_{d-f} \cdot \prod_{i=1}^{f} G_i$$

with $G_i \in U(G)$ for some $f \in \{0, 1, \ldots, d\}$ and $K \triangleleft G * S_d$ implies $f = 0$ or $f = d$ because otherwise we would get a contradiction to (i) by choosing $G = S_1$. If $f = d$ we have to show that $G_i \simeq G_j$, $1 \le i, j \le f$. Let G_i operate on $\{a_i\} \times B$ and G_j on $\{a_j\} \times B$, where $a_i, a_j \in A$. If we can show that for each $z \in G_i$ there is a $z' \in G_j$ such that for all $b \in B$ we have $z \cdot b = z' \cdot b$ then we are finished because then it follows that G_i is isomorphic to a subgroup of G_j, and the assertion follows by interchanging G_i and G_j. Thus let $z \in G_i$ and let t_{ij} be the transposition of a_i and a_j in S_d. Let $\underline{1}|_B$ denote the mapping $A \to G$ which maps each element of A onto the identity element of G and let $\underline{z}$ denote any mapping $A \to G$ with $\underline{z}(a_i) = z$. Then $K \triangleleft G * S_d$ implies that there exists an element $(f, \underline{z}') \in K$ such that for all $b \in B$

$$(t_{ij}, \underline{1}|_B)(Id_A, \underline{z})(t_{ij}, \underline{1}|_B)^{-1}(a_j, b) = (f, \underline{z}')(a_j, b).$$

The left hand side of this equation equals to $(a_j, \underline{z}(a_i) \cdot b)$ and, hence, the right hand side equals to $(a_j, \underline{z}'(a_j) \cdot b)$. It follows that $\underline{z}'(a_j) \in G_j$ and for all $b \in B, \underline{z}'(a_j) \cdot b = \underline{z}(a_i) \cdot b = z \cdot b$. Hence, we can choose z' as $\underline{z}'(a_j)$ which proves the assertion. We now know that $K \in NU(G * S_d)$ implies $K = G * S_d$ or $K = G'^d$ with $G' \in U(G * S_d)$. It remains to show that in the latter case $G' \triangleleft G$. As $G * S_d$ contains G^d as subgroup it follows that $G'^d \triangleleft G * S_d$ implies $G'^d \triangleleft G^d$ and hence $G' \triangleleft G$.

For the other inclusion of (iii) we have to show that $G' \in NU(G)$ implies $G'^d \in NU(G * S_d)$. As $G'^d \in U(G * S_d)$, it remains to be shown that $G'^d \triangleleft G * S_d$. Thus let $(f, \underline{h}) \in G * S_d$ and $(Id_A, \underline{h}') \in G'^d$. Denoting $(f, \underline{h})(Id_A, \underline{h}')(f, \underline{h})^{-1}$ by z we have to show that there exists an element $z' \in G'^d$ with $z \cdot (a, b) = z' \cdot (a, b)$ for all $(a, b) \in A \times B$.

But

$$z \cdot (a, b) = (f, \underline{h})(Id_A, \underline{h}')(f^{-1}(a), (\underline{h}(f^{-1}(a)))^{-1}(b))$$
$$= (f, \underline{h})(f^{-1}(a), \underline{h}'(f^{-1}(a))(\underline{h}(f^{-1}(a)))^{-1}(b))$$
$$= (a, \underline{h}(f^{-1}(a))\underline{h}'(f^{-1}(a))(\underline{h}(f^{-1}(a)))^{-1}(b)).$$

As $\underline{h}(f^{-1}(a)) \in G$ and $\underline{h}'(f^{-1}(a)) \in G'$ the assumption $G' \triangleleft G$ implies that there is a $h'' \in G'$ with $\underline{h}(f^{-1}(a))\underline{h}'(f^{-1}(a))(\underline{h}(f^{-1}(a)))^{-1}(b) = h''(b)$ for all $b \in B$. Setting $\underline{h}''(a) := \underline{h}(f^{-1}(a))\underline{h}'(f^{-1}(a))(\underline{h}(f^{-1}(a)))^{-1}$ it follows that $\underline{h}''(a) \in G'$ for all $a \in A$ and that z' can be chosen as $(Id_A, \underline{h}'')$.

The preceding lemma has recursively determined the structure of those unwindings of the Galois group that are also normal subgroups. We now define a subclass of $I(T_1, T_2)$ and show that it corresponds to $NU(Gal)$.

DEFINITION 3.2. Let (T_1, T_2) be a tree extension with Galois group Gal. A tree $T = (V, E) \in I(T_1, T_2)$ is a normal subtree of T_2 if $v \in V$ implies $Gal \cdot v \subset V$.

COROLLARY 3.1. Let (T_1, T_2) with $T_2 = (V_2, E_2)$ be a tree extension with Galois group Gal and Galois correspondence (t,g). Let $T = (V,E)$ be a normal subtree of T_2. Then $t(g(T)) = T$ if and only if for all $v \in V$ and for all $w \in \bigcup_{v \in V} Adj(v,T_2) - V$ we have $|Adj(v,T_2) \cap Gal \cdot w| \neq 1$.

PROOF. "if": Let $v \in V$ and $\{v,w\} \in E_2$ be arbitrarily chosen. If $w \in V$ then $Gal \cdot w \subset V$ and hence equation (*) of Theorem 3.1 is not satisfied. If $w \notin V$ then $V \cap Gal\, w = \emptyset$.
Hence

$$|(Adj(v,T_2) \cap Gal \cdot w) - V| = |Adj(v,T_2) \cap Gal \cdot w|$$
$$> 1,$$

and thus $\{v,w\}$ does not satisfy equation (*) of Theorem 3.1.

"only if": As $w \notin V$ we have $V \cap Gal \cdot w = \emptyset$. Hence, $c := |(Adj(v,T_2) \cap Gal \cdot w) - V| = |Adj(v,T_2) \cap Gal \cdot w|$. As $w \in Adj(v,T_2) \cap Gal \cdot w$ the assertion follows from Theorem 3.1. (ii).

THEOREM 3.2. Let (T_1,T_2) be a tree extension with Galois group Gal and Galois correspondence (t,g).

(i) If $T = (V,E)$ is a normal subtree of T_2 then $g(T) \in NU(Gal)$.

(ii) If $G \in NU(Gal)$ then $t(G)$ is a normal subtree of T_2.

(iii) Let T be a normal subtree of T_2. Then $Gal/g(T) \simeq Gal(T/T_1)$.

PROOF. It follows with Lemma 1.4 that it suffices to consider the case $T_1 = K_1$. Let $V(T_1) = \{r\}$ and let $T^1, \ldots, T^n$ be the component subtrees of the forest $< V(T_2) - \{r\} >$. Then $Gal = \prod_{i=1}^{c}(G_i * S_{d_i}) \cdot S_1$ where each G_i is isomorphic to the automorphism group of some T^j rooted at the unique point v_j adjacent to r. We use induction on the height $h(T_2)$ of T_2. The case $h(T_2) = 1$ is trivial.

(i) Denote by $\hat{T}^i$ the tree $< V(T^i) \cap V >$. By assumption T is a normal subtree of T_2. Hence, if $\hat{T}^i$ is not empty, then $\hat{T}^i$ is a normal subtree of T^i with respect to the tree extension $((\{v_i\}, \emptyset), T^i)$. Thus $g(T) = \prod_{i=1}^{c}(G_i * S_{d_i - f_i} \cdot G_i'^{f_i}) \cdot S_1$ with $f_i \in \{0, d_i\}$, where $G_i'\ Gal(T^i|\hat{T}^i)$. If $\hat{T}^i$ is not empty, then by the induction hypothesis $G_i' \in NU(G_i)$, $1 \leq i \leq c$, which proves the assertion.

(ii) As $G \in NU(Gal)$ we have

$$G = \prod_{i=1}^{c}(G_i * S_{d_i - f_i} \cdot G_i'^{f_i}) \cdot S_1$$

with $f_i \in \{0, d_i\}$. We can assume that $d_i \neq 1$, $1 \leq i \leq c$. If $f_i = 0$ then $t(G)$ contains no point of the corresponding trees T^j. If $f_i = d_i$ then $t(G)$ contains the f_i subtrees $\tau_j(G_i')$ of certain tree extensions $((\{v_j\}, \emptyset), T^j)$ which have Galois group G_i and Galois correspondences (τ_j, γ_j). by the induction hypothesis $G'_i \in NU(g_i)$ implies that $\tau_j(G_i')$ is a normal subtree of T^j. As $t(G)$ also contains r it follows that $t(G)$ is a normal subtree of T_2.

(iii) As T is normal we can define a group homomorphism $\rho: Gal(T_2/T_1) \to Gal(T/T_1)$ where for each $h \in Gal(T_2/T_1)$ we denote by $\rho(h)$ the restriction of h to T. As ρ is surjective and $\ker \rho = g(T)$ the assertion follows from the first isomorphism theorem (of group theory).

4. A Solvability Criterion

Algorithmic problems whose Galois groups are those of tree extensions (T_1, T_2) suggest that the solvability of the problem by an elementary operation corresponds to the extensibility of the tree T_1 to the tree T_2 by adjoining stars according to certain rules, see [4]. In the following we consider this situation from the viewpoint of Galois theory.

By a *star*, we mean a rooted tree of height 1. A rooted tree T is specified by the triple (V, E, r) where (V, E) is a tree and $r \in V$ is the root. The star having two points is denoted by $K_{2,r}$. The formula $M = [3a, 4b]$ means that the *multiset* M consists of 3 copies of element a and 4 copies of element b. Thus $M - [a, 2b, c] = [2a, 2b]$, $M \cup [a, 2b, c] = [4a, 6b, c]$ and $2M = [6a, 8b]$.

DEFINITION 4.1. Let (T_1, T_2) be a tree extension and T a star, where $T_i = (V_i, E_i)$, $i = 1, 2$, and $T = (V, E, r)$. Let $v \in V_1$. We define the tree $T_1(v \to T)$ to be (V', E') where $V' = V_1 \cup (V - \{r\})$ and $E' = E_1 \cup \{\{v, w\} | \{r, w\} \in E\}$. We say that $T_1(v \to T)$ is *admissible* if

(i) $T_1(v \to T) \in (T_1, T_2)$ and

(ii) $V - \{r\}) \cap \bigcup_{w \in Adj(v, T_2) - (V - \{r\})} Gal \cdot w = \emptyset$.

The following definition makes precise what we mean by extending a tree T_1 to a tree T_2 by adjoining stars from a given multiset according to certain rules.

DEFINITION 4.2. A *firmament* is a multiset of stars not containing $K_{2,r}$. We say that a firmament F *solves* a tree extension (T_1, T_2) with

$T_1 \neq T_2$ if there exists a sequence $(T_{1,0}, \ldots, T_{1,n})$ of trees and a sequence $(T^1, \ldots, T^n)$ of stars with

(i) $T_{1,i} = T_{1,i-1}(v \to T^i)$ for some point v in $T_{1,i-1}$, $1 \leq i \leq n$;

(ii) $T_{1,i}$ is admissible, $1 \leq i \leq n$;

(iii) $\bigcup_{i=1}^{n} \{T^i\} = F \cup [(n-|F|)K_{2,r}]$;

(iv) $T_{1,0} = T_1$ and $T_{1,n} = T_2$.

Thus a firmament F solves a tree extension (T_1, T_2) if one can extend T_1 to T_2 by successively adjoining to T_1 exactly the elements of F and possibly copies of $K_{2,r}$ such that the intermediate results are admissible trees. One can think of other criteria of the solvability of a tree extension by a firmament [6] but we shall only derive a criterion that formally resembles the solvability criterion by radical extensions of the Galois theory of field extensions. The following definition is the first step into this direction.

DEFINITION 4.3. Let G be a tree group. A normal series

$$G = G_1 \trianglelefteq G_2 \trianglelefteq \ldots \trianglelefteq G_n = E$$

of G is a *normal unwinding series*, briefly n.u.s., of G if $G_i \in U(G)$, $1 \leq i \leq n$. The quotient multiset. $Q(G)$ of the n.u.s. $\underline{G} = (G_i)_{1 \leq i \leq n}$ of the tree group G is the multiset $[G_i/G_{i-1} | 1 \leq i \leq n]$ where the quotient groups G_i/G_{i-1} are considered as abstract groups. A n.u.s. $\underline{G}$ of G is *maximal* if it does not permit a proper refinement that is a n.u.s. of G.

LEMMA 4.1. Let $\underline{G} = (G_1, \ldots, G_n)$ be a normal unwinding series of a tree group G. Then $\underline{G}$ is maximal if and only if for any $i \in \{1, \ldots, n\}$ the groups G_{i-1} and G_i differ in exactly one factor $H * S_d$ of the product making up G_{i-1} and this factor is changed into the product H^d in G_i.

PROOF. "only if": Let $\underline{G}$ be maximal, $G_{i-1} = \prod_{j=1}^{c} G_{i-1,j}$, $G_i = \prod_{j=1}^{c} G_{ij}$ where the $G_{i-1,j}$ are not direct products of tree groups and the numbering is such that $G_{i-1,j} \geq G_{ij}$, $1 \leq j \leq c$. At least for one $l \in \{1, \ldots, c\}$ we must have $G_{i-1,l} \neq G_{i,1}$ because otherwise we would have $G_{i-1} = G_i$. If for another $l' \neq l$ we have $G_{i-1,l'} \neq G_{i,l'}$ then $K := \prod_{\substack{j=1 \\ j \neq l'}}^{c} G_{ij} \cdot G_{i-1,l'}$ satisfies $K \in U(G)$ and $G_{i-1} \trianglelefteq K \trianglelefteq G_i$, a contradiction to the maximality of $\underline{G}$. Hence, $G_{i-1,l}$ is the only factor in G_{i-1} that is changed in G_i. As $G_{i-1,l} = H * S_d$ the conditions $G_{i-1} \trianglelefteq G_i$ and $G_i \in U(G)$

imply by Lemma 3.2 that $G_{i-1,l}$ is replaced in G_i by H'^d where $H' \in NU(H)$.

If $H' \neq H$ then $K := \prod_{\substack{j=1 \\ j \neq l}}^{c} G_{i-1,l} \cdot H^d$ satisfies $K \in U(G)$ and $G_{i-1} \trianglelefteq K \trianglelefteq G_i$, a contradiction to the maximality of $\underline{G}$. Hence, $H' = H$.

"if": Let $G_{i-1} \trianglelefteq K \trianglelefteq G_i$. As G_{i-1} differs from G_i in a single factor $G_{i-1,l}$ it follows that K differs from G_{i-1} only in $G_{i-1,l'}$ too. As $G_{i-1,1} = H * S_d$ the conditions $G_{i-1} \trianglelefteq K$ and $K \in U(G)$ imply that $G_{i-1,l}$ is changed into H'^d in K where $H' \in NU(H)$. As $G_{i-1,l}$ is changed into H^d in G_i the condition $K \trianglelefteq G_i$ implies $H' \trianglelefteq H$. But $H' \in NU(H)$ implies $H' \leq H$. Hence, $H' = H$ and $K = G_i$.

The following lemma is a strengthening of the Jordan-Hölder theorem valid for tree groups.

LEMMA 4.2. Two maximal normal unwinding series $\underline{G}_1$ and $\underline{G}_2$ of a tree group G are equivalent, i.e. $Q(\underline{G}_1) = Q(\underline{G}_2)$.

PROOF. The assertion is equivalent to: $Q(\underline{G})$ is independent of the choice of a maximal n.u.s. $\underline{G}$ of G. In order to prove this we use induction on the maximal number h of nestings of wreath products in a direct factor of G. If $h = 0$ then $G = \prod_{i=1}^{c} S_{d_i}$ and the assertion is trivially verified. Assume that it is proved for all h smaller than n and let G have $h = n$. Let $G = \prod_{i=1}^{c} H_i * S_{d_i}$. Then

$$Q(\underline{G}) = \bigcup_{i=1}^{c} (d_i Q(\underline{H}_i) \cup S_{d_i})$$

for any maximal n.u.s. $\underline{G}$ of G and the assertion follows because by the induction hypothesis $\overline{Q(H_i)}$ is independent of the choice of a maximal n.u.s. $\underline{H}_i$ of H_i.

By the preceding lemma the quotient multiset $Q(\underline{G})$ depends only on the tree group G and thus will be denoted by $Q(G)$. If F is a firmament we denote by $Aut(F)$ the multiset $[Aut(T) | T \in F]$ where $Aut(T)$ is the abstract group underlying the automorphism group of the rooted tree T.

THEOREM 4.1. (solvability criterion) A firmament F solves a tree extension (T_1, T_2) if and only if $Aut(F) = Q(Gal(T_2/T_1))$.

PROOF. It follows with Lemma 1.4 that it suffices to consider the case $T_1 = K_1$, i.e. T_2 is a rooted tree. We use induction on the height h of T_2. The case $h = 1$ is trivial. Assume that the assertion is proved for all h smaller than n and let T_2 be a rooted tree of height n with root r.

Let $T^1, \ldots, T^m$ be the component trees of the forest $<V(T_2) - \{r\}>$. Then $Gal(T_2/T_1) = \overset{c}{\underset{i=1}{\Pi}}(G_i * S_{d_i}) \cdot S_1$ where the G_i are isomorphic to the automorphism groups of certain T^j rooted at their unique point v_j adjacent to r.

Also

$$Q(Gal(T_2/T_1)) = \overset{c}{\underset{i=1}{\cup}}(d_i Q(G_i) \cup \{S_{d_i}\})$$

$$= \overset{m}{\underset{j=1}{\cup}} Q(Gal(T^j/(\{v_j\}, \varnothing))) \cup \overset{c}{\underset{i=1}{\cup}} \{S_{d_i}\}.$$

Let $St(k)$ denote the star having $k+1$ points; thus $St(2) = K_{2,r}$. Then F solves (T_1, T_2) iff $F = F' \cup \overset{m}{\underset{j=1}{\cup}} F_j$ where $F' = \overset{c}{\underset{i=1}{\cup}} [ST(d_i)]$ and F_j solves T^j. The assertion now follows from the induction hypothesis.

References

[1] P.M. Cohn, *Algebra*, Vol. 2. John Wiley, Chichester (1979).
[2] E. Engeler, Lower bounds by Galois theory. *Astérisque*. 38/39 (1976) 45-52.
[3] E. Engeler, Generalized Galois theory and its application to complexity. *Theoret. Comput. Sci.* 13 (1981) 271-293.
[4] G. Gati, Some elements of a Galois theory of the structure and complexity of the tree automorphism problem. *Theoret. Comput. Sci.* 14 (1981) 95-109.
[5] F. Harary, *Graph Theory*. Addison-Wesley, Reading, Mass. (1969).
[6] F. Harary, personal communication (1980).
[7] G. Pólya, Kombinatorische Anzahlbestimmungen für Gruppen, Graphen und chemische Verbindungen. *Acta Math.* 68 (1937) 145 - 254.
[8] A. Robinson, *Introduction to Model Theory and the Metamathematics of Algebra*. North-Holland, Amsterdam (1963).

Real Graph Polynomials

C.D. Godsil

ABSTRACT

We study the properties of three polynomials associated with graphs, having in common the feature that their zeros are always real. Recent results on these polynomials are discussed and some open problems concerning them are presented.

1. Introduction

Here three polynomials associated with graphs are discussed. The first of these is the characteristic polynomial of the adjacency matrix of a graph. The second is a family of polynomials introduced by Rolf Schneider in [19], where they are used to give a proof of Aleksandrov's inequalities. The third is the matchings polynomial, which is a generating function for the number of matchings with k edges in a given graph ($k = 0, 1, \ldots$).

One common feature of these polynomials is that their zeros are always real. The polynomials of Schneider include as a special case the characteristic polynomial and it will be seen that the characteristic and matchings polynomials are also closely related. Our hope is that the reader(s) of this paper will become convinced that collectively and individually these polynomials are worth further study.

The contents of the remainder of this paper can be summarized as follows. In Section 2 we consider for subsequent use in Section 4, the relation between the characteristic polynomial and walks in a graph. The material is not new, although the presentation is. In Section 3 we consider the polynomials introduced by Schneider and show how they can be used to derive some results of Richard Stanley. (The brevity of this section is forced on me by the fact that I know of nothing more about these polynomials.) In Section 4 a number of recent results concerning the matchings polynomials are surveyed. The final section lists some selected problems concerning these polynomials.

2. The Characteristic Polynomial

Let G be a graph with adjacency matrix A. The characteristic polynomial $\phi(G, x)$ of G is just $det(xI-A)$. This polynomial is well known, our discussion here is intended chiefly to review some results needed in Section 4. Further information can be found in the book [3] and, at less length in [1].

What we do need is the relation between $\phi(G, x)$ and the walks in G. A *walk* of length r in G is a sequence $v_0,\ldots,v_r$ of vertices from G such that consecutive vertices are adjacent. It is *closed* if $v_0 = v_r$. If $v_i \in V(G)$ then $C_i(G, x)$ is the ordinary generating function for the number of closed walks of length k in G with v_i as first vertex ($k = 0, 1,\ldots$). We define $C(G, x)$ to be:

$$\sum_{i \in V(G)} C_i(G, x).$$

Thus $C(G,x)$ enumerates all closed walks in G and $C_i(G,x)$ enumerates those with initial vertex i.

2.1. THEOREM. We have

$$C_i(G, x^{-1}) = x\phi(G\backslash i, x)/\phi(G, x) \qquad (1)$$

and

$$C(G, x^{-1}) = x\phi'(G, x)/\phi(G, x). \qquad (2)$$

The first identity is basically a disguised form of the usual expression for the inverse of a matrix, since $C_i(G, x)$ is just the i^{TH} diagonal entry of $(I - xA)^{-1}$. To the best of my knowledge (1) first appeared explicitly as equation (18) in [15]. The second identity follows from the first because

$$\phi'(G, x) = \sum_{i \in V(G)} \phi(G\backslash i, x). \qquad (3)$$

The coefficient of x^k in $C(G, x)$ is $tr(A^k)$. Hence (2) expresses, in terms of generating functions. Newton's relations between the coefficients of a polynomial and the sums of powers of its zeros.

Both these results are important and useful. Examples of their application will be found in [10], [11] and [15]. (Kasteleyn's paper [15] contains a great deal of interesting material and is worthy of being studied very thoroughly.) By way of an application of Theorem 2.1 we will derive the so-called "feasibility conditions" for distance regular graphs.

One preliminary concept is required. Let G be a graph and let π be a partition of its vertex set. Suppose π has components

$\pi_1, \ldots, \pi_r$. We say π is *equitable* if, for all $i, j = 1, \ldots, r$, the number of vertices in π_j adjacent to a given vertex x in π_i is independent of the choice of x in π_i. We denote this number by q_{ij} and call $Q(G, \pi) = (q_{ij})$ the *quotient matrix* of G. The discrete partition of $V(G)$ (with each component a single vertex) is always equitable. The subgraph induced by a single component of an equitable partition is necessarily a regular graph. The orbits of any group of automorphisms of G form an equitable partition.

Let $\Delta(v) = (\Delta_0, \ldots, \Delta_r)$ be the partition of G which has its ith component the set of vertices in G at distance i from the vertex v in G. Then G is *distance regular* if for each vertex v, Δ_v is equitable and the quotient matrix $Q(G, \Delta(v))$ is independent of the choice of v. A simple example of a distance regular graph is Petersen's graph.

Suppose $\pi = (\pi_1, \ldots, \pi_r)$ is an equitable partition of G such that π_i is a single vertex v. Let $Q = Q(G, v)$. Then we may view Q as the adjacency matrix of a directed multigraph M, with vertices $w_1, \ldots, w_r$ corresponding to the components of π. We will use v to denote the vertex w_1 in M. A simple induction argument shows that the number of closed walks of length k in G which start at v equals the number of closed walks in M which start at v. Consequently

$$\frac{\phi(G\backslash v, x)}{\phi(G, x)} = \frac{\phi(M\backslash v, x)}{\phi(M, x)}. \tag{4}$$

Now suppose $\phi(G\backslash v)$ is independent of the choice of v (e.g., if G is vertex-transitive or distance-regular). Then if $n = |V(G)|$ it follows using (3) that $n\phi(G\backslash v) = \phi'(G)$, whence

$$\frac{\phi'(G, x)}{\phi(G, x)} = \frac{n\phi(M\backslash v, x)}{\phi(M, x)}. \tag{5}$$

The left side of (4) is a rational function and its residue at a pole θ is equal to the multiplicity of θ as a zero of $\phi(G, x)$. Consequently the residue of $n\phi(M\backslash v)/\phi(M)$ at each of its poles must be an integer.

The point here is that, given M (from which n may be recovered) it is possible to compute the multiplicity of each zero of $\phi(G, x)$. Thus we have a criterion for determining whether a given M is a quotient matrix with respect to an equitable partition of some graph - namely the multiplicities we compute must all be non-negative integers. This condition is called the "feasibility condition". It has been used many times, particularly in establishing the non-existence of many cages and Moore graphs (for a discussion of this see e.g. [1]).

The usual derivation of this result only works for distance regular graphs, whereas the argument just used applies whenever $\phi(G\backslash v)$ is independent of v. An even more general form is established in [11] using a superficially different proof.

It follows from (4) that a graph with irreducible characteristic polynomial can have no equitable partition other than the discrete one. Consequently the automorphism group of such a graph must be trivial. The fact that irreducible characteristic polynomial implies trivial automorphism group was first noted by A. Mowshowitz. (See Theorem 2.10 in [3].) Tutte has shown that if $\phi(G, x)$ is irreducible then G is reconstructible [21]. (For a different proof, see [11].)

3. A Generalization of $\phi(G, x)$

The motivation for this section is provided by Section 2 of Stanley's paper [20], where the following result is established.

3.1. THEOREM. Let M be a unimodular matroid of rank m. Let $M_1, \ldots, M_r (2 \leq r \leq m)$ be a partition of the elements of M. Let $a_3, \ldots, a_r$ be non-negative integers summing to $m - l$. Let f_k be the number of bases of M containing k elements in M_1, $l - k$ in M_2 and a_i in M_i ($i = 3, \ldots, r$). Then the sequence $f_k / \binom{l}{k}$ ($k = 0, 1, \ldots, l$) is log-concave.

A non-negative sequence $g_1, \ldots, g_l$ is *log-concave* if $g_k^2 \geq g_{k-1} g_{k+1}$. For fixed l the binomial coefficients $\binom{l}{k}$ form a log-concave sequence. The term-wise (or Hadamard) product of two log-concave sequences is log-concave, so Theorem 3.1 implies that the numbers f_k form a log-concave sequence. Stanley's proof of the above result uses the Aleksandrov-Fenchel inequalities for mixed volumes. However he notes that if M is the cycle matroid of a graph then, in an important special case, the log-concavity of the numbers $f_k / \binom{l}{k}$ is implied by the fact that the polynomial $f(M, x) = \sum_{k=0}^{l} f_k x^k$ has only real zeros.

This follows from Newton's inequalities, which assert that if $\sum_{k=0}^{l} a_k x^k$ has only real zeros and $b_k = a_k / \binom{l}{k}$ then $b_k^2 \geq b_{k-1} b_{k+1}$. (We do not need to assume here that the a_k are non-negative.) Stanley asks whether, for an arbitrary unimodular (= regular) matroid, the polynomial $f(M, x)$ has only real zeros. It turns out that this is indeed the case and that this can be proved using a result due to Schneider [19], which he employs to prove Aleksandrov's inequalities.

3.2. THEOREM. Let $A, B_1, \ldots, B_r$ be symmetric $m \times m$ matrices where $r < m$ and the B_i are positive semidefinite. Let $a_3,\ldots,a_r$ be non-negative integers summing to $m - t$ and let $\tau_l(x,y_1)$ be the coefficient of $y_2^{t-l} y_3^{a_3} \cdots y_r^{a_r}$ in

$$det(xA + y_1B_1 + \cdots + y_rB_r).$$

Then the zeros of $\tau_l(x, 1)$ are real and are interlaced by the zeros of $\tau_{l-1}(x, 1)$ $(l = 1,\ldots,m)$.

By *interlaced* we mean that between any two of the l zeros of $\tau_l(x, 1)$ there is a zero of $\tau_{l-1}(x, 1)$. The most important special case of this result occurs when A is also positive semidefinite. For then it is possible to show that the coefficients of $\tau_l(x, 1)$ are non-negative. So if $\tau_l(x, 1)$ is $\sum_{k=0}^{l} f_k x^k$ then the sequence $g_k = f_k / \binom{l}{k}$ $(k = 0,\ldots,l)$ is log-concave. The resulting inequalities $g_k^2 \geq g_{k-1} g_{k+1}$ are Aleksandrov's inequalities.

If, in the theorem, we take B_1 to be the identity matrix, $a_3,\ldots,a_r$ to be zero and $t = m$, then $\tau_m(-1,x)$ is just the characteristic polynomial of A. This explains the title of this section.

We now indicate how Theorem 3.2 implies Theorem 3.1 and, in particular, that $f(M, x)$ has only real zeros. Let B be an $m \times n$ matrix with the property that any $m \times m$ submatrix has determinant $0, 1$ or -1. Let $E_1,\ldots,E_r$ be $m \times m$ diagonal matrices with entries 0 or 1 such that $E_1 + \cdots + E_r$ is the identity matrix. We consider the coefficient of $x_1^{a_1} \cdots x_r^{a_r}$ in

$$det(B(x_1E_1 + \cdots + x_rE_r)B^t) \tag{1}$$

where $a_1 + \cdots a_r = m$.

The matrices $E_1,\ldots,E_r$ determine a partition of the columns of B. From the Binet-Cauchy identity, the coefficient in which we are interested is the sum of terms of the form $det(CC^t)$, where C is an $m \times m$ submatrix of B obtaining by taking (for $i = 1,\ldots,r$) a_i of the columns of B determined by E_i, i.e., a_i distinct non-zero columns from BE_i. By our hypothesis on the $m \times m$ submatrices of B we see that $det(CC^t) = (detC)^2$ is either 0 or 1. Thus our coefficient is the number of non-singular $m \times m$ submatrices of B. If we consider the linear matroid formed by the columns of B, this is the number of bases of M with a_i elements chosen from the subset of M determined by E_i.

On the other hand the determinant (1) is the determinant of

$$x_1 D_1 + \cdots + x_r D_r,$$

where $D_i = BE_i B^t$ is a symmetric positive semidefinite matrix. Applying 3.2 with $D_1 = A$, $D_2 = B_1$ etc. yields the conclusion that $\sum_{k=0}^{t} f_k x^k$ has real zeros. Thus we obtain 3.1.

It will perhaps help if we give one simple application of Theorem 3.1 (due to Stanley). Let G be a graph and let S be a subset of its edges. Assume G has n vertices and f_k be the number of spanning trees of G with exactly k edges in S. Then the sequence $f_k / \binom{n-1}{k}$ is log-concave (and the polynomial $\sum_{k=0}^{n-1} f_k x^k$ has only real zeros).

4. The Matchings Polynomial

A matching in a graph G is a set of k edges, no two of which have a vertex in common. Let $p(G, k)$ denote the number of matchings in G with exactly k edges and let $\nu = \nu(G)$ be the maximum number of edges in a matching from G. We define the matchings polynomial $\mu(G, x)$ to be

$$\sum_{k=0}^{\nu} p(G, k) (-1)^k x^{n-2k},$$

where $n = |V(G)|$. By a standard convention $p(G, 0) = 1$, hence $\mu(G, x)$ is a monic polynomial. This polynomial is surveyed in [9] and the reader is directed to this source for further information.

One of the most surprising properties of the matchings polynomial of a graph is that its zeros are always real. This was first proved by Heilmann and Lieb in [13]. This also follows from the next result which is proved in [6].

4.1. THEOREM Let G be a graph and let r be a vertex in G. Then there is a tree T with a vertex w such that $\mu(G, x)$ divides $\mu(T, x)$ and

$$\frac{\mu(G \backslash v, x)}{\mu(G, x)} = \frac{\mu(T \backslash w, x)}{\mu(T, x)}.$$

The tree T is obtained by taking the (simple) paths in G which start at r as vertices, and declaring two such paths to be adjacent if one is a maximal proper subpath of the other. The vertex w corresponds to the trivial path formed by the vertex v. Given the definition of T, the proof of Theorem 4.1 is straightforward. It depends chiefly on the following properties of the matchings polynomial:

(a) the matchings polynomial of a disconnected graph is equal to the product of the matchings polynomials of its components, and

(b) if the graph $\hat{H}$ is obtained from the graph H by taking a new vertex and joining it to each vertex in a specified subset S of $V(H)$ then

$$\mu(\hat{H}, x) = x\mu(H, x) - \sum_{v \in S} \mu(H\backslash v, x).$$

The importance of Theorem 4.1 stems from the fact that, for any tree T, $\phi(T, x) = \mu(T, x)$. (This follows from the usual expression for the coefficients of the characteristic polynomial of a graph - see Proposition 7.3 in [1] or Theorem 1.2 in [3]). In particular the zeros of $\mu(G, x)$ must be real, since the zeros of $\mu(T, x) = \phi(T,x)$ are real.

From our discussion in Section 2 we also recognize that $\mu(T\backslash w, x)/\mu(T, x)$ is essentially the generating function for the number of closed walks of length k in T which start at w. Given the definition of T it is easy to see that any closed walk in T starting at w determines a closed walk in G starting at v. Thus $\mu(G\backslash v)/\mu(G)$ can be viewed as the generating function for a class of closed walks in G starting at v. A precise definition of the walks in G enumerated by $\mu(G\backslash v)/\mu(G)$ is given in [6], the interested reader will find further details there.

From the definition of $\mu(G, x)$ and the fact that its zeros are real we see that the zeros of

$$\sum_{k=0}^{m} p(G, k) x^{m-k},$$

where m is the size of the largest matching in G, are real and negative. Consequently the sequence of numbers $p(G, k)/\binom{m}{k}$, and hence the numbers $p(G, k)$, form a log-concave sequence. (This was first noted by Heilmann and Lieb in [13].)

If G consists of m independent edges then $p(G, k)$ is just $\binom{m}{k}$. Thus the number of edges in a matching chosen at random from G (with all matchings equally probable) is a binomial random variable. By the DeMoivre-Laplace limit theorem, when m is large this random variable is approximately normally distributed. A surprising generalization of this result holds true. Let G_m ($m = 1, 2, \ldots$) be a sequence of graphs, each regular of valency Δ and such that $V(G_m)$ increases with m. Then the probability that a matching chosen at random from G_m has k edges is asymptotically normally distributed. For a proof of this, or at least of more general results which immediately yield this, we refer the reader to [8]. Here we note only that the proof makes use of the fact, first noted by Heilmann and Lieb in [13], that the largest zero $\mu(G, x)$ is less than $2\sqrt{\Delta-1}$, where Δ is the maximum valency of a vertex in G.

The next group of results indicates an interesting connection between $\mu(G, x)$ and the paths in G. We begin with an identity which is a special case of Theorem 6.3 from [13].

4.2. LEMMA. Let u, v be distinct vertices in the graph G. Then

$$\mu(G\backslash u)\mu(G\backslash v) - \mu(G)\mu(G\backslash uv) = \sum_P \mu(G\backslash P)^2 \qquad (1)$$

where the sum is over all paths in G joining u to v and $G\backslash P$ is obtained by deleting all the vertices in P from G).

Now in analogy to equation (3) of Section 2 we have

$$\frac{d}{dx} \mu(G, x) = \sum_{v \in V(G)} \mu(G\backslash v, x) \qquad (2)$$

This is easily proved from the definition of $\mu(G, x)$. It follows that summing (1) over all vertices v in G, with the convention that $\mu(G\backslash uu) = 0$, we get

$$\mu(G\backslash u)\mu'(G) - \mu(G)\mu'(G\backslash u) = \sum_P \mu(G\backslash P)^2 \qquad (3)$$

where our sum is now over all paths in G starting at v. Summing (3) over u now yields

$$\mu'(G)^2 - \mu(G)\mu''(G) = \sum_P \mu(G\backslash P)^2 \qquad (4)$$

with the sum over all paths in G.

If θ is a real number let $m(\theta, G)$ be the multiplicity of θ as a zero of $\mu(G, x)$. Any real zero of the sum in (3) must be a zero of $\mu(G\backslash P, x)$, for each path P in G. Hence we find that $m(\theta, G\backslash P) \geq m(\theta, G) - 1$ for any zero θ of $\mu(G, x)$. As a consequence we have immediately:

4.3. LEMMA. The maximum multiplicity of a zero of $\mu(G, x)$ is less than or equal to the number of disjoint paths needed to cover the vertices of G.

Another consequence of (3) above is the following result, essentially due to Heilmann and Lieb [13].

4.4. LEMMA. The number of distinct zeros of $\mu(G, x)$ is at least equal to the number of vertices in a longest path in G.

PROOF. The number of real zeros of the RHS of (4) is clearly at most twice the degree of $\mu(G\backslash P, x)$, where P is a longest path in G. On the other hand if θ is a zero of $\mu(G, x)$ then it is a zero of the LHS of (4.4) with multiplicity at least $2(m(\theta, G) - 1)$ so we get

$$2\sum_\theta (m(\theta, G) - 1) \leq 2|V(G\backslash P)|$$

where the sum is over the zeros of $\mu(G, x)$. If $\mu(G)$ has k distinct zeros it follows that $|V(G)| - k$ is not greater than $|V(G)| - |V(P)|$ and so $|V(P)| \leq k$.

These results indicate a situation where the behaviour of $\mu(G, x)$ differs markedly from that of $\phi(G, x)$. For it is known (see e.g., Theorem 4.8 in [10]) that the characteristic polynomial of a vertex-transitive graph with at least three vertices has multiple zeros. But all known vertex-transitive graphs have Hamilton paths and hence, by Lemma 4.4, their matchings polynomials have only simple zeros. It seems highly likely the matchings polynomial of a vertex-transitive graph must have simple zeros, but I have not been able to prove this.

Perhaps the best way to motivate the final part of this section is to offer:

4.5. THEOREM. The number of perfect matchings in the complement of the graph G is

$$(2\pi)^{-1/2} \int_{-\infty}^{\infty} \mu(G, x)\exp(-x^2/2)dx.$$

For the proof of this the reader is referred to [7]. Theorem 4.5 has some interesting consequences. Let $\overline{G}$ denote the complement of G. The number of matchings in the complement of $G \cup K_{n-2k}$ ($n = |V(G)|$) is easily seen to be $(n-2k)! \, p(\overline{G}, k)$. Hence

$$(n - 2k)!p(\overline{G}, k) = \int_{-\infty}^{\infty} \mu(G, x)\mu((K_{n-2k}, x)\exp(-x^2/2)dx \quad (5)$$

This implies that $\mu(G, x)$ determines $\mu(\overline{G}, x)$, as noted also in [16: Problem 5.18] and [22]. Taking $G = K_n$ and $m = n - 2k$ in this identity yields:

$$(2\pi)^{-1/2}\int_{-\infty}^{\infty} \mu(K_n, x)\, \mu(K_m, x)\exp(-x^2/2)\,dx = \begin{cases} m! & m = n \\ 0 & m \neq n \end{cases}.$$

It follows from this that $\mu(K_n, x)$ coincides with the Hermite polynomial of degree n. (For further details on the connections between $\mu(G, x)$ and some classical families of orthogonal polynomials see Section III in [13], and [9].) Using the above orthogonality relation it is easy to verify that

$$\mu(\overline{G}, x) = \sum_{k=0}^{\left[\frac{n}{2}\right]} p(G, k)\, \mu(K_{n-2k}, x). \tag{6}$$

One consequence of this is that the coefficients in the expansion of x^n in terms of Hermite polynomials are the same, up to sign, as the coefficients of the Hermite polynomial of degree n. (Set $G = K_n$ in (6)).

Now suppose G is a subgraph of the complete bipartite graph $K_{m,n}$, where $m \leq n$. Define the *rook polynomial* $r(G, x)$ by

$$r(G, x) = \sum_{k=0}^{m} (-1)^k\, p(G, k)\, x^{m-k}.$$

Let $G*$ be the subgraph of $K_{m,n}$ formed by the edges not in G (i.e., the bipartite complement of G). Then if $m = n$ we find that the number of perfect matchings in $G*$ is

$$\int_0^\infty r(G, x)\, e^{-x}\,dx.$$

If, in general, we set $s = n - m$, then in analogy to (5) we have

$$\int_0^\infty r(G, x) r(K_{l,l+s}, x) x^s e^{-x}\, dx = \begin{cases} l!(l+s)!\, p(G*, m-l) & l \leq m \\ 0 & l > m \end{cases}.$$

Here we view $G \cup K_{l,l+s}$ as a subgraph of $K_{n+l, n+l}$. The product being integrated with respect to $e^{-x}dx$ is just $r(G \cup K_{l,l+s}, x)$. Setting $G = K_{m,n}$ yields that the polynomials $\mu(K_{m,m+s}, x)$ are orthogonal with respect to the measure $x^s e^{-x}dx$ and so forces their identification with the generalized Laguerre polynomials.

Note finally that our "rook polynomial" is basically the same as the rook polynomial studied in Riordan's book [18]. In particular we can say that a rook polynomial is just the matchings polynomial of a bipartite graph.

5. Open Problems

Those graphs with all eigenvalues greater than or equal to -2 and which are not line graphs have been determined by Cameron et al [2]. A.J. Hoffman has shown that a graph where each vertex has "large" valency and with all eigenvalues greater than $-1-\sqrt{2}$ must have all eigenvalues greater than or equal to -2.

5.1. PROBLEM. Determine the graphs with least eigenvalues in the interval $(-1-\sqrt{2}, -2)$.

(Perhaps this problem is not perfectly posed, but anyone who reads [2] will see the type of result to be aimed at.) Incidentally the value of "large" above is determined by Ramsey Theory and so is finite only in the technical sense.

I. Gutman has proposed that the sum of the absolute values of the eigenvalues of a graph be called its *energy,* since in Quantum Chemistry this sum can be so interpreted. He has raised the following:

5.2. PROBLEM. Show that, amongst all graphs with n vertices, the complete graph has the greatest energy.

The most important graphs in chemistry are planar and bipartite. Hence it would also be interesting to know the maximum value of the energy of graphs having one or both of these properties. Gutman has shown, in [12], that amongst trees with n vertices the path has maximal energy. The reader is referred to this paper for further information on this topic.

Let G be a graph and let θ be a zero of $\mu(G, x)$. Call a vertex v in G θ-*critical* if the multiplicity of θ in $\mu(G\backslash v)$ is less than its multiplicity in $\mu(G)$, i.e., if $m(\theta, G\backslash u) < m(\theta, G)$.

5.3. PROBLEM. Is it true that if each vertex in a graph G is $\theta-$ critical then $m(\theta, G) = 1$?

This does hold when $\theta = 0$ because then $m(\theta, G)$ is just the number of vertices not in a maximum matching and so the graph G would be *factor-critical.* In this case it is known that each maximum matching misses just one vertex (see e.g. [16: Problem 7.26]). The result also holds if G is a tree. This is proved, in another context, by A. Neumaier [17]. Given the obvious relevance of trees (see Theorem 4.1) this is strong evidence, and the reader might wonder why 5.3 is stated as a problem and not a conjecture.

However a positive solution would have very strong consequences. First it would imply that if G was vertex-transitive then the zeros of $\mu(G, x)$ are simple (it follows from equation (2) in Section 4 that if G is vertex-transitive then all vertices are θ-critical - otherwise θ would have the sample multiplicity in $\mu(G, x)$ and $\mu'(G, x)$!). More importantly it would yield a generalization of Tutte's criterion for the existence of perfect matchings. (The result would be that if $m(\theta, G) = k > 1$ then G has a cutset C whose removal left $|C| + k$ components, each of which has θ as a simple zero of its matchings polynomials.)

One piece of information which should be mentioned is that the zeros of $\mu(G\backslash v, x)$ interlace those of $\mu(G, x)$. (This follows from Theorem 4.1, but was first proved in [13]). Using this, together with 4.1 and 4.2, I can show that if $m(G, \theta) = k > 1$ then G has an induced subgraph with k components, each of which is θ-critical with regard to every vertex and has θ as a zero with multiplicity one. Details of this will appear elsewhere.

References

[1] N. Biggs, *Algebraic Graph Theory*, Cambridge University Press, Cambridge, 1974.

[2] P.J. Cameron, J.M. Goethals, J.J. Seidel and E.E. Shult, Line graphs, root systems and elliptic geometry. *J. Algebra*, 43 (1976), 305-327.

[3] D.M. Cvetković, M. Doob and H. Sachs, *Spectra of Graphs*, Academic Press, N.Y., 1980.

[4] P. Erdös, and M. Simonovits, Compactness results in extremal graph theory, Combinatorica, 2 (1982), 275-288.

[5] R.J. Faudree and B.D. McKay, The number of walks of length a power of three in a graph, to appear.

[6] C.D. Godsil, Matchings and walks in graphs, *J. Graph Theory*, 5 (1981), 285-297.

[7] C.D. Godsil, Hermite polynomials and a duality relation for the matchings polynomials, *Combinatorica*, 1 (1981), 257-262.

[8] C.D. Godsil, Matching behaviour is asymptotically normal, *Combinatorica*, 1 (1981), 369-376.

[9] C.D. Godsil and I. Gutman, On the theory of the matching polynomial, *J. Graph Theory* 5 (1981), 137-144.

[10] C.D. Godsil and B.D. McKay, Feasibility conditions for the existence of walk-regular graphs. *Linear Algebra and Appl.*, 30 (1980), 51-61.

[11] C.D. Godsil and B.D. McKay, Spectral conditions for the reconstructibility of a graph. *J. Combinatorial Theory B*, 30 (1981), 285-289.

[12] I. Gutman, Acyclic systems with extremal Huckel π-electron energy. *Theoret. Chim. Acta*. 45 (1977), 79-87.

[13] Ole J. Heilmann and Eliott H. Lieb, Theory of monomer-dimer systems, *Commun. Math. Physics*, 25 (1972), 190-232.

[14] A.J. Hoffman, On graphs whose least eigenvalue exceeds $-1-\sqrt{2}$ *Linear Algebra and Appl.* 16 (1977), 153-165.

[15] P.W. Kasteleyn, Graph theory and crystal physics, in *Graph Theory and*

Theoretical Physics, ed F. Harary, Academic Press, N.Y., 1967.

[16] L. Lovasz, *Combinatorial Problems and Exercises*, North-Holland, Amsterdam, 1979.

[17] A. Neumaier, The second largest eigenvalue of a tree, *Linear Algebra and its Appl.*, 46 (1982), 9-25.

[18] J. Riordan, *An Introduction to Combinatorial Analysis*, Wiley, N.Y., 1958.

[19] Rolf Schneider, On A.D. Aleksandrov's inequalities for mixed discriminants, *J. Math. Mech.* 15 (1966), 285-290.

[20] Richard P. Stanley, Two combinatorial applications on the Aleksandrov-Fenchel inequalities. *J. Combinatorial Theory* A, 31 (1981), 56-65.

[21] W.T. Tutte, All the king's horses, in *Graph Theory and Related Topics* (eds. J.A. Bondy and U.S.R. Murty) Academic Press, N.Y., 1979.

[22] T. Zaslavsky, Complementary matching vectors and the uniform matching extension property, *Europ. J. Combinatorics*, 2 (1981), 91-103.

On the Minimal Cut Problem

Mark K. Goldberg
Robert Gardner

ABSTRACT

This paper is devoted to the problem of partitioning the vertex set of a graph into two equally sized subsets (to within one element) in such a way that the number of edges cut is minimal. A sequence of $n \log n$ - time approximation algorithms is presented. It is proved that for almost all graphs each member of the sequence produces a partition close to the best solution.

1. Introduction

This paper is devoted to the problem of partitioning the vertex set of a graph $G(V,E)$ into two disjoint subsets V_1 and V_2 in such a way that

(a) $|V_1| = \lfloor \frac{1}{2}|V| \rfloor$, $|V_2| = \lceil \frac{1}{2}|V| \rceil$;

(b) the number of the edges that have one end in V_1 and one end in V_2 is minimal.

The problem and its diverse generalizations were studied in [1], [5], [9], [10]. It is known [6] that the problem is NP-complete, making unlikely the existence of a polynomial algorithm producing an exact solution for every graph. However, in the areas of its practical applications such as computer program segmentation [8], or layout of electronic circuits on computer boards ([3], [4], [7]), the problem is particularly important for graphs on many thousands of vertices. In such a situation, one cannot afford even a quadratic algorithm.

In this paper we present a sequence of nlogn-algorithms that produce good approximate solutions. It is proved that for almost all graphs without parallel edges, every algorithm of the sequence constructs a partition close to the best solution.

Throughout the paper the following notations will be used. Given a graph G, $V = V(G)$ and $E = E(G)$ denote respectively the set of vertices and the set of edges of G; $n(G) = |V(G)|$, $p(G) = |E(G)|$. We permit graphs to have parallel edges; however, loops are not allowed. If $x, y \in V(G)$ then $d(x)$ is the degree of x; $d(x,y)$ is the number of the edges joining x and y. If $A, B \subset V$ then $p(A,B)$ denotes the number of edges of the form $(x,y)(x \in A, y \in B)$.

We will consider only those partitions of $V(G)$ that satisfy the condition (a) above. It is clear that if $n = n(G)$ is even (resp. odd) then the number of different partitions is $\frac{1}{2}\binom{n}{\frac{n}{2}}$ (resp. $\binom{n}{\frac{n-1}{2}}$). If τ is a partition of $V(G)$ then $c_\tau(G)$ denotes the number of edges cut by τ; $c(G) = \min_\tau c_\tau(G)$, $c(n,p) = \max_G c(G)$ where G ranges over the set of all graph with n vertices and p edges.

Let $\mathbf{G}(n,p) = \{G \mid n(G) = n, p(G) = p\}$ where G has no parallel edges. Then $K(n,p) = |\mathbf{G}(n,p)|$; $K(n,p,\alpha) = |\{G : G \in \mathbf{G}(n,p)$ and $c(G) \leq \alpha p(G)\}|$

All concepts we do not explain here can be found in [2].

2. Upper and Lower Bounds

We start with the following simple bound

THEOREM 1. *For every graph $G(V,E)$ with $|V| = n$ and $|E| = p$*

$$c(G) \leq \begin{cases} \dfrac{p \cdot n}{2(n-1)} & \text{if } n \text{ is even,} \\ \dfrac{p(n+1)}{2n} & \text{if } n \text{ is odd} \end{cases}.$$

PROOF. First let n be even. Then

$$\frac{1}{2}\binom{n}{\frac{n}{2}} c(G) \leq \sum_\sigma c_\sigma(G) \qquad (*)$$

where σ ranges over the set of all partitions of G.

Straightforward calculations show that in the right side of $(*)$ the contribution of each edge is exactly $\binom{n-2}{\frac{n}{2}-1}$, therefore

$$\frac{1}{2}\binom{n}{\frac{n}{2}}c(G) \le \binom{n-2}{\frac{n}{2}-1}p(G)$$

The result follows.

In the case of odd n instead of (*) we have

$$\binom{n}{\frac{n-1}{2}}c(G) \le \sum_\sigma c(G)$$

but now the contribution of each edge is equal to $2\binom{n-2}{\frac{n-1}{2}-1}$ resulting in

$$c(G) \le \frac{1}{2}(1 + \frac{1}{n})p.$$

A lower bound for $c(n,p)$ is derived by considering the following graph $G^*_{n,p}$.

Let s be the largest integer such that $p \ge s(n-1) - \frac{s(s-1)}{2}$ and let $r = p - s(n-1) + \frac{s(s-1)}{2}$. It can be easily proved that $0 \le r \le n-s-2$. Now, the set of vertices V and the set of edges E of $G^*_{n,p}$ are defined by

$$V = \{x_1, x_2, \ldots, x_n\}$$
$$E = \{(x_i, x_j) \mid i = 1, \ldots, s; j = 1, \ldots, n(i \ne j)\} \cup$$
$$\{(x_{s+1}, x_j) \mid j = s+2, \ldots, s+r+1\}.$$

Thus, if $r = 0$ then $G^*_{n,p}$ contains exactly s vertices of degree $n-1$ and their deletion produces an empty graph on $n-s$ vertices. If $r > 0$ then $G^*_{n,p} - \{x_1, \ldots, x_s\}$ is a graph with $n-s$ vertices and r edges $(x_{s+1}, x_{s+2}), \ldots, (x_{s+1}, x_{s+r+1})$.

Routine calculations show that the minimal partition cuts the set $\{x_1, \ldots, x_s\}$ into two parts of sizes $\lfloor\frac{s}{2}\rfloor, \lceil\frac{s}{2}\rceil$ and

$$c(G^*_{n,p}) = \{\frac{s(2n-s)}{4}\} + (r - \lceil\frac{(n-s)/2}{2}\rceil - 1)^+$$

where $\{z\}$ denotes the integer nearest to z and $(z)^+ = \frac{z + |z|}{2}$.

Since $p = \frac{s(2n-s)}{2} - \frac{s-2r}{2}$ one can see that $G^*_{n,p}$ provides a lower bound for $c(n,p)$ that is very close to the upper bound proved above.

CONJECTURE. For all n and p
$$c(n,p) = c(G^*_{n,p}).$$

From the proof of theorem 1 it clearly follows that if $c(G)$ is close to $\frac{1}{2}p$ then for most partitions σ the value $c_\sigma(G)$ is close to $\frac{1}{2}p$ as well. It may seem from this that such graphs are rather rare. However, this is not the case. As was proved in [8] if $p/n \to \infty$ almost all graphs are such that for every partition $c_\sigma(G)$ is asymptotically $\frac{1}{2}p$. As a matter of fact, in [8] one can find essentially stronger results. The next theorem can be considered as an illustration of them. We give here an independent and simple proof of the theorem.

THEOREM 2. For any positive $\alpha < \frac{1}{2}$ there exists a β such that if $p > \beta n$ and $n \to \infty$ then
$$\frac{K(n,p,\alpha)}{K(n,p)} \to 0.$$

PROOF. Let us fix some set V of n vertices and some partition $\sigma = (V_1, V_2)$. Also, let $R(n,p,i)$ (resp. $R_\sigma(n,p,i)$) denote the number of graphs on V with p edges and $c(G) = i (resp. c_\sigma(G) = i)$. Then $K(n,p,\alpha) = \sum_{i \leq \alpha p} R(n,p,i) < 2^n \sum_{i \leq \alpha p} R_\sigma(n,p,i)$.

Let us assume that n is even (the odd n case is similar). Then
$$K(n,p,\alpha) < 2^n \cdot \sum_{i \leq \alpha p} \binom{\frac{n^2}{4}}{i} \binom{\frac{n^2}{4} - \frac{n}{2}}{p - i}$$

Since for any $i < \frac{1}{2}p$
$$\binom{\frac{n^2}{4}}{i-1} \binom{\frac{n^2}{4} - \frac{n}{2}}{p - i + 1} < \binom{\frac{n^2}{4}}{i} \binom{\frac{n^2}{4} - \frac{n}{2}}{p - i},$$

$$\frac{K(n,p,\alpha)}{K(n,p)} < \frac{p \cdot 2^n \binom{\frac{n^2}{4}}{q}\binom{\frac{n^2}{4}-\frac{n}{2}}{p-q}}{\binom{\frac{n^2}{2}-\frac{n}{2}}{p}}, \text{ where } q = \lfloor \alpha p \rfloor.$$

Then

$$\frac{\binom{\frac{1}{4}n^2}{q}\binom{\frac{1}{4}n^2-\frac{1}{2}n}{p-q}}{\binom{\frac{1}{2}n^2-\frac{1}{2}n}{p}} = \frac{\binom{\frac{1}{4}n^2}{q}\binom{\frac{1}{4}n^2}{p-q}}{\binom{\frac{1}{4}n^2}{\frac{1}{2}p}^2} \cdot$$

$$\frac{\binom{\frac{1}{4}n^2}{\frac{1}{2}p}^2}{\binom{\frac{1}{2}n^2}{p}} \cdot \frac{\binom{\frac{1}{2}n^2}{p}\binom{\frac{1}{4}n^2-\frac{1}{2}n}{p-q}}{\binom{\frac{1}{2}n^2-\frac{1}{2}n}{p}\binom{\frac{1}{4}n^2}{p-q}}$$

$$= \prod_{i=1}^{\frac{1}{2}p-q} \left(\frac{q+i}{\frac{1}{4}n^2-q+1-i} \cdot \frac{\frac{1}{4}n^2-\frac{1}{2}p+1-i}{\frac{1}{2}p+i} \right) \cdot \frac{\binom{p}{\frac{1}{2}p}\binom{\frac{1}{2}n^2-p}{\frac{1}{4}n^2-\frac{1}{2}p}}{\binom{\frac{1}{2}n^2}{\frac{1}{4}n^2}} \cdot$$

$$< \prod_{j=1}^{\frac{n}{2}} \left(\frac{\frac{1}{2}n^2 - j + 1}{\frac{1}{4}n^2 - j + 1} \cdot \frac{\frac{1}{4}n^2 - p + q - j + 1}{\frac{1}{2}n^2 - p - j + 1} \right)$$

$$< \prod_{i=1}^{\frac{1}{2}p-q} \left(\frac{q+i}{\frac{1}{2}p+i} \cdot \frac{\frac{1}{4}n^2 - \frac{1}{2}p + 1 - i}{\frac{1}{4}n^2 - q + 1 - i} \right) \cdot \prod_{j=1}^{\frac{n}{2}} \left(\frac{\frac{1}{2}n^2 - j + 1}{\frac{1}{4}n^2 - j + 1} \right) < \left(\frac{\frac{1}{2}p}{p-q} \right)^{\frac{1}{2}p-q}.$$

$$2^{\frac{n}{2}} \left(1 + \frac{1}{n}\right)^{\frac{n}{2}} < \sqrt{e} \cdot 2^{\frac{n}{2}} \cdot \left(\frac{1}{2(1-\alpha)} \right)^{p(\frac{1}{2}-\alpha)}$$

The result follows for $p > \beta n$, where β is such that

$$2^{\frac{3}{2}} \cdot \left(\frac{1}{2(1-\alpha)} \right)^{\beta(\frac{1}{2}-\alpha)} < 1.$$

3. Algorithms

We now describe a sequence $\{Cut(2k)\}$ of $n \log n$ algorithms that produces approximate solutions to the problem. Each algorithm of the sequence reduces the problem for the original graph to that for a smaller graph and does so several times until the final graph is sufficiently small to be treated by an exhaustive search procedure. We do not specify this procedure. Let S_{2k} denote an arbitrary algorithm which for every input $[G(V,E); A; B]$ where $|V| \leq 2k$, A and B are disjoint subset of V and $|A|, |B| \leq \frac{1}{2}|V|$, produces a partition $\tau = (V_1, V_2)$ such that $A \subset V_1$, $B \subset V_2$ and, for which, with these restrictions, the number of the edges cut by τ is minimal. Since k is assumed to be bounded the running time of S_{2k} can be taken as a constant.

We start with the descriptions of two procedures.

Procedure CONTRACTION (2k)

Input: Graph $G(V,E)$ on $2kq$ vertices (k,q are integers)
Output: Graph $G'(V',E')$ on $2q$ vertices.
1. Let $x_1, x_2, \ldots, x_{2kq}$ be the order of the vertex set of G. Form subgraphs $G_1, G_2, \ldots, G_q$ induced on subsets

$X_1 = (x_1, \ldots, x_{2k})$,
$X_2 = (x_{2k+1}, \ldots, x_{4k}), \ldots, X_q = (x_{2kq-2k+1}, \ldots, x_{2kq})$.

2. Apply S_{2k} to $[G_i;\emptyset;\emptyset]$. Let $\tau_i = (Y_{2i-1}, Y_{2i})$ be the resulting partition of G_i ($i = 1, 2, \ldots, q$).
3. Form graph $G'(V', E')$ on $2q$ vertices $x_1', \ldots, x_{2q}'$ defining two vertices x_i' and x_j' to be joined by m edges iff there are exactly m edges between Y_i and Y_j. Stop. □

Procedure EXTENSION

Input: Graph $G(V, E)$; ordered set $R = \{\bar{x}_1, \bar{x}_2, \ldots, \bar{x}_r\}$ ($R \subset V; r \leq 2k-1$); partition $(\bar{V}_1, \bar{V}_2)$ of $G-R$.
Output: Partition (V_1, V_2) of G.

1. If r is even go to 5 otherwise proceed to 2.
2. Put $H = G - \{\bar{x}_1\}$.
3. Apply EXTENSION to H; let $(\bar{U}_1, \bar{U}_2)$ be the resulting partition of H.
4. Calculate $d(\bar{x}_1, \bar{U}_1)$ and $d(\bar{x}_1, \bar{U}_2)$. If $d(\bar{x}_1, \bar{U}_1) \leq d(\bar{x}_1, \bar{U}_2)$ then put $V_1 = U_1, V_2 = U_2 \cup \{x_1\}$, otherwise put $V_1 = U_1 \cup \{x_1\}, V_2 = U_2$. Stop.
5. Form graph F on the vertices $\bar{x}_1, \bar{x}_2, \ldots, \bar{x}_r, \bar{x}_{r+1}, \bar{x}_{r+2}$ by contracting subsets $\bar{V}_1, \bar{V}_2$ into $\bar{x}_{r+1}, \bar{x}_{r+2}$ respectively.
6. Apply S_{2k} to $[F; \{\bar{x}_{r+1}\}; \{\bar{x}_{r+2}\}]$. Let (W_1, W_2) be the resulting partition of F.
7. Construct partition (V_1, V_2) of G from (W_1, W_2) by substituting $\bar{V}_1$ and $\bar{V}_2$ for $\bar{x}_{r+1}$ and $\bar{x}_{r+2}$ respectively. Stop. □

Algorithm CUT ($2k$)

Input: Graph $G(V, E)$.
Output: Partition $\tau(V_1, V_2)$ of G; the number c of the edges cut by τ.

1. Put $H = G(V, E)$.
2. If $n(H) \leq 2k$ apply S_{2k}. Stop.
3. Calculate the remainder r of $n = \nu(H)$ modulo $2k$. If $r > 0$ go to 4 otherwise go to 7.
4. Construct sequence $\bar{x}_1, \bar{x}_2, \ldots, \bar{x}_r$ of vertices by the rule:

   (i) $\bar{x}_1$ is a vertex of the maximal degree; (ii) $\bar{x}_{i+1}$ is such that $d^*(\bar{x}_{i+1}) - d(\bar{x}_i, \bar{x}_{i+1})(n-i-1)$ is maximal; where $d^*(\bar{x}_{i+1})$ is

the degree of $\bar{x}_{i+1}$ in $F = H - \{x_1, \ldots, x_{i-1}\}$.
5. Apply CUT(2k) to $H' = H - \{\bar{x}_1, \ldots, \bar{x}_r\}$; let τ be the resulting partition.
6. Apply EXTENSION to $[H; \bar{x}_1, \bar{x}_2, \ldots, \bar{x}_r; \tau]$. Stop.
7. Apply CONTRACTION(2k). Let $\bar{H}$ be the output.
8. Apply CUT(2k) to $\bar{H}$; let $\bar{\tau}$ be the resulting partition.
9. Restore the partition τ of H from $\bar{\tau}$ be the resulting partition.
10. Restore the partition τ of H from $\bar{\tau}$ by replacing vertices of $\bar{H}$ with the corresponding sets of H. Print. Stop.

4. Running Time and Accuracy

It is not hard to see that the running time of CUT(2k) is $O(n\log n)$ for every fixed k. Consider:

1) Step 4 in CUT(2k) can be performed in linear time.
2) CUT(2k) calls itself recursively, each time on a graph half size of the previous. Therefore, $\log n$ calls are made to CUT(2k).
3) EXTENSION and CONTRACTION (2k) are evidently linear.

DEFINITION. A partition τ of a graph G with n vertices and p edges is called admissible if

$$c_\tau(G) \leq \begin{cases} \dfrac{pn}{2(n-1)} & \text{if } n \text{ is even} \\ \dfrac{p(n+1)}{2n} & \text{if } n \text{ is odd} \end{cases}.$$

LEMMA. Let the input partition of EXTENSION be admissible and let the sequence $\bar{x}_1, \ldots, \bar{x}_r$ of vertices be chosen according to the rule from Step 4 of CUT(2k). Then the output partition is also admissible.

PROOF. It is sufficient to consider the cases $r = 1$ and r even. We will prove the case r even only since the case $r = 1$ is much simpler and can be done in a similar way.

Let us consider $\frac{1}{2}r$ graphs defined by

$$G_1 = G, \quad G_{i+1} = G_i - \{\bar{x}_{2i-1}, \bar{x}_{2i}\} \quad (i = 1, \ldots, \frac{r}{2} - 1).$$

From the rule in step 4 of CUT(2k) it follows that given G_i and $\bar{x}_{2i-1}$, $\bar{x}_{2i}$ maximizes the quantity $d^*(\bar{x}_{2i}) - d(\bar{x}_{2k}, \bar{x}_{2i-1})(n(G_i) - 2)$ in G_i, where $d^*(x)$ denotes the degree of x in the graph G_i.

Therefore if we prove the lemma in the case $r = 2$, then each of the graphs $G_{r/2}, \ldots, G_2, G_1$ gets an admissible partition and therefore Step 6 in EXTENSION produces an admissible partition too.

Now, let $\{G(V,E); \bar{x}_1, \bar{x}_2; \sigma = (\bar{V}_1, \bar{V}_2)\}$ be the input of EXTENSION, let $\bar{x}_2$ be such that $d(\bar{x}_2) - d(\bar{x}_1,\bar{x}_2)(n-2)$ is maximal, and also let $\tau = (V_1, V_2)$ be the output partition.
Then

$$c_\tau \le c_\sigma(G - \{\bar{x}_1,\bar{x}_2\}) + d(\bar{x}_1,\bar{x}_2) + \frac{d(\bar{x}_1) - d(\bar{x}_1,\bar{x}_2) + d(\bar{x}_2) - d(\bar{x}_1,\bar{x}_2)}{2}$$

$$\le \frac{1}{2}p(G - \{\bar{x}_1,\bar{x}_2\}) \cdot \frac{n-2}{n-3} + \frac{d(\bar{x}_1) + d(\bar{x}_2)}{2}$$

$$= \frac{1}{2}(p(G) - d(\bar{x}_1) - d(\bar{x}_2) + d(\bar{x}_1,\bar{x}_2))\frac{n-2}{n-3} + \frac{d(\bar{x}_1) + d(\bar{x}_2)}{2}$$

$$= \frac{1}{2}p(G) \cdot \frac{n}{n-1} + \frac{1}{2(n-3)}(\frac{2p(G)}{n-1} + d(\bar{x}_1,\bar{x}_2)(n-2) - d(\bar{x}_1) - d(\bar{x}_2)).$$

Thus, the result would follow from the inequality

$$\frac{2p(G)}{n-1} + d(\bar{x}_1,\bar{x}_2)(n-2) - d(\bar{x}_1) - d(\bar{x}_2) \le 0.$$

If it fails then $\forall x \ne x_1$

$$\frac{2p(G)}{n-1} + d(\bar{x}_1,x)(n-2) - d(\bar{x}_1) - d(x) > 0$$

due to maximality of $d(\bar{x}_2) - d(\bar{x}_1,\bar{x}_2)(n-2)$.
Therefore,

$$\frac{2p(G)}{n-1}(n-1) + (n-2)\sum_{x \ne x_1} d(\bar{x}_1,x) > (n-1)d(\bar{x}_1) + \sum_{x \ne x_1} d(x)$$

implying

$$2p(G) + (n-2)d(\bar{x}_1) > (n-1)d(\bar{x}_1) + 2p - d(\bar{x}_1)$$

which is not possible.

Now we are ready to prove the following.

THEOREM 3. *If τ is the output partition of CUT(2k) applied to a graph $G(V,E)$ then τ is admissible.*

PROOF. It will be carried out by induction on the number Θ of calls of the procedure $S(2k)$.

If $\Theta = 1$ then $n(G) \le 2k$ and the result follows from Theorem 1.

Let $\Theta > 1$ and let H be the graph to which S_{2k} was applied the last time. The number $n(H)$ is apparently $\leq 2k$. Because of the lemma we can assume that H is the last term of a sequence of graphs $F_1, F_2, \ldots, F_t = H$ such that F_{i+1} is constructed from F_i by applying CONTRACTION(2k) $(i = 1, \ldots, t-1)$. Thus $n(H)$ is even. Let $z_1, z_2, \ldots, z_{2d}$ be the set of the vertices of H ordered in such a way that z_{2i-1}, z_{2i} correspond to the classes of the same partition $\tau_i = (Y_{2i-1}, Y_{2i})$ occurring in the course of applying CONTRACTION(2k) $(i = 1, \ldots, d)$.

Let us denote by Z_j the set of vertices of F_1 contracted into z_j $(j = 1, \ldots, 2d)$. Because of the induction assumption the partition $\sigma_i = (Z_{2i-1}, Z_{2i})$ of the subgraph T_i induced on $Z_{2i-1} \cup Z_{2i}$ is admissible $(i = 1, \ldots, d)$. Therefore, if $m = |Z_j|$ $(j = 1, \ldots, 2d)$, $p_i = p(T_i)$, and $c_i = c_{\sigma_i}(T_i)$ then

$$c_i \leq \frac{1}{2}p_i\left(1 + \frac{1}{2m-1}\right) = \frac{p_i \cdot m}{2m-1} \quad (i = 1, \ldots, d).$$

Let us assume for simplicity that d is even. (The odd d-case is similar). If $p_{ij} = p(Z_{2i-1} \cup Z_{2i}, Z_{2j-1} \cup Z_{2j})$ $(i \neq j)$ then every partition of F_1, that does not cut $Z_{2i-1} \cup Z_{2i}$ for all $i = 1, \ldots, d$ can be interpreted as a partition of H. If one of these partitions is admissible, an admissible partition of $\overline{H}_1$ will be returned. Therefore, assuming that none of such partitions is admissible we come to the following inequality.

$$\frac{1}{2}\left[\sum_{\substack{i,j=1 \\ i \neq j}}^{d} p_{i,j}\right]\left(1 + \frac{1}{d-1}\right) > \frac{1}{2}p\left(1 + \frac{1}{2dm-1}\right),$$

or

$$\frac{2 \cdot m \cdot p \cdot (d-1)}{2dm - 1} < \bar{p} \qquad (*)$$

where $\bar{p} = \sum_{\substack{i,j=1 \\ i \neq j}}^{d} p_i, j$.

Let us consider partitions of F_1 that cut every set $Z_{2i-1} \cup Z_{2i}$ into the parts Z_{2i-1} and Z_{2i} $(i = 1, \ldots, d)$. It is clear that
(a) every such partition can be originated from some partition of H,
(b) the minimal partition of such a kind cuts not more than $\frac{1}{2}\bar{p} + \sum_{i=1}^{d} c_i$ edges. So,

$$c_\tau(F_1) \le \frac{1}{2}\bar{p} + \sum_{i=1}^{d} c_i \le \frac{1}{2}\bar{p} + \frac{m}{2m-1}\sum_{i=1}^{d} p_i$$

$$= \frac{1}{2}\bar{p} + \frac{m}{2m-1}(p-\bar{p}) = \frac{m}{2m-1}p - \bar{p}\frac{1}{2(2m-1)}.$$

Using (*) we come to

$$c_\tau(F_1) < \frac{m}{2m-1}p - \frac{m(d-1)}{(2dm-1)(2m-1)} \cdot p = \frac{m}{2m-1}p\left(1 - \frac{(d-1)}{2dm-1}\right)$$

$$= \frac{mp(2dm-d)}{(2m-1)(2dm-1)} = \frac{mdp}{2dm-1}.$$

The proof is complete.

Acknowledgement

The authors would like to thank Prof. Gary Miller for his valuable remarks, and the referee for many helpful suggestions.

References

[1] Earl R. Barnes, An algorithm for partitioning the nodes of a graph, IBM Research Report, 1981.

[2] J.A. Bondy, U.S.R. Murty, Graph theory with applications, North Holland, New York, 1980.

[3] Michael Burstein, Partitioning of VLSI networks, IBM Research Report, 1982.

[4] L.I. Corrigan, "A Placement Capability Based on Partitioning", Proc. 16th Design Automation Conf., June 1979, pp. 406-413.

[5] W.E. Donath, A.J. Hoffman, "Lower Bounds for the Partitioning of Graphs", IBM Journal of Research and Development, Vol. 17, No. 5, 1973, pp. 420-425.

[6] M.R. Garey, D.S. Johnson, "Computers and Intractability", W.M. Freeman and Co., 1979, pp. 209-210.

[7] A.D. Friedman, P.K. Menon, Theory and design of switching circuits, Bell Telephone Laboratories, Inc. and Computer Sciences Press, Inc. 1975.

[8] B.W. Kernighan, "Some Graph Partitioning Problems Related to Program Segmentation", Ph.D. Thesis, Princeton University, January 1969.

[9] B.W. Kernighan, S. Lin, "An Efficient Heuristic Procedure for Partitioning Graphs", Bell Systems Technical Journal, vol. 49, 1970, pp. 291-307.

[10] V.P. Kosyrev, A.D. Korshynov, On the size of a cut in a random graph, Promlemy Kibernetiki, 1974 (Russian).

On Isometric Embeddings of Graphs

R. L. Graham

ABSTRACT

For a finite connected undirected graph $G=(V,E)$ one can associate a metric d_G by defining $d_G(x,y)$ to be the number of edges in a shortest path between x and y, where x and y are any two vertices in V, the vertex set of G. If (M,d_M) is an arbitrary metric space we say that an embedding $\lambda:V\to M$ is *isometric* if for all $x, y \in V$, $d_M(\lambda(x),\lambda(y))=d_G(x,y)$.

In this note we survey recent results on this topic, especially in the case in which M is formed from the cartesian product of graphs.

1. Introduction

With a finite connected undirected graph* $G = (V,E)$ one can associate a metric $d_G: V \times V \to \mathbb{N}$ (the set of nonnegative integers) by defining $d_G(x,y)$ to be the number of edges in the shortest path between x and y for all $x,y \in V$, the vertex set of G. If (M,d_M) is an arbitrary metric space we say that an embedding $\lambda:V\to M$ is *isometric* † if for all $x,y \in V$,

$$(1) \qquad d_M(\lambda(x),\lambda(y)) = d_G(x,y).$$

We denote this by $G \stackrel{d}{\to} M$.

A fair number of papers have appeared in the past few years which deal with various properties of graphs which have isometric embeddings in certain metric and semi-metric§ spaces (e.g., see [As1],

---
*In general, we follow the terminology of [BM].
†In the literature this is also sometimes said to be "distance preserving."
§i.e., the triangle inequality may fail (in French, écart).

[As2], [As3], [AD 1], [AD12], [ADzl], [ADz2], [Av1], [Av2], [Dew], [Dez1], [Dez2], [DR], [Dj], [F], [GP1], [GP2], [HL], [GP2], [HL], [K1], [K2], [K3]).

In this note we will describe some very recent work in this subject, and in particular, when the "host" metric space is itself derived from a graph.

We should remark here that many of the results we discuss actually apply to general metric and semi-metric spaces. However, we will usually restrict ourselves to metrics induced by graphs.

2. Some Background

Some of the early interest in questions of this type were motivated by the investigation of routing algorithms in data networks ([P], [GP1], [BGK]). Several of the first results involved the (semi-metric) space S consisting of the three symbols 0, 1 and *, with the distance d_S given by:

$$d_S(x,y) = \begin{cases} 1 & \text{if } x=0, y=1 \text{ or } x=1, y=1, \\ 0 & \text{otherwise} \end{cases}$$

In general, if (M, d_M) is a (semi)-metric space there is a natural metric d_M on the cartesian product M^n given by:

For $\bar{x} = (x_1, \ldots, x_n)$, $\bar{y} = (y_1, \ldots, y_n) \in M^n$,

$$(2) \quad d_{M^n}(x,y) := \sum_{k=1}^{n} d_M(x_k, y_k).$$

FACT ([GPI]). Every finite metric space M can be embedded isometrically into S^N for a suitable integer $N = N(M)$.

DEFINITION. For a graph G, denote by $N(G)$ the *least* integer N so that $G \stackrel{d}{\to} S^N$.

Let $D(G)$ denote the distance matrix of G. That is, if $V = V(G) = \{v_1, \ldots, v_n\}$ then $D(G) = (d_{ij})$ is the n by n matrix defined by setting $d_{ij} = d_G(v_i, v_j)$. Observe that since $D(G)$ is real and symmetric then the eigenvalues of $D(G)$ are real. Denote the number of positive, negative and zero eigenvalues of $D(G)$ by $n_+(G), n_-(G)$ and $n_0(G)$, respectively.

A basic result concerning $N(G)$ is the following.

THEOREM ([GPI]). For every (connected) graph G,

(3) $\quad N(G) \geq \max\{n_+(G), n_-(G)\}.$

PROOF: Suppose $\lambda: G \to S^N$ is an isometry (= isometric embedding). Thus, since

$$d_{ij} = \sum_{k=1}^{N} d_s(\lambda(v_i)_k, \lambda(v_j)_k)$$

where

$$(x) = (\lambda(x)_1, \ldots, \lambda(x)_N)$$

then by the definition of d_S, we have the basic decomposition

(4) $\quad \sum_{i,j} d_{ij} x_i x_j = \sum_{k=1}^{N} \left(\sum_{a \in A_k} x_a\right)\left(\sum_{b \in B_k} x_b\right)$

where

$$A_k = \{a: \lambda(v_a)_k = 0\},$$
$$B_k = \{b: \lambda(v_b)_k = 1\}$$

and the x_k are indeterminates.

We can rewrite (4) as

(4′) $\quad \sum_{i,j} d_{ij} x_i x_j = \frac{1}{4} \sum_{k=1}^{N} \left\{ \left(\sum_{a \in A_k} x_a + \sum_{b \in B_k} x_b\right)^2 - \left(\sum_{a \in A_k} x_a - \sum_{b \in B_k} x_b\right)^2 \right\}$

The significance of (4′) is that it represents a decomposition of the quadratic form $\sum_{i,j} d_{ij} x_i x_j$ into a sum and difference of squares. Thus, by Sylvester's "law of inertia", $N = \geq n_+(G)$ and $N \geq n_-(G)$, i.e.,

$$N \geq \max\{n_+(G), n_-(G)\}$$

and the theorem is proved. □

In [GP1], [GP2], $N(G)$ was determined for a number of classes of graphs. In particular, it was shown that for complete graphs K_n, trees T_n and cycles C_n, with n vertices:

$$N(K_n) = n-1,$$
$$N(T_n) = n-1,$$

$$N(C_n) = \begin{cases} n-1 \text{ for } n \text{ odd}, \\ \dfrac{n}{2} \text{ for } n \text{ even}. \end{cases}$$

On the basis of evidence of this type, it was conjectured in [GP1] that for every graph G with n vertices,

(5) $\quad N(G) \leq n-1$

This was finally very recently established by P. Winkler [Win1] (improving earlier estimates of Yao [Y]).

We point out that the assertion $N(K_n) = n-1$ has the following equivalent combinatorial interpretation.

FACT ([GP1]). Suppose the edge set of K_n is decomposed into t disjoint complete bipartite subgraphs. Then $t \geq n-1$.

At present no purely combinatorial proof of this is known. However, Tverberg [Tv] has recently given the following nice algebraic argument.

Denote the hypothesized bipartite graphs' vertex sets by A_k and B_k, $1 \leq k \leq t$. Thus,

(6) $\quad \displaystyle\sum_{1 \leq i \leq j \leq n} x_i x_j = \sum_{k=1}^{t} \left(\sum_{a \in A_k} x_a \right) \left(\sum_{b \in B_k} x_b \right)$

Consider the following system of $t+1$ homogeneous linear equations in the n variables x_i:

$$\sum_{a \in A_k} x_a = 0, \quad 1 \leq k \leq t,$$

and

$$\sum_{i=1}^{n} x_i = 0.$$

If $(y_1, \ldots, y_n)$ is any solution to this system then we must have

$$0 = \left\{ \sum_{i=1}^{n} y_i \right\}^2 = \sum_{i=1}^{n} y_i^2 + 2 \sum_{i<j} y_i y_j$$

$$= \sum_{i=1}^{n} y_i^2 + 2 \sum_{k=1}^{t} \left(\sum_{a \in A_k} x_a \right) \left(\sum_{b \in B_k} x_b \right)$$

$$= \sum_{i=1}^{n} y_i^2$$

i.e., $y_i = 0$ for all i. Hence, the number of equations in the system must be as large as the number of variables, i.e., $t+1 \leq n$. □

3. A Geometric Interpretation of the Tree Theorem

For a tree T_n with n vertices, the equality $N(T_n) = n-1$ relied on the fact that any tree T has $n_+(T) = 1$ (and consequently, $n_-(T) = n-1$ since $D(T_n)$ has trace 0). This follows from the surprising fact that det $D(T_n)$ depends only on n and is otherwise independent of the structure of T_n. Specifically, we have:

THEOREM ([GP1]). *For a tree T_n with n vertices*

(7) $\qquad \det D(T_n) = (-1)^{n-1}(n-1)2^{n-2}.$

A straightforward proof is not difficult. One can simply (arbitrarily) choose a root in T_n and successively subtract the row and column of each "son" from those of his "father," always selecting the unprocessed vertices furthest from the root. At the termination of this process the resulting matrix $D' = (d'_{ij})$ satisfies

$$d'_{ij} = \begin{cases} 1 & \text{if } i = 1, j \neq 1 \text{ or } j = 1, i \neq 1, \\ -2 & \text{if } i = 1, j \neq 1, \\ 0 & \text{otherwise} \end{cases}$$

and (7) follows easily.

However, one strongly suspects from the form of (7) that something deeper must be involved. The fact that the factor $n-1$ represents the number of edges of T_n was observed in [GHH] where the following result was proved.

Let $cof(G)$ denote the sum of the cofactors of the matrix $D(G)$ for a graph of G.

THEOREM ([GHH]). *If G has blocks* $G_1, \ldots, G_r$ then*

$$cof(G) = \prod_{k=1}^{r} cof(G_k) \qquad (i)$$

---

*i.e., maximal 2-connected subgraphs

$$\det D(G) = \sum_{k=1}^{r} \det D(G_k) \prod_{i \neq k} cof(G_i) \qquad (ii)$$

Note that when all $cof(G_k) \neq 0$ then (ii) can be rewritten in the attractive form

$$\frac{\det D(G)}{cof(G)} = \sum_{k=1}^{r} \frac{\det D(G_k)}{cof(G_k)} \qquad (ii)'$$

Thus, for T_n, all blocks consist of a single edge K_2 which has $\det D(K_2) = -1$, $cof(K_2) = -2$ and consequently (7) follows at once.

The next result we give generalizes (7) and gives a geometrical explanation for the factor 2^{n-2}. Consider the set Q^n of vertices of the (usual) unit n-cube in $\mathbb{R}^n$, i.e.,

$$Q^n = \{0,1\}^n = \{\bar{a} = (a_1, \ldots, a_n) : a_k = 0 \text{ or } 1, 1 \leq k \leq n\}.$$

There is a natural metric d_H on Q^n, called the Hamming metric, given by

$$d_H((a_1, \ldots, a_n), (b_1, \ldots, b_n)) = \sum_{k=1}^{n} |a_k - b_k|.$$

i.e., the distance between $\bar{a}$ and $\bar{b}$ in Q^n is just equal to the number of coordinate positions in which they differ. Let us call a set $S \subseteq Q^n$ *full-dimensional* if the convex hull of S has positive n-dimensional volume.

THEOREM 1. If $\{\bar{a}_0, \ldots, \bar{a}_n\}$ is any full-dimensional set of $n+1$ points of Q^n then

$$\det(d_H(\bar{a}_i, \bar{a}_j)) = (-1)^n n \cdot 2^{n-1} \qquad (8)$$

PROOF: For $\bar{a}_i = (a_{i1}, \ldots, a_{in})$ write

$$a_{ik} = \frac{1}{2} + \frac{1}{2}\alpha_{ik}$$

where $\alpha_{ik} = \pm 1$. Thus

$$d_H(\bar{a}_i, \bar{a}_j) = \sum_{k=1}^{n} |a_{ik} - a_{jk}| \qquad (9)$$

$$= \frac{1}{2} \sum_{k=1}^{n} |\alpha_{ik} - \alpha_{jk}|$$

$$= \frac{1}{2} \sum_{k=1}^{n} (1 - \alpha_{ik}\alpha_{jk})$$

$$= \frac{1}{2}(n - \bar{\alpha}_i \cdot \bar{\alpha}_j)$$

where $\bar{\alpha}_i \cdot \bar{\alpha}_j$ denotes the inner product of the vectors $\bar{\alpha}_i = (\alpha_{i1}, \ldots, \alpha_{in})$ and $\bar{\alpha}_j = (\alpha_{j1}, \ldots, \alpha_{jn})$. It follows from elementary linear algebra that for any square matrix $M = (m_{ij})$, if J is the matrix with all entries equal to 1 then

(10) $\det(M + xJ) = \det M + x \det(m_{ij} - m_{1j} - m_{i1} + m_{11})$.

Thus, from (9) we have

(11) $\det(d_H(a_i, a_j)) = \det\left[\frac{1}{2}(n - \bar{\alpha}_i \cdot \bar{\alpha}_j)\right]$

$$= \frac{1}{(-2)^{n+1}} \det(\bar{\alpha}_i \cdot \bar{\alpha}_j - n)$$

$$= \frac{1}{(-2)^{n+1}} \{\det(\bar{\alpha}_i \cdot \bar{\alpha}_j) - n \det(\bar{\alpha}_i \cdot \bar{\alpha}_j - \bar{\alpha}_0 \cdot \bar{\alpha}_j - \bar{\alpha}_i \cdot \bar{\alpha}_0 + \bar{\alpha}_0 \cdot \bar{\alpha}_0)\}$$

$$= \frac{1}{(-2)^{n+1}} \{\det(\bar{\alpha}_i \cdot \bar{\alpha}_j) - n \det((\bar{\alpha}_i - \bar{\alpha}_0) \cdot (\bar{\alpha}_j - \bar{\alpha}_0))\}$$

The determinants which appear in (11) of the form $\det(\bar{x}_i \cdot \bar{x}_j)$, called *Gramians*, occur frequently in linear algebra. One of their particularly useful properties is the following.

FACT (see [Ga]). For a set of vectors $\bar{x}_1, \ldots, \bar{x}_n$ in $\mathbb{R}^n$, the Gramian $\det(\bar{x}_i \cdot \bar{x}_j)$ is just the square of the (m-dimensional) volume of the parallelepiped spanned by the $\bar{x}_k$.

In particular, in (11) since the $n+1$ $\bar{\alpha}_k$'s all lie in $\mathbb{R}^n$ then $\det(\bar{\alpha}_i \cdot \bar{\alpha}_j) = 0$. On the other hand, the vectors $\bar{\alpha}_k - \bar{\alpha}_0$ by hypothesis span an n-dimensional space. It is clear that the parallelepiped they span has width 2 in each dimension (since all $\alpha_{ik} = \pm 1$), and consequently, has volume 2^n.

Thus, continuing (11), we obtain

$$\det d_H(\bar{a}_i, \bar{a}_j) = \frac{1}{(-2)^{n+1}} (0 - n \cdot (2^n)^2)$$

$$= (-1)^n n \cdot 2^{n-1}$$

and the theorem is proved. □

Symmetric differences. It is often natural to interpret binary n-tuples as characteristic functions on the set $[n] = \{1, \ldots, n\}$ so that each

$\bar{a} = (a_1, \ldots, a_n)$ corresponds to a subset $S(\bar{a}) \subseteq [n]$ by

$$k \in S(\bar{a}) <=> a_k = 1.$$

With this association it is easy to see that

$$d_H(\bar{a}, \bar{b}) = |S(\bar{a}) \Delta S(\bar{b})|$$

where $X \Delta Y$ denotes the symmetric difference $X/Y \cup Y/X$ of X and Y.

Let us say that a family of subsets of $[n]$ is full-dimensional if the corresponding n-tuples are. We can restate Theorem 1 in these terms.

THEOREM 1'. *Suppose* $\{S_0, \ldots, S_n\}$ *is a full-dimensional family of subsets of* $[n]$. *Then*

$$\det(|S_i \Delta S_j|) = (-1)^n n \cdot 2^{n-1}.$$

The advantage of this formulation is that it can be readily extended to the following more general situation. Suppose μ is a discrete measure on $2^{[n]}$, i.e.,

$$\mu(k) \geq 0, \ k \in [n]$$

$$\mu(X) = \sum_{x \in X} \mu(x), \ X \subseteq [n].$$

THEOREM 2. *Suppose* $\{S_0, \ldots, S_n\}$ *is a full-dimensional family of subsets of* $[n]$. *Then*

(12) $$\det(\mu(S_i \Delta S_j)) = (-1)^n 2^{n-1} \sum_k \mu(k) \prod_k \mu(k).$$

The proof of (12), which we will not give here, depends on an extension of $(ii)'$ appearing in [GHH]. Of course, when μ is just the counting measure, i.e., $\mu(k) = 1$ for all $k \in [n]$, then (12) reduces to the previous result.

4. Embedding in the m-Cube.

The cartesian product K_2^n of K_2 (the complete graph on two vertices) is usually called the n-cube in the graph theory literature. The induced metric $d_{K_2^n}$ on K_2^n is just the Hamming metric d_H. While we have seen earlier that every graph G embeds isometrically in $\{0, 1, *\}^N = S^N$ for some N this is certainly not true for K_2^N. The problem of characterizing those G for which $G^d K_2^m$ for some m was settled by the following result of Djoković. First we need a definition.

For two vertices x and y on a graph G define $N(x,y) := \{z \in V(G): d_G(x,z) < d_G(y,z)\}$ (i.e., $N(x,y)$ is the set of points nearer to x than to y).

THEOREM (Djoković [Dj]). $G \xrightarrow{d} K_2^m$ for some m if and only if

(i) G is bipartite,
(ii) For each edge $\{x,y\}$ of G, if $a,b \in N(x,y)$ and $d_G(a,c) + d_G(c,b) = d_G(a,b)$ then $c \in N(x,y)$. (Of course, this also applies to $N(y,x)$).

What (ii) says is that $N(x,y)$ is closed under taking shortest paths. Note that because of (i), $N(x,y) \cup N(y,x)$ is actually a partition of the vertices of G whenever $\{x,y\}$ is an edge.

In [Dj], Djoković introduces the following equivalence relation θ on the edge set $E(G)$ of G by defining:

$$\{x,y\} = e \stackrel{\theta}{\sim} e' <=> e' \text{ intersects both } N(x,y) \text{ and } N(y,x).$$

The corresponding set $E(G)/\theta$ of θ-equivalence classes has the following property. Denote by $dim(G)$ the least m such that $G \xrightarrow{d} K_2^m$ (when G satisfies (i) and (ii)).

THEOREM (Djoković [Dj])

$$dim(G) = card(E(G)/\theta).$$

In fact, as pointed out by P. Winkler [Win2], this approach actually shows that any isometry of G into K_2^m can only have $m = dim(G)$ (for a larger m none of the additional coordinates are used) and furthermore, the embedding is unique up to symmetries of the m-cube.

The next result ties $dim(G)$ directly to the distance matrix $D(G)$.

THEOREM 3.

(13) $\quad dim(G) = n_-(G)$

PROOF: First, recall from (3) that

$$N(G) \geq n_-(G).$$

Next, we claim

$$G \xrightarrow{d} K_2^m => n_+(G) = 1.$$

This can be seen by observing (as was done in [BG]) in (4') that when no *'s are used, $A_k \cup B_k$ is then a partition of $V(G)$ and consequently, the quadratic form $\sum_{i,j} d_{ij} x_i x_j$ is expressible as a sum of *one* positive square and some negative squares. This implies $n_+(G) = 1$.

Suppose $\lambda: G \to K_2^m$ is isometry.

CLAIM: rank $(D(G)) = m+1$.
PROOF OF CLAIM: On one hand
$$rank(D(G) = n_-(G) + n_+(G)$$
$$= n_-(G) + 1$$
$$\leq N(G) + 1$$
$$\leq m + 1$$

On the other hand, since G is connected there must exist $v_0, v_1, \ldots, v_m \in V(G)$ such that the set $\{\lambda(v_0), \lambda(v_1), \ldots, \lambda(v_m)\}$ is full-dimensional in K_2^m. Thus the submatrix

$$(d_G(v_i, v_j)) = (d_H(\lambda(v_i), \lambda(v_j)))$$

is nonsingular (by Theorem 1) and so,

$$rank(D(G)) \geq m+1$$

Consequently,

$$rank(D(G)) = m+1.$$

which proves the claim, and

$$n_-(G) = m = N(G) = dim(G)$$

which proves the theorem. □

We note that it follows from these considerations, for example, that $G \xrightarrow{d} K_2^m$ then $det(D(G)) \neq 0$ iff G is a tree.

5. Embedding in Products of Graphs.

A natural extension of the questions raised in the preceding section is to attempt to characterize those G for which $G \xrightarrow{d} H^m$ or $G \xrightarrow{d} H_1 \times \cdots \times H_m$ hold, for various choices of H and $H_1, \ldots, H_m$. Unfortunately, our knowledge for these more general questions is rather incomplete at present. In this section we will mention several results (without proof) which are known.

To begin with, we should observe that no graph H can be a "universal host" (i.e., such that every graph G embeds isometrically into some power of H) since, for example, it is not hard to show that

$$G \xrightarrow{d} H^m \Longrightarrow \chi(G) \leq \chi(H)$$

where χ denotes chromatic number. (Compare with condition (i) of

Djokovič's first theorem.) One reason that it was possible for $G \stackrel{d}{\to} S^N$ to hold for any G is that $S = \{0,1,*\}$ is not a metric space, since

$$d_S(0,1) = 1 > 0 = d_S(0,*) + d_S(*,1)$$

A more substantial restriction on G is given by the following considerations. For $x,y \in V(G)$ define the set

$$S(x,y) = \{z \in V(G): d_G(x,z) = d_G(y,z)\}$$

(the set of points of G at the *same* distance from x and y). With $N(x,y)$ and $N(y,x)$ defined as before, define the relation $\hat{\theta}$ on $E(G)$ as follows: If $e = \{x,y\}$ and e' are edges of G then

$$e \stackrel{\hat{\theta}}{\sim} e' \iff e' \text{ intersects at least two of the sets } N(x,y), N(y,x), S(x,y)$$

It can be checked that $\hat{\theta}$ is symmetric and reflexive but not transitive (consider a 5-cycle).

Define θ to be the *transitive closure* of $\hat{\theta}$. Note that θ is actually the same equivalence relation given by Djokovič in the case that G is bipartite.

A basic result concerning θ is the following.

FACT. Suppose $\lambda: G \stackrel{d}{\to} H_1 \times \cdots \times H_m$. If $\{x,y\} \stackrel{\theta}{\sim} \{x',y'\}$ and $\lambda(x)$ and $\lambda(y)$ differ only in their k^{th} components then $\lambda(x')$ and $\lambda(y')$ differ only in their k^{th} components.

Of course, if $\{x,y\}$ is an edge then $d_G(x,y) = 1$ and so $\lambda(x)$ and $\lambda(y)$ can only differ in one component. This fact can be used to prove the following result.

THEOREM 4. If $G \stackrel{d}{\to} H^m$ then each connected component C of G induced by the θ-equivalence classes must satisfy $C \stackrel{d}{\to} H$.

This is a rather strong restriction. It implies, for example, that if $G \stackrel{d}{\to} K_3^m$ then all such C are either K_2's or K_3's. Let us call G *prime* if $G \stackrel{d}{\to} H_1 \times \cdots \times H_m \implies G \stackrel{d}{\to} H_k$ for some k. The preceding results can be used to prove the following.

THEOREM 5. If G has a single θ-equivalence class then G is prime.

For example, it is easy to check that this is the case for the 5-cycle C_5 although it seems to be a rather delicate condition in general. For example, for the graphs G and $G' = G-\{e\}$ shown in Figure 1, G has a single θ-equivalence and so, is prime while G' has *three* θ-equivalence classes and in fact, embeds isometrically in K_3^3. (We have shown the appropriate images next to each vertex, where we take $V(K_3) = \{0,1,2\}$).

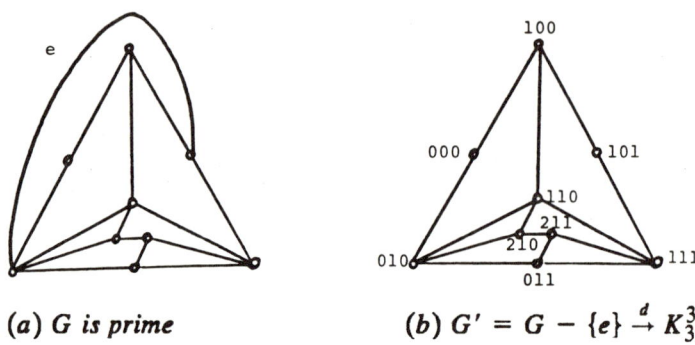

(a) G is prime   (b) $G' = G - \{e\} \xrightarrow{d} K_3^3$

Figure 1

6. Concluding Remarks

The problem of embedding graphs isometrically into other graphs is a special case the more general topic of embedding (finite) metric spaces isometrically into other metric (or semi-metric) spaces. This topic has an extensive literature, some of which can be found in [ADz1], [ADz2], [Av3], [K2], [K3]. Of course, many of these more general results impinge on our studies. For example, it follows from these considerations that if $G \xrightarrow{d} K_3^m$ (or indeed, if $G \xrightarrow{d} H^m$ for any graph H with at most four vertices) then $n_+(G) = 1$. The reason for this is as follows.

Lets say that an n by n distance matrix $D = (d_{ij})$ is of *negative type* if

(14) $x_1 + \cdots + x_n = 0,\ x_k \in \mathbb{R} => \sum_{i,j} d_{ij} x_i x_j \leq 0$.

Similarly, call D *hypermetric* if

(15) $x_1 + \cdots + x_n = 1,\ x_k \in \mathbb{Z} => \sum_{i,j} d_{ij} x_i x_j \leq 0$.

Although (14) and (15) are similar, (15) is actually much stronger. Not only does it imply (14) but also that the space actually satisfies the triangle inequality (and many stronger ones), something that (14) does not do. It is not hard to show that

$$D \text{ is of negative type} \Rightarrow n_+(D) = 1.$$

No example is currently known of a graph G for which $n_+(G) = 1$ and $D(G)$ is *not* of negative type although such graphs undoubtedly exist.

It turns out that the properties of hypermetricity and negative type are preserved under taking products and isometric subsets. Thus,

$$K_3 \text{ is of negative type } (easy \ to \ check)$$

$$\Rightarrow K_3^m \text{ is of negative type}$$

$$\Rightarrow G \stackrel{d}{\hookrightarrow} K_3^m \text{ is of negative type}$$

$$\Rightarrow n_+(G) = 1$$

An interesting related question is the following. Suppose X is a semi-metric space with distance matrix C. Let $D^{(k)}$ denote the distance matrix corresponding to the product space X^k. As just remarked, if X is of negative type then so is X^k and consequently $n_+(D^{(k)}) = 1$ for any k. Does the converse hold? In other words, does $n_+(D^{(k)}) = 1$ for all k imply that X is of negative type?

It was conjectured at one time by Deza [Dez2] that hypermetricity was a sufficient condition for isometric embeddability into $l_1 (i.e., \mathbb{R}^m \text{ with } d(\bar{x}, \bar{y}) = \sum_i |x_i - y_i|)$. This was shown *not* to be the case by Avis [Av3] (see also Assouad [As1]), who proved that the graph $K_7 - P_3$ is hypermetric but not isometrically embeddable into l_1. However, it is true [Dez1], [K1] that hypermetricity is a necessary condition for l_1-embeddability. In the same spirit it is easy to show that the graph $K_3 + \{e\}$ (an edge) is of negative type but not hypermetric (see [AsD2]).

Finally, we arrive at the graph $K_{3,2}$, which is exceptional in several respects. Since $n_+(K_{3,2}) = 2$, $K_{3,2}$ is not of negative type and therefore not isometrically embeddable into any K_2^m or even $\mathbb{R}^n$. In fact, it is not even a *subgraph* of K_2^m. It also turns out that

$$N(K_{3,2}) = 4 > max(n_+(K_{3,2}), n_-(K_{3,2})) = 3.$$

showing that equality does not have to hold in (3).

At present no necessary condition is known for a graph to be l_1-embeddable, hypermetric or of negative type. It would seem fruitful to study the characteristic polynomials of the associated distance matrices of various spaces rather than just the signs of the eigenvalues. This has been initiated for trees in [EGG] and [GL]. It seems quite likely that our understanding of this whole general area would increase substantially if the corresponding results were known for more general graphs, e.g., those $G \xrightarrow{d} K_2^m$.

Acknowledgements

The author wishes to acknowledge the valuable discussions he has had on the preceding topics with Peter Winkler and Hans Witsenhausen.

Added in proof: The converse of Theorem 5 has now been proved by Peter Winkler. Also, H.J. Landau has shown that $n_+(D^{(2)})=1$ already complies that the underlying space X is of negative type.

References

[As1] P. Assouad, Un espace hypermétrique non plongeable dans un espace L^1, C. R. Acad. Sci. Paris, 285 (ser A) (1977), 361-363.

[As2] ········· Plongements isométrique dans L^1: aspect analytique, Séminaire d'Initiation à l'Analyse 1979-1980 (Paris 6) exposé no. 14.

[As3] ········· Sur les inequalités valides dans L^1 (preprint).

[ADl1] P. Assouad and C. Delorme, Graphes plongeables dans L^1, C. R. Acad. Sci. Paris 291 (1980), 369-372.

[ADl2] P. Assouad and C. Delorme, Distances sur les graphes et plongements dans L^1, I,II (preprints).

[Adz1] P. Assouad and M. Deza, Espaces métriques plongeables dans un hypercube, Annals of Disc. Math 8 (1980), 197-210.

[ADz2] ···················· Metric subspaces of L^1 (preprint).

[Av1] D. Avis, Hamming metrics and facets of the Hamming cone, Tech. Report SOCS - 78.4 School of Comp. Sci., McGill Univ, 1978.

[Av2] ······· Extremal metrics induced by graphs, Annals of Dis. Math. 8 (1980), 217-220.

[Av3] ······· Hypermetric spaces and the Hamming cone, Canad. J. Math. 33 (1981), 795-802.

[BG] I. F. Blake and J. H. Gilchrist, Addresses for graphs, IEEE Trans. on Inf. Th., vol. IT-19, (1973), 683-688.

[B] F. T. Boesch, Properties of the distance matrix of a tree, Quart. Appl. Math. 16 (1969), 607-609.

[BM] J. A. Bondy and U. S. R. Murty, Graph Theory and Applications, American Elsevier Publishing Company, New York, 1976.

[BGK] L. H. Brandenburg, B. Gopinath and R. P. Kurshan, On the addressing problem of loop switching, Bell Sys. Tech. Jour. 51 (1972), 1445-1469.

[Dew] A. K. Dewdney, The embedding dimension of a graph, Ars Combinatoria 9

(1980), 77-90.

[Dez1] M. Deza (M. E. Tylkin), On Hamming geometry of unitary cubes, Doklady Akad. Nauk. SSR 134 (1960), 1037-1040 (Russian).

[Dez2] ------------------- Realizability of matrices of distances in unitary cubes, Problemy kibernetiki 7 (1962), 31-42 (Russian).

[DR] M. Deza and I. G. Rosenberg, Intersection and distance patterns, Tech. report, Centre de Recherches Mathématique, Univ. de Montreal, 1977.

[Dj] D. Z. Djoković, Distance preserving subgraphs of hypercubes, J. Comb. Th. (B) 14 (1973), 263-267.

[EGG] M. Edelberg, M. R. Garey and R. L. Graham, On the distance matrix of a tree, Dis. Math 14 (1976), 23-29.

[F] V. Firsov, Isometric embedding of a graph in a Boolean cube, Kibernetica 1 (1965), 95-96 (Russian).

[Ga] F. R. Gantmacher, The Theory of Matrices, vol. I, Chelsea Pub. Co., New York 1959.

[Go] A. J. Goldman, Realizing the distance matrix of a graph, J. Res. Nat. Bur. Standards B70, (1966), 153-154.

[GP1] R. L. Graham and H. O. Pollak, On the addressing problem for loop switching, Bell Sys. Tech. Jour. 50 (1971), 2495-2519.

[GP2] --------------------------- On embedding graphs in squashed cubes, Graph Theory and Applications, in Lecture Notes in Math. No. 303, Springer-Verlag, New York, 1972.

[GHH] R. L. Graham, A. J. Hoffman and H. Hosoya, On the distance matrix of a directed graph, Jour. Graph Th. 1 (1977), 85-88.

[GL] R. L. Graham and L. Lovász, Distance matrix polynomials of trees, Advances in Math. 29 (1978), 60-88.

[HY] S. L. Hakimi and S. S. Yau, Distance matrix of a graph and its realizability, Quart. Appl. Math. 12 (1965), 305-317.

[HL] I. Havel and P. Liebl, Embedding the polytomic tree into the n-cube, Čas. pěst mat. 98 (1973), 307-314.

[HMG] H. Hosoya, M. Murakimi and M. Gotoh, Distance polynomial and characterization of a graph, Natur. Sci. Rep. Ochanomizu Univ. 24 (1973), 27-34.

[K1] J. B. Kelly, Metric inequalities and symmetric differences, in : Inequalities II, O. Shisha, editor, Acad. Press, New York, 1970, pp. 193-212.

[K2] ---------- Hypermetric spaces and metric transforms, in : Inequalities II, O. Shisha, editor, Acad. Press, New York, 1972, pp. 149-159.

[K3] ---------- Hypermetric spaces, in : Lecture Notes in Math. No. 490, Springer-Verlag, New York, 1975, pp. 17-31.

[PH] A. N. Patrinos and S. L. Hakimi, The distance matrix of a graph and its tree realization, Quart. Appl. Math. 19 (1972), 255-269.

[P] J. R. Pierce, Network for block switching of data, Bell Sys. Tech. Jour. 51 (1972), 1133-1145.

[R] R. M. Robinson, Prob. 3705, Amer. Math. Monthly 41 (1934), 581.

[S] G. Szegö, Sol. to Prob. 3705, Amer. Math. Monthly 43 (1936), 246-259.

[Tv]   H. Tverberg, On the decomposition of K_n into complete graphs, J. Graph. Th 6 (1982), 493-494.

[Win1] P. M. Winkler, Proof of the squashed cube conjecture (to appear in Combinatorica).

[Win2] ------------ (personal communication).

[Wit] H. S. Witsenhausen (personal communication).

[Y]   A. C-C. Yao, On the loop switching addressing problem, SIAM Jour. on Computing 7 (1978), 515-523.

The Maximum Weighted Cycle-Packing Problem and its Relation to the Chinese Postman Problem

Meigu Guan

ABSTRACT

We call a set of cycles $C=\{c_1,c_2,\ldots,c_s\}$ in a connected undirected graph a cycle-packing of G if any two distinct cycles in C are edge-disjoint. An optimization problem concerning cycle-packing is proposed in this paper, i.e., to find a cycle-packing with maximum total weight for a given weighted graph G. It is proved that this problem is actually equivalent to the Chinese postman problem, so it can be solved by a polynomial bounded algorithm.

Most problems concerning cycle-packing and cycle-covering of a graph seem to be very difficult. In this paper, however, we investigate a problem of this kind which can be solved by a polynomial bounded algorithm.

Definition 1: Let $G = (V,E)$ be an undirected connected graph, and let $C = \{C_1, C_2, \ldots, C_s\}$ be a set of cycles of G.

(1) If C_i and C_j are edge-disjoint for any $i \neq j$, then we call C a cycle-packing of G.

(2) If for every $e \in E$, there exists a cycle C_i, such that $e \in C_i$, then we call C a cycle-covering of G.

Furthermore, if G is weighted with a non-negative weight $w(e)$ for each $e \in E$ and if $C = \{C_1, C_2, \ldots, C_s\}$ is either a cycle-packing or a cycle-covering of G, then the weight of C is defined to be

$$w(C) = \sum_{i=1}^{s} w(C_i)$$

where $w(C_i)$ is the sum of the weights of all edges belonging to C_i.

The following two problems concerning cycle-packing and cycle-covering may be very difficult:

PROBLEM 1 (maximum cardinality cycle-packing problem): Find a cycle-packing of G with maximum number of cycles.

PROBLEM 2 (minimum cardinality cycle-covering problem): Find a cycle-covering of G with minimum number of cycles.

In the sequel, we consider two similar problems:

PROBLEM 1^0 (maximum weighted cycle-packing problem): Find a cycle-packing of G with maximum weight.

PROBLEM 2^0 (minimum weighted cycle-covering problem): Find a cycle-covering of G with minimum weight.

We shall prove that problem 1^0 is actually equivalent to the Chinese postman problem (see [1], [4]) and obtain an algorithm for finding a maximum weighted cycle-packing. And we shall discuss briefly problem 2^0.

LEMMA 1: Let $G = (V,E)$ be a connected graph, and let $\overline{G}$ be the (connected) graph obtained from G by duplicating the edges of some $E_0 \subset E$. Then $E - E_0$ admits a cycle-decomposition $C = \{C_1,\ldots,C_s\}$ for some $s \geq 1$ (i.e. C is a cycle-packing of G), if and only if $\overline{G}$ is Eulerian.

The proof of Lemma 1 can be left to the reader.

THEOREM 1: Let $C = \{C_1, C_2, \ldots, C_s\}$ be a cycle-packing of a weighted connected graph $G = (V,E)$, and let $\overline{G}$ and $E_0 = \{e | e \in E, e \notin C_j, j = 1,2,\ldots,s\}$ have the same meanings as in Lemma 1. Then C is a cycle-packing of G with maximum weight if and only if $\overline{G}$ is an Eulerian supergraph of G with minimum weight.

PROOF: First, suppose $C = \{C_1, C_2, \ldots, C_s\}$ is a cycle-packing with maximum weight. We want to prove that the corresponding $\overline{G}$ is an Eulerian supergraph of G with minimum weight. Let G' be an Eulerian supergraph of G obtained by duplicating the elements of some $E' \subset E$. Then by Lemma 1, $E - E'$ can be expressed as a cycle-packing C' of G. Since C is a maximum weighted cycle-packing, we have

$$w(C') \leq w(C).$$

On the other hand, we clearly have

$$2w(G) = w(C) + w(\overline{G})$$
$$= w(C') + w(G').$$

It follows that

$$w(\overline{G}) \le w(G').$$

Thus we have proved that among all the Eulerian supergraphs of G obtained by the same type of construction employed to obtain $\overline{G}$, G' respectively, $\overline{G}$ has minimum weight. However, it was proved in [4] that if G^0 is an Eulerian supergraph of G such that some edge of G corresponds to more than two edges of G^0, then G^0 cannot be a minimum weighted Eulerian supergraph of G. So $\overline{G}$ is an Eulerian supergraph of G with minimum weight.

Similarly, we can prove that C is a maximum cycle-packing if $\overline{G}$ is an Eulerian supergraph of G with minimum weight. We omit the details here.

Theorem 1 shows that the maximum weighted cycle-packing problem is actually equivalent to the Chinese postman problem. Furthermore, the proof given above indicates that one can find a maximum weighted cycle-packing of a graph G by the following algorithm:

STEP 1: Find an Eulerian supergraph $\overline{G}$ of G with minimum weight (i.e. solve the Chinese postman problem for graph G). Let E_0 be the set of edges which are duplicated in $\overline{G}$.

STEP 2: Decompose $E - E_0$ into an edge-disjoint cycles $C_1, C_2, \ldots, C_s$. Then $C = \{C_1, C_2, \ldots, C_s\}$ is a desired maximum weighted cycle-packing of G.

Since step 1 can be done with a polynomial bounded algorithms [1], and step 2 is much easier, we have proved that the maximum weighted cycle-packing problem belongs to class P; i.e. it can be solved by a polynomial bounded algorithm.

Let us turn to problem 2^0; ie. that of finding a cyle-covering of G with minimum weight. Obviously, a connected graph G has a cycle-covering if and only if it is bridgeless (i.e. 2-connected), we may therefore assume that G is bridgeless when discussing the cycle-covering problem.

It is easy to see that any cycle-covering of G corresponds to an Eulerian supergraph of G. But the converse is not true. Itai and Rodeh showed in [3] that some Eulerian supergraph G' (obtained by the above construction) of the Peterson graph cannot be considered as obtained from a cycle-covering of G, because when one decomposes G' into edge-disjoint cycles, there always exists a cycle consisting of an edge and its duplicated edge, which is not a cycle of the original graph.

However, it has been proved (see [2]) that if G is planar, any Eulerian supergraph of G (obtained by the above construction) can always be considered as obtained from a cycle-covering of G. For details, see [2].

Acknowledgement:

This paper was written when the author was on leave at the University of Waterloo. The hospitality of the University, especially that of the Department of Combinatorics and Optimization, is greatly appreciated.

References

[1] J. Edmonds and E. L. Johnson, Matching, Euler tours and the Chinese postman, Mathematical Programming 5 (1973), 88-124.

[2] Meigu Guan and H. Fleischner, On the minimum cycle-covering problem for planar graphs, Research Report CORR 83-3, Faculty of Mathematics, University of Waterloo (1983).

[3] A. Itai and M. Rodeh, Covering a graph by circuits, Automata, Languages and Programming (G. Goos and J. Hartmanis, ed.), Fifth Colloquium, Udine, July 1978, 289-299.

[4] Kwan Mei-ko (Meigu Guan), Graphic programming using odd or even points, Chinese Mathematics 1 (1960), 273-277.

Digraphs with Converse and Line Digraph Isomorphic

R. L. Hemminger*

ABSTRACT

A characterization of finite digraphs D such that the line digraph of D is isomorphic to the converse of D is given.

1. Digraphs with $L(D) \approx D'$

In this paper we characterize those finite digraphs D with the property that $L(D) \approx D'$, the converse of D. We allow D to have loops but since we want D' to be a line digraph, it (and hence D) can have no multiple arcs. Hereafter we will use the term digraph to mean a finite digraph without multiple arcs. We will let (a,b) denote the ordered pair of vertices a and b from D and set $((a,b), (c,d)) \in E(L(D))$ iff $(a,b), (c,d) \in E(D)$ and $b = c$. We first record a simple but useful observation.

LEMMA 1. For every digraph D, $L(D') \approx L(D)'$.

PROOF: The mapping σ, given by $\sigma(a,b)=(b,a)$, establishes the desired isomorphism.

COROLLARY 2: If D is a digraph with $L(D) \approx D'$, then $L^2(D) \approx D$.

PROOF: We have $L^2(D) \approx L(D') \approx L(D)' \approx (D')' \approx D$.

Digraphs D with $L^2(D) \approx D$ have been characterized in [1], so at this point we review the pertinent material.

If, for some positive integers n and k, we have $L^{n+k}(D) \approx L^n(D)$, then we say that the digraph D is *periodic*, and the smallest value of k for which this is true is called its *period*. An important concept in this discussion is that of an *out-tree* which is a directed tree with one vertex (called the *root*) which can reach any other vertex

---
* Written while the author was visiting the Universität für Bildungswissenschaften, Klagenfurt, Austria.

in the tree by a directed path; an *in-tree* is the converse of an out-tree.

If D is a digraph with a single directed circuit $C: v_0, v_1, \ldots, v_m, v_0$, then we define its *basic configuration* to be that subdigraph formed by the union of all directed paths to and from vertices of C. Let A_i be the subdigraph with root v_i consisting of the union of all directed paths from v_i that are otherwise disjoint from C; B_i is similarly defined as the corresponding in-subdigraph rooted at v_i. Since C is the only directed circuit in D, we see that $A_i \cap B_j \subseteq \{v_i\} \cap \{v_j\}$ and that the basic configuration of D is C plus the A_i and B_i. If we also have that the A_i are all out-trees and are pairwise disjoint and that the B_i are all in-trees and pairwise disjoint, then we call the basic configuration of D an *eddy digraph* of D. Clearly each of the line digraph iterates of D have a single circuit and it is not difficult to see (but somewhat messy) that for all sufficiently large n, the basic configuration of $L^n(D)$ is an eddy digraph. The cyclic (modulo m) sequences $\{A_i\}$ and $\{B_i\}$ are called the *out-tree* and *in-tree sequences* of D. A critical observation is that the line digraph of an eddy digraph has the same sequences with the in-trees advanced one vertex on the circuit relative to the out-trees. All of the above is illustrated in Figure 1 where the basic configurations of D and its iterates are given. Note that those of $L^2(D)$ and $L^3(D)$ are eddy digraphs.

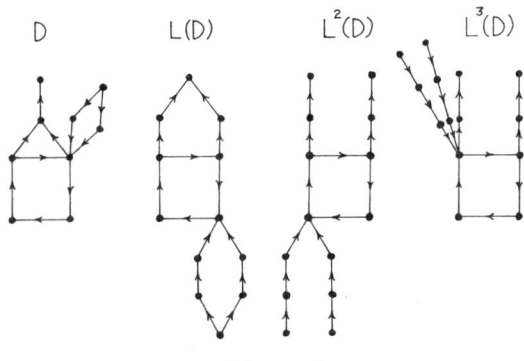

Figure 1

We define the *out-tree index* of the unicyclic digraph D to be the minimum positive integer r for which $A_{i+r} \approx A_i$ for all i; the *in-tree index* is defined similarly. The main result in [1] is that the period of a unicyclic digraph D is the greatest common divisor of its out-tree and in-tree indices.

The reader should note that the digraphs labelled L(D), $L^2(D)$ and $L^3(D)$ in Figure 1 are only the basic configurations of those digraphs. The description of the rest of $L^m(D)$ when D is a unicyclic digraph such that $L^t(D) \approx D$ for some t is given in [1]. It is independent of t and is determined uniquely from the eddy digraph E in D by certain generalizations of the Heuchenne condition that characterizes line digraphs. We will call this subdigraph so determined by E the *Heuchenne component* determined by E. The Heuchenne components determined by the eddy digraphs of $L^2(D)$ and $L^3(D)$ in Figure 1 are given in Figure 2.

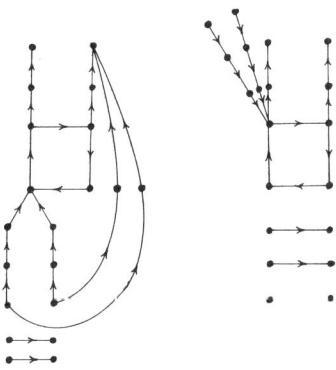

Figure 2

In the remainder of the paper D will denote a digraph with the property that $L(D) \approx D'$. We will denote eddy subdigraphs of D by E's and Heuchenne components of D by H's. We need the following facts from [1].

(1) D is of period one or two and consists of the disjoint union of Heuchenne components. Each such Heuchenne component consists of a unicyclic weakly connected component of D plus certain associated weakly connected tree components of D. Note that a Heuchenne component of D need not have period one or two; for example, we could have $D = H \cup L(H) \cup L^2(H)$ of period one with H a Heuchenne component of D of period three. However, the Heuchenne component is still determined uniquely by its eddy subdigraph and the generalized Heuchenne conditions.

(2) For an eddy digraph E and for all sufficiently large n, $L^n(E)$ is the Heuchenne component determined by the eddy digraph in $L^n(E)$. This is true for all $n \geq 1$ if we replace E by the Heuchenne component it determines.

(3) All line digraph iterates $L^n(E)$, $n \geq 0$, of an eddy digraph E have the same out-tree sequence and the same in-tree sequence. Moreover, two eddy digraphs E_1 and E_2 have the same out-tree sequence and the same in-tree sequence if and only if there are nonnegative integers m and n such that $L^m(E_1) \approx L^n(E_2)$.

(4) Two eddy digraphs are isomorphic if and only if their line digraphs are isomorphic. This follows easily since every source (sink) in an eddy digraph is of in(out)-degree one. Moreover, the result carries over to the Heuchenne components determined by two eddy digraphs since the Heuchenne components are uniquely determined by the eddy digraphs.

(5) The Heuchenne components of D' are the converses of the Heuchenne components of D.

In the following let H be a Heuchenne component of D and let $\hat{H}$ be a subdigraph of D containing H such that $L(\hat{H}) \approx \hat{H}'$.

So $\hat{H}$ is of period one or two and consists of a disjoint union of Heuchenne components of D. We will say that $L^m(H)$ *is in* $\hat{H}$ if $L^m(H)$ is isomorphic to a Heuchenne component of $\hat{H}$. Note that, by (2), $L^m(H)$ is a Heuchenne component.

LEMMA 3: If H is a Heuchenne component of $\hat{H}$, then $L^m(H)$ is in $\hat{H}$ if and only if $L^{m+1}(H')$ is in $\hat{H}$.

PROOF: By (1) and (2) we see that the Heuchenne components of $L(\hat{H})$ are the line digraph images of the Heuchenne components in $\hat{H}$. So we have, by (4), that $L^m(H)$ is in $\hat{H}$ if and only if $L^{m+1}(H)$ is in $L(\hat{H})$. But $L(\hat{H}) \approx \hat{H}'$. The result follows by (5) and Lemma 1.

COROLLARY 4: Let H be of period t. Then

(a) $L^{2m}(H)$ is in $\hat{H}$ for all $m \geq 0$, and

(b) if $L^m(H)$ is not in $\hat{H}$ for some $m > 0$, then m is odd, t is even and, for all $r \geq 0$, $L^{m+2r}(H)$ is not in $\hat{H}$.

PROOF:

(a) $L(H')$ is in $\hat{H}$ by Lemma 3 and by a second application of that lemma we have that $L^2(H)$ is in $\hat{H}$. The claim follows by induction.

(b) Follows from (a).

LEMMA 5: Let H be of period t. Then

(a) if $L^m(H)$ and $L^n(H)$ both occur in $\hat{H}$ with $0 \leq m < n < t$, then they are nonisomorphic and hence disjoint (that is, the Heuchenne components to which they are isomorphic are disjoint),

(b) if t is even and $L^{2m-1}(H')$ and $L^{2n}(H)$ both occur in $\hat{H}$ with $0 \leq 2m-1$, $2n < t$, then they are nonisomorphic and hence disjoint, unless $t = 2k$ with k odd, and $L^k(H) \approx H'$.

PROOF:

(a) If $L^m(H) \approx L^n(H)$, then $L^k(H) \approx H$ for some k, $1 \leq k < t$ which is a contradiction. It follows from (1) and (2) that they are disjoint.

(b) Since t is even, one of the two numbers between 1 and $t-1$ that the two differences of $2m-1$ and $2n$ are congruent to modulo t, is less than or equal to $\frac{t}{2}$. If k is that number, then k is odd. If we also have $L^{2m-1}(H') \approx L^{2n}(H)$, then $L^k(H) \approx H'$. But then $L^{2k}(H) \approx H$ which means that $t = 2k$. That completes the proof of the lemma.

Let H be of period t. Lemma 5 suggests the definition of the following digraphs. Let H_1 denote a digraph isomorphic to the disjoint union of H, L(H), $L^2(H),\ldots,L^{t-1}(H)$; let H_2 denote the part of H_1 involving the even iterates of the line digraph ($H = L^0(H)$); and let H_3 denote the part of H_1 involving the odd iterates of the line digraph. Thus H_1 is the disjoint union of H_2 and H_3, $L(H_1) \approx H_1$ and when t is even, $H_3 \approx L(H_2)$ and $H_2 \approx L(H_3)$.

LEMMA 6: Let H be of period t. Then

(a) H_2 is isomorphic to a subdigraph of $\hat{H}$,

(b) if $\overline{H}$ is isomorphic to the disjoint union of H_1 and H_1', then $L(\overline{H}) \approx \overline{H}'$,

(c) if $H' \approx L^n(H)$ for some n, $0 \leq n < t$, and $\overline{H} = H_1$, then $L(\overline{H}) \approx \overline{H}'$,

(d) if $t = 2m$ and $\overline{H}$ is isomorphic to the disjoint union of H_2 and H'_3, then $L(\overline{H}) \approx \overline{H}'$, and

(e) if $t = 2m$ with m odd, $H' \approx L^m(H)$ and $\overline{H} = H_2$, then $L(\overline{H}) \approx \overline{H}'$.

PROOF:

(a) This is immediate from Corollary 4(a) and Lemma 5(a).

(b) In this case we have

$$L(H_1 \cup H_1') \approx L(H_1) \cup L(H_1)' \approx H_1 \cup H_1' \approx (H_1 \cup H_1')'.$$

The point of (c) and (d) is that sometimes $H_1 \cup H_1'$ contains more than is needed. A further refinement of (c) and (d) is given by (e).

(c) If $H' \approx L^n(H)$, then $L^r(H') \approx L^{n+r}(H)$, $0 \leq r < t$, and so $H_1' \approx H_1$. Consequently, $L(H_1) \approx H_1 \approx H_1'$ as desired.

(d) Noting the comment preceding the lemma, we have, since t is even,

$$L(H_2 \cup H_3') \approx L(H_2) \cup L(H_3)' \approx H_3 \cup H_2' \approx (H_2 \cup H_3')'.$$

(e) In this case we have $H_3 \approx H_2'$ since $L^{m+2r}(H) \approx L^{2r}(H')$, $0 \leq r < m$. But we also have $H_3 \approx L(H_2)$ so the conclusion follows.

The following lemma furnishes us with a converse to Lemma 6.

LEMMA 7: Let H be a Heuchenne component of period t and let $\hat{H}$ be a minimal subdigraph of D containing H and having $L(\hat{H}) \approx \hat{H}'$. Then

(a) if t is even, say $t = 2m$, then $\hat{H}$ is as $\overline{H}$ in Lemma 6(e) or (d), depending on whether $H' \approx L^m(H)$ or not.

(b) if t is odd, then $\hat{H}$ is as $\overline{H}$ in Lemma 6 (c) or (b), depending on whether $H' \approx L^n(H)$ for some n, $0 \leq n < t$, or not.

PROOF:

(a) Let $t = 2m$. By Lemma 6(a), $H_2 \subseteq \hat{H}$ and so, by Lemma 3, $L^{2r-1}(H') \subseteq \hat{H}$, $1 \leq r \leq m$. Thus, by Lemma 5(a), $H_3' \subseteq \hat{H}$ and, by Lemma 5(b), H_2 and H_3' are disjoint unless m is odd and $L^m(H) \approx H'$. In the former case $\hat{H}$ contains the disjoint union of H_2 and H_3' and so equals it by Lemma 6(d). In the latter case, $\hat{H}$ equals H_2 by Lemma 6(e).

(b) By Lemma 5(a), $H_1 \subseteq \hat{H}$ and so, by Lemma 3, $L(H_1') \subseteq \hat{H}$. Now, by (1) and (2), H_1' and H_1 are disjoint if and only if $L^r(H') \not\approx L^n(H)$ for all r and n, $0 \leq r, n < t$, or equivalently, if and only if $H' \not\approx L^n(H)$ for all n, $0 \leq n < t$.

We consider the two cases.

(i) $H' \not\approx L^n(H)$ for all n, $0 \leq n < t$. Then $\hat{H}$ contains a subdigraph $\overline{H}$ isomorphic to the disjoint union of H_1 and H_1' and so, from Lemma 6(b), we see that $\hat{H} \approx \overline{H}$.

(ii) $H' \approx L^n(H)$ for some n, $0 \leq n < t$. Then, by Lemma 6(c), $L(H_1) \approx H_1'$ and so $\hat{H} \approx H_1$.

Combining Lemmas 6 and 7 we have the desired characterization.

THEOREM: For a finite digraph D, $L(D) \approx D'$ if and only if the Heuchenne components of D can be partitioned into sets of the following forms where H always represent a Heuchenne component of period t:

(1) $\{H, H', L(H), L(H)', \ldots, L^{t-1}(H), L^{t-1}(H)'\}$,
(2) $\{H, L(H), L^2(H), \ldots, L^{t-1}(H)\}$ where $H' \approx L^n(H)$ for some n, $0 \le n < t$,
(3) $\{H, L^2(H), \ldots, L^{2m-2}(H)\}$ where $L^m(H) \approx H'$ with $t = 2m$, m odd, or
(4) $\{H, L(H)', L^2(H), L^3(H)', \ldots, L^{2m-2}(H), L^{2m-1}(H)'\}$ where $t = 2m$.

Note that the set in (2) is H_1 while that in (1) is $H_1 \cup H_1'$. And the set in (3) is H_2 while that in (4) is $H_2 \cup H_3'$. Thus if t is even *and* $H' \approx L^n(H)$, then the sets in (2) and (4) are the same while the set in (1) contains two copies of the set in (2). If in addition, $t = 2m$ and m is odd, then the set in (2) contains two copies of the set in (3).

The theorem and proof also supply an efficient algorithm for testing a digraph D for $L(D) \approx D'$: First find the directed circuits in D and the weakly connected components containing them; the basic configuration of each directed circuit must be an eddy digraph if $L(D) \approx D'$. From each eddy digraph we can produce the Heuchenne component it determines and test if the weakly connected component containing the determining directed circuit is correct. Finally we can see if D minus the weakly connected unicyclic components is isomorphic to the union of the forests that must be associated with each eddy digraph. We note that the isomorphisms checks required can be done efficiently since only forests are involved.

References
[1] R. L. Hemminger, Digraphs with periodic line digraphs, *Studia Sci. Math. Hungar.* 9 (1974) 27-31: MR52#2948.

Efficient Algorithms for the Maximum Weight Clique Problem on Circular-Arc Graphs and Circle Graphs *

Wen-Lian Hsu

ABSTRACT

Circle graphs and circular-arc graphs are the intersection graphs of chords and arcs in a circle. In this paper we present algorithms for finding maximum weight cliques in these graphs. The running times of the algorithms are $O(n^2 \log n)$ for circle graphs and $O(n^3)$ for circular-arc graphs. Our algorithms are based on the scanning of appropriate endpoint sequences and efficient bookkeeping of results for subproblems.

1. Introduction

Let $G = (V,E)$ be a *simple graph*, i.e., a finite, undirected, loopless graph without multiple edges. Let V and E denote the vertex and edge set of G, respectively. Denote the cardinality of V by n. A *clique* is a complete subgraph. An *independent set* is a subset $P \subseteq V$ such that $v_i, v_j \in P$ implies $(v_i, v_j) \notin E$. The complement $\bar{G}$ of G is the graph $\bar{G} = (V, \bar{E})$, where $\bar{E} = \{(x,y) | x,y \in V, x \neq y \text{ and } (\underline{x},y) \notin E\}$. It is easy to see that an independent set in G is a clique in $\bar{G}$ and vice versa. A *weighted* graph is a simple graph with weights on its vertices.

A graph $G = (V,E)$ is called a *circular-arc graph* if there is a one-to-one correspondence between V and a set S of arcs in a circle such that two vertices are adjacent if and only if the corresponding arcs have a nonempty intersection. S is called an *intersection model* for G. If S is a family of intervals on a real line, G is called an *interval graph*. If S is a family of chords in a circle, G is called a *circle graph*. Clearly, every interval graph is a circular-arc graph since we can

---
* This research was supported in part by the National Science Foundation under Grant ECS-8105989.

represent the intervals by arcs on a circle. Figure 1 gives a circular-arc graph and its corresponding intersection model.

Figure 1. A circular-arc graph and its arc representation

This graph is not an interval graph since we cannot represent the cycle $X_1X_2X_3X_4$ by intervals on the real line. An equivalent description of a circle graph (which is also called an *overlap graph*) is that (see [1]) each vertex of the graph corresponds to an interval on a real line and each edge corresponds to two intervals overlapping without one being completely contained in the other (i.e., strictly overlapping).

These classes of graphs have drawn considerable attention in recent years [1,2,3,4,5,7,8]. Tucker [8] has shown that recognizing circular-arc graphs can be done in $O(n^3)$ time. Gavril has given a polynomial algorithm for finding a maximum clique in a circular-arc graph. His algorithm involves a procedure for finding a maximum independent set of a bipartite graph, which takes $O(n^{2.5})$ time in the cardinality case [6], and $O(n^3)$ time in the weighted case (using the maximum flow algorithm). Since this procedure is called n times in his maximum clique algorithm, the total complexity is $O(n^{3.5})$ in the cardinality case and $O(n^4)$ in the weighted case. In this paper we present an $O(n^3)$ algorithm for the weighted problem assuming that the arc representation of a circular-arc graph is given. The maximum clique problem on circle graphs can also be done in polynomial time once an interval representation is given. Gavril has given an $O(n^3)$ algorithm for the weighted case: Buckingham [1], Rotem and Urrutia [7] have, independently, given $O(n^2)$ algorithms for the cardinality case. We present in Section 2 an $O(n^2 \log n)$ algorithm for the weighted case.

Our clique finding algorithm for circular-arc graphs is much more complicated than that for interval graphs. This is essentially due to the fact that in using the interval model for circular-arc graphs, there are arcs "crossing" both ends of the interval and it becomes harder to count things by scanning the endpoint sequence "only" once. Gavril's approach to constructing a maximum clique in the neighborhood of each arc exploits very little of the structure of circular-arc graphs. We develop an algorithm based on the scanning of appropriate sets of arcs in a certain order. Using our approach, constructing a maximum clique for the entire graph is about the same order of complexity as constructing a maximum clique in the neighbourhood of each arc.

2. An $O(n^2 \log n)$ Maximum Weight Clique Algorithm for Circle Graphs

In this section, we assume an interval representation of a circle graph is given; namely, each vertex of the graph corresponds to an interval on a real line and two vertices are connected by an edge if and only if the corresponding intervals strictly overlap. An example of a clique is shown in Figure 2.

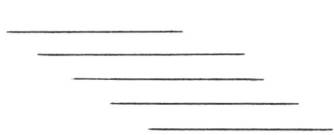

Figure 2. A clique in a circle graph (overlap graph)

Without loss of generality, assume all endpoints of the intervals are distinct. For each clique, let us call the interval with the smallest left endpoint the *"beginning interval"* of the clique. Our maximum weight clique algorithm goes as follows. For each interval, we compute a maximum weight clique beginning with this interval. A maximum weight clique of the graph can then be chosen as the one with the maximum weight from among these cliques.

Label the intervals as $1, \ldots, n$ according to their ascending left endpoint order. Denote by x_i, y_i, the coordinate of the left, right endpoint of i, respectively. Let the weight of an interval i be $wt(i)$. In the following we describe an $O(n \log n)$ algorithm which finds a maximum weight clique beginning with interval 1. Clearly, the same method can be used to find maximum weight cliques beginning with any interval.

First of all we determine the interval s such that $x_s < y_1$ and $x_{s+1} > y_1$. We only have to consider those intervals $2, \ldots, s$ which overlap with interval 1. Let us call the interval with the largest left endpoint in a clique the *"ending interval"* of that clique. Our algorithm scans the intervals $2, \ldots, s$ from left to right. For each interval i it scans, the weight (denoted by $\lambda(i)$) of a maximum clique beginning at interval 1 and ending at interval i is computed. After the sth interval is scanned, a maximum weight clique beginning at interval 1 can then be selected.

The computation of $\lambda(i)$ can be carried out by a dynamic programming algorithm. Let $\lambda(1) = wt(1)$ and $\lambda(i) = 0$ for those interval i with $y_i < y_1$.

LEMMA 1. $\lambda(i) = wt(i) + \max\{\lambda(j) | 1 \le j < i\ \&\ y_1 \le y_j < y_i\}$ for all i s.t. $y_i > y_1$.

PROOF. Clearly, if a maximum weight clique beginning at interval 1 and ending at interval i_m consists of the intervals $1 = i_0, i_1, \ldots, i_m$ (where $i_0 < i_1 < \cdots < i_m$), then $\{i_0, i_1, \ldots, i_{m-1}\}$ is a maximum weight clique beginning at interval 1 and ending at interval i_{m-1}. Thus $\lambda(i_m) = wt(i_m) + \lambda(i_{m-1})$. It is also clear that $\lambda(i_m) \ge wt(i_m) + \lambda(j)$ for each j such that $1 \le j \le i_m$ and $y_1 \le y_j \le y_{i_m}$.

Hence, to find $\lambda(i)$ we just have to compare $\lambda(j)$ for all intervals $j < i$ such that $y_j < y_i$. This process can be speeded up by the following observation:

LEMMA 2. If $j_1 < j_2$, $y_{j_1} > y_{j_2}$ and $\lambda(j_1) \le \lambda(j_2)$, then deleting the information $\lambda(j_1)$ will not change our computation of the weight of a maximum clique beginning with interval 1.

PROOF. By Lemma 1, $\lambda(j_1)$ will possibly be considered only when we compute $\lambda(i)$ for some interval i such that $i > j_1$ and $y_i > y_{j_1}$. However, after $\lambda(j_2)$ has been computed we need not consider $\lambda(j_1)$ for any later calculation of $\lambda(i)$ with $i > j_2$ and $y_i > y_{j_1}$.

We say j_2 dominates j_1 w.r.t. $\lambda(j)$'s when they satisfy the assumptions in Lemma 2. Hence in the set $\{\lambda(j) | 1 \le j < i, y_1 \le y_j\}$ we can delete dominated $\lambda(j)$'s and order the remaining undominated subset $\{\lambda(j) | 1 \le j < i, y_1 \le y_j, \lambda(j) > \max_{1 \le j' < j, y_1 \le y_{j'}} \lambda(j')\}$ into $\lambda(1) = \lambda(i_0) < \lambda(i_1) < \cdots < \lambda(i_m)$ where $i_0 < i_1 < \cdots < i_m$ and $y_{i_0} < y_{i_1} < \cdots < y_{i_m}$. In computing $\lambda(i)$ we determine the integer k such that $y_{i_k} < y_i < y_{i_{k+1}}$ and let $\lambda(i) = wt(i) + \lambda(i_k)$. This formula conforms to Lemma 1 since $\lambda(i_k) = \max\{\lambda(j) | 1 \le j < i\ \&\ y_1 \le y_j < y_i\}$. The presence of the new interval i might create some other dominated intervals $\{j | i_{k+1} \le j \le i_m, \lambda(j) < \lambda(i)\}$. This set can easily be identified and deleted by comparing $\lambda(i)$ and $\lambda(i_{k+1}), \ldots$ one by one. Hence the insertion of $\lambda(i)$ and following deletions give us a new set of undominated $\lambda(j)'s$ arranged in ascending $\lambda(i_j)$, y_{i_j} and the same procedure repeats for the next interval $i + 1$. The only work involved at the ith iteration is determining the integer k, which takes at most $O(\log n)$ time and deleting $\lambda(j)'s$ dominated by interval i. The former operation has to be repeated at most n times; the latter deletions totaled at most $O(n)$ times. Hence the entire procedure of finding a maximum weight clique beginning at interval 1 takes at most $O(n \log n)$ time. The running time of the maximum weight clique algorithm on circle graphs is therefore bounded by $O(n^2 \log n)$.

3. An Analysis of Cliques in a Circular-Arc Graph

Without loss of generality, assume that all endpoints of the n arcs are distinct. Let the length of the circle be 1, we can assume all arcs have lengths less than 1. Arbitrarily choose an arc from the collection. Starting clockwise from the left endpoint of this arc, we can label all endpoints as $1, 2, \ldots, 2n$. Specify arbitrarily an order for the arcs and denote arc i by (x_i, y_i), where x_i is the label of its left endpoint and y_i, the label of its right endpoint. Note that x_i can be larger than y_i, in which case the arc (x_i, y_i) goes across $x_i, x_{i+1}, \ldots, 2n, 1, \cdots, y_i$. We say that a label x is *between* y and z if either (i) $y < x < z$ or (ii) $y > z$ & ($x > y$ or $x < z$).

Two arcs (a_i, b_i), (a_j, b_j) are said to *intersect strictly* if a_i is between a_j, b_j and b_j is between a_i, b_i. Two strictly intersecting arcs i and j are said to form a *Type I pair* if together they do not cover the whole circle; otherwise, they are said to form a *Type II pair*. These two types are depicted in Figure 3.

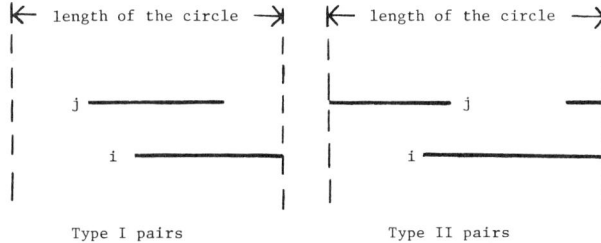

Figure 3. Two types of pairs in P

The next lemma gives a classification of the cliques in a circular-arc graph.

LEMMA 3. Let $(a_1,b_1), \ldots, (a_l,b_l)$ be a collection A of arcs which form a clique in a circular-arc graph G. Then one of the following two conditions is true:

(i) There exists an arc (a_i,b_i) which is properly contained in every other arc in A.

(ii) There exist two strictly intersecting arcs (a_i,b_i), (a_j,b_j) in A such that no arc of A is properly contained in either one of them and for every other arc (a_k,b_k) in A exactly one of the following conditions is true:

   (a) (a_k,b_k) properly contains (a_i,b_i)
   (b) a_k is between a_j,a_i; b_k is between b_j,b_i
   (c) a_k is between a_i,b_j; b_k is between b_i,a_k
   (d) a_k is between b_j,b_i and also between b_j,a_j; b_k is between a_j,a_k
   (e) b_i is between b_j,a_j; a_k is between b_i,a_i; b_k is between b_j,b_i (note: in this case arcs i and j must form a Type I pair).

PROOF. Let B be the subcollection of A that contains all arcs which do not properly contain any other arc of A. If B contains a single arc (a_i,b_i), then condition (i) holds. Hence assume B contains more than one arc. If there exist a Type II pair i,j in B then choose this pair. Otherwise pick any arc i in B. Starting from the left endpoint of i, we can find an arc j in B whose left endpoint is the leftmost in the counterclockwise direction before b_i. If a_j is between a_i and b_i, then we interchange the indices i and j. Note that there is no arc in B with its left endpoint between b_i,a_j and its right endpoint between a_i,b_j.

The proof for condition (ii) does not depend on whether i and j form a Type I pair or a Type II pair except that condition (ii) (e) is possible only for Type I pairs.

Now consider an arc (a_k,b_k) in A such that $k \neq i$, $k \neq j$. Suppose (a_k,b_k) does not properly contain (a_i,b_i). consider the following cases:

CASE 1. a_k is between a_j,a_i. Since (a_k,b_k) cannot be properly contained in (a_j,b_j) we have b_k between b_j,a_k. Since (a_k,b_k) does not properly contain (a_i,b_i) we conclude that b_k is between b_j,b_i. This implies condition (b).

CASE 2. a_k is between a_i,b_j. Since (a_k,b_k) cannot be properly contained in (a_i,b_i), we have b_k between b_i and a_k. This implies condition (c).

CASE 3. a_k is between b_j, b_i and also between b_i, a_k. Since (a_k, b_k) must intersect (a_j, b_j), we have b_k between a_j, a_k. This implies condition (d).

CASE 4. None of the above cases apply. This can only happen when i, j form a Type I pair and a_k is between b_i, a_j. Since (a_k, b_k) must intersect (a_i, b_i), we have b_k between a_i, a_k. However, b_k cannot be between a_i, b_j because then (a_k, b_k) would contain some $(a_s, b_s) \in B$ (could be itself) with a_s between b_i, a_j and b_s between a_i, b_j, which is impossible by the selection of (a_i, b_i) and (a_j, b_j). Hence b_k is between b_j, a_k, but this implies (a_k, b_k) properly contains (a_j, b_j). Since (a_k, b_k) does not properly contain (a_i, b_i), we have b_k between b_j, b_i which is condition (e).

By Lemma 3, a clique in a circular-arc graph satisfies either condition (i) or condition (ii). To find a maximum weight clique in G it suffices to find a maximum weight clique among all cliques satisfying condition (i) and another one among all cliques satisfying condition (ii). This is discussed in the next section.

4. An $O(n^3)$ Maximum Weight Clique Algorithm for Circular-arc Graphs

Consider a circular-arc graph G represented by its arc intersection model. First of all, we construct a "containment graph" H from G with directed edges. Each vertex of H corresponds to an arc in G. An edge (i, j) directed from i to j appears in H iff arc i is properly contained in arc j. H is a transitive graph. The number of arcs which properly contain an arc i is equal to the out-degree of i. Thus, it is not difficult to find a maximum weight clique among all cliques satisfying condition (i) of Lemma 3.

Next, we calculate the weight of a maximum clique satisfying condition (ii) of Lemma 3. A clique for which condition (ii) applies is said to be *generated* by its two arcs (a_i, b_i), (a_j, b_j). It should be noted that such a clique does not contain any arc which is properly contained in (a_i, b_i) or (a_j, b_j) and its weight is certainly bounded by a maximum weight clique generated by (a_i, b_i), (a_j, b_j). Since there is no advanced information about which pair of arcs will generate a maximum weight clique satisfying condition (ii), we simply consider the set P of all eligible pairs of arcs (i.e., arcs (x_i, y_i), (x_j, y_j) that strictly intersect) and compute a maximum weight clique generated by each such pair.

For each pair (i, j) in P, we first compute the number of arcs satisfying one of conditions (a), (b), (c), (d) and (e). Arcs satisfying (a) or (e) can be determined by the containment graph H. Arcs satisfying (b), (c) and (d) can be found by searching through every arc in G, which takes $O(n)$ time. Denote the total weight of arcs satisfying (d) by $\lambda_d(i,j)$. Since there can be at most $O(n^2)$ pairs in P, this part of the algorithm takes at most $O(n^3)$ time. (It can be carried out in $O(n^2 \log n)$ time by using a more sophisticated data structure. However the remaining part still takes $O(n^3)$ time.) It remains to show that the remaining part can also be done in $O(n^3)$ time.

Further analyzing condition (ii), one can see that every arc satisfying condition (a), (c) or (d) will intersect with any other arc satisfying condition (ii), but arcs satisfying (b) might not intersect arcs satisfying (d). Hence we still have to make a careful selection with regard to which arc satisfying (b) or (d) should be included in order to form a clique with the maximum weight. The optimal selection from arcs satisfying (b) or (d) will be determined separately for different types of pairs in P.

Let us first consider Type I pairs. For each arc i we determine an optimal selection for each type I pair (i,j) in P. Denote the total weight of arcs selected (including the weight of j but not i) by $\lambda(i,j)$. It should be noted that once the selection of arcs satisfying (b) is determined, the corresponding set of arcs satisfying (d) which can be included in a clique is also determined. Hence it is sufficient to consider the optimal selection of arcs satisfying (b). We will use a dynamic programming procedure which recursively finds $\lambda(i,j)$ for all arc j such that (i,j) is a Type I pair in P. Each Type I pair (i,j) satisfies that x_j is between y_i, x_i. Therefore we can arrange all these arcs into a list $L_i = \{i_1, \ldots, i_m\}$ according to the counterclockwise order of x_{i_t}. Figure 4 shows an example of possible arc j forming Type I pair (i,j) with arc i.

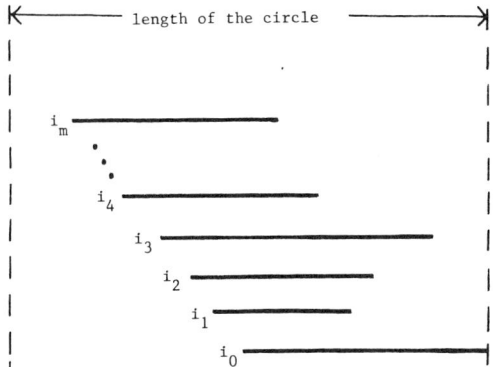

Figure 4. An example of arcs between Type I pair (i_0, i_m)

We now describe the recursive formula used in our algorithm. Suppose we have computed $\lambda(i, i_t)$ $t = 1, \ldots, k-1$ and want to determine $\lambda(i, i_k)$. For each $t \in \{1, \ldots, k-1\}$ such that y_t is between y_k, y_i, let $\alpha(i_t)$ be the total weight of arcs in $\{i_{t+1}, \cdots, i_{k-1}\}$ whose right endpoints are between y_{i_t}, y_i (or equivalently, which properly contain the arc i_t).

LEMMA 4.

$$\lambda(i, i_k) = \max \begin{Bmatrix} \max\ [\lambda(i, i_t) + \alpha(i_t) + \lambda_d(i_t, i_k), \lambda_d(i, i_k)] \\ t \in \{1, \ldots, k-1\} \\ y_t \text{ is between } y_k, y_i \end{Bmatrix}$$

PROOF. Consider an optimal selection $S = \{i, i_1', i_2', \ldots, i_l', i_k\}$ (ordered by left endpoints). If $l = 0$, then $\lambda(i, i_k) = \lambda_d(i, i_k)$. Otherwise proceed as follows. Starting from i_l', we can find the first arc, say i_q', such that it does not properly contain any other arc in the selection S. Since S is an optimal selection for the pair (i, i_k), the following must be true:

1. The set $\{i, i_1', \ldots, i_{q-1}', i_q'\}$ must be an optimal selection for the pair (i, i_q'). This optimal selection results in a total weight $\lambda(i, i_q')$.

2. The set $\{i_{q+1}', \ldots, i_l'\}$ consists of all those arcs in the ordered subset $\{i_q'(=i_r), i_{r+1}, \ldots, i_{k-1}\}$ which properly contain the arc i_q'. The total weight of this set is $\alpha(i_q')$.

3. The set of arcs satisfying (d) for (i, k) whose left endpoints are between $y_{i_k}, y_{i_q'}$ is included in the clique. The total weight of these arcs is $\lambda_d(i_q', i_k)$. The total weight of the optimal selection

S is thus $\lambda(i,i_q') + \alpha(i_q') + \lambda_d(i_q',i_k)$, which must be $\max_{t \in \{1,\ldots,k-1\}}[\lambda(i,i_t) + d_k(i_t) + \lambda_d(i_t,i_k)]$.

Our algorithm for computing $\lambda(i,j)$ on Type I pairs goes as follows. Start with an arc i. scan the list L_i from left to right. Whenever an arc i_k is scanned, execute the following:

1. For each $t \in \{1,\ldots,k-1\}$ such that y_{i_t} is between y_{i_k}, y_i, calculate $\lambda(i,i_t) + \alpha(i_t) + \lambda_d(i_t,i_k)$. Take the maximum of these numbers and $\lambda_d(i,i_k)$ to be $(\lambda(i,i_k)$.

2. For each $t \in \{1,\ldots,k-1\}$ such that y_t is between x_i, y_k, augment $\alpha(t)$ by $wt(i_k)$.

Next we consider Type II pairs. For each arc j we determine an optimal selection for each type II pair (i,j) in P. Again, denote the total weight of arcs selected (including the weight of i but not j) by $\lambda(i,j)$. Since it is possible that the same clique can be generated both by a Type I pair and by a Type II pair, we will now be concerned with those cliques which cannot be generated by Type I pairs.

LEMMA 5. If a clique contains two arcs i,j such that (i,j) is a Type I pair in P and both i and j do not properly contain any other arc in the clique, then this clique can be generated by a Type I pair in P by condition (ii) of Lemma 3.

PROOF. Consider the subset B of arcs which do not properly contain any other arc of the clique. Then $i,j \in B$. Starting from the left endpoint of i, we search for an arc k in B whose left endpoint is the leftmost in the counterclockwise direction before y_i. The existence of such an arc k is ensured by the existence of the arc j. (i,k) is a Type I pair and the clique can be generated by (i,k).

Now, if an optimal selection for a Type II pair (i,j) contains an arc k satisfying condition (d) with y_k between x_j, x_i, then we claim that the corresponding clique can be generated by a Type I pair. First, the two arcs j and k do not cover the whole circle. Since the clique does not contain any arc properly contained in arc j, there must be an arc k' (could be k itself) contained in k which does not properly contain any other arc in the clique. clearly, (j,k') is a Type I pair in P. By Lemma 5, this clique can be generated by a Type I pair according to condition (ii) of Lemma 3. Hence we do not have to consider the inclusion of arcs satisfying condition (d) with its right endpoint between x_j, x_i. This leads us to consider arcs satisfying (b) and arcs satisfying (d) with their right endpoints between x_i and their left endpoints. However, these two types of arcs always intersect and there is no need to make any selection. Therefore, for each Type II pair (i,j), the clique we need to consider includes arcs satisfying (a), (b), (c), (e) together with those arcs satisfying (d) with their right endpoints between x_i and their left endpoints.

Having determined an optimal selection for each Type I pair, formed the clique for each Type II pair as above and calculated the maximum total weight of arcs containing the same arc i for each arc i, we simply choose the clique with the maximum weight as our maximum weight clique.

5. Conclusion

In this paper we present two algorithms for finding maximum weight cliques on circle graphs and circular-arc graphs. Both algorithms make use of dynamic programming techniques. It remains to be seen whether more sophisticated data representation could produce faster algorithms. For the case of circle graphs, we have not been able to reduce the complexity of the weighted case down to $O(n^2)$. We suspect that the weighted case is, indeed, harder than the cardinality case, which can be done in $O(n^2)$ time. For circular-arc graphs, there are some steps which can be carried out in $O(n^2 \log n)$ time, but the dominating step is selecting arcs satisfying conditions (b) or (d) for Type I pairs, which takes $O(n^3)$ time.

References

[1] M.A. Buckingham, "Efficient stable set and clique finding algorithms for overlap graphs," Department of Computer Science, New York University, New York, NY 10012, 1981.

[2] M.R. Garey, D.S. Johnson, G.L. Miller and C.H. Papadimitriou, "The complexity of coloring circular arcs and chords," *SIAM Journal on Algebraic and Discrete Methods*, 1(1980), pp. 216-228.

[3] F. Gavril, "Algorithms on circular-arc graphs," *Networks*, Vol. 4, 1974, pp. 357-369.

[4] F. Gavril, "Algorithms for a maximum clique and a maximum independent set of a circle graph," *Networks*, 3(1973), pp. 261-273.

[5] U.I. Gupta, D.T. Lee and J. Y-T. Leung, "Efficient algorithms for interval graphs and circular-arc graphs," accepted for publication in *Networks*, 1981.

[6] J. Hopcroft and R.M. Karp, "An $n^{5/2}$ algorithm for maximum matching in bipartite graphs ," *SIAM Journal on Computing*, 2(4), 1973, pp. 225-231.

[7] D. Rotem and J. Urrutia, "Finding maximum cliques in circle graphs," *Networks*, 11(1981), pp. 269-278.

[8] A. Tucker, "An efficient test for circular-arc graphs," *SIAM Journal on Computing* 9(1), 1980, pp. 1-24.

On Some Algebraic Properties of Graphs

F. Jaeger

ABSTRACT

This study is motivated firstly by a number of results (essentially due to Shank, Rosenstiehl and Read) relating some algebraic properties of planar graphs and geometric properties of their left-right paths, and secondly by the representation of planar graphs in terms of bipartite circle graphs given by De Fraysseix. Bringing the two theories together, we immediately obtain interesting properties of bipartite circle graphs. Our purpose here is to present a generalization of these properties to all circle graphs. The general results have various connections with previous work by different authors.

I. Introduction.

We begin in Section II with general definitions on graphs, binary spaces, and embeddings of graphs on surfaces. Section III introduces the main concepts studied in this paper - chord diagrams, circle graphs, plane graphs and their left-right paths - and their mutual connections. Section IV presents the main results together with the known properties of planar graphs which they generalize. Finally, Section V, VI and VII are devoted to different applications of the main results.

II. Definitions.

1) **Graphs:** We consider only finite undirected graphs. Loops and multiple edges are allowed. A graph is *simple* if it has no loops and no multiple edges. The vertex-set (respectively: edge-set) of the graph G is denoted by $V(G)$ (respectively: $E(G)$). We write $G = (V,E)$ to state that $V(G) = V$ and $E(G) = E$. For other definitions on graphs the

reader should refer to [1].

2) Spaces: Let X be a finite set. $P(X)$ is the set of subsets of X. For A,B in $P(X)$ we denote by $A+B$ the symmetric difference of A and B. For $A \in P(X)$ and $\alpha \in GF(2)$ let: $\alpha A = \emptyset$ if $\alpha = 0$, $\alpha A = A$ if $\alpha = 1$. $P(X)$ together with the two above defined operations is a vector space over $GF(2)$.

Let $(A_i, i \in I)$ be a family of subsets of X. We denote by $<A_i, i \in I>$ the subspace of $P(X)$ generated by this family. For the sake of simplicity, for every x in X, the subset $\{x\}$ of X will be also denoted by x. The meaning will be clear from the context. $\{x/x \in X\}$ is a base of $P(X)$, so that $dim P(X) = |X|$. The associated scalar product (non singular symmetric bilinear form) is defined for A,B in $P(X)$ by: $A.B = 0$ if $|A \cap B|$ is even, $A.B = 1$ otherwise. Thus for instance, for $x \in X$, $A \in P(X)$: $x.A = 1$ iff $x \in A$.

Let $\mathbf{F}$ be a subspace of $P(X)$. The set: $\{A \in P(X)/\forall B \in \mathbf{F}, A.B = 0\}$ is a subspace of $P(X)$. It is called *the subspace of $P(X)$ orthogonal to* $\mathbf{F}$ and is denoted by $\mathbf{F}^{\perp}$. A *space on X* is any subspace of $P(X)$. A space $\mathbf{F}$ on X and a space $\mathbf{F}'$ on X' are *isomorphic* when there exists a bijection $\phi: X \to X'$ such that $\mathbf{F}' = \{\phi(A)/A \in \mathbf{F}\}$.

REMARK: Any result on spaces can be reformulated in terms of binary matroids (see [24], Chapter 10).

3) Cycle spaces and cocycle spaces of graphs: Let $G = (V,E)$ be a graph. The *boundary* of an edge e with ends v, v' is $\partial_G(e) = v + v'$ (so that $\partial_G(e) = \emptyset$ if e is a loop). For $F \subseteq E$ the boundary of F is $\partial_G(F) = \sum_{e \in F} \partial_G(e)$. Thus ∂_G is a linear mapping from $P(E)$ to $P(V)$. The kernel of ∂_G is denoted by $\mathbf{C}(G)$. The space $\mathbf{C}(G)$ on E is the *cycle space of G*. The *coboundary* of a vertex v is $\omega_G(v) = \{e \in E/v \in \partial_G(e)\}$. For $W \subseteq V$ the coboundary of W is $\omega_G(W) = \sum_{v \in W} \omega_G(v)$. Thus ω_G is a linear mapping from $P(V)$ to $P(E)$. The image of ω_G is denoted by $\mathbf{K}(G)$. The space $\mathbf{K}(G)$ on E is the *cocycle space of G*.

It is easy to show that $\mathbf{K}(G) = \mathbf{C}(G)^{\perp}$. A space is called *graphic* (respectively: *cographic*) if it is isomorphic to the cycle (respectively: cocycle) space of some graph. A space is called *planar* if it is isomorphic to the cycle space of some planar graph. By Whitney's duality theorem, a space is planar iff it is both graphic and cographic.

4) Graphs embedded on surfaces: We now present a description of embeddings of graphs on surfaces (not necessarily orientable) which is the most convenient for our purposes. This description is attributed to J. Edmonds (private communication) in [9]. See also [18], [23]. In the sequel, "embedding" stands for "2-cell embedding".

Let $G = (V,E)$ be a connected graph. We shall consider each edge e as subdivided into two *half-edges*, one for each end of e (this approach is presented in [10], where half-edges are called "brins"). For every half-edge h we denote by βh the corresponding edge and by αh the other half-edge of βh. Thus α is a fixed-point-free involution, whose orbits can be identified with the edges of G.

For v in V, a *rotation of* v is a cyclic permutation Π_v on the set of half-edges incident to v. A *rotation system* is a family $\Pi = (\Pi_v, v \in V)$ where Π_v is a rotation of v. Π is identified with the permutation on the set of half-edges whose cycles are the different Π_v's ($v \in V$). An *embedding scheme* is a pair (Π, λ), where Π is a rotation system and λ is a mapping from E to $GF(2)$.

Let us call *walk-unit* each pair (h,b), where h is a half-edge and $b \in GF(2)$. Let W be the set of walk-units. The embedding scheme (Π, λ) defines a permutation σ on W as follows. For (h,b) in W, $\sigma(h,b) = (h',b')$, where:

$$h' = \alpha \Pi h \text{ if } b = 0, \ \alpha \Pi^{-1} h \text{ if } b = 1$$

$$b' = b + \lambda(\beta h').$$

Each orbit of σ can be ordered in a sequence $(h_1,b_1),(h_2,b_2)$,...,(h_k,b_k) with $(h_{i+1},b_{i+1}) = \sigma(h_i,b_i)$ ($i = 1,...,k$) (the subscripts are to read modulo k). Then it is clear from the definition of σ that $(\beta h_1, \beta h_2, \ldots, \beta h_k)$ is the sequence of edges of a closed walk of G. Intuitively, this closed walk is obtained by starting on the edge βh_1, traveling towards the end-vertex v_1 corresponding to h_1, then turning at v_1 according to the rotation of v_1 if our "behaviour" is normal ($b_1 = 0$) and in the opposite way otherwise. This leads us on the edge βh_2, traveling towards the end-vertex v_2 corresponding to h_2, and our behaviour at v_2 will change ($b_2 \neq b_1$) if and only if $\lambda(\beta h_2) = 1$. We go on in this way until we come back to the situation corresponding to the initial walk-unit (h_1,b_1).

A closed walk derived as above from an orbit of σ will be called a (Π, λ)-*walk*. Two (Π, λ)-walks deriving from the same orbit of σ will be considered the same. It is easy to check that if $(h_1,b_1),(h_2,b_2)$,...,(h_k,b_k) is an orbit O of σ, then $(\alpha h_k, 1+b_{k-1})$, $(\alpha h_{k-1}, 1+b_{k-2})$,...,$(\alpha h_2, 1+b_1)$, $(\alpha h_1, 1+b_k)$ is also an orbit of σ, distinct from the first one, which we denote by $\bar{O}$. Clearly $\bar{\bar{O}} = O$, and we say that $O, \bar{O}$ are mutually *opposite*. In this case the corresponding (Π, λ)-walks are also opposite, and we say that they form an *opposite pair*.

We now select arbitrarily one member of each opposite pair of (Π, λ)-walks: let $F_1, \ldots, F_r$ be these (Π, λ)-walks. Clearly each edge appears twice in the list $F_1, \ldots, F_r$. More precisely, it appears either twice in one of the F_i's and not in the others, or once in two different F_i's and not in the others. It is then not difficult to show that the F_i's are the oriented boundaries of faces of an embedding of G on some surface S, with an arbitrary orientation specified on each face. Changing the orientation on a face amounts to replace the corresponding F_i by the opposite (Π, λ)-walk. It can be shown that every embedding of G on a surface can be obtained in this way.

The surface S will be *orientable* if one can choose the orientations of the faces such that each edge is taken in both directions in the associated boundaries. The following result appears in a slightly different form in [23] (Theorem 5).

PROPOSITION 1: S is orientable iff $\{e \in E/\lambda(e) = 1\} \in \mathbf{K}(G)$.

REMARK: The case where λ is identically zero corresponds to the classical description of orientable embeddings ([4]).

III. Chord Diagrams, Plane Graphs and Circle Graphs.

1) Chord diagrams: We define a *chord diagram* to be a pair (R, H), where R is a cubic Hamiltonian graph and H is a Hamiltonian cycle of R for which we choose one of the two possible circular orientations as "clockwise". The edges of H are called *circle edges*, the other edges of R are called *chords*.

A *representation of* (R, H) is a drawing of R in the plane (vertices map onto distinct points, edges map onto Jordan curves with appropriate ends) such that H is a circle (the "clockwise" orientation of H being the clockwise orientation of this circle) and chords are straight line segments. In the sequel we shall assume that every chord diagram has been given such a representation. This will allow us to deal at the same time with graphical and geometrical properties of (R, H).

2) Circle graphs: Two chords c_1, c_2 of the chord diagram (R,H) *cross* iff each one of the two paths joining in H the end-vertices of c_1 contains an end-vertex of c_2. Equivalently, c_1 and c_2 cross iff their images in the representation of (R,H) cross in the geometrical sense (this is independent of the representation we have chosen).

The *cross-graph* of (R,H) is the simple graph with the chords of (R,H) as vertices, two vertices being adjacent iff the corresponding chords cross. A *circle graph* is a simple graph which is isomorphic to the cross-graph of some chord diagram. Circle graphs are also called chord intersection graphs, or chord graphs. One can find in [5], [8] a survey of this class of graphs.

3) Bipartite chord diagrams and plane graphs: A chord diagram is *bipartite* when its cross-graph is bipartite. Let (R,H) be a bipartite chord diagram with cross-graph $G = (V,E)$. Let $\{V_1, V_2\}$ be a partition of V which is a coloring of G in two colors.

We can embed the cubic Hamiltonian graph R in the plane as follows: H is drawn as a circle. The chords of V_1 are drawn as Jordan curves inside H. The chords of V_2 are drawn as Jordan curves outside H. An example is given in Figure 1.

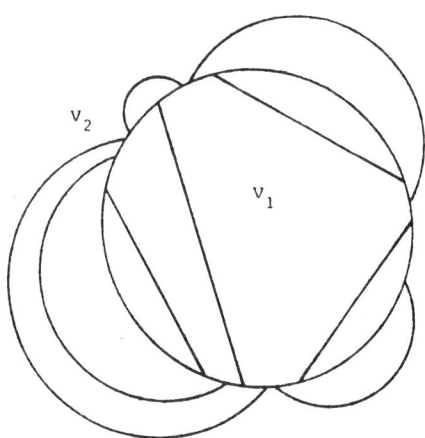

Figure 1

We now present a construction due to H. de Fraysseix [6]. The interior of H is partitioned into regions $Q_1, \ldots, Q_n$. We place a vertex q_i in the interior of the region Q_i ($i = 1, \ldots, n$). For every chord v_1 of V_1 separating the regions Q_i and Q_j we draw an edge v_1^* joining q_i and q_j and crossing v_1. Then every chord v_2 of V_2 with ends on the boundaries of the regions Q_k and Q_l is extended into an edge v_2^* with ends q_k, q_l. We obtain a connected plane graph M with vertex-set $q_1, \ldots, q_n$. The edges of M corresponding to chords of V_1 form a spanning tree T of M. We shall denote by $\bar{T}$ the associated cotree $E(M) - T$.

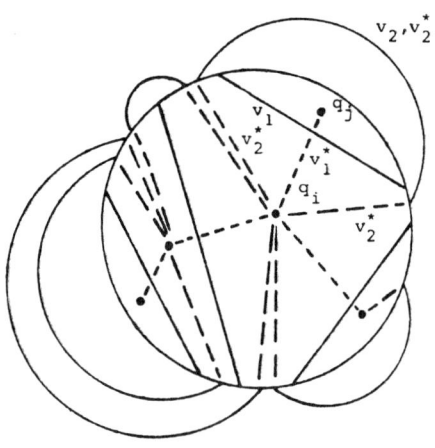

Figure 2

Conversely every connected plane graph M can be obtained in the above way from a bipartite chord diagram.

An interesting aspect of this representation of plane graphs is that it gives a convenient representation of their cycle and cocycle spaces. Indeed, for v in V, denote by N_v the *neighbourhood* of v in the cross-graph G, i.e. the set of vertices of G adjacent to v. We identify the chord x of (R, H) (equivalently the vertex x of the cross-graph G) with the corresponding edge x^* of M. V_1 is identified to T and V_2 to $\bar{T}$. It is then easy to see that for $v_2 \in V_2$, $v_2 + N_{v_2}$ corresponds to the cycle closed by v_2^* in the tree T (usually called the fundamental cycle of v_2^*

in T). Similarly for $v_1 \in V_1$, $v_1+N_{v_1}$ corresponds to the fundamental cocycle of v^*_1 in $\bar{T}$. Then $<v_2+N_{v_2}, v_2 \in V_2>$ is identified with $C(M)$ and $<v_1+N_{v_1}, v_1 \in V_1>$ is identified to $K(M)$.

4) Left-right paths: The *left-right paths* of M can be defined as the (Π,λ)-walks of M, where Π is the rotation system corresponding to the embedding of M in the plane, and λ is identically 1. Thus they are obtained by traveling on edges and turning alternately on left and on right at vertices.

We are now looking for a description of the left-right paths of M in terms of the bipartite chord diagram (R,H). This is the purpose of the remaining of this section, where for convenience we study general chord diagrams first, and then specialize to bipartite chord diagrams.

5) Labelled chord diagrams: A *labelled chord diagram* is a triple (R,H,ϵ) where (R,H) is a chord diagram and ϵ is a mapping from the set of chords of (R,H) to $GF(2)$. To each labelled chord diagram (R,H,ϵ) corresponds a unique embedding scheme (ρ,λ) of the cubic graph R defined as follows:

- ρ is the rotation system which assigns to every vertex the clockwise rotation around this vertex (given by the representation of (R,H) in the plane).

- λ is the mapping from $E(R)$ to $GF(2)$ which takes the value 0 on the circle edges and is equal to ϵ on the set of chords. To be short, we call the (ρ,λ)-walks of R *the ϵ-walks of (R,H,ϵ)*.

These ϵ-walks are obtained as follows: travel on a first chord, with a first specified behaviour. Turn left or right when arriving on the circle according to whether your behavior is normal or not. Leave the circle on the first encountered chord (because the behavior does not change on circle edges), and go on in this way, changing behavior only when traveling along a chord on which ϵ takes the value 1, until you come back to the initial situation. There are also two opposite ϵ-walks which do not use any chord: they consist of the cycle H taken in some direction and they will be called the *outer ϵ-walks*. The other ϵ-walks are called *inner*.

6) Representation of left-right paths. When (R,H) is bipartite, with associated plane map M, if ϵ is identically equal to 1, it is easy to see that the inner ϵ-walks of (R,H,ϵ) are in one-to-one correspondence with the left-right paths of M. More precisely, the sequences of edges of left-right paths of M are the sequences of chords of the inner ϵ-walks.

IV - Algebraic Properties of ϵ-Walks.

1) Introduction. We consider a labelled chord diagram (R,H,ϵ). We denote by V its set of chords, by $G = (V,E)$ its cross-graph, and by N_v the neighbourhood in G of the vertex v. We select one member of each opposite pair of inner ϵ-walks of (R,H,ϵ): let $F_1, \ldots, F_r$ be these ϵ-walks. Recall that each chord appears either twice in one of the F_i's and not in the others, or once in two different F_i's and not in the others. We define the *core of* F_i $(i = 1,\ldots,r)$ as the set of chords which appear exactly once in F_i. We denote this set by C_i.

We now present properties of ϵ-walks which generalize the known properties of left-right paths.

2) Cores and neighbourhoods.

2.1) GENERAL CASE

PROPOSITION 2: $\forall i \in \{1,\ldots,r\}$, $\forall v \in V$: $C_i \cdot (\epsilon(v)v + N_v) = 0$.

PROOF. The closed walk F_i is a sequence $(a_1, s_1, \ldots, a_j, s_j, \ldots, a_k, s_k)$ where the a_j's are edges of R and the s_j's are vertices of R. Let B_i be the set of edges of R which appear exactly once in F_i: B_i contains all the circle edges of F_i, together with the chords belonging to the core C_i. We have:

$$\sum_{j=1}^{k} \partial_R(a_j) = (s_k + s_1) + (s_1 + s_2) + \ldots + (s_{k-1} + s_k) = \emptyset.$$

Hence $\partial_R(B_i) = \emptyset$ or equivalently $B_i \in C(R)$.

Now let $v \in V$ be a chord. We associate to v a cocycle of R, denoted by θ_v, defined as follows. We choose one of the two paths joining in H the two end-vertices of v. Let P be:

- the set of inner vertices of this path if $\epsilon(v) = 0$;
- the set of inner vertices of this path, plus one arbitrary end-vertex if $\epsilon(v) = 1$. Then θ_v is the cocycle $\omega_R(P)$. It is clear that the chord-edges of θ_v are just the elements of $\epsilon(v)v + N_v$. The general situation is described in Figure 3.

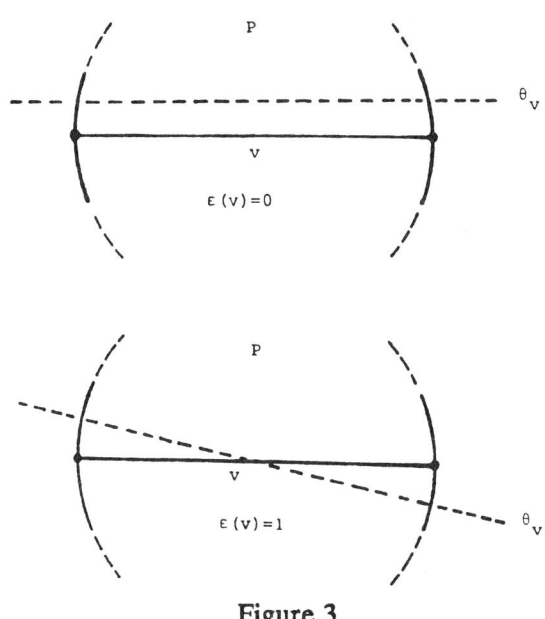

Figure 3

Let Q be the set of circle edges of θ_v. Clearly Q either is empty or consists of exactly two edges, according to whether P is empty or not. Now we have

$$B_i \cdot \theta_v = 0 \text{ (because } B_i \in C(R), \theta_v \in K(R)).$$

Since $\theta_v \cap V = \epsilon(v)v + N_v$, we obtain:

$$B_i \cdot (\epsilon(v)v + N_v) = B_i \cdot Q,$$

or equivalently:

$$C_i \cdot (\epsilon(v)v + N_v) = B_i \cdot Q.$$

The last member of this equality is 0. This is clear if $Q = \emptyset$. Otherwise the result follows immediately from the definition of ϵ-walks. □

2.2) BIPARTITE CASE. Assume that (R,H) is bipartite with associated plane map M and ϵ is identically equal to 1. As we have seen in Section III.6, to each inner ϵ-walk F_i corresponds a left-right path P_i of M such that the sequence of edges of P_i is the sequence of chords of F_i. We define the *core of* P_i as the set of edges of M which occur exactly once in P_i. Thus the core of P_i is identified with C_i.

Finally, we recall that a *bicycle of M* (see [16]) is an element of the space $C(M) \cap K(M)$ (the *bicycle space* of M).

Then Proposition 2, together with the representation of $C(M)$ and $K(M)$ obtained in Section III.3, immediately gives:

PROPOSITION 2': The core of any left-right path of M is a bicycle.

3) The tripartition theorem.

3.1) GENERAL CASE. Consider an ϵ-walk $F_i = (a_1, s_1, \ldots, a_j, s_j, \ldots, a_k, s_k)$ $(a_j \in E(R), \; s_j \in V(R) \; \forall j \in \{1, \ldots, k\})$. Assume that the chord v appears twice in F_i: we may write $a_{j_1} = a_{j_2} = v$ for some distinct j_1, j_2 in $\{1, \ldots, k\}$. Then F_i can be partitioned into two closed walks, each one with first and last vertex $s_{j_1} - 1$. We select arbitrarily one of these closed walks and denote it by F_i^y. We denote by C_i^y the set of chords which appear exactly once in F_i^y.

PROPOSITION 3: Let x be a chord.

(a) If x appears in two different ϵ-walks, one of which is F_i, then $x \in C_i$ and: $\epsilon(x)x + N_x = \sum_{v \in C_i - x} \epsilon(v)v + N_v$.

(b) If x appears twice in F_i and is taken twice with the same direction, then $x \in C_i^x$ and: $[1 + \epsilon(x)]x + N_x = \sum_{v \in C_i^x - x} \epsilon(v)v + N_v$.

(c) If x appears twice in F_i and is taken in the two opposite directions, then $x \notin C_i^x$ and: $x = \sum_{v \in C_i^x} \epsilon(v)v + N_v$.

PROOF.

(a) If x appears in two different ϵ-walks, one of which is F_i, then $x \in C_i$. Let A be the matrix over $GF(2)$, with rows and columns indexed by V, such that for every v, v' in V:

$$A_v^{v'} = (\epsilon(v)v + N_v) \cdot v'.$$

Thus A is obtained from the usual adjacency matrix of G by the assignment of the value $\epsilon(v)$ to the vth diagonal element for each v in V. By Proposition 2, $\forall v \in V$: $C_i \cdot (\epsilon(v)v + N_v) = 0$, and by the symmetry of A we have:

$$\sum_{v \in C_i} \epsilon(v)v + N_v = \emptyset,$$

which is equivalent to the required equality.

(b) Assume that $F_i = (a_1, s_1, \ldots, a_k, s_k)$, with $a_{j_1} = a_{j_2} = x$. Since x is taken twice with the same direction $s_{j_1-1} = s_{j_2-1}$ and $s_{j_1} = s_{j_2}$. We may assume that $F_i^x = (a_{j_1}, s_{j_1}, \ldots, a_{j_2-1}, s_{j_2-1})$. Clearly $x \in C_i^x$. The situation is described in Figure 4.

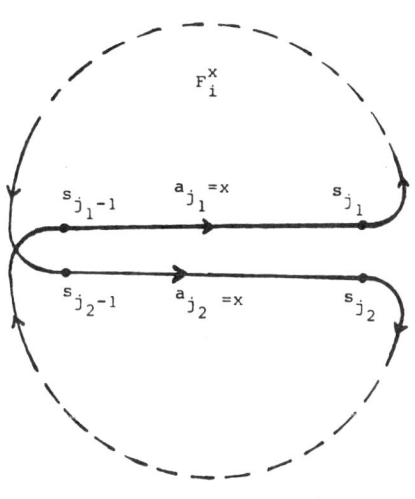

Figure 4

Let ϵ' be the mapping from V to $GF(2)$ defined by: $\forall v \in V-x$, $\epsilon'(v) = \epsilon(v)$; $\epsilon'(x) = 1 + \epsilon(x)$. Each ϵ-walk F_j with $j \neq i$ is also an ϵ'-walk. Moreover it is easy to see that, by replacing ϵ by ϵ', F_i is split into two ϵ'-walks, one of which is F_i^x. We may now apply the result just obtained for case (a) to the labelled chord diagram (R, H, ϵ'): we obtain the required equality.

(c) Assume as above that $F_i = (a_1, s_1, \ldots, a_k, s_k)$, $a_{j_1} = a_{j_2} = x$. Since x is taken in the two opposite directions, we have now: $s_{j_1} = s_{j_2-1}$ and $s_{j_1-1} = s_{j_2}$. We may assume that $F_i^x = (a_{j_1}, s_{j_1}, \ldots, s_{j_2-1}, a_{j_2}, s_{j_2})$. Clearly $x \notin C_i^x$. The situation is described in Figure 5.

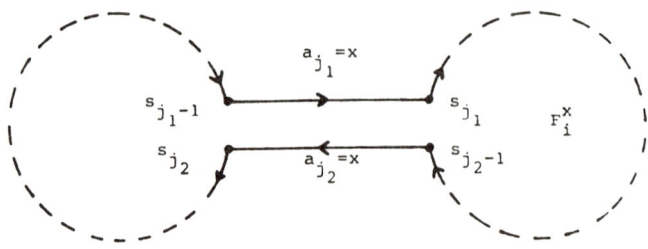

Figure 5

Replace the chord x by two new chords x_1, x_2 parallel to x and sufficiently close to x. We obtain a new chord diagram (R', H'). The new set of chords is $V' = (V-x) \cup \{x_1, x_2\}$. Let ϵ' be the mapping from V' to $GF(2)$ defined by: $\forall v \in V-x$, $\epsilon'(v) = \epsilon(v)$; $\epsilon'(x_1) = \epsilon'(x_2) = 1$. Each ϵ-walk F_j with $j \neq i$ is also an ϵ'-walk of (R', H', ϵ'). Moreover, it is easy to check that we have now two new ϵ'-walks replacing F_i, one of which has core $C_i^x \cup \{x_1, x_2\}$.

Let G' be the cross-graph of (R', H'). For $w \in V'$, N'_w denotes the neighbourhood of w in G'. We have:

$$N'_{x_1} = N'_{x_2} = N_x \cdot \forall v \in V-x, N'_v = N_v \text{ if } x \notin N_v;$$

$$N'_v = (N_v - x) \cup \{x_1, x_2\} \text{ if } x \in N_v.$$

Applying (a) to the core $C_i^x \cup \{x_1, x_2\}$ in (R', H', ϵ'), we obtain:

$$\epsilon'(x_1)x_1 + N'_{x_1} + \epsilon'(x_2)x_2 + N'_{x_2} = \sum_{v \in C_i^x} \epsilon'(v)v + N'_v.$$

This may be rewritten as:

$$x_1 + x_2 = \sum_{v \in C_i^x} \epsilon(v)v + N'_v.$$

The definition of N'_v for $v \in V-x$ gives immediately:
$$x = \sum_{v \in C_i^x} \epsilon(v)v + N_v \square$$

REMARK. One can show that for every simple graph $G = (V,E)$ every mapping ϵ from V to $GF(2)$ and every element x of V, the space $<\epsilon(v)v+N_v, \ v \in V-x>$ contains exactly one of the following elements: $\epsilon(x)x+N_x$, $[1+\epsilon(x)]x+N_x$ and x. Thus Proposition 3 provides a geometric interpretation of this algebraic result in the case of circle graphs.

3.2) BIPARTITE CASE. The notations are those of Section IV.2.2. Thus if the edge x of M appears twice in the left-right path P_i, the set of edges which appear exactly once in P_i between the two occurrences of x is identified with $C_i^x - x$.

We now derive from Proposition 3 the following edge tripartition theorem of P. Rosenstiehl ([19], Proposition 1):

PROPOSITION 3': If the edge x appears:
(i)   Exactly once in P_i, C_i is a bicycle which contains x.
(ii) Twice in P_i with the same direction, $C_i^x - x$ is a cocycle and $C_i^x \cup x$ is a cycle.
(iii) Twice in P_i with opposite directions, $C_i^x - x$ is a cycle and $C_i^x \cup x$ is a cocycle.

PROOF. Note first that (i) immediately follows from Proposition 2'. For cases (ii) and (iii): we are in situation (b) or (c) of Proposition 3, so that we may write:
$$x = \sum_{v \in C_i^x} v + N_v.$$

One can easily deduce from this (using the same argument as in the proof of Proposition 3(a)) the following property:

for $v \in V$, $C_i^x \cdot (v+N_v) = 1$ if and only if $v = x$.

Equivalently:

$$C_i^x \cdot (v+N_v) = x \cdot v \quad (\forall v \in V).$$

It follows that:

$$(C_i^x+x) \cdot (v+N_v) = x \cdot N_v \quad (\forall v \in V).$$

Using the description of $C(M)$ and $K(M)$ given in Section III.3, we obtain:

- C_i^x is a cocycle if $x \in V_1$ and a cycle if $x \in V_2$.
- C_i^x+x is a cycle if $x \in V_1$ and a cocycle if $x \in V_2$.

Now it is easy to see geometrically that:

- if $x \in V_1$, the edge x appears twice with the same direction in P_i iff the chord x appears twice with opposite directions in F_i;
- if $x \in V_2$, the edge x appears twice with the same direction in P_i iff the chord x appears twice with the same direction in F_i.

Hence in situation (ii) either $x \in V_1$ and we are in situation (c) (so that $x \notin C_i^x$) or $x \in V_2$ and we are in situation (b) (so that $x \in C_i^x$). Similarly in situation (iii) either $x \in V_1$ and we are in situation (b) (so that $x \in C_i^x$) or $x \in V_2$ and we are in situation (c) (so that $x \notin C_i^x$). In both cases the properties stated in Proposition 3' are easily verified. □

4) Core graph and core space.

4.1) GENERAL CASE. Recall that the inner ϵ-walks $F_1, \ldots, F_r$, together with one of the two opposite outer ϵ-walks, are the face-boundaries of an embedding of the graph R on some surface S. If we contract every edge of H, we obtain a new graph L with exactly one vertex. Every edge of L is a loop which can be identified with a chord of R. The clockwise orientation of H gives naturally a rotation Π of the unique vertex of L. We may as well consider Π as a rotation system of L, and then (Π,ϵ) defines an embedding scheme of L. It is clear that (Π,ϵ) defines an embedding of L on S, the face-boundaries of which are the intersections of the inner ϵ-walks $F_1, \ldots, F_r$ with the set of chords.

We denote by D the dual-graph of L on S, which is a connected graph with r vertices, one for each face of L. For $i = 1,\ldots,r$, the core C_i of F_i corresponds by duality to the set of edges of D with exactly one end equal to the ith vertex of D. It follows that the space $<C_i, i = 1,\ldots,r>$ (which we shall call the *core space* of (R,H,ϵ)) is isomorphic to the cocycle space of D. For this reason D will be called the *core-graph* of (R,H,ϵ).

PROPOSITION 4: The core space is the subspace of $P(V)$ orthogonal to $<\epsilon(v)v+N_v, v \in V>$.

PROOF. Let $Z = <\epsilon(v)v+N_v, v \in V>$. It follows from Proposition 2 that $<C_i, i = 1,\ldots,r> \subseteq Z^\perp$. Moreover, the isomorphism from the core space to $K(D)$ gives:

$$dim <C_i, i = 1,\ldots,r> = dim\ K(D).$$

Since D is connected, $dim\ K(D) = |V(D)| - 1 = r-1$. Thus it is enough to prove:

(i) $dim\ Z^\perp \leq r-1$.

We do this by induction on r.

For $r = 1$, situation (a) cannot occur in Proposition 3. Then it follows from cases (b) and (c) of this Proposition that for every x in V, $x \in Z$. Hence $Z^\perp = \{\varnothing\}$ and (i) holds.

If $r > 1$: there exists a chord x of type (a). Delete this chord from the labelled chord diagram (R,H,ϵ): this corresponds to a merging of the two ϵ-walks containing x into a single one, from which x is deleted. Equivalently this corresponds to a contraction of the edge of the core graph D dual to the edge x. Thus the number of opposite pairs of inner ϵ-walks of the new chord diagram is $r' = r-1$.

On the other hand, the new cross-graph G' is the subgraph of G induced by $V-x$. The new space to consider is the space $Z' = <\epsilon(v)v + (N_v-x), v \in V-x>$ on $V-x$. It is clear then that $dim\ Z' \leq dim\ Z$. Equivalently:

$$|V-x| - dim\ (Z')^\perp \leq |V| - dim\ (Z)^\perp$$

so that:

$$dim\ (Z)^\perp \leq dim(Z')^\perp + 1.$$

By our induction hypothesis: $dim\ (Z')^\perp \leq r'-1 = r-2$. Property (i) follows. □

4.2) BIPARTITE CASE. With the notations of Section IV.2.2, Proposition 4 together with well known properties of cocycle spaces of graphs immediately gives the following result of Rosenstiehl ([19], Corollary 1) and Shank ([22], Theorem 2):

PROPOSITION 4': The cores of any $r-1$ of the P_i's form a basis of the bicycle space.

V. Inversion of Adjacency Matrices of Labelled Circle Graphs.

A *labelled graph* is a pair (G,ϵ), where $G = (V,E)$ is a simple graph and ϵ is a mapping from V to $GF(2)$. The *adjacency matrix of* (G,ϵ) is the square symmetric matrix A over $GF(2)$ with rows and columns indexed by V, defined as follows:

For every v,v' in V: $A_v^{v'} = (\epsilon(v)v+N_v) \cdot v'$.

(This concept already appears in the proof of Proposition 3(a)).

In this section we present a geometric interpretation of inversion of adjacency matrices of labelled graphs in the case of circle graphs.

PROPOSITION 5: Let (G,ϵ) be a labelled circle graph with non-singular adjacency matrix A. Then the inverse matrix A^{-1} is also an adjacency matrix of some labelled circle graph.

PROOF: G is the cross-graph of some chord diagram (R,H). We consider the labelled chord diagram (R,H,ϵ). Since A is non-singular, by Proposition 4, the core space has dimension zero. Hence there is only one inner ϵ-walk F_1 (with a specified direction).

It follows from Proposition 3 that for every chord x:

$$x = \sum_{v \in C_1^x} \epsilon(v)v+N_v. \tag{i}$$

Let B be the square matrix over $GF(2)$, with rows and columns indexed by V, defined as follows:

For every v,x in V: $B_v^x = v \cdot C_1^x$.

In other words, the xth column of B is the representative vector of C_1^x. Then the property (i) is easily seen to be equivalent to: $AB = I$, where I is the ($|V| \times |V|$) identity matrix over $GF(2)$. Thus $B = A^{-1}$.

Each chord appears twice in F_1. Hence one can construct a chord diagram (R',H') with the same set of chords as (R,H), i.e. V, such that the sequence of chords encountered in a cyclic walk around H' is identical to the sequence of chords of F_1. Let G' be the cross-graph of (R',H'), and let N'_x denote the neighbourhood of the chord x in G'. Define the mapping ϵ' from V to $GF(2)$ by: $\epsilon'(x) = 1$ if x belongs to C_1^x, $\epsilon'(x) = 0$ otherwise. It is now easy to check that: $\forall x \in V$, $C_1^x = \epsilon'(x)x+N'_x$. Hence A^{-1} is the adjacency matrix of (G',ϵ'). □

VI. Embeddings of Loop-Graphs.

A *loop-graph* is a connected graph, all edges of which are loops, so that it has exactly one vertex. Let L be such a graph, and let (Π,ϵ) be an embedding scheme of L. (Π,ϵ) defines an embedding of L on some surface S. Let l,l' be two loops of L. Denote by l_1,l_2 (respectively: l'_1,l'_2) the two halves of l (respectively: l'). We say that l,l' *cross* if, in the cyclic sequence determined by Π, the four corresponding half-edges occur in the order l_1, l'_1, l_2, l'_2 or l_1, l'_2, l_2, l'_1. We denote by G the simple graph whose vertices are the loops of L, two loops being adjacent if they cross. We denote by A the adjacency matrix of the labelled graph (G,ϵ).

PROPOSITION 6: If ϵ is identically zero, S is orientable with orientable genus $g(S) = \frac{1}{2}(\text{rank } A)$. Otherwise S is non-orientable with non-orientable genus $\tilde{g}(S) = \text{rank } A$.

PROOF: First note that the cocycle space of a loop-graph reduces to $\{\emptyset\}$. Hence by Proposition 1, S is orientable iff ϵ is identically zero. By expanding the unique vertex of L into a cycle corresponding to the cyclic sequence determined by Π, it is easy to obtain a chord diagram (R,H) such that:

- the cross-graph of (R,H) is G
- the faces of the embedding of L correspond one-to-one to opposite pairs of inner ϵ-walks of (R,H,ϵ).

It then follows from Proposition 4 that the number f of faces of the embedding satisfies:

$$\text{dim Ker } A = f - 1.$$

Hence:

$$\text{rank } A = |E(L)| - f+1 = |E(L)| - f - |V(L)| + 2.$$

The result now follows from the Euler-Poincaré formula. □

REMARKS:
(1) This theorem was obtained in a different way by Brahana ([2], section 26). Related work appears in [3] and [7].
(2) S is the sphere iff rank $A = 0$, i.e. A is the zero matrix. Then no two chords cross and the core graph is easily seen to be a tree.
(3) Proposition 6 implies that if ϵ is identically zero, rank A is even.

More generally, this is true for all symmetric matrices A over $GF(2)$ with zero diagonal.

VII. On Graphic Spaces and Planar Spaces.

1) A characterization of graphic spaces: The following result appears in [11] (see also [12] for related work):

PROPOSITION 7: A space is graphic iff it is of the form $<\epsilon(v)v+N_v, v \in V>$ for some labelled circle graph (G, ϵ).

PROOF: If (G, ϵ) is a labelled circle graph, $<\epsilon(v)v+N_v, v \in V>$ is graphic by Proposition 4.

Conversely, consider a graphic space of the form $\mathbf{C}(D)$ for some graph D. We may assume that D is connected. It is known that D has an embedding on some surface S with exactly one face ([13], Chapter IV, Theorem 5 and [18], Theorem 13). The dual of D on S is a loopgraph L. By expanding the unique vertex of L into a cycle in the appropriate way, we obtain a labelled chord diagram (R, H, ϵ), the core space of which is isomorphic to $\mathbf{K}(D)$ (See section IV. 4.1). Then if G is the cross-graph of (R, H), it follows from Proposition 4 that $\mathbf{C}(D)$ is isomorphic to $<\epsilon(v)v+N_v, v \in V>$. □

Similarly one can prove:

PROPOSITION 8: A space is isomorphic to the cycle space of a connected graph embeddable with exactly one face on some orientable surface iff it is of the form $<N_v, v \in V>$ for some circle graph G.

Connected graphs embeddable with exactly one face on some orientable surface are characterized by the following combinatorial property obtained by Xuong [14]: there exists a cotree, all connected components of which have an even number of edges. For a survey of related problems, see [15], [17].

2) A characterization of planar spaces without bicycles. Let F be a space on a set X. F is said to be *without bicycles* if $\mathbf{F} \cap \mathbf{F}^\perp = \{\varnothing\}$. In this case, every subset Y of X can be written uniquely in the form $Y = \gamma(Y) + \theta(Y)$, with $\gamma(Y) \in \mathbf{F}$, $\theta(Y) \in \mathbf{F}^\perp$. It is easy to show (see [16], theorem 2.5) that if x, x' are distinct elements of X, x belongs to $\gamma(x')$ if and only if x' belongs to $\gamma(x)$. In this case, x and x' are said to be *interlaced*. The *interlacement graph of* F is the simple graph with vertex-set X, where two elements of X are adjacent if and only if they are interlaced in $\mathbf{F}$ (see for instance [21]).

The following result appears in [6] and extends the characterization of planar graphs without bicycles by an "algebraic diagonal" presented in [16] (Theorem 10.1) and [20] (Theorem 13).

PROPOSITION 9: *A space without bicycles is planar iff its interlacement graph is a circle graph.*

PROOF: Let $\mathbf{F}$ be a space on a set X, without bicycles. Let G be the interlacement graph of $\mathbf{F}$. With the notations given above, let ϵ be the mapping from X to $GF(2)$ defined by:

$$\epsilon(x) = 1 \text{ if } x \in \gamma(x), \text{ and } \epsilon(x) = 0 \text{ if } x \in \theta(x).$$

Then clearly for every x in X:

$$\gamma(x) = \epsilon(x)x + N_x, \quad \theta(x) = [1+\epsilon(x)]x + N_x,$$

where N_x denotes the neighbourhood of x in G. It follows that $\mathbf{F} = <\epsilon(x)x+N_x, \ x \in X>$ and $\mathbf{F}^\perp = <[1+\epsilon(x)]x+N_x, \ x \in X>$. Now if G is a circle graph, by Proposition 7, $\mathbf{F}$ and $\mathbf{F}^\perp$ are graphic, so that $\mathbf{F}$ is planar.

Conversely, if $\mathbf{F}$ is planar, we consider a plane graph M such that $\mathbf{F} = C(M)$, and a bipartite chord diagram (R,H) from which M is obtained as in Section III.3. By Proposition 4', M has only one left-right path P_1. It immediately follows from Proposition 3' that for every x in X, N_x is identical to $C_1^x - x$.

By the same argument as in the proof of Proposition 5, we conclude that G is a circle graph. □

References.

[1] C. Berge, *Graphes et Hypergraphes*, Dunod, Paris, 1974.

[2] H.R. Brahana, Systems of circuits on two-dimensional manifolds, *Annals of Math.* 23, 1921, 144-168.

[3] M. Cohn, A. Lempel, Cycle decomposition by disjoint transpositions, *J. of Combinatorial Theory* (A) Vol. 13 (no. 1) 1972, 83-89.

[4] J. Edmonds, A combinatorial representation for polyhedral surfaces, *Notices Am. Math. Soc.* 7 (1960), 646.

[5] J.C. Fournier, Graphes de cordes, hypergraphes de chaines d'un arbre et matroïdes graphiques, *Actes du Colloque "Algèbre Appliquée et Combinatoire"*, Grenoble, Juin 1978, C. Benzaken editor, 164-171.

[6] H. De Fraysseix, Local complementation and interlacement graphs, *Discrete Mathematics*. Vol. 33 (no. 1) 1981, 29-35.

[7] R. Goldstein, E. Turner, Applications of topological graph theory to Group theory, *Math. Zeit.* 165 (1979), 1-10.

[8] M.C. Golumbic, *Algorithmic graph theory and perfect graphs,* Academic Press, New York, 1980, Chapter 11.

[9] G. Haggard, Edmonds characterization of Disc Embeddings, in: *Proceedings of the 8th Southeastern Conference on Combinatorics, Graph Theory, and Computing,* Utilitas Mathematica, Winnipeg, 1977, 291-302.

[10] A. Jacques, *Constellations et propriétés algébriques des Graphes Topologiques,* Thesis, Paris, 1969.

[11] F. Jaeger, Graphes de cordes et espaces graphiques, *European Journal of Combinatorics,* to appear.

[12] F. Jaeger, Symmetric representations of binary matroids, *Colloque International sur la Théorie des Graphes et la Combinatoire,* Luminy, June 1981, to appear.

[13] Nguyen Huy Xuong, *Sur quelques problèmes d'immersion d'un graphe dans une surface,* Thesis, Université Scientifique et Médicale de Grenoble, 1977.

[14] Nguyen Huy Xuong, How to determine the maximum genus of a graph, *J. of Combinatorial Theory* (B) Vol. 26 (no. 2) 1979, 217-225.

[15] Nguyen Huy Xuong, C. Payan, Sur un théorème min-max en Théorie des Graphes, d'après L. Nebesky, *Colloque International sur la Théorie des Graphes et la Combinatoire,* Luminy, June 1981, to appear.

[16] R.C. Read, P. Rosenstiehl, On the principal edge tripartition of a graph, *Annals of Discrete Math.* 3 (1978), 195-226.

[17] R.D. Ringeisen, Survey of results on the maximum genus of a graph, *J. of Graph Th.* 3 (1979), 1-13.

[18] G. Ringel, The Combinatorial Map Color Theorem, *J. of Graph Th.* 1 (1977), 141-155.

[19] P. Rosenstiehl, Bicycles et diagonales des graphes planaires, *Cahiers du C.E.R.O.,* Vol. 17, no. 2-3-4 (1975), 365-383.

[20] P. Rosenstiehl, Mot et vecteur d'un graphe mêlé, *Journées de Combinatoire et Informatique,* Bermond et Cori editors, Bordeaux, June 1975, 317-328.

[21] P. Rosenstiehl, Les graphes d'entrelacement d'un graphe, in: *Problèmes combinatoires et théorie des graphes,* Colloque International du CNRS, Orsay, July 1976, 359-362.

[22] H. Shank, The theory of left-right paths, in: *Combinatorial Mathematics, III,* Lecture Notes in Mathematics 452, Springer Verlag, Berlin, 1975, 42-54.

[23] S. Stahl, Generalized embedding schemes, *J. of Graph Th.* 2 (1978), 41-52.

[24] D.J.A. Welsh, *Matroid Theory,* Academic Press, London, 1976.

The Tutte Polynomial of a Morphism of Matroids. II Activities of Orientations

Michel Las Vergnas

ABSTRACT

We define the activities of an oriented matroid with respect to its orientation. The main result expresses the Tutte polynomial of an oriented matroid (and more generally of a certain two-variable specialization of the Tutte polynomial of an oriented matroid perspective) in terms of the activities of all its reorientations. We discuss various corollaries and applications of this theorem to graphs, arrangements of hyperplanes and convex polytopes.

1. Introduction

The present paper takes place among several works whose common origin is a theorem established by R.P. Stanley [21] in 1973: *The number of acyclic orientations of an undirected graph G is equal to $(-1)^{|V(G)|} \chi(G;-1)$, where $\chi(G;q)$ is the chromatic polynomial of G.*

First generalizations of this theorem have been independently given by T. Brylawski and D. Lucas to regular matroids [4], by T. Zaslavsky to arrangements of hyperplanes in $\mathbb{R}^l$ [24] and by M. Las Vergnas to oriented matroids [11] (see also [15]). The relationship between these theorems, mentioned here in increasing order of generality, will be briefly discussed in Section 4 below.

For oriented matroids we have:

THEOREM A [11]: The number of acyclic reorientations of an oriented matroid M is equal to $t(M;2,0)$, where $t(M;\zeta,\eta)$ is the Tutte polynomial of M (see Section 2 for notations and terminology).

Theorem A generalizes to oriented matroid perspectives:

THEOREM B [13]: The number of acyclic/totally cyclic reorientations of an oriented matroid perspective P is equal to $t(P;0,0,1)$, where $t(P;\zeta,\eta,\xi)$ is the Tutte polynomial of P.

Theorem B, which contains Theorem A as a specialization, unifies several counting theorems on graphs and arrangements of hyperplanes (see Section 4).

Our purpose in the present paper is to prove a generalization of Theorem B dealing with the coefficients of $t(P;\zeta,\eta,1)$. The main theorem, Theorem 3.1, gives a combinatorial interpretation of these coefficients in terms of activities of reorientations of the oriented matroid perspective P. Theorem 3.1 was announced in [14] and quoted in the survey [16] (Th. 7.1). We discuss applications of these theorems in Section 4. We present in Section 5 a further generalization.

2. Terminology and Notations

Basic definitions, properties and examples of oriented matroids used in this paper may be found in [2]. For the reader's convenience we recall briefly the main ones.

Oriented matroids abstract the sign properties of linear dependence in vector spaces over ordered fields (or division rings). A *signed circuit* X of an *oriented matroid* M on a (finite) set E is a signed subset of E with a *positive* part X^+ and a *negative* part X^- ($X = X^+ \cup X^-$, $X^+ \cap X^- = \varnothing$). The *opposite* of X, denoted by $-X$, is defined by $(-X)^+ = X^-$, $(-X)^- = X^+$. The collection $\mathbf{O}$ of signed circuits of an oriented matroid is characterized by the following properties:

(i) $\varnothing \notin \mathbf{O}$; $X \in \mathbf{O}$ implies $-X \in \mathbf{O}$; $X, Y \in \mathbf{O}$ and $X \subseteq Y$ as sets imply $X = Y$ or $X = -Y$ as signed sets,

(ii) *(elimination property)* for all $X, Y \in \mathbf{O}$, $e \in X^+ \cap Y^-$ and $x \in (X \setminus Y) \cup (Y \setminus X) \cup (X^+ \cap Y^+) \cup (X^- \cap Y^-)$ there is $Z \in \mathbf{O}$ such that $x \in Z$, $Z^+ \subseteq (X^+ \cup Y^+) \setminus \{e\}$ and $Z^- \subseteq (X^- \cup Y^-) \setminus \{e\}$.

Clearly the subsets of E underlying the signed circuits of M constitute the circuits of a matroid on E.

Let M be an oriented matroid on E and $e \in E$. We denote by $M \setminus e$ resp. M/e the oriented matroid obtained from M by *deleting* resp. *contracting* e. The signed circuits of $M \setminus e$ are all the signed circuits of M not containing e. The signed circuits of M/e are all the non empty signed subsets of the form $X \setminus \{e\}$, X a signed circuit of M, minimal with respect to inclusion as sets.

LEMMA 2.1 [2] Let X be a signed circuit of M and $x \in X \setminus \{e\}$. Then there is a signed circuit X' of M/e such that $x \in X'$, $X'^+ \subseteq X^+$ and $X'^- \subseteq X^-$.

There is a unique way to sign the cocircuits of an oriented matroid

M such that the following *orthogonality property* holds: for any signed circuit X and signed cocircuit Y of M if $X \cap Y$ is non empty then both $(X^+ \cap Y^+) \cup (X^- \cap Y^-)$ and $(X^+ \cap Y^-) \cup (X^- \cap Y^+)$ are non empty. These signed cocircuits define an oriented matroid: the *orthogonal* oriented matroid M^* of M. We have $(M^*)^* = M$, $(M \backslash e)^* = M^*/e$.

A signed set X is *positive* if $X^- = \emptyset$.

LEMMA 2.2 [2] An element $e \in E$ is either in a positive circuit of M or in a positive cocircuit, but not in both.

Let $\mathbf{O}$ be the collection of signed circuits of M. For $A \subseteq E$ we denote by $_{\overline{A}}M$ the oriented matroid obtained from M by *sign reversal* on A: the signed circuits of $_{\overline{A}}M$ are the signed sets $_{\overline{A}}X$, $X \in \mathbf{O}$, where $(_{\overline{A}}X)^+ = (X^+ \backslash A) \cup (X^- \cap A)$ and $(_{\overline{A}}X)^- = (X^- \backslash A) \cup (X^+ \cap A)$. We say that $_{\overline{A}}M$ is a reorientation of M. We write $_{\overline{e}}M$ instead of $_{\overline{\{e\}}}M$.

A *matroid perspective* $P: M \to M'$ on a set E constituted by two matroids M, M' on E such that every circuit of M is a union of circuits of M' (with no substantial loss of generality matroid perspectives correspond to *strong maps* of matroids, see [23 Chap. 17]). An *oriented matroid perspective* $P: M \to M'$ on E is constituted by two oriented matroids M, M' on E such that every signed circuit X of M is a union of circuits X' of M' such that $X'^+ \subseteq X^+$ and $X'^- \subseteq X^-$ [13]. Oriented matroid perspectives abstract linear maps of vector spaces over ordered fields or division rings.

We say that an oriented matroid perspective $P: M \to M'$ is *acyclic/totally cyclic* if M is acyclic and M' totally cyclic. We denote by $_{\overline{A}}P$ the reorientation $_{\overline{A}}M \to _{\overline{A}}M'$.

We recall that the *Tutte polynomial* $t(M)$ of a matroid M is the polynomial in two variables defined by

$$t(M; \zeta, \eta) = \sum_{X \subseteq E} (\zeta - 1)^{r(M) - r_M(X)} (\eta - 1)^{|X| - r_M(X)}$$

where $r_M(X)$ denotes the rank of X in M and $r(M)$ the rank of M (see [21 Chap. 12]). The *Crapo invariant* $\beta(M)$ *is the coefficient of* ζ *in* $t(M; \zeta, \eta)$.

The *Tutte polynomial* of a matroid perspective $P: M \to M'$ is a polynomial in three variables $t(P; \zeta, \eta, \xi)$ [12] [16] [17]. We will only use here the specialization $t_1(P; \zeta, \eta) = t(P; \zeta, \eta, 1)$. We have

$$t_1(M, M'; \zeta, \eta) = \sum_{X \subseteq E} (\zeta - 1)^{r(M') - r_{M'}(X)} (\eta - 1)^{|X| - r_M(X)}$$

$t_1(P)$ satisfies inductive relations similar to those satisfied by the

Tutte polynomial of a matroid:

(i) if $e \in E$ is neither an isthmus of M' nor a loop of M we have $t_1(P) = t_1(P\backslash e) + t_1(P/e)$ ($P\backslash e$ denotes the matroid perspective $M\backslash e \to M'\backslash e$)

(ii) if $e \in E$ is an isthmus of M' (hence necessarily an isthmus of M) we have $t_1(P) = \zeta t_1(P\backslash e)$

(iii) if $e \in E$ is a loop of M (hence necessarily a loop of M') we have $t_1(P) = \eta t_1(P\backslash e)$

(iv) $t_1(\emptyset) = 1$

3. Activities of Orientations. The Main Theorem

Let E be a finite (totally) ordered set and M be an oriented matroid on E. We define the *external* resp. *internal activity* of M with respect to the ordering, denoted by $o(M)$ resp. $o^*(M)$, as the number of elements of E which are the least element of some positive circuit (resp. positive cocircuit) of $M^{(1)}$. Clearly $o^*(M) = o(M^*)$.

THEOREM 3.1 Let $M \to M'$ be an oriented matroid perspective on a (finite) set E with a total ordering. We have

$$t(M,M';\zeta,\eta,1) = \sum_{A \subseteq E} \left(\frac{1}{2}\zeta\right)^{o^*(\frac{1}{A}M')} \left(\frac{1}{2}\eta\right)^{o(\frac{1}{A}M)}$$

The main step of the proof of Theorem 3.1 is the following lemma:

LEMMA 3.2 Let $M \to M'$ be an oriented matroid perspectve on a set E with a total ordering. Let e be the greatest element of E. Then if e is neither an isthmus of M' nor a loop of M we have either $o^*(M') = o^*(M\backslash e)$, $o(M) = o(M\backslash e)$, $o^*(\frac{1}{e}M') = o^*(M'/e)$, $o(\frac{1}{e}M) = o(M'\backslash e)$ or $o^*(M') = o^*(M'/e)$, $o(M) = o(M/e)$, $o^*(\frac{1}{e}M') = o^*(M\backslash e)$,

---

(1) (The ideas of activities of an orientation (in the case of a directed graph) goes back to a paper of G. Berman [1]. We conjectured Theorem 3.1 as a possible generalization of Berman's Theorem 4.1 to oriented matroids and oriented matroid perspectives (also generalizing [13] Th. 1). It should be pointed out that the above definition of activities is different from that of Berman (in fact it arose during a conversation with C. Greene in May 1977, a conversation that we gratefully acknowledge). The author was for some time superficially convinced that the two definitions were equivalent in some sense, and thus that Theorem 3.1 indeed generalized [1 Th. 4.1] (see [14]). However, a recent careful checking of [1] has shown that, whereas many of the ideas remain basic to the subject (see [18]), the proof of [1 Th. 4.1] is incomplete. Furthermore, in view of several examples, it appears that the theorem as stated does not hold in general).

$o(\frac{-}{e}M) = o(M\backslash e)$.

PROOF. Let $x \in E$. We set $o(M;x)=1$ if x is the least element of some positive circuit of M, $o(M;x)=0$ otherwise. Clearly $o(M) = \sum_{x \in E} o(M;x)$. Similarly we set $o^*(M';x)=1$ if x is the least element of some positive cocircuit of M', $o^*(M';x)=0$ otherwise.

We break the proof of Lemma 3.2 into several steps. Let $x \in E\backslash\{e\}$.

(1) $o(M;x)=1$ or $o(\frac{-}{e}M;x)=1$ imply $o(M/e;x)=1$

Let X be a positive circuit of M (resp. $\frac{-}{e}M$) with least element x. By Lemma 2.1 there is a positive circuit X' of $M/e = (\frac{-}{e}M)/e$ such that $x \in X' \subseteq X$, hence $o(M/e;x)=1$.

(2) $o(M;x)=1$ and $o(\frac{-}{e}M;x)=1$ imply $o(M\backslash e;x)=1$

There exist signed circuits X and Y of M both with least element x such that $X^- = \emptyset$ and $Y^- \subseteq \{e\}$. Suppose $o(M\backslash e;x)=0$: Then necessarily $e \in X^+ \cap Y^-$, hence by elimination there exists a positive circuit Z such that $x \in Z \subseteq (X \cup Y)\backslash\{e\}$. We have thus $o(M\backslash e;x)=1$, a contradiction.

(3) $o(M\backslash e;x)=1$ implies $o(M;x)=o(\frac{-}{e}M;x)=1$

The proof is immediate.

(4) $o(M/e;x)=1$ implies $o(M;x)=1$ or $o(\frac{-}{e}M;x)=1$

Let X' be a positive circuit of M/e with least element x. There exists a signed circuit X of M such that $X' = X\backslash\{e\}$. Since e is the greatest element of E, x is the least element of X. We have $X^- \subseteq \{e\}$, hence X is a positive circuit of M, or $\frac{-}{e}X$ is a positive circuit of $\frac{-}{e}M$.

(5) $o(M;x)=0(M\backslash e;x)$ if and only if $o(\frac{-}{e}M;x)=o(M/e;x)$.

Suppose $o(M;x)=o(M\backslash e;x)$ and $o(\frac{-}{e}M;x) \neq o(M/e;x)$. We have $o(\frac{-}{e}M;x)=0$ (since $o(\frac{-}{e}M;x)=1$ implies $o(M/e;x)=1$ by (1)) and $o(M/e;x)=1$. Then $o(M;x)=1$ by (4), hence $o(M\backslash e;x)=1$, which contradicts (3) (since $o(\frac{-}{e}M;x)=0$).

Conversely suppose $o(\frac{-}{e}M;x)=o(M/e;x)$ and $o(M;x) \neq o(M\backslash e;x)$. We have $o(M\backslash e;x)=0$ (since $o(M\backslash e;x)=1$ implies $o(M;x)=1$ by (3)) and $o(M;x)=1$. Then $o(M/e;x)=1$ by (1), hence $o(\frac{-}{e}M;x)=1$. But this, together with $o(M;x)=1$ contradicts (2) (since $o(M\backslash e;x)=0$).

(6) *If for some* $y \in E\backslash\{e\}$ *we have* $o(M;y) \neq o(\frac{-}{e}M;y)$ *and* $o(M;y)=o(M\backslash e;y)$, *then* $o(M;x)=o(M\backslash e;x)$ *for all* $x \in E\backslash\{e\}$.

Suppose $o(M;x) \neq o(M\backslash e;x)$ for some $x \in E\backslash\{e,y\}$. Then $o(M\backslash e;x)=0$ and $o(M;x)=1$ (since $o(M\backslash e;x)=1$ implies $o(M;x)=1$ by (3)). Hence there is a positive circuit X of M with least element x, and necessarily $e \in X$. On the other hand $o(M\backslash e;y)=0$ (since $o(M\backslash e;y)=1$ implies $o(M;y)=o(\overline{}_e M;y)$ by (3)), hence $o(M;y)=0$, $o(\overline{}_e M;y)=1$ and there is a signed circuit Y of M with least element y such that $Y^- \subseteq \{e\}$. Necessarily $e \in X^+ \cap Y^-$, since $o(M\backslash e;x)=o(M\backslash e;y)=0$. Set $z=\text{Min}(x,y)$. By the elimination property there is a positive circuit Z of M such that $z \in Z \subseteq (X \cup Y)\backslash\{e\}$. Clearly z is the least element of Z. Hence $o(M\backslash e;z)=1$, a contradiction.

(7) $o(M;x)=o(\overline{}_e M;x)$ implies

$$o(M;x)=o(\overline{}_e M;x)=o(M\backslash e;x)=o(M/e;x)$$

Suppose $o(M;x)=o(\overline{}_e M;x)$. If $o(M;x)=0$, then $o(M\backslash e;x)=0$ by (3). If $o(M;x)=1$, then $o(M\backslash e;x)=1$ by (2). In both cases $o(M;x)=o(M\backslash e;x)$. Now $o(\overline{}_e M;x)=o(M/e;x)$ by (5).

(8) *We have* $o(M;x)=o(M\backslash e;x)$ *and* $o(\overline{}_e M;x)=o(M/e;x)$ *for all* $x \in E\backslash\{e\}$, *or* $o(M;x)=o(M/e;x)$ *and* $o(\overline{}_e M;x)=o(M\backslash e;x)$ *for all* $x \in E\backslash\{e\}$.

If $o(M;x)=o(\overline{}_e M;x)$ for all $x \in E\backslash\{e\}$, then $o(M;x)=o(\overline{}_e M;x)=o(M\backslash e;x)=o(M/e;x)$ for all $x \in E\backslash\{e\}$ by (7). Suppose there exists $y \in E\backslash\{e\}$ such that $o(M;y) \neq o(\overline{}_e M;y)$. We have $o(M;y)=o(M\backslash e;y)$ or $o(\overline{}_e M;y)=o(M\backslash e;y)$. Suppose for instance $o(M;y)=o(M\backslash e;y)$. We have $o(M;x)=o(M\backslash e;x)$ for all $x \in E\backslash\{e\}$ by (6), then $o(\overline{}_e M;x)=o(M/;x)$ for all $x \in E\backslash\{e\}$ by (5).

(9) $o(M;x)=1$ *implies* $o^*(M';x)=0$.

Suppose $o(M;x)=1$: there is a positive circuit X of M with least element x. Since $M \to M'$ is an oriented matroid perspective there is a positive circuit X' of M' such that $x \in X' \subseteq X$, hence there is no positive cocircuit of M' containing x by Lemma 2.2, implying $o^*(M';x)=0$.

(10) *If* $o(M;x) \neq o(\overline{}_e M;x)$, *then* $o(M;x)=o(M\backslash e;x)$ *implies* $o^*(M';x)=o^*(M'\backslash e;x)$.

Suppose $o^*(M';x) \neq o^*(M'\backslash e;x)$ under the hypothesis of (10). By (1^*) (i.e. by (1) applied to M^*) $o^*(M';x)=1$ implies $o^*(M'\backslash e;x)=1$. Hence $o^*(M';x)=0$ and $o^*(M'\backslash e;x)=1$. Now by (4^*) $o^*(M';x)=1$ or $o^*(\overline{}_e M';x)=1$. Hence $o^*(\overline{}_e M';x)=1$. Then by (9) $o(\overline{}_e M;x)=0$. It follows that $o(M;x)=1$, hence $o(M\backslash e;x)=1$, by hypothesis, contradicting $o(\frac{-}{e}M;x)=0$ by (3).

(11) We have either $o^*(M';x)=o^*(M'\backslash e;x)$, $o(M;x)=o(M\backslash e;x)$, $o^*(\frac{-}{e}M';x)=o^*(M'/e;x)$ and $o(\frac{-}{e}M;x)=o(M/e;x)$ for all $x \in E\backslash\{e\}$, or $o^*(M';x)=o^*(M'/e;x)$, $o(M;x)=o(M/e;x)$, $o^*(\frac{-}{e}M';x)=o^*(M\backslash e;x)$ and $o(\frac{-}{e}M;x)=o(M\backslash e;x)$ for all $x \in E\backslash\{e\}$.

If $o^*(M';x)=o^*(\frac{-}{e}M;x)$ for all $x \in E\backslash\{e\}$, or $o(M;x)=o(\frac{-}{e}M;x)$ for all $x \in E\backslash\{e\}$, then (11) follows clearly from (7), (7^*) and (8), (8^*). If there is $x \in E\backslash\{e\}$ such that $o^*(M';x) \neq o^*(\frac{-}{e}M';x)$ and $o(M;x) \neq o(\frac{-}{e}M;x)$ then (11) follows from (8), (8^*) and (10).

The remaining possibility is that for all $x \in E\backslash\{e\}$ we have either $o^*(M';x) \neq o^*(\frac{-}{e}M;x)$, $o(M;x)=o(\frac{-}{e}M;x)$ or $o^*(M';x)=o^*(\frac{-}{e}M';x)$, $o(M;x) \neq o(\frac{-}{e}M;x)$, and both cases occur. Replacing if necessary $M \to M'$ by $\frac{-}{e}M \to \frac{-}{e}M'$ we may suppose notations such that $o(M;y)=o(M\backslash e;y)$ for some $y \in E\backslash\{e\}$ with $o(M;y) \neq o(\frac{-}{e}M;y)$. Then by (6) $o(M;x)=o(M\backslash e;x)$ for all $x \in E\backslash\{e\}$.

If $x \in E\backslash\{e\}$ is such that $o(M;x) \neq o(\frac{-}{e}M;x)$ we have $o^*(M';x)=o^*(M\backslash e;x)$ by (10).

Consider now $x \in E\backslash\{e\}$ such that $o(M;x)=o(\frac{-}{e}M;x)$. We have $o^*(M';x) \neq o^*(\frac{-}{e}M';x)$ by our hypothesis. If $o^*(M';x)=0$ we have $o^*(\frac{-}{e}M';x)=1$ hence there is a cocircuit X' of M' with $X'^- = \{e\}$. By our hypothesis there is $y \in E\backslash\{e\}$ with $o(M;y)=o(M\backslash e;y) \neq o(\frac{-}{e}M;y)$: by (3) we have $o(M\backslash e;y)=0$, hence $o(M;y)=0$, $o(\frac{-}{e}M;y)=1$, implying the existence of a circuit Y of M with $Y^- = \{e\}$. Since $M \to M'$ there is a circuit Y' of M' with $Y'^- = \{e\}$. Then X' and Y' contradict the orthogonality property. Therefore $o^*(M',x)=1$ and $o^*(\frac{-}{e}M';x)=0$. Now by (1^*) $o^*(M'\backslash e;x)=1$, hence $o^*(M';x)=o^*(M'\backslash e;x)$.

Thus for all $x \in E\backslash\{e\}$ we have $o^*(M';x)=o^*(M'\backslash e;x)$ and $o(M;x)=o(M\backslash e;x)$. Then (11) follows from (5), (5^*) (or (8), (8^*)).

(12) Lemma 3.2 is obtained by summing up the equalities in (11) for all $x \in E\backslash\{e\}$, and by observing that $o^*(M';e)=o^*(\frac{-}{e}M';e)=o(M;e)=o(\frac{-}{e}M;e)=0$, since e is the greatest element of E and is neither an isthmus of M' nor a loop of M. □

Let $P: M \to M'$ be an oriented matroid perspective on a set E with a total ordering. For i,j two non-negative integers let $o_{ij}(P)$ denote the number of subsets $A \subseteq E$ such that $o^*(\frac{}{A}M') = i$ and $o(\frac{}{A}M) = j$.

LEMMA 3.3 With the above notations, let e be the greatest element of E. Then we have

(i) if e is neither an isthmus of M' nor a loop of M
$$o_{ij}(P) = o_{ij}(P \backslash e) + o_{ij}(P/e)$$

(ii) if e is an isthmus of M' (hence also of M) $o_{ij}(P) = 2o_{i-1,j}(P \backslash e)$.

(iii) if e is a loop of M (hence also of M') $o_{ij}(P) = 2o_{i,j-1}(P \backslash e)$.

PROOF. Write for short $(o^*, o)(P) = (o^*(M'), o(M))$. Then for $A \subseteq E \backslash \{e\}$, if e is neither an isthmus of M' nor a loop of M, we have by Lemma 3.2

$$\{(o^*, o)(\tfrac{}{A}P), (o^*, o)(\tfrac{}{A \cup \{e\}}P)\} = \{(o^*, o)(\tfrac{}{A}(P \backslash e)), (o^*, o)(\tfrac{}{A}(P \backslash e))\}$$

The collection of these equalities for $A \subseteq E \backslash \{e\}$, clearly implies (i). The proofs of (ii) and (iii) are straightforward and left to the reader. □

PROOF OF THEOREM 3.1. The derivation of Theorem 3.1 from Lemma 3.3 consists in straightforward substitutions in the inductive relations satisfied by $t_1(P)$ (see Section 2). We leave these verifications to the reader. □

We state as a corollary the important particular case of an oriented matroid:

COROLLARY 3.4. Let M be an oriented matroid on a (finite) set E with a total ordering. We have

$$t(M; \zeta, \eta) = \sum_{A \subseteq E} \left(\frac{1}{2}\zeta\right)^{o^*(\frac{}{A}M)} \left(\frac{1}{2}\eta\right)^{o(\frac{}{A}M)}$$

PROOF. Corollary 3.4 is obtained by applying Theorem 3.1 to $P: M \to M$: Clearly P is an oriented matroid perspective and we have immediately from the definitions (see Section 2) $t(P; \zeta, \eta, \xi) = t(M; \zeta, \eta)$. □

4. Corollaries and Applications

We give in this section some corollaries and applications of Theorem 3.1.

We note that Theorem A and B are specializations of Theorem 3.1:

Let M be an oriented matroid on E. Consider any ordering of E. By Corollary 3.4 we have (formally) $t(M; 2, 0) = \sum_{A \subseteq E} 1^{o^*(\frac{}{A}M)} 0^{o(\frac{}{A}M)}$. In this sum a term is 1 if $o(_{\overline{A}}M)=0$ i.e. if $_{\overline{A}}M$ is acyclic, and is 0 otherwise. Hence Theorem A.

Let $P:M\to M'$ be an oriented matroid perspective on E. Consider any ordering of E. By Lemma 2.2 a reorientation $_{\overline{A}}P$ of P, $A\subseteq E$, is acyclic/totally cyclic if and only if $o^*(_{\overline{A}}M')=o(_{\overline{A}}M)=0$. Hence by Theorem 3.1 the constant term of $t(P;\zeta,\eta,1)$ i.e. $t(P;0,0,1)$ counts the number of acyclic/totally cyclic reorientations of P. Theorem B contains Theorem A (see [13]).

Theorem A generalizes directly Stanley's theorem: Let G be a graph and $\vec{G}$ be any orientation of G. Let M be the oriented cycle matroid of $\vec{G}$ [2]. Clearly the acyclic orientations of G are in 1-1 correspondence with the acyclic reorientations of M, hence by Theorem A the number of acyclic reorientations of G is equal to $t(M;2,0)$. Now the chromatic polynomial of G is given by $\chi(G;q)=(-1)^{|V|}(-q)^{c(G)}t(\mathcal{C}(G);1-q,0)$, where V is the vertex-set of G, $c(G)$ the number of connected components of G and $\mathcal{C}(G)$ the nonoriented cycle matroid of G underlying M [23 Chap. 15]. In particular we have $t(\mathcal{C}(G);2,0)=(-1)^{|V|}(G;-1)$.

Proposition 4.5 of [4] is the particular case of Theorem A when M is the oriented matroid of the signed linear dependencies of the columns of a totally unimodular matrix (i.e. M is regular).

Zaslavski's Theorem A of [24] counts the number of regions determined by hyperplanes in $I\!R^n$. This theorem is equivalent to the particular case of Theorem A when M is the oriented matroid of the signed linear dependencies of hyperplanes of $I\!R^n$. We have

LEMMA 4.1 Let h_e $e \in E$ be linear forms over $I\!R^n$ and M be the oriented matroid on E of the signed linear dependencies over $I\!R$ of these forms [2]. Then there is a 1-1 correspondence between the regions of $I\!R^n$ determined by the hyperplanes $h_e=0$ $e \in E$ and the acyclic reorientations of M.

Lemma 4.1 is a particular case of a correspondence between topes of a sphere-system representing an oriented matroid (see below) and acyclic reorientations of this matroid. A direct proof as a straightforward application of classical results of the theory of linear inequalities (see for instance [9]) is as follows:

We observe that the regions of $I\!R^n$ determined by the hyperplanes $h_e=0$ $e\in E$ are in 1-1 correspondence with the subsets $A\subseteq E$ such that the system of inequalities $\epsilon_e h_e > 0$ $e\in E$, where $\epsilon_e=-1$ for $e\in A$, $\epsilon_e=1$ for $e\in E\setminus A$, has a solution. Now by [9 Th. 2.9] either the inequalities $\epsilon_e e h_e > 0$ have a solution in $I\!R^n$ or there are $\lambda_e \in I\!R$, not all zero, such that $\epsilon_e \lambda_e \geq 0$ for all $e \in E$ and $\sum_{e \in E} \lambda_e \epsilon_e h_e = 0$ (.e. M has a positive circuit), but not both.

By lemma 4.1 and Theorem A the number of regions of $\mathbb{R}^n$ determined by the hyperplanes $h_e = 0$ $e \in E$ is equal to $t(M;2,0)$, which is Zaslavsky's theorem. The counting of regions determined by affine hyperplanes of $\mathbb{R}^n$ (i.e. not necessarily containing 0) can be reduced to the case of hyperplanes containing 0 by considering the suspension in $\mathbb{R}^{n+1}$ (see below). We quote two other equivalent attractive formulations:

Let E be a finite set of points in $\mathbb{R}^n$. An unordered pair $A, E\backslash A$ where $A \subsetneq E$, is a *non-Radon partition* of E if there is an affine hyperplane H of $\mathbb{R}^n$ such that all points of A are in one of the two open half-spaces determined by H and all the points of $E\backslash A$ in the other. Let M be the matroid of affine dependencies of E over $\mathbb{R}$ The number of non-Radon partitions of E is equal to $\frac{1}{2}t(M;2,0)$ [3] [23]. The application of Theorem A is straightforward: Let now M denote the oriented matroid of signed affine dependencies of E over R. By Hahn-Banach Theorem $A, E\backslash A$ is a non-Radon partition of E if and only if there is no signed circuit X of M with $X^+ \subseteq A$ and $X^- E\backslash A$, or, equivalently, if and only if the reorientation $_{\overline{A}}M$ of M is acyclic.

Let v_e $e \in E$ be a set of vectors of $\mathbb{R}^n$. The convex hull Z of the set of 2^n points x of R^n defined by $= \sum_{e \in E} \epsilon_e v_e$, where $\epsilon_e = \pm 1$ is called a *zonotope*. The vertices of Z are in 1-1 correspondence with the regions determined by the hyperplanes normal to the v_e's [20], hence in 1-1 correspondence with the acyclic reorientations of the oriented matroid M of the signed linear dependencies of the v_e's: they are also counted by $t(M;2,0)$ [10] [24].

Another theorem of Zaslavsky [24] counts the number of bounded regions determined by affine hyperplanes in $\mathbb{R}^n$. This theorem can be derived from Theorem B as follows:

Let H_e $e \in E$ be affine hyperplanes of $\mathbb{R}^n$, with equations $h_e = b_e$ where h_e is a linear form over $\mathbb{R}^n$ and $b_e \in \mathbb{R}$. We denote by M the oriented matroid on $E \cup \{\infty\}$ of the signed linear dependencies over $\mathbb{R}$ of the $|E|+1$ linear forms over $\mathbb{R}^{n+1}$: $\bar{h}_e = h_e - b_e x_{n+1}$ $e \in E$ and $\bar{h}_\infty = x_{n+1}$. Let $M' = M\backslash\infty \oplus 0(\infty)$ be the direct sum of M/∞ and the rank zero matroid on $\{\infty\}$ (M' is the oriented matroid of the signed linear dependencies of the $|E|+1$ linear forms over $\mathbb{R}^n$ h_e $e \in E$ and $h_\infty = 0$). Clearly $P:M \to M'$ is an oriented matroid perspective.

LEMMA 4.2 The bounded regions of $\mathbb{R}^n$ determined by the hyperplanes H_e $e \in E$ are in 1-1 correspondence with the acyclic/totally cyclic reorientations of P.

As above the regions determined by the H_e's are in 1-1 correspondence with the subsets $A \subsetneq E$ such that the system of inequalities $\epsilon_e h_e > b_e$ $e \in E$ has a solution, where $\epsilon_e = -1$ for $e \in A$, $\epsilon_e = 1$ otherwise. Furthermore a region is bounded if and only if the system $\epsilon_e h_e \geq 0$ $e \in E$ has only the zero solution. The first system has a solution in $I\!R^n$ if and only if the homogeneous system $\epsilon_e \bar{h}_e > 0$ $e \in E$ and $\bar{h}_\infty > 0$ and has a solution in $I\!R^{n+1}$, so as above if and only if $_{\overline{A}}M$ is acyclic. The second system has only the zero solution if and only if there are $\mu_e \in I\!R$ for all $e \in E$ such that $\epsilon_e \mu_e > 0$ and $\sum_{e \in E} \mu_e \epsilon_e h_e = 0$ [9 Th. 2.9] i.e. if and only if $_{\overline{A}}M'$ is totally cyclic.

By Lemma 4.2 and Theorem B the number of bounded regions determined by the H_e's is thus $\frac{1}{2} t(P;0,0,1)$ ($\frac{1}{2}$ since to each $A \subsetneq E$ such that $_{\overline{A}}P$ is acyclic/totally cyclic corresponds $A' = (E \backslash A) \cup \{\infty\}$ with the same property). It follows easily from the definition of $t(P)$ that $t(M, M/\infty \oplus O(\infty); 0, 0, 1) = 2\beta(M)$ (see [13]), hence Zaslavsky's theorem.

We have shown in [15] that an acyclic oriented matroid may be considered as a generalization of a convex polytope. In this context Zaslavsky's theorem is generalized by Statement (c) of [5] p.157: *the number of acyclic reorientations of an oriented matroid M such that a given element is an internal point is equal to* $2\beta(M)$. More generally we have the following corollary of Theorem B: *the number of acyclic reorientations of M such that no facet contains a given flat X of M is equal to* $t(M, M/x \oplus O(X); 0, 0, 1)$ ([5] Statement (b) p.154). We refer the reader to [5] for some other statements of a similar flavour.

The Representation Theorem for oriented matroids due to Folkman-Lawrence [8] and strengthened by Edmonds-Mandel [19] states that any oriented matroid can be represented by a system of topological spheres of $I\!R^n$ satisfying certain regularity conditions, generalizing the representation by hyperplanes of an oriented matroid coordinatizable over $I\!R$ In this representation the regions determined by the topological spheres, or *topes*, correspond bijectively to the acyclic reorientations of the oriented matroid. Thus, using this representation, Theorems A, B and 3.1 have equivalent dual forms. for instance, Theorem A may be stated: *the number of topes in a sphere system representing an oriented matroid M is equal to* $t(M; 2, 0)$, and the above consequence of Theorem B: *the number of topes whose boundaries do not meet a given member of the sphere system is equal to*

$2\beta(M)$.

We conclude this brief survey by two standard examples of oriented matroid perspectives arising from graphs [13] [16] [17]:

1) Let $G=(V,E)$ be a graph and $V=V_1+V_2+\ldots+V_k$ be a partition of V into k non-empty subsets. Let $G'=(V',E)$ be the graph with vertex-set $V'=\{v'_1,v'_2,\ldots,v'_k\}$ and same edge-set E such that $e \in E$ joins v'_i and v'_j in G' if and only if e joins V_i and V_j in G (in other words G' is obtained from G by identifying vertices with respect to the given partition). Clearly $C(G) \rightarrow C(G')$ is a matroid perspective. Furthermore if G is directed and G' is given the corresponding orientation, then $C(G) \rightarrow C(G')$ is an oriented matroid perspective, where $C(G)$ and $C(G')$ denote now the oriented cycle matroids of G and G'.

The particular case when G' is obtained from G by identifying the two vertices u,v of a given edge is specially interesting. As easily seen an acyclic/totally cyclic reorientation of $C(G) \rightarrow C(G')$ corresponds to an acyclic orientation of G with u unique source and v unique sink or with v unique source and u unique sink. Thus, using Theorem B we get a theorem of Greene-Zaslavsky [10]: *the number of acyclic orientations of G with u unique source and v unique sink is equal to* $\beta(C(G))$. We mention that Stanley's theorem and the above result can be derived from counting regions in $\mathbb{R}^n$ by associating with an edge v_iv_j of G the hyperplane $x_i=x_j$ of $R^{|V|}$.

2) Let G and G^* be two connected graphs dually imbedded in a surface. We know from homology theory (see for instance [7]) that $\mathbb{B}(G^*) \rightarrow C(G)$ is a matroid perspective, where $\mathbb{B}(G^*)$ denotes the cocycle matroid of G^*. Furthermore if the surface is orientable and G, G^* are directed such that all rotations defined by directed edges e, e^* in correspondence by the duality are consistent, then $\mathbb{B}(G^*) \rightarrow C(G)$ is an oriented matroid perspective. By Theorem B we have

COROLLARY 4.3 *Let G, G^* be two connected graphs dually imbedded in an orientable surface. The number of orientations of G such that both G and G^* with the dual orientation are strongly connected is equal to* $t(\mathbb{B}(G^*),C(G);0,0,1)$.

5. Further Properties

Let M be a matroid on a (totally) ordered set E and B be a base of M. The *external activity* $\epsilon(B)$ resp. *internal activity* $\iota(B)$ of B is the number of elements $e \in E\setminus B$ resp. $e \in B$ which are the least element with respect to the ordering of the unique circuit resp. cocircuit of M contained in $B\cup\{e\}$ resp. in $(E\setminus B)\cup\{e\}$. The definition of activities is due to Tutte [22] in the case of graphs and to Crapo [6] in the general matroid case.

THEOREM [22] [6] : We have

$$t(M;\zeta,\eta)=\sum_{i,j}b_{ij}\zeta^i\eta^j$$

where b_{ij} is the number of bases B of M with $\iota(B)=i$ and $\epsilon(B)=j$. (see [16] [17] for a generalization of this theorem to matroid perspectives).

Restricting our attention to oriented matroids rather than oriented matroid perspectives, for the sake of simplicity, we observe that Corollary 3.4 may equivalently be stated

THEOREM: For any oriented matroid M on a (totally) ordered set E we have

$$t(M;\zeta,\eta)=\sum_{i,j}2^{-i-j}o_{ij}\zeta^i\eta^j$$

where o_{ij} is the number of subsets A of E such that $o^*(_A^-M)=i$ and $o(_A^-M)=j$.

Comparing these two expressions we get the remarkable equality:

COROLLARY 5.1

$$o_{ij}=2^{i+j}b_{ij}$$

This equality suggests a natural question: Is it possible to describe a (canonical) correspondence between reorientations and bases of an oriented matroid consistent with these relations? More explicitly we want to associate with any subset A of E a base $B=f(A)$ of M such that $\iota(B)=o^*(_A^-M)$ and $\epsilon(B)=o(_A^-M)$, the correspondence being such that for any base B of M there are exactly $2^{\iota(B)+\epsilon(B)}$ subsets $A\subseteq E$ such that $f(A)=B$.

The answer to the above question is positive: we construct a canonical correspondence with the required properties in [18].

References

[1] G. Berman, The dichromate and orientations of a graph, Canad. J. Math., 29 (1977) 947-956.

[2] R. Bland & M. Las Vergnas, Orientability of matroids J. Combinatorial Theory Ser. B, 24 (1978) 94-123.

[3] T. Brylawski, A combinatorial perspective on the Radon convexity theorem, Geometriae Dedicata, 5 (1976) 459-466.

[4] T. Brylawski & D. Lucas, Uniquely representable combinatorial geometries, in: Teorie Combinatorie (vol. 1), B. Segre ed., Accademia Nazionale dei Lincei, Roma 1976, 83-108.

[5] R. Cordovil, M. Las Vergnas & A. Mandel, Euler's relation, Möbius functions

and matroid indentities, Geometriae Dedicata, 12 (1982) 147-162.
[6] H. H. Crapo, The Tutte polynomial, Aequationes Math., 3 (1969) 211-229.
[7] J. Edmonds, On the surface duality of linear graphs, J. Res. Nat. Bur. Stand., 69 B (1965) 121-123.
[8] J. Folkman & J. Lawrence, Oriented matroids, J. Combinatorial Theory Ser. B, 25 (1978) 199-236.
[9] D. Gale, *The theory of linar economic models*, McGraw-Hill, New York-Toronot-London 1960.
[10] C. Greene & T. Zaslavsky, On the interpretations of Whitney numbers through arrangements of hyperplanes, zonotopes, non-Radon partitions and orientations of graphs, to appear.
[11] M Las Vergnas, Matroïdes orientables, C.R. Acad. Sci. Paris Sér. A, 280 (1975) 61-64.
[12] M Las Vergnas, Extensions normales d'un matroïde, polynome de Tutte d'un morphisme, C.R. Acad. Sci. Paris Sér. A, 280 (1975) 1479-1482.
[13] M Las Vergnas, Acyclic and totally cyclic orientations of combinatorial geometries, Discrete Math., 20 (1977) 51-61.
[14] M. Las Vergnas, Sur les activités des orientations d'un géométrie combinatoire, Actes Colloque Math. Discrètes: codes et hypergraphes, Bruxelles 1978, Cahiers Centre Etudes Recherche Opérationnelle (Bruxelles), 20 (1978) 293-300.
[15] M. Las Vergnas, Convexity in oriented matroids, J. Combinatorial Theory Ser. B, 29 (1980) 231-243.
[16] M. Las Vergnas, On the Tutte polynomial of a morphism of matroids, Annals of Discrete Math., 8 (1980) 7-20.
[17] M. Las Vergnas, The Tutte polynomial of a morphism of matroids I, J. Combinatorial Theory Ser. A, to appear.
[18] M. Las Vergnas, Bases and orientations in graphs and matroids, in preparation.
[19] A. Mandel, Topology of oriented matroids, Thesis, Waterloo 1982.
[20] P. McMullen, On zonotopes, Tans. Amer. Math. Soc., 159 (1971) 91-109.
[21] R. Stanley, Acyclic orientations of graphs, Discrete Math., 5 (1973) 171-178.
[22] W. T. Tutte, A contribution to the theory of chromatics polynomials, Canad. J. Math., 6 (1954) 80-91.
[23] D. J. A. Welsh, *Matroid theory*, Academic Press, London 1976.
[24] T. Zaslavsky, Facing up to arrangements: face-count formulas for partitions of spaces by hyperplanes, Memoirs Amer. Math. Soc., No 154 (1975).

The Cycle-Chromatic Number of a Hypergraph and an Inequality of Lovász

Li Wei-xuan

ABSTRACT

The cycle-chromatic number of a hypergraph H, denoted by $c(H)$, is defined to be the minimum number of colours needed to colour the vertices of H such that no cycle in H^2 is monochromatic. Let $H=(X;E_1,E_2,\ldots,E_m)$ be a hypergraph with n vertices and p components. It is proved that

$$\sum_{i=1}^{m}|E_i| \leq n+(m-p)c(H).$$

By this inequality, we deduce some sufficient conditions for the following inequality, due to Lovász, to hold:

$$\sum_{i=1}^{m}|E_i| \leq n+(m-p)r(H)$$

where $r(H)$ is the rank of H.

Let $H=(X;E_1,E_2,\ldots,E_m)$ be a hypergraph with n vertices and p components. The *square* of H is defined to be the hypergraph

$$H^2 = (E_i \cap E_j;\ i,j = 1,2,\ldots,m\ E_i \cap E_j \neq \emptyset,\ i \neq j).$$

Set $r=\max_{i \neq j} |E_i \cap E_j|$, i.e., the rank of H^2. An inequality obtained by L. Lovász [6] and extended by P. Hansen and M. Las Vergnas [3] is as follows:

THEOREM 1. If H has no cycles of length more than two, then

$$\sum_{i=1}^{m}|E_i| \leq n+(m-p)r. \tag{1}$$

Recently, Theorem 1 was generalized in several directions by B.D. Acharya and M. Las Vergnas [1], L. Lovász [7], C.Q. Zhang and W.X. Li [8], and Q.M. He [4]. In this present paper, by considering the connection between the cycle-chromatic number of H and Inequality (1), some further sufficient conditions for (1) are developed. Thereby, the result in [8] will be deduced in a more natural way. For the terms and symbols not specified here, see [2].

The *cycle-chromatic number* of H, denoted by c, is defined to be the minimum number of colours needed to colour the vertices of H such that no cycle in H is monochromatic. Since two vertices in the same edge of H^2 form a cycle of H of length 2, it is easy to see that a colouring of X such that no cycle of H is monochromatic induces a strong colouring of H^2, i.e., no two vertices in the same edge of H^2 have the same colour. Hence

$$c \geq s \geq r, \tag{2}$$

where $s = \gamma(H^2)$ is the strong chromatic number of H^2.

THEOREM 2. With the notations defined above, we have

$$\sum_{i=1}^{m} |E_i| \leq n + (m-p)c. \tag{3}$$

If $c > n/\tau(H)$, where $\tau(H)$ is the transversal number of H, and every component of H has at least two edges, then (3) holds strictly.

PROOF. For a given colouring f of X with the colour set $\{1, 2, \ldots, c\}$ such that no cycle in H is monochromatic, define a graph G as follows: The vertex set of G is

$$X(G) = X \cup \{y_{it}; \ i=1,2,\ldots,m, \ t=1,2,\ldots,c\}.$$

There is an edge joining a vertex $v \in X$ to a vertex y_{it} if and only if $v \in E_i$ and $f(v) = t$. It is evident that G has $n + mc$ vertices and $\sum_{i=1}^{m} |E_i|$ edges. Observing that $y_{i_1 t_1}$ and $y_{i_2 t_2}$ are of the same component of G only if E_{i_1} and E_{i_2} are of the same component of H and $t_1 = t_2$, we see that G has at least pc components. On the other hand, we claim that G is a forest. In fact, if there were a cycle $x_1 y_{i_1 t_1} x_2 y_{i_2 t_2} \cdots x_q y_{i_q t_q} x_1$ in G, then $f(x_1) = t_1 = f(x_2) = t_2 = \ldots = f(x_q) = t_q$ and $x_1 E_{i_1} x_2 E_{i_2} \ldots x_q E_{i_q} x_1$ would be a monochromatic cycle in H, contradicting the hypothesis. Therefore (3) follows by the well-known equation concerning the number of vertices, edges and components of a forest.

If the equality holds in (3), then there are exactly pc components in G. Hence, each vertex y_{it} in G is adjacent to a vertex $v \in X$. Thus, for every colour class C_t, the set of vertices coloured by the colour t,

$$C_t \cap E_i \neq \emptyset \quad i=1,2,\ldots,m.$$

In other words, each colour class is a transversal set of H. So that,

$$n = \sum_{t=1}^{c}|C_t| \geq c\tau(H).$$

The last assertion of the theorem follows.

It deserves to be noted that (3) gives a lower bound other than (2) for the cycle-chromatic number of H, i.e.,

$$c \geq \frac{\sum_{i=1}^{m}|E_i| - n}{m - p}. \tag{3'}$$

The following corollary is immediate.

COROLLARY 3. If $c=r$, then (1) holds.

We note that the condition given in Corollary 3 is not necessary. A counterexample is as follows:

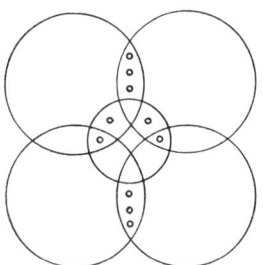

Figure 1.

By (2), the condition $c=r$ is equivalent to $c=s$ and $s=r$. It will be convenient to consider the latter two conditions instead of the former. First, we have the following result:

THEOREM 4. If every cycle in H has two vertices belonging to at least two common edges, then $c=s$.

PROOF. Without loss of generality, we may suppose that every vertex of H belongs to at least two edges of H. Consider a strong colouring of H^2. Since every cycle of H has two vertices belonging to the same edge of H^2, and hence they are coloured by different colours, no cycle of H is monochromatic in this colouring. So that, $c \leq s$. By (2), we conclude that $c=s$, completing the proof.

If the 2-section $H_{(2)}$ is triangulated, i.e., every elementary cycle of $H_{(2)}$ with length at least 4 has a chord, then the condition in

Theorem 4 can be restricted to cycles of length 3.

THEOREM 5. If $H_{(2)}$ is a triangulated graph and every cycle in H of length 3 has two vertices belonging to at least two common edges, then $c=s$.

PROOF. Consider a strong colouring f of H^2. Assume that there is a monochromatic cycle in H. Since $H_{(2)}$ is triangulated, we can find a monochromatic triangle xyz in $H_{(2)}$ such that $x,y \in E$, $y,z \in E'$ and $z,x \in E''$ with $E' \neq E''$, where E, E' and E'' are edges of H. If $E \neq E'$ and $E \neq E''$, then $xEyE'zE''x$ is a cycle of length 3. Since x,y,z are coloured by the same colour, no two of them are contained in two common edges, contradicting the hypothesis. Otherwise, we have that either E' or E'' contains all the three vertices. Hence, we have $z,x \in E' \cap E''$ or $y,z \in E' \cap E''$, contradicting that f is a strong colouring of H^2. The theorem is proved.

We observe that, if H is conformal, then each cycle of length 3 is contained in an edge of H. A conformal hypergraph with triangulated 2-section is called a *T-hypergraph* in [3]. Then we have the following consequence:

COROLLARY 6. If H is a T-hypergraph, then $c=s$.

It is regretful that, though a T-hypergraph H satisfies (1), as shown in [1], and $c=s$, it may fail to satisfy $s=r$. A counterexample is shown in Figure 2.

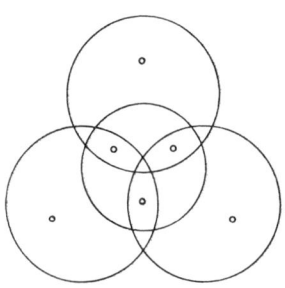

Figure 2

However, we can show that, if H is a T-hypergraph and H^2 is uniform, then H has $s=r$.

Let $L_w(H)$ be a simple edge-weighted graph with its vertex set being the family of edges of H, and E_i and E_j being joined by an edge of weight w in $L_w(H)$ if and only if $|E_i \cap E_j| = w > 0$. A parameter $w(H)$ was defined by M. Lewin [5] as the maximum weight of a spanning forest of $L_w(H)$. Obviously,

$$w(H) \leq (m-p)r. \tag{4}$$

If H^2 in r-uniform, then the equality holds in (4). In [1], the *cyclomatic number* of H is defined to be

$$\mu(H) = \sum_{i=1}^{m} |E_i| - n - w(H), \tag{5}$$

and it is proved that $\mu(H) \geq 0$, and $\mu(H) = 0$ if and only if H is a T-hypergraph. By these results, we have the following theorem:

THEOREM 7. If H^2 is r-uniform, then the following statements are equivalent:

(i) $c = r$

(ii) $\sum_{i=1}^{m} |E_i| \leq n + (m-p)r.$

(iii) $\sum_{i=1}^{m} |E_i| = n + (m-p)r.$

(iv) H is a T-hypergraph.

PROOF. It is clear that (i) implies (ii) by Corollary 3, and (ii) implies (iii) by $w(H) = (m-p)r$ and

$$0 \geq \sum_{i=1}^{m} |E_i| - n - (m-p)r = \sum_{i=1}^{m} |E_i| - n - w(H) = \mu(H) \geq 0.$$

On the other hand, this inequality shows $\mu(H) = 0$, and then H is a T-hypergraph. By a result in [1], (iv) implies (ii). Hence, we see that (ii) is equivalent to (iv).

Now suppose that (iii) is valid. Since (iii) implies (ii), and (ii) implies (iv), by Corollary 6, it is sufficient to prove that $s = r$. Observing that H is conformal, if $F_{ij} = E_i \cap E_j$ and $F_{ik} = E_i \cap E_k$ are edges of H^2, then there is an edge E of H containing F_{ij}, F_{ik} and $F_{jk} = E_j \cap E_k$. By

$$r = |E \cap E_j| \geq |F_{jk} \cup F_{ij}| \geq r,$$

we have $|F_{jk} \cup F_{ij}| = r$ and then $F_{jk} \subseteq F_{ij}$. Similarly, $F_{jk} \subseteq F_{ik}$.

If $F_{ij} \cap F_{ik} \neq \emptyset$, then $F_{jk} \neq \emptyset$. Since $|F_{jk}| = |F_{ij}| = |F_{ik}| = r$, we see that $F_{ij} = F_{jk} = F_{ik}$. Therefore, we conclude that for any pair of edges F' and F'' in H^2, either $F' = F''$ of $F' \cap F'' = \emptyset$. Then $s = r$ follows by the following obvious proposition.

LEMMA 8. If any pair of edges of H^2 is either equal or disjoint, then $s = r$.

We note further that, if H^2 is γ-perfect, or blanced, or odd-cycle-free, then $s = r$ [2, Chapter 20]. However, since H^2 is, in general, more complicated than H, we would prefer to develop a simpler, though weaker, condition which is imposed on H.

THEOREM 9. If H has no odd cycle, then $s=r$.

PROOF. By the above observation, we need only to show that H^2 has no odd cycles.

Denote an edge $E_i \cap E_j$ in H^2 by F_{ij}. Assume that, to the contrary, there is an odd cycle

$$C' = x_1 F_{i_1 i_2} x_2 F_{i_3 i_4} \ldots x_q F_{i_{2q-1} i_{2q}} x_1$$

in H^2. If $i_1, i_2, \ldots, i_{2q-1}$ are different, then

$$C = x_1 E_{i_1} x_2 E_{i_3} \ldots x_q E_{i_{2q-1}} x_1$$

is an odd cycle in H, contradicting the hypothesis. Otherwise, we may suppose that $i_1, i_2, \ldots, i_{2l}$ are different and $i_{2l+1} = i_1$. If $l=1$, then $x_1 E_{i_2} x_2 E_{i_4} x_3 E_{i_1} x_1$ is an odd cycle in H; if $l>1$, then either

$$C_1 = x_1 E_{i_2} x_2 E_{i_3} x_3 E_{i_5} \ldots x_l E_{i_{2l-1}} x_{l+1} E_i x_1 \quad (l \text{ even})$$

or

$$C_2 = x_2 E_{i_3} x_3 E_{i_5} \ldots x_l E_{i_{2l-1}} x_{l+1} E_{i_1} x_2 \quad (l \text{ odd})$$

is an odd cycle in H, contradicting the hypothesis. The proof is completed.

Combining a condition for $c=s$ with another one for $s=r$, we will get a sufficient condition for (6). In particular, Theorem 4 and Theorem 9 yield the following result which appeared in [8]:

COROLLARY 10. If H has no odd cycles and every cycle in H has two vertices contained in at least two common edges, then (1) holds.

By this corollary, Theorem 1 follows readily.

Acknowledgment.

The author would like to thank Professor C. Berge for his helpful suggestions and Mr. Zhang Cunquan for his kind cooperation.

References

[1] B. Devadas Acharya, M. Las Vergnas, Hypergraphs with cyclomatic number zero, triangulated graphs and an inequality, to appear in *J. Combinatorial Theory*.

[2] C. Berge, *Graphes et Hypergraphes*, Dunod, Paris, 1970.

[3] P. Hansen, M. Las Vergnas, On a property of hypergraphs with no cycles of length greater than two, In *Hypergraph Seminar*, ed. C. Berge, D. K. Ray-Chaudhuri, Springer-Verlag, Berlin, 1974 99-101.

[4] He Qimei, The cyclomatic number of a hypergraph and an inequality of Lovás, (Chinese), to appear in *Acta Mathematicae Applagatae Sinica*.

[5] M. Lewin, On hypergraphs without significant cycles, *J. Combinatorial Theory* B, 20 (1976), 80-83.

[6] L. Lovász, Graphs and set-systems, in *Beitrage zur Graphentheorie*, ed. H. Sachs, H. S. Voss, H. Walther, Teubner, Leipzig, 1968, 99-106.

[7] L. Lovász, *Combinatorial Problems and Exercises*, Problem 13.2, North-Holland, Amsterdam, 1979.

[8] Zhang Cunquan, Li Weixuan, On a property of hypergraphs, submitted to *J. Combinatorial Theory B*.

On k-Critically n-Connected Graphs.

W. Mader

ABSTRACT

This paper gives a survey of results and open problems on k-critically n-connected graphs.

A graph G is called *k-critically n-connected* or *(n,k)-critical* or simply an *(n,k)-graph*, $(n \geq k \geq 1)$, if for all $V' \subseteq V(G)$ with $|V'| \leq k$, we have $\mu(G-V') = n - |V'|$, where $\mu(G)$ denotes the connectivity number of G. (We shall always assume $\mu(G)$ to be finite.) This concept was introduced by *Maurer* and *Slater* in [12], and it generalizes the well known concept of a critically n-connected graph (i.e. $(n,1)$-graph). First let us consider some examples. The complete graph K_{n+1} is an (n,n)-graph, and it is not difficult to see that it is the only finite (n,n)-graph (proposition 2.1 in [12]). Let S_k arise from K_{2k+2} by deleting the edges of a 1-factor of K_{2k+2}. Then S_k is a $(2k,k)$-graph, but it is not $(k+1)$-critical. Let Z_m denote the integers modulo m and N_m the positive integers less or equal to m. The graph $H_m(k) := (Z_m, \{[x, x+\kappa] : x \in Z_m \wedge \kappa \in N_k\})$ is easily recognized as a $(2k,2)$-graph for all integers m and k with $\frac{m}{2} > k \geq 2$. Define the graph $W_m(k)$ by $V(W_m(k)) := \{(i_1, \ldots, i_k) : \bigwedge_{\kappa \in N_k} i_\kappa \in N_m\}$ and $E(W_m(k)) := \{[(i_1, \ldots, i_k), (j_1, \ldots, j_k)] : \bigvee_{\kappa \in N_k} i_\kappa \neq j_\kappa \wedge \bigvee_{\kappa \in N_k} i_\kappa = j_\kappa\}$. It is checked in [9] that $W_m(k)$ is an $(m^k - (m-1)^k - 1, k)$-graph for all $m \geq 2$ and $k \geq 2$, with the only exception $m = k = 2$.

We need some further notations and definitions. For $X \subseteq V(G)$, we define $N(X;G) := \{y \in V(G) - X : \bigvee_{x \in X} [x,y] \in E(G)\}$ and $E(X;G) := \{[x,y] \in E(G) : x \in X \wedge y \in V(G) - X\}$; for a subgraph H and a vertex x of G, we write $E(H;G)$ and $E(x;G)$ instead of $E(V(H);G)$ and $E(\{x\};G)$, respectively. As usual, $\lambda(G)$ denotes the edge-connectivity of G. - Let G be any graph and for every $x \in V(G)$, let H_x be given a graph. We can choose an H_x' isomorphic to H_x such that $V(H_x') \cap V(H_y') = \emptyset$ for all vertices $x \neq y$ of G. Then the

lexicographic product $G[H_x]$ is defined by $V(G[H_x]) := \bigcup_{x \in V(G)} V(H_x')$
and
$E(G[H_x]) := \bigcup_{x \in V(G)} E(H_x') \cup \bigcup_{[x,y] \in E(G)} \{[x',y'] : x' \in V(H_x') \wedge y' \in V(H_y')\}$.
We say, $G[H_x]$ arises from G by replacing $x \in V(G)$ by H_x. If all H_x are isomorphic to H, we write $G[H]$ instead of $G[H_x]$. - A *least separating set* T of G is a separating vertex set T of G with $|T| = \mu(G)$. If T is a least separating set of G, we call any union of at least one, but not all components of $G-T$ a *fragment of* G (relative to T). A fragment of G with the least number of vertices is called an *atom of* G. Of course, every non-complete graph has fragments and at least one atom. The following property of atoms is fundamental.

THEOREM 1 [8]. *Let A be an atom of the finite graph G and let T be a least separating set of G with $T \cap A \neq \emptyset$. Then $A \subsetneq T$ and $|A| \leq \dfrac{\mu(G)}{2}$.*

The commonest (n,k)-graphs are, of course, the $(n,1)$-graphs, which are simply called critically n-connected. They have been first considered by *Chartrand, Kaugars* and *Lick* who proved the following:

THEOREM 2 (Chartrand, Kaugars, Lick [1]). *Every critically n-connected, finite graph $G(n \geq 2)$ contains a vertex z of degree $d(z;G) \leq \dfrac{3n}{2} - 1$.*

Using theorem 1, an easy proof of theorem 2 is given in [8]. - There is a lot of examples which show that, in general, a critically n-connected graph does not contain a vertex of degree less than $\lfloor \dfrac{3n}{2} \rfloor - 1$. For instance, let us consider the graph $C_l(n,m)$ for integers $l \geq 4$ and $1 \leq m \leq \dfrac{n}{2}$ which arises from a circuit of length l by replacing exactly two adjacent vertices by K_m and all other vertices by K_{n-m}. These graphs are critically n-connected for all admissible l,n,m, and for $m = \dfrac{n}{2}$, their minimum degree is equal to $\dfrac{3n}{2} - 1$. But these graphs have more than one vertex of degree $\leq \dfrac{3n}{2} - 1$, and it was proved by Hamidoune in [5] that an $(n,1)$-graph necessarily contains even two vertices of degree $\leq \dfrac{3n}{2} - 1$. In general, for $n \geq 3$, an $(n,1)$-graph does not have more than two vertices of degree $\leq \dfrac{3n}{2} - 1$, as the graphs $C_l(n,1)$ show. But the two vertices of degree $\leq \dfrac{3n}{2} - 1$ in $C_l(n,1)$ have degree equal to n, and it was proven quite recently by *Veldman* in [14] that it is so not by chance. It is possible to generalize these results in the following way.

THEOREM 3 [10]. *Every finite, critically n-connected, non-complete graph G has two (vertex-) disjoint fragments F with* $|F| \le \frac{n}{2}$. *(For all $x \in V(F)$, $d(x;G) \le \frac{3n}{2} - 1$.)*

There are many infinite, critically n-connected graphs (notice that theorem 2 does not hold for them), but it is shown in [9] that every $(n,2)$-graph is finite, or equivalently, that every infinite graph G contains vertices $x \ne y$ with $\mu(G - \{x,y\}) \ge \mu(G) - 1$. It is possible to improve this result:

THEOREM 4 [11]. *If G is an infinite n-connected graph, then there is an infinite subset $X \subseteq V(G)$ such that $\mu(G - X') \ge n-1$ for all $X' \subseteq X$.*

Every $(n,2)$-graph is finite, but there is an infinite number of different $(n,2)$-graphs, for instance, the $(2k,2)$-graphs $H_m(k)$ for all $2 \le k < \frac{m}{2}$. This situation changes, when we consider $(n,3)$-graphs.

THEOREM 5 [9]. *Every $(n,3)$-graph has less than $6n^2$ vertices.*

I don't know if a bound quadratic in n is best possible, but the 3-critical graphs $W_m(3)$ show that the best estimation would be of order $n^{\frac{3}{2}}$, because $\left(\frac{\mu(W_m(3))}{3}\right)^{\frac{3}{2}} < |W_m(3)| < \mu(W_m(3))^{\frac{3}{2}}$. But if we assume $k > \lceil\frac{n}{2}\rceil$, there is a much better bound for the order of an (n,k)-graph.

THEOREM 6 (*Hamidoune* [6]). *If G is an $(n, \lceil\frac{n}{2}\rceil + 1)$-graph, then $|G| \le \frac{3n}{2}$.*

This theorem was a rather successful attack on the following challenging conjecture.

CONJECTURE 1 (*Slater* [12]). *There is no non-complete (n,k)-graph for $k > \frac{n}{2}$.*

This conjecture is settled only for all $n \le 10$ and all even $n \le 18$. It is possible also to show that there is no non-complete (n,k)-graph for $k > \frac{2}{3}n$. For a proof of conjecture 1, it would be enough to prove it for all odd n, because a non-complete $(2k + 2, k+2)$-graph furnishes a non-complete $(2k+1, k+1)$-graph by deleting a vertex. Therefore, the following conjecture, which I proved for $k = 3$, is stronger than Slater's one.

CONJECTURE 1a. *For $k \geq 3$, the only non-complete $(2k,k)$-graph is S_k.*

For $k = 2$, there is an infinite number of $(2k,k)$-graphs as the graphs $H_m(k)$ show. But these graphs $H_m(2)$ are almost all $(4,2)$-graphs.

THEOREM 7 [9]. *The following graphs are all $(4,2)$-graphs: $H_m(2)$ for $m \geq 5$, $W_3(2) = L(K_{3,3})$, and $L(K_{4,4}$ -1-factor).*

($K_{n,n} := K_2[\overline{K}_n]$ is a complete bipartite graph, and $L(G)$ denotes the line graph of G.)

I had announced this theorem in [9] p. 145, l. 17, but I have not yet published the troublesome proof. Now there is no need to give a direct proof of this result, because in the meantime a more general theorem has appeared, from which theorem 7 is easily deduced. For stating this nice result of M. Fontet, we need some further concepts.

A graph G is called *minimally n-connected*, if $\mu(G) = n$, but $\mu(G-e) < n$ for every edge $e \in E(G)$, and it is called *n-contraction-critical*, if it is minimally n-connected, and contracting any edge e of G, we get a graph G_e with $\mu(G_e) < n$. For a non-complete graph G of connectivity n, the last condition is equivalent to saying that for every pair of adjacent vertices x and y, there is a least separating set T of G containing x and y. Therefore, every minimally n-connected $(n,2)$-graph is n-contraction-critical (in particular, the graphs $H_m\left(\frac{n}{2}\right)$ for n even are n-contraction-critical), but not vice versa. Let us call a graph G $(n + \frac{1}{2})$-*edge-connected*, if $\lambda(G) \geq n$ and for every separating edge set $T \subseteq E(G)$ with $|T| = n$, $G-T$ has an isolated vertex. Let M denote the class of all finite, 3-regular, $\frac{7}{2}$-edge-connected graphs. ($\frac{7}{2}$-edge-connected is equivalent to 3-edge-connected and at least 4-cyclically connected.) It is easy to see that for $G \in \mathbf{M}$, $L(G)$ is 4-contraction-critical. As an analogue to *Tutte's* theorem for 3-connected graphs, M. Fontet discovered the nice fact that these graphs $L(G)$ and the graphs $H_m(2)$ are all 4-contraction-critical graphs.

THEOREM 8 (*Fontet* [3] and [4]) *. *The finite, 4-contraction-critical graphs are exactly the graphs $H_m(2)$ for all $m \geq 5$ and the graphs $L(G)$, where G is any 3-regular, $\frac{7}{2}$-edge-connected, finite graph.*

(These two classes of graphs have only the graph $H_6(2) = L(K_4)$ in common.)

For handling the graphs $L(G)$, it is necessary to find a construction for all $G \in \mathbf{M}$. It is well known that every 3-regular, 3-edge-connected, finite graph is constructable from K_4 by successive application of the following operation: choose two different edges k_1 and k_2, subdivide k_i by a new vertex z_i for $i = 1, 2$ ($z_1 \neq z_2$), and add the edge $[z_1, z_2]$. If we apply this operation to a graph $G \in \mathbf{M}$, choosing the edges k_1 and k_2 disjoint, we get again a graph from $\mathbf{M}$, and it is possible to get all elements of $\mathbf{M}$ in this way, starting from K_4 and the 1-dimensional skeleton Q of a 3-dimensional cube.

THEOREM 9 (*Fontet* [3] and [4]) †. *Every finite, 3-regular, $\frac{7}{2}$-edge-connected graph is obtained from K_4 or $Q := K_{4,4}$ -(1-factor) by successive application of the following procedure P: choose any disjoint edges k_1 and k_2, subdivide them by new vertices $z_1 \neq z_2$, respectively, and add the edge $[z_1, z_2]$.*

This theorem was proved in [4] by induction, first verifying it for all graphs with not more than 10 vertices by inspection. We shall give a more elegant proof for theorem 9 at the end of this paper. Now we will show how theorems 8 and 9 imply theorem 7.

First let us remark that the only graph $G \in \mathbf{M}$ which contains a triangle Δ is the graph K_4. ($|E(\Delta; G)| = 3$ as G is 3-regular, hence $|G - V(\Delta)| = 1$ as G is $\frac{7}{2}$-edge-connected.) Let the graph $\overline{G}$ arise from a $G \in \mathbf{M}$ by operation P; $k_1 = [u_1, v_1]$, $k_2 = [u_2, v_2]$, z_1, z_2 being as in theorem 9, $k_i' = [u_i, z_i]$ and $k_i'' = [z_i, v_i]$ may arise from k_i by the subdivision for $i = 1, 2$. Let us assume $L(\overline{G})$ to be (4,2)-critical. We prove that then $G \simeq K_4$. For this we show that the assumption $|G| > 4$ leads to a contradiction.

Then there is an $x_1 \in V(G) - \{u_1, v_1, u_2, v_2\}$. Since G does not contain a triangle, there is an edge $k = [x_1, x_2] \in E(G)$ with $x_2 \neq u_1, v_1, u_2, v_2$. $L(\overline{G})$ being (4,2)-critical, there is a minimal

---

* As I learned at this Waterloo conference from C. *Thomassen*, this result (or a similar one) has been rediscovered recently by *Martinov*.
† *Fontet* stated this theorem in another, but equivalent form.

separation set $T \supseteq \{[x_1,x_2], [z_1,z_2]\}$ of four edges in $\overline{G}$ such that both components of $\overline{G}-T$ contain an edge. For $i = 1,2$, C_i denotes the component of $\overline{G}-T$ containing x_i. $E(C_i) \cap E(x_i;\overline{G}) \neq \emptyset$ since $|C_i| \geq 2$ for $i = 1,2$ and $E(x_1;\overline{G}) \cup E(x_2,\overline{G}) \subseteq E(G)$ by choice of k. Let T' arise from $T - \{[z_1,z_2]\}$ by replacing the edges of $T \cap \{k_1',k_1'',k_2',k_2''\}$ by k_1, k_2, respectively. Then $|T'| \leq 3$ and T' separates G into two components containing edges, contradicting the $\frac{7}{2}$-edge-connectivity of G.

As we have shown now, $L(\overline{G})$ (4,2)-critical implies $G \simeq K_4$ and hence $\overline{G} \simeq K_{3,3}$. By theorem 9, therefore, only graphs $K_4, K_{3,3}$, and Q from M can have a (4,2)-critical line graph, and these graphs do. Hence, by theorem 8, the only minimally 4-connected, 2-critical graphs are the graphs $H_m(2)$ for all $m \geq 5$, $L(K_{3,3})$, and $L(Q)$. On the other hand, it is easily checked that a graph obtained by adding an edge to any of these graphs is not (4,2)-critical. This proves theorem 7.

There are some further conjectures which are stronger than conjecture 1. Let us consider the graph $G := K_{k+1}[H]$, where H consists of two disjoint edges. Then G is a $(4k,k)$-graph of minimum degree $4k+1$. But I should conjecture that every $(4k,k+1)$-graph has a vertex of degree $4k$.

CONJECTURE 2. *Every finite* $(n, \lfloor \frac{n}{4} \rfloor + 1)$*-graph contains a vertex of degree* n.

This conjecture which I verified for all $n \leq 10$ implies conjecture 1 in the following way: if there were a non-complete $(2k+1,k+1)$-graph G, then $G[K_2]$ would be a $(4k+2,k+1)$-graph without vertices of degree $4k+2$. The following conjecture of *Entringer* and *Slater* is obviously stronger yet than conjecture 2.

CONJECTURE 3 (*Entringer* and *Slater* [2]). *If A is an atom of the finite (n,k)-graph G, then* $|A| \leq \frac{n}{2k}$.

For $k = 1$, this upper bound for $|A|$ follows from theorem 1. For $k = 2$, conjecture 3 was proved recently, as we shall see after conjecture 4. For $k > \frac{n}{2}$, it is equivalent to conjecture 1. For $k = \lfloor \frac{n}{2} \rfloor \geq 1$, conjecture 3 is equivalent to the assertion that every finite $(n, \lfloor \frac{n}{2} \rfloor)$-graph contains a vertex of degree n. It is no problem to prove this and it was done in [2]. This result was improved in [7] by showing that every $(n, \lfloor \frac{n}{2} \rfloor)$-graph has at least two vertices of degree

n, but I am sure that this is best possible only for $n = 3$ (cf. also conjecture 1a).

As we have seen in the paragraph before theorem 3, a finite $(n,1)$-graph, in general, does not contain more than two disjoint fragments F with $|F| \le \frac{n}{2}$. But disregarding the order of the fragments, we can always find four (vertex-) disjoint fragments in a finite, non-complete $(n,1)$-graph, which is proved in [10]. The $(n,1)$-graphs $C_l(n,m)$ with $l \ge 4$ and $\frac{n}{2} > m \ge 1$ (cf. the paragraph behind theorem 2) show that there are not more than 4 disjoint fragments, in general. For $n = 2$, our result furnishes a theorem of *Nebeský* [13], saying that every finite, non-complete $(2,1)$-graph has 4 vertices of degree 2, because an (inclusion-) minimal fragment of a $(2,1)$-graph consists obviously only of one vertex. Perhaps, it is possible to generalize these facts to (n,k)-graphs with $k > 1$.

CONJECTURE 4. *Every finite, non-complete (n,k)-graph has a (vertex-) disjoint system of $2k+2$ fragments.*

The $(2mk,k)$-graphs $S_k[K_m]$ show that this would be best possible for every k. We shall prove now that conjecture 4 for $k-1$ implies conjecture 3 for k. Let A be an atom of a finite (n,k)-graph G (hence G non-complete). It is easily deduced from theorem 1 that $G-A$ is a non-complete $(n - |A|, k-1)$-graph. Hence by conjecture 4, there are disjoint fragments $F_1, \ldots, F_{2k}$ in $G-A$. As $\mu(G-A) = \mu(G) - |A|$, F_κ is also a fragment of G and $N(F_\kappa;G) \supseteq A$ for all $\kappa \in N_{2k}$. Hence $V(F_\kappa) \cap N(A;G) \ne \emptyset$ for all $\kappa \in N_{2k}$. If $V(F_\kappa) \subseteq N(A;G)$, then $|F_\kappa| \ge |A|$, because A is an atom and F_κ is a fragment of G. If $V(F_\kappa) - N(A;G) \ne \emptyset$, then $N(F_\kappa;G-A) \cup (V(F_\kappa) \cap N(A;G))$ is a separating set of G, hence $|V(F_\kappa) \cap N(A;G)| \ge |A|$, as $N(F_\kappa;G-A) \cup A$ is least separating set. Hence $n = |N(A;G)| \ge \sum_{\kappa=1}^{2k} |V(F_\kappa) \cap N(A;G)| \ge 2k|A|$, that means $|A| \le \frac{n}{2k}$, as claimed.

Having proved conjecture 4 for $k = 1$ in [10], we know, therefore, that the atoms of $(n,2)$-graphs have at most $\frac{n}{4}$ vertices. By the way, this fact implies immediately conjecture 2 for all $n \le 9$.

Finally, we will mention yet another open question for (n,k)-graphs. It is easy to see that every $(n,3)$-graph has diameter less or equal to 4 (Corollary 4 in [9]). But I don't know any $(n,3)$-graph of diameter 3 or 4. If there are any (which I should suspect), given any $n \ge 6$, what is the least integer $k(n)$ such that every non-complete

$(n,k(n))$-graph has diameter 2 ? The $(2m,2)$-graphs $H_n(m)$ show $k(2m) \geq 3$. Conjecture 2 implies immediately $k(n) \leq \lfloor \frac{n-2}{4} \rfloor + 3$. (From conjecture 1a we know $k(6) = 3$.)

Let us turn now to the proof of theorem 9. First two lemmas. A component C of a graph may be called trivial, if $|C| = 1$. For $k = [x,y] \in E(G)$, $V(k) := \{x,y\}$.

(a) Let G be a 3-regular, 3-edge-connected graph and $T \subseteq E(G)$ be a separating set with $|T| \leq 4$. Then $G-T$ has exactly two components. If C is a non-trivial component of $G-T$ and $|T| = 3$, then $|V(C) \cap \bigcup_{k \in T} V(k)| = 3$.

(b) Let T be a minimal separating set of $G \in \mathbf{M}$ consisting of four edges. If C is a component of $G-T$ with $|V(C) \cap \bigcup_{k \in T} V(k)| \leq 3$, then $|C| = 2$.

We prove only (b). T being minimal separating and G 3-regular, $|C| \geq 2$ holds. If there is a $z \in V(C)$ incident to two edges k_1 and k_2 of T, then $T' := (E(z;G) \cup T) - \{k_1,k_2\}$ is a separating set with $|T'| = 3$. G being $\frac{7}{2}$-edge-connected, a component of $G-T'$ is trivial, that means $C-z$ is trivial, hence $|C| = 2$.

Let us consider now any $G \in \mathbf{M}$ with $|G| \geq 5$ and choose any $k = [x_1,x_2] \in E(G)$. For $i = 1,2$, $E(x_i;G) = \{k,k_i = [x_i,y_i], k_i' = [x_i,y_i']\}$. G having no triangles (cf. the deduction of theorem 7), $\{y_1,y_1'\} \cap \{y_2,y_2'\} = \emptyset$ and $[y_i,y_i'] \notin E(G)$ for $i = 1,2$. Reversing the procedure described in theorem 9, we define $G_k := (G - \{x_1,x_2\}) \cup \{[y_1,y_1'],[y_2,y_2']\}$. Of course, G_k is 3-regular. For $T \subseteq E(G_k)$ we define $P(T) \subseteq E(G) - \{k,k_1',k_2'\}$ by $P(T) - \{k_1,k_2\} = T - \{[y_i,y_i'] : i = 1,2\}$ and for $i = 1,2, k_i \in P(T)$ if and only if $[y_i,y_i'] \in T$. If $T \subseteq E(G_k)$ is a separating set of G_k, then $T' := P(T) \cup \{k\}$ is a separating set of G and every non-trivial component of $G_k - T$ furnishes a non-trivial component of $G-T'$. Hence G_k is 3-edge-connected, as G_k is 3-regular and G is $\frac{7}{2}$-edge-connected. If G_k is not $\frac{7}{2}$-edge-connected, there is a 3-element separating set $T \subseteq E(G_k)$ such that $G_k - T$ has no isolated vertex. By (a), therefore, the edges of T are disjoint, hence the edges of $P(T)$ are disjoint, but then also the edges of $T' := P(T) \cup \{k\}$ are disjoint by (b), as $G-T'$ has two non-trivial components, hence T' is minimal separating by the $\frac{7}{2}$-edge-connectivity of G.

Let us assume now that G is not constructable from any $G^* \in M$ by procedure P described in theorem 9. As we have seen above, then for every $k \in E(G)$, there is a separating edge set T_k containing k consisting of four disjoint edges. Of course, both the components C_k^1 and C_k^2 of $G - T_k$ are non-trivial and T_k is minimal separating. Let be $L := \{C: C$ induced subgraph of G such that $E(C;G)$ consists of four disjoint edges $\}$. Consider any $C \in L$. Then $\lambda(C) \geq 2$, because the edges of $E(C;G)$ are disjoint and G is $\frac{7}{2}$-edge-connected. If $T \subseteq E(C)$ is a separating edge set of C consisting of two disjoint edges and C' is a component of $C - T$, then $|E(C';G)| = 4$, hence $C' \in L$ or $|C'| = 2$ by (b). Choose a minimal element C_o^1 of L and denote $T_o := E(C_o^1;G)$ and $C_o^2 := G - V(C_o^1)$; then also $C_o^2 \in L$. Consider any $k \in E(C_o^1)$. Of course, $C_k^i \cap C_o^1 \neq \emptyset$ for $i = 1,2$ and by minimality of C_o^1, also $C_k^i \cap C_o^2 \neq \emptyset$ for $i = 1,2$. Hence $T_k \cap E(C_o^i)$ separates C_o^i for $i = 1,2$ and, therefore, $|T_k \cap E(C_o^i)| = 2$ for $i = 1,2$. The components of $C_o^i - T_k$ are $C_o^i \cap C_k^1$ and $C_o^i \cap C_k^2$, and from above we get $|E(C_o^i \cap C_k^1;G)| = 4$ and $C_o^i \cap C_k^1 \in L$ or $|C_o^i \cap C_k^1| = 2$. By minimality of C_o^1, therefore, $|C_o^1 \cap C_k^i| = 2$ for $i = 1,2$, say, $V(C_o^1 \cap C_k^i) = \{x_1^i, x_2^i\}$. Then C_o^1 is a quadrangle, say $C_o^1 = (\{x_j^i : i,j \in N_2\}, \{[x_1^1, x_2^1], [x_i^1, x_i^2] : i \in N_2\})$. Let $e_j^i = [x_j^i, y_j^i]$ be the edge of $E(x_j^i;G) \cap T_o$ for $i,j \in N_2$ (see figure 1).

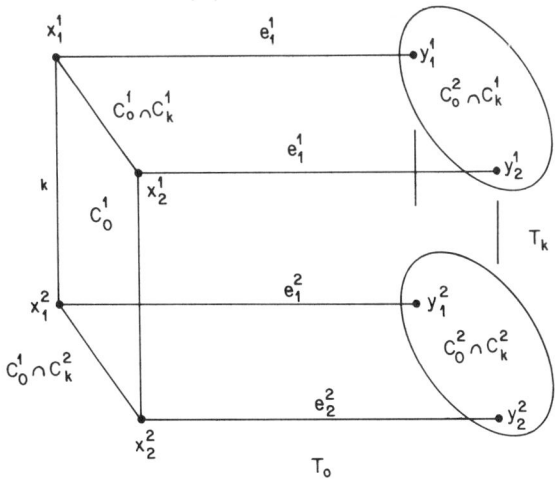

Figure 1.

Let us consider now $T_{k'}, C_{k'}^1, C_{k'}^2$ for $k' := [x_1^1, x_2^1]$; say, $x_i^1 \in C_{k'}^i$ for $i = 1,2$. Using the properties proved for T_k, C_k^1, C_k^2, we see

$[x_i^1, x_i^2] \in E(C_{k'}^i)$, hence $\{y_i^1, y_i^2\} \subseteq V(C_{k'}^i)$, therefore, $|E(C_{k'}^i \cap C_o^2) \cap T_k| = 1$ for $i = 1,2$, say, $E(C_{k'}^i \cap C_o^2) \cap T_k = \{k_i\}$. Correspondingly, we have $|E(C_k^i \cap C_o^2) \cap T_{k'}| = 1$, say, $E(C_k^i \cap C_o^2) \cap T_{k'} = \{k_i'\}$ for $i = 1,2$. Then $E(C_o^2 \cap C_k^i \cap C_{k'}^i; G) = \{e_j^i, k_j, k_i'\}$. G being $\frac{7}{2}$-edge-connected, this implies $V(C_o^2 \cap C_k^i \cap C_{k'}^j) = \{y_j^i\}$ for all $i,j \in N_2$, and G is the 1-dimensional skeleton Q of a 3-dimensional cube.

References

[1] G. Chartrand, A. Kaugars and D.R. Lick, Critically n-connected graphs, Proc. Amer. Math. Soc. 32(1972), 63-68.

[2] R.C. Entringer and P.J. Slater, A note on k-critically n-connected graphs, Proc. Amer. Math. Soc. 66(1977), 372-375.

[3] M. Fontet, Graphes 4-essentiels, C.R. Acad. Sc. Paris, t.287(1978), Série A 289-290.

[4] M. Fontet, Connectivité des graphes automorphismes des cartes: propriétés et algorithmes, Thèse d'Etat, Université P. et M. Curie, Paris 1979.

[5] Y.O. Hamidoune, On critically h-connected simple graphs, Discrete Math. 32(1980), 257-262.

[6] Y.O. Hamidoune, On multiply critically h-connected graphs, J. Combinatorial Theory (B) 30(1981), 108-112.

[7] Y.O. Hamidoune, On a conjecture of Entringer and Slater, Discrete Math. 41(1982), 323-326.

[8] W. Mader, Eine Eigenschaft der Atome endlicher Graphen, Arch. Math. 22(1971), 333-336.

[9] W. Mader, Endlichkeitssätze für k-kritische Graphen, Math. Ann. 229(1977), 143-153.

[10] W. Mader, Disjunkte Fragmente in kritisch n-fach zusammenhängenden Graphen, submitted to European Journal of Combinatorics.

[11] W. Mader, On infinite n-connected graphs, submitted to Combinatorica.

[12] St. B. Maurer and P.J. Slater, On k-critical, n-connected graphs, Discrete Math. 20(1977), 255-262.

[13] L. Nebeský, On induced subgraphs of a block, J. Graph Theory 1(1977), 69-74.

[14] H.J. Veldman, A necessary condition for a graph to be critically h-connected, submitted to Monatshefte für Mathematik.

Graph Width and Well-Quasi-Ordering: a Survey

Neil Robertson
*P. D. Seymour**

ABSTRACT

The concept of the tree-width of a graph appears to be new, and has some interesting applications. We have used it to show, for example, that for any infinite set of planar graphs, some one must be isomorphic to a minor of another, and that for any fixed forest H, there is a good algorithm to test if an arbitrary graph has a minor isomorphic to H. In this paper we survey our methods and results. Full details will be given elsewhere.

1. Introduction

We are concerned here with problems about minors of graphs. (Throughout, all graphs are finite, but may have loops or multiple edges.) A graph H is a *minor* of a graph G if H can be obtained from G by the use of vertex-deletion, edge-deletion and edge-contraction. Nowhere in this paper will it be necessary to distinguish between isomorphic graphs, and so we shall (loosely) regard isomorphic graphs as the same. Thus we shall speak of graphs having K_4 as a minor rather than having a minor isomorphic to K_4, etc.

The "minor" relation should be distinguished from the topological minor relation, commonly used in the theory of well-quasi-ordering. We say that H is a *topological minor* of G if there is a subgraph of G which is isomorphic to a subdivision of H (where a *subdivision* of H means a graph obtained from H by repeatedly replacing edges by pairs of edges in series). If H is a topological minor in G then it is a minor of G, but the converse is not always true; for example, the Petersen

---

*Research partially supported by NSF grant MCS 8103440

graph has K_5 as a minor, but not as a topological minor.

Probably the most well-known theorem about graph minors is the Kuratowski-Wagner theorem.

(1.1) *G is planar if and only if it does not have K_5 or $K_{3,3}$ as a minor.*

Actually, this is usually formulated in terms of topological minors as

(1.2) *G is planar if and only if it does not have K_5 or $K_{3,3}$ as a topological minor.*

But the form (1.1) is more appropriate to this paper. It can be restated as

(1.3) *The minor-minimal graphs which are not planar are K_5 and $K_{3,3}$.*

[A graph is *minor-minimal* with a property if it has the property but no other minor of it does.]

This has an analogue for the projective plane, due to Archdeacon, Glover, Huneke and Wang [1,3].

(1.4) *There are precisely 35 minor-minimal graphs not embeddable in the projective plane.*

(Again, this theorem is more usually formulated using topological minors where there are 103 minimal graphs). It has been a popular conjecture that the corresponding list for any surface should be finite: and indeed, Archdeacon and Huneke [2] have recently shown

(1.5) *For any non-orientable surface S, the list of minor-minimal graphs not embeddable in S is finite.*

But there is a much more general conjecture, probably due to K. Wagner (unpublished), which can be expressed as

(1.6) (Conjecture) *For any property of graphs, the list of minor-minimal graphs without the property is finite.*

This form of Wagner's conjecture exhibits the connection with (1.5); but it is more naturally stated as

(1.7) (Conjecture) *If C is a set of graphs such that no member of C is a minor of another, then C must be finite.*

The equivalence with (1.6) is easy. This generalizes a well-known conjecture of Vazsonyi (see [4], conjecture 2), the following.

(1.8) (Conjecture) *If C is a set of graphs with all vertex valencies at most 3, and no member of C is a topological minor of another, then C must be finite.*

(1.7) implies (1.8), because if H has no vertex with valency > 3, then H is a minor of G if and only if H is a topological minor of G. The reason why the valency restriction is necessary in (1.8) is that otherwise there are counterexamples; for example, if C_n^2 denotes the result of replacing each edge of the n-vertex circuit by two parallel edges, then the set

$$\left\{C_3^2, C_4^2, C_5^2, \ldots\right\}$$

is infinite, but no member of it is a topological minor of another.

A quasi-ordered set is a *well-quasi-order* if it has no infinite antichain and no infinite descending chain. Wagner's conjecture may be viewed as asserting that the "minor" quasi-order of graphs is a well-quasi-order. The theory of well-quasi-orderings is an established branch of order theory. But we do not discuss the general theory of well-quasi-ordering here, although our proofs in [10] do use some of its results.

Let us say that a set of graphs is an *antichain* if no member of it is a minor of another (thus, it is an antichain in the "minor" order of the set of all graphs). Wagner's conjecture is

(1.9) (Conjecture) *Every antichain of graphs is finite.*

Some partial results towards this are known. For example, Kruskal [4] proved

(1.10) *Every antichain of trees is finite.*
(Indeed, Kruskal proved a stronger form of this, using topological minors.) Mader [5] proved

(1.11) *Every antichain of graphs with no $K_5 - e$ minor is finite* (where $K_5 - e$ is K_5 with one edge missing) and

(1.12) *Every antichain of graphs with a bounded number of vertex-disjoint circuits is finite.*

We have a common generalization of these results, using the concept of the tree-width of a graph (which we shall define later), the following.

(1.13) *Every antichain of graphs of bounded tree-width is finite.*

It is easy to show that this implies the three earlier results. Furthermore, we can use (1.13) and some structure theorems about tree-width to deduce

(1.14) *Every antichain of planar graphs is finite*
and

(1.15) *Every antichain containing a forest is finite.*

These results are one reason for the introduction of the new invariant "tree-width." In the next section we describe another.

2. Minor Algorithms

Given two arbitrary graphs G and H, the problem "is H a minor of G?" is easily seen to be NP-complete. However, for fixed H, the problem appears to become much more tractable. For example, for any simple graph H with at most 5 vertices, there is a good algorithm to test if G has H as a minor ("good" meaning, as usual, with running time bounded above by a polynomial in the size G). We conjecture that this is always the case.

(2.1) (Conjecture) *For any fixed graph H, there is a good algorithm to test if an arbitrary graph G has H as a minor.*

Our results on tree-width yield the following.

(2.2) *(2.1) is true if G is restricted to be planar*

and

(2.3) *(2.1) is true if H is a forest.*

These results are both consequences of our structure theorems about tree-width ((3.8) and (3.9) below) and the following.

(2.4) *For any fixed graph H and integer k, there is a good algorithm which when applied to an arbitrary graph G, outputs either*

    (i) *that G has tree-width $> k$ (and hence path-width $> k$)*

    (ii) *that H is a minor of G*

or   (iii) *that H is not a minor of G.*

3. Tree-width and Path-width

We say a graph G has *tree-width* $\leq k$ if there is a tree T and vertex-disjoint graphs $G_v (v \in V(T))$, such that each G_v has at most $k + 1$ vertices, and such that by suitably identifying pairwise some vertices of G_u with some of G_v, for each adjacent pair of vertices u, v of T, we can obtain G. G has *path-width* $\leq k$ if the tree T is a path.

The following results are easy to prove.

(3.1) *If H is a minor of G then the tree-width of H is not greater than the tree-width of G. (A similar result holds for path-width.)*

(3.2) *The complete graph K_n has tree-width and path-width both $n - 1$.*

(3.3) *If G is simple and connected, then it has tree-width ≤ 1 if and only if it has no K_3 minor, that is, it is a tree.*

(3.4) *If G is 2-connected, then it has tree-width ≤ 2 if and only if it has no K_4 minor, that is, it is a series-parallel graph.*

(This is easy to prove by induction; or one can use the cleavage unit tree structure for non-separable graphs, described for example in [12]. Series-parallel graphs are the non-separable graphs in which every cleavage unit is a "polygon" or "bond".)

(3.5) *If G is simple and connected, then it has path width ≤ 1 if and only if it is a caterpillar.*

(3.6) *The binary tree of depth n (that is, n + 1 generations, each father having two sons) has path-width $\frac{\lceil n \rceil}{2}$.*

(The fact that this tree has path-width $\leq \frac{\lceil n \rceil}{2}$ may be proved by induction on n, if the following inductive statement is used: it has a path-decomposition of width $\leq \frac{\lceil n \rceil}{2}$, and if n is odd it had such a decomposition in which every subgraph used contains the "root.")

(3.7) *The $n \times n$ grid (the graph of the chessboard with n^2 squares) has tree-width n.*

Our two structure theorems for width are as follows.

(3.8) *For any planar graph H there is a number k such that every planar graph G without H as a minor has tree-width $\leq k$.*

(3.9) *For any forest H, there is a number k such that every graph without H as a minor has path-width $\leq k$ (and hence tree-width $\leq k$).*

The number k produced by out proof of (3.9) is huge, involving an iterated exponential, $|V(H)|$ exponentiated by itself $3|V(H)|$ times. But the fact that such a number exists at all is sufficient for our purposes; we do not need it to be small.

It is interesting that in another sense, (3.9) *is* sharp. For if H is not a forest, there are graphs with arbitrarily large path-width not containing H as a minor (for example, the binary trees), and so the conclusion of (3.9) does not hold for any H except forests. (3.8) is sharp in the same way; for the conclusion of (3.8) is false for every nonplanar H, since there are planar graphs with arbitrarily high tree-width, such as the grids. We conjecture the following strengthening of (3.8).

(3.10) (Conjecture) *For any planar graph H there is a number k such that every graph without H as a minor has tree-width $\leq k$.*

Added in proof, July 1983: we have now proved conjecture (3.10). The proof will appear in [11].

4. Applications of the Structure Theorems

In this section we explain how (1.14), (1.15), (2.2) and (2.3) are derived from (1.13), (2.4), (3.8) and (3.9).

PROOFS OF (1.14), (1.15) These are similar. If C is an antichain of planar graphs, choose $H \in C$ and then no other graph in C has H as a minor. By (3.8), the graphs in $C - \{H\}$ have bounded tree-width, and

$C - \{H\}$ is an antichain; and so by (1.13), $C - \{H\}$ is finite. This proves (1.14). To prove (1.15), we suppose $H \in C$ is a forest, and argue similarly using (3.9) and (1.13).

PROOFS OF (2.2), (2.3) Again, these are similar. To prove (2.2), let H be a fixed planar graph. Let k be as in (3.8). Given an arbitrary planar graph G, we apply (2.4) to G, H, k. Each of the three possible outcomes determines whether H is a minor of G. For (2.3) we argue similarly, using (2.4) and (3.9).

There is another pleasing application of these results, as follows.

(4.1) *For any integer k, there is a good algorithm to test if an arbitrary graph G has tree-width $\leq k$.*

PROOF. There are only finitely many minor-minimal graphs with tree-width $> k$, by (1.13), since they form an antichain and they all have tree-width $k + 1$. Let them be $H_1, \ldots, H_n$ say. Given an arbitrary graph G, we apply (2.4) with $H = H_i$ for $i = 1, 2, \ldots, n$ in turn. If for some i, the application of (2.4) yields that G has tree-width $> k$, we are done. Otherwise, we decide either that G contains one of $H_1, \ldots, H_n$ as a minor, or that it does not. But G has tree-width $> k$ if and only if it contains one of $H_1, \ldots, H_n$ as a minor (only "if" is clear; "if" follows from (3.1)), and so again we are done.

There is a similar result for path-width.

5. Conclusion

We do not describe here the proofs of (1.13), (2.4), (3.8) and (3.9), because they are all rather technical; the reader is referred to [10], [8], [9] and [7] respectively. Suffice it to say that to prove (1.13), we observe that graphs of tree-width $\leq k$ are "tree-shaped" and we are able to adapt Nash-Williams' proof [6] of Kruskal's theorem to apply to "tree-shaped graphs"; and to prove (2.4) we use the fact that if G has tree-width $\leq k$, it must have a vertex cutset of size $\leq k + 1$ dividing G into two parts each containing at most $\frac{2}{3}$ of the vertices, and a "divide and conquer" method can be used.

Finally, there clearly remain a number of open problems. The most fundamental of these is, what can be said about the structure of the graphs not containing a fixed graph as a minor? If we could find an extension of (3.8), (3.9) and (3.10) to answer this, we could hope to use it to solve (1.7) and (2.1) in general. We propose the following. [G is the *sum* of G_1 and G_2 if it can be obtained from G_1 and G_2 by choosing a clique from each (of the same size), deleting the edges of the cliques, and pairwise identifying the vertices of one clique with those of the other. The *completion* of a set of graphs is the smallest set including it, closed under sums. A *k-vertex extension* of a graph G is a graph H such that there exist k vertices of H the deletion of which yields G.]

(5.1) (Conjecture) *For any graph H there is a number k and a surface S such that every graph not containing H as a minor is in the completion of k-vertex extensions of graphs embeddable in S.*

If this could be proved, we would have a chance of proving (1.7); for

(a) it may be possible to adapt the proof of (1.14) to prove that for any fixed surface S, any antichain of graphs embeddable in S must be finite;

(b) there are standard vertex-labelling methods which may be adequate to handle k-vertex extensions, and

(c) the method of proof of (1.13) may extend to deal with completions in general (rather than just graphs of tree-width $\leq k$, which are the graphs in the completion of the set of graphs with at most $k + 1$ vertices).

But evidently there is still a long way to go.

Acknowledgement

We would like to thank J.B. Kruskal for his helpful comments on the presentation of this paper.

References

[1] D. Archdeacon, *A Kuratowski Theorem for the Projective Plane*. Thesis, The Ohio State University (1980).

[2] D. Archdeacon and P. Huneke, "On irreducible graphs for non-orientable surfaces", preprint (1980).

[3] H. Glover, P. Huneke and C.S. Wang, "103 graphs that are irreducible for the projective plane," *J. Combinatorial Theory* (B) 27 (1979), 332-370.

[4] J. Kruskal, "Well-quasi-ordering, the tree theorem, and Vazsonyi's conjecture," *Trans. Amer. Math. Soc.* 95 (1960), 210-225.

[5] W. Mader, "Wohlquasiordnete Klassen endlicher Graphen," *J. Combinatorial Theory* (B), 12 (1972), 105-122.

[6] C. St.J.A. Nash-Williams, "On well-quasi-ordering finite trees," *Proc. Comb. Phil. Soc.* 59 (1963), 833-835.

[7] Neil Robertson and P.D. Seymour, "Graph minors I: Excluding a forest," *J. Combinatorial Theory* (B), to appear.

[8] Neil Robertson and P.D. Seymour, "Graph minors II: Algorithmic aspects of tree-width," submitted.

[9] Neil Robertson and P.D. Seymour, "Graph minors III: Planar tree-width,"

submitted.

[10] Neil Robertson and P.D. Seymour, "Graph minors IV: Tree-width and well-quasi-ordering," submitted.

[11] Neil Robertson and P.D. Seymour "Graph minors V: excluding a planar graph," submitted.

[12] W.T. Tutte, *Connectivity in Graphs*. Mathematical Expositions, No. 15 (Univ. Toronto Press, Toronto, Ontario; Oxford Univ. Press, London, 1966), chapter 11.

Cographic Subspaces of Cocycle Spaces For Line Graphs of Complete Graphs

Ching-Hsien Shih

ABSTRACT

A vector space over $GF(2)$ is called cographic if it is the cocycle space of some graph. The codimension-k cographic subspaces of a $k+2$-connected cographic space are known, and all the special cases are derived from the cocycle space for the line graph of the complete graph with four vertices. In this paper we classify the cographic subspaces of cocycle spaces for line graphs of complete graphs.

1. Introduction

Let G be a graph and $K(G)$ be the cocyle space of G. In order to study the cographic subspaces of $K(G)$, we need to find the relations between graphs G and H such that $E(H)=E(G)$ and $K(H)$ is a subspace of $K(G)$. Theorem 1 in [S] gives these relations when $K(H)$ is a codimension-1 subspace of $K(G)$. In this paper Theorem 5 gives these relations when $K(H)$ is a codimension-k subspace of $K(G)$ ($k \geq 2$) and G is $(k+2)$-connected. We conjecture that there is a similar theorem when G is $(k+1)$-connected and $k \geq 5$. We found that all exceptional cases of theorem 5 are derived from $L(K_4)$ and all those of the conjecture are derived from $L(K_5)$. That motivated the study of the cographic subspaces of $K(L(K_n))$.

MAIN THEOREM. Let H be a graph such that $E(H)=E(L(K_n))$ and $K(H)$ is a proper subspace of $K(L(K_n))$, then either

(i) there exists a graph G', obtained from $L(K_n)$ by identifying two distinct vertices, such that $K(H) \subseteq K(G')$, or

(ii) $H=(n-2)K_n$.

2. Definitions and known theorems

We assume that G is a connected graph in this paper. Let S be a subset of $V(G)$.

DEFINITION 1. $\delta_G(S) = \{e : e \in E(G), e$ has one end in S and one end not in $S\}$.

DEFINITION 2. The cocyle space of G is $\mathbf{K}(G) = \{\delta_G(S) : S \subseteq V(G)\}$.

DEFINITION 3. The cycle space of G is $\mathbf{C}(G) = \{E(H) : H$ is a subgraph of G, all vertices of which have even degree $\}$.

An element of a cycle space is called a cycle and an element of a cocycle space is called a cocycle. The power set $\mathbf{P}(E(G))$ can be considered as a vector space over $GF(2)$, where the addition is symmetric difference. A subspace of $\mathbf{P}(E(G))$ is called graphic (cographic), if it is the cycle space (cocycle space) of some graph. For $a, b \in \mathbf{P}(E(G))$, we say that a is orthogonal to b if $|a \cap b| \equiv 0 \pmod{2}$. For a subspace S of $\mathbf{P}(E(G))$, let $S^\perp$ denote the collection of elements of $\mathbf{P}(E(G))$ which are orthogonal to every element of S. We note that:

THEOREM 1. $\mathbf{K}(G)$ and $\mathbf{C}(G)$ are subspaces of $\mathbf{P}(E(G))$ and $K(G)^\perp = \mathbf{C}(G)$. Furthermore, $\dim(\mathbf{K}(G)) = |V(G)| - 1$. (see [H]).

Let K_n be the complete graph on n vertices, $L(K_n)$ be the line graph of K_n, and mK_n be the graph formed by replacing each edge of K_n with exactly m edges in parallel. The following two theorems given without proof state basic facts about K_n and $L(K_n)$ used later.

THEOREM 2. If $\mathbf{K}(H) \subseteq \mathbf{K}(K_n)$ and $E(H) = E(K_n)$ for $n \geq 5$ then either

(i) there exists a graph G', obtained from K_n by identifying two distinct vertices, such that $\mathbf{K}(H) \subseteq \mathbf{K}(G')$, or

(ii) $H = K_n$. (See [J]).

THEOREM 3. $L(K_n)$ is $2(n-2)$-connected and $|V(L(K_n))| = \frac{1}{2}n(n-1)$.

The following somewhat technical theorems are proved in [S] and give the motivation of the paper. We assume G has no loops or multiple edges and $E(G) = E(H)$.

THEOREM 4. If G is $(k+3)$-connected and $\mathbf{K}(H)$ is a codimension-k subspace of $\mathbf{K}(G)$ then H can be obtained from G by identifying vertices.

If G is a graph and X is a set disjoint from $V(G)$ then $G + X$ denotes the graph obtained from G by adding the elements of X as vertices and joining every vertex of G to every element of X by a single edge. As usual, when $X=\{x\}$ is a singleton, we write $G+X=G+x$.

THEOREM 5. If G is $(k+2)$-connected, $\mathbf{K}(H)$ is a codimension-k subspace of $\mathbf{K}(G)$ and $k \geq 2$ then one of the following holds.

(i) There is a graph G', obtained from G by identifying two distinct vertices, such that $\mathbf{K}(H) \subsetneq \mathbf{K}(G')$.

(ii) There exists disjoint graph G_1, G_2 and distinct vertices $x_1, x_2, \ldots, x_{k+1}, x_{k+2} \in V(G_1)$, $y_1, y_2, \ldots, y_{k+2} \in V(G_2)$ such that one of the following holds.

(a) G is obtained from G_1, G_2 and $k+1$ edges $e_1, e_2, \ldots, e_{k+1}$ by identifying one end of e_i with x_i, the other end with y_i, for $i=1,2,\ldots,k+1$, and identifying x_{k+2} with y_{k+2}. H is obtained from G_1, G_2 and $e_1, e_2, \ldots, e_{k+1}$ by identifying one end of e_i with x_{k+2}, the other end with y_{k+2}, and identifying x_i with y_i, for $i = 1,2,\ldots,k+1$. (See Figure 1.)

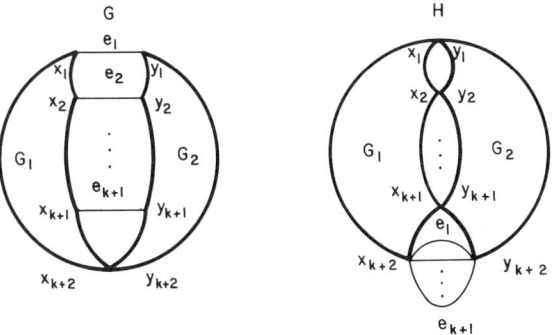

Figure 1

(b) G is obtained from G_1, G_2 and k edges $e_1, e_2, \ldots, e_k$ by identifying one end of e_i with x_i, the other end with y_i, for $i=1,2,\ldots,k$, and identifying x_{k+1} with y_{k+1} and x_{k+2} with y_{k+2}. H is obtained from G_1, G_2 and $e_1, e_2, \ldots, e_k$ by identifying one end of e_i with x_{k+1} and y_{k+2}, the other end of e_i with x_{k+2} and y_{k+1}, and identifying x_i with y_i for $i=1,2,\ldots,k$. (See Figure 2.)

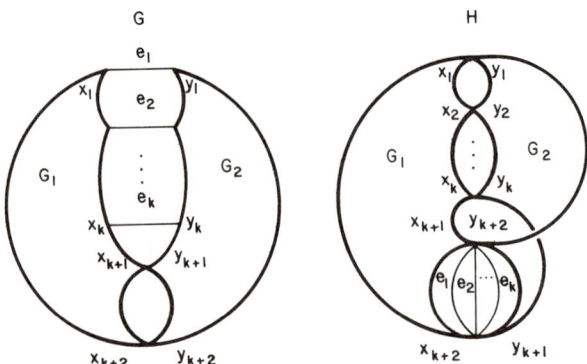

Figure 2

(iii) $k=2$, $G=L(K_4)$, $H=2K_4$ and either

(a) the edge with ends $\{i,j\}$, $\{j,k\}$ in G has ends i,k in H for $i,j,k \in \{1,2,3,4\}$ or

(b) the edge with ends $\{i,j\}$, $\{j,k\}$ in G has ends i,k in H for $\{i,j\} \neq \{1,2\} \neq \{j,k\}$, the edge with ends $\{1,2\}$, $\{1,k\}$ in G has ends $1,k$ in H, and the edge with ends $\{1,2\}$, $\{2,k\}$ in G has ends $2,k$ in H for $k \in \{3,4\}$.

In (b) we may add an edge with ends $\{1,2\}$, $\{3,4\}$ in G and ends 3,4 in H.

(iv) $k=3$, $G=L(K_4)+t$, $H=3K_4$ and the ends in H of edges of G not incident to t are defined as in (iii) (a) and the edge with ends t, $\{i,j\}$ in G has ends i,j in H for $i,j \in \{1,2,3,4\}$.

(v) $k=4$, $G=L(K_4)+\{t,t'\}$, $H=4K_4$ and the end in H of edges of G not incident to t' are defined as in (iv) and the edge with ends t', $\{i,j\}$ in G has ends h, m in H for $i,j \in \{1,2,3,4\}$. Here $\{h,m\}=\{1,2,3,4\}-\{i,j\}$.

CONJECTURE 1. If G is $(k+1)$-connected, $K(H)$ is a codimension-k subspace of $K(G)$ and $k \geq 5$ then either one of (i), (ii) in Theorem 5 is true (in (ii) we may delete e_1) or one of the following holds.

(iii) $k=5$, $G=L(K_5)$, $H=3K_5$ and either

(a) the edge with ends $\{i,j\}$, $\{j,k\}$ in G has ends i,k in H for $i,j,k \in \{1,2,3,4,5\}$, or

(b) the edge with ends $\{i,j\}$, $\{j,k\}$ in G has ends i,k in H for $\{i,j\} \neq \{1,2\} \neq \{j,k\}$, the edge with ends $\{1,2\}$, $\{1,k\}$ in G has ends $1,k$ in H, and the edge with ends $\{1,2\}$, $\{2,k\}$ in G has ends $2,k$ in H for $k \in \{3,4,5\}$.

In (b) we may add an edge with ends $\{1,2,\}$, $\{i,j\}$ in G and ends i,j in H for every $i,j \in \{3,4,5\}$. We may delete the edge with ends $\{i,k\}$, $\{j,k\}$ in G, if we have added two edges with ends $\{1,2,\}$, $\{i,k\}$, and $\{1,2\}$, $\{j,k\}$ in G for some $\{i,j,k\}=\{3,4,5\}$.

(iv) $k=6$, $G=L(K_5)+t$, $H=4K_5$ and the ends in H of edges of G not incident to t are defined as in (iii) (a) and the edge with ends t, $\{i,j\}$ in G has ends i,j in H for $i,j \in \{1,2,3,4,5\}$.

In Theorem 5 and Conjecture 1, alternative (i) and (ii) are the general cases, the exceptional cases of Theorem 5 are derived from $L(K_4)$, and the exceptional cases of Conjecture 1 are derived from $L(K_5)$.

3. Lemmas

In order to find the cographic subspaces of $\mathbf{K}(L(K_n))$ for $n \geq 4$ we need the following lemmas.

LEMMA 1. If $\mathbf{K}(H) \subseteq \mathbf{K}(G)$ and G' is the graph obtained from G by identifying vertices x and y, then there exists a path with ends x,y in G whose edge set is a cycle in H if and only if $\mathbf{K}(H) \subseteq \mathbf{K}(G')$.

PROOF. If $\mathbf{K}(H) \subseteq \mathbf{K}(G')$, then by Theorem 1, $C(G') \subseteq C(H)$. The edge set of any path with ends x,y in G is certainly a cycle in G' hence a cycle in H.

Conversely, assume that there exists a path P in G with ends x,y whose edge set is a cycle in H. It is easy to see that $C(G) \subseteq C(G')$. By Theorem 1, the codimension is one. So $C(G')$ is generated by $C(G)$ and $E(P)$. By Theorem 1, $C(G) \subseteq C(H)$. Since $E(P) \in C(H)$, $C(G') \subseteq C(H)$. By Theorem 1 again, we have $\mathbf{K}(H) \subseteq \mathbf{K}(G')$.

REMARK. It follows that the existence of a path with ends x and y in G whose edge set is a cycle in H implies that the edge set of any path with ends x and y in G is a cycle in H. An interesting special case is when a non-loop-edge of G becomes a loop of H.

LEMMA 2. If $\mathbf{K}(H) \subseteq \mathbf{K}(G)$ then either

(i) there exists a graph G', obtained from G by identifying two vertices, such that $\mathbf{K}(H) \subseteq \mathbf{K}(G')$, or

(ii) the edge set of every triangle in G is the edge set of a triangle in H.

REMARK. We note that (i) and (ii) can occur simultaneously.

PROOF. Since the edge set of a triangle in G is a cycle in G it is also a cycle in H. If this cycle is not a triangle, it contains a loop. Then (i) is true by Lemma 1.

For two graphs G,H, with $E(G)=E(H)$, we call $A \subsetneq V(G)$ *faithful* if it can be identified with a subset of $V(H)$ such that the following property holds: if both ends x,y of the edge e in G are in A then e has also ends x,y in H. Let $G|S$ denote the induced graph of G on a set S. Here S is a subset of $V(G)$ or $E(G)$ depending on the context.

LEMMA 3. If $K(H) \subseteq K(G)$, $A \subsetneq V(G)$, A is faithful and $G|A$ is connected, then either

  (i) there exists a graph G', obtained from G by identifying two vertices, such that $K(H) \subseteq K(G')$, or

  (ii) for any $v \in V(G)-A$ such that v is joined to at least four vertices of A in G, $A \cup \{v\}$ is faithful.

PROOF. Assume (i) does not hold, and let $a_1, a_2,\ldots,a_n$ be all the vertices of A joined to v by edges, say $e_1, e_2,\ldots,e_n$ respectively. Consider a cycle in G formed by e_1, e_2 and a path in $G|A$ joining a_1 and a_2. Since the path is still a path in H joining a_1 and a_2 and since $C(G) \subseteq C(H)$, $\{e_1,e_2\}$ is a path in H with ends a_1,a_2. Since (i) does not hold, e_1 and e_2 are not loops in H. Thus $\{e_1,e_2\}$ is a simple path in H with ends a_1,a_2. It follows that e_1 has exactly one end in $\{a_1,a_2\}$. Similarly, e_1 has exactly one end in $\{a_1,a_i\}$ for $i=3,\ldots,n$. Since $n \geq 4$ this implies that a_1 is an end of e_1 in H. Let b be the other end of e_1 in H. If $b \in A$, let P be a path in $G|A$ joining a_1 and b. P is still a path in H joining a_1 and b. Thus P together with e_1 forms a path in G joining v and b, and a cycle in H. By Lemma 1, this would imply (i): contradiction. Thus $b \notin A$, e_1 has ends a_1,b, and e_2 has ends a_2,b. Similarly, the ends of e_i are a_i and b for $i=2,3,\ldots,n$. This implies that we may identify v with b and $A \cup \{v\}$ is faithful.

4. The Main Theorem.

MAIN THEOREM. Let $G=L(K_n)$. If $K(H) \subseteq_+ K(G)$ for $n \geq 4$ and H is connected then either

  (i) there exists G', obtained from G by identifying two distinct vertices, such that $K(H) \subseteq K(G')$, or

  (ii) $H=(n-2)K_n$ is such that either

    (a) the edge with ends $\{i,j\}$, $\{j,k\}$ in G has ends i,k in H

for $i,j,k \in \{1,2,\ldots,n\}$, or

(b) the edge with ends $\{i,j\}$, $\{j,k\}$ in G has ends i,k in H for $\{i,j\} \neq \{1,2\} \neq \{j,k\}$, the edge with ends $\{1,2\}$, $\{1,k\}$ in G has ends $1,k$ in H, and the edge with ends $\{1,2\}$, $\{2,k\}$ in G has ends $2,k$ in H for $k=3,4,\ldots,n$.

PROOF. Assume (i) does not hold. Then by Lemma 1, a path in $L(K_n)$ cannot be a cycle in H.

Case 1. ($n=4$). By Theorem 3, $L(K_4)$ is 4-connected with six vertices. If the codimension is less than 3 by applying Theorem 4 or Theorem 5 we will have the conclusion. If the codimension is greater than 3 then Theorem 1 implies that $|V(H)| \leq 2$. Applying Lemma 2, by considering any triangle in $L(K_4)$, we have a contradiction. Therefore the codimension is 3. This implies that $|V(H)|=3$. Let $e_1, e_2, \ldots, e_5$ be edges of G having $\{1,2\}$ and $\{1,3\}$, $\{1,2\}$ and $\{1,4\}$, $\{1,3\}$ and $\{1,4\}$, $\{3,4\}$ and $\{1,4\}$, $\{2,4\}$ and $\{1,4\}$ as their ends respectively. Then e_1, e_2 and e_3 form a triangle in G. By Lemma 2, they form a triangle in H. By Lemma 1, e_4 cannot have the same ends in H as e_2 or e_3. But $|V(H)|=3$ implies that e_4 has the same ends as e_1. Similarly, e_5 has the same ends as e_1. This implies that e_4 and e_5 have the same ends, which contradicts Lemma 1

Case 2. ($n=5$). By Theorem 3, $L(K_5)$ is 6-connected and $|V(L(K_5))|=10$. By Theorem 4 and Theorem 5, either the conclusion is true or the codimension is greater than 4. Therefore $|V(H)| \leq 5$. Let G_1 be the $L(K_4)$ subgraph of $L(K_5)$ induced by $\{\{i,j\} : i,j \in \{1,2,3,4\}\}$. Let H_1 be the subgraph of H induced by $E(G_1)$. Then

$$\mathbf{K}(H_1) = \{\theta \cap E(G_1) : \theta \in \mathbf{K}(H)\} \subseteq \{\theta \cap E(G_1) : \theta \in \mathbf{K}(G)\} = \mathbf{K}(G_1).$$

Since (i) does not hold, by Lemma 1, no path in G_1 is a cycle of H_1. Again by Lemma 1, there does not exist G_1', obtained from G_1 by identifying two distinct vertices, such that $\mathbf{K}(H_1) \subseteq \mathbf{K}(G_1')$. Applying the discussion of case 1, we see that (ii) is true for H_1 and G_1. Thus, $|V(H)| \geq |V(H_1)| = 4$ so that $|V(H)|=4$ or $|V(H)|=5$ and $V(H_1) = \{1,2,3,4\}$.

Let e_1, e_2, e_3 be edges with ends $\{5,4\}$ and $\{3,4\}$, $\{5,3\}$ and $\{3,4\}$, $\{2,4\}$, and $\{3,4\}$. Since H_1 and G_1 satisfy (ii), e_3 has ends 2 and 3 in H. Since e_1 and e_3 belong to a triangle of $L(K_5)$, Lemma 2 implies that exactly one of 2 and 3 is an end of e_1 in H. Similarly, by considering the triangle defined by $\{5,4\}$, $\{3,4\}$, and $\{1,4\}$, exactly one of 1 and 3 is an end of e_1 in H. Similarly, exactly one of 2 and 4 is an end of e_2 in H, and exactly one of 1 and 4 is an end of e_2 in H.

Case 2.a. ($|V(H)|=4$). Then $V(H) = \{1,2,3,4\}$ and the ends of e_1 in H are either 1 and 2 or 3 and 4, and the ends of e_2 in H are either 1 and 2 or 3 and 4. But e_1 and e_2 are two edges of a triangle in $L(K_5)$, and so by Lemma 2, e_1 and e_2 are two edges of a triangle in H. We have a contradiction.

Case 2.b. ($|V(H)| = 5$). Let 5 be the vertex of H not in H_1. Then the ends of e_1 in H are 1 and 2, 3 and 4, or 3 and 5, and the ends of e_2 in H are 1 and 2, 4 and 3, or 4 and 5.

Let e_4, e_5, e_6 be the edges in $L(K_5)$ with ends $\{5,3\}$ and $\{5,4\}$, $\{5,3\}$ and $\{5,2\}$, and $\{5,4\}$ and $\{5,2\}$. Since e_1, e_2 and e_4 form a triangle in G, by Lemma 2, they form a triangle in H. If the ends of e_1 in H are 1 and 2, then e_1 and e_2 cannot be two edges of a triangle in H, we have a contradiction. If the ends of e_1 in H are 3 and 4 then the ends of e_2 must be 4 and 5 and the ends of e_4 are 5 and 3. If the ends of e_1 in H are 3 and 5 then either the ends of e_2 are 3 and 4, and the ends of e_4 are 5 and 4, or the ends of e_2 are 4 and 5 and the ends of e_4 are 3 and 4. Thus either the ends of e_4 in H are 3 and 4 or 5 is an end of e_4.

If the ends of e_4 in H are 3 and 4 then it is not hard to show that (ii) holds by applying Lemma 2 and choosing the proper triangles.

We assume (ii) does not hold. Then 5 is an end of e_4. Similarly, 5 is an end of e_5 and e_6. By Lemma 2, e_4, e_5, e_6 for a triangle in H. We have a contradiction.

Case 3. ($n \geq 6$). Let $A_i = \{\{i,j\} : j \in \{1,2,\ldots,n\} - \{i\}\}$, for $i \in \{1,2,\ldots,n\}$, $G_i = G|A_i$, $G_{i,j} = G|A_i \cup A_j$, $H_i = H|E(G_i)$, and $H_{i,j} = H|E(G_{i,j})$. It is obvious that $K(H_{i,j})$ is a subspace of $K(G_{i,j})$. Let $d_{i,j}$ denote the corresponding codimension. We also know that $G_i \approx K_{n-1}$ and $K(H_i) \subseteq K(G_i)$. Therefore Theorem 2 together with the fact that (i) does not hold, implies that $H_i \approx G_i$. So, $d_{i,j} = \dim(K(G_{i,j})) - \dim(K(H_{i,j})) \leq \dim(K(G_{i,j})) - \dim(K(H_i)) = n-2$. Since (i) does not hold, $H_{i,j}$ is not obtained from $G_{i,j}$ by identifying vertices. It is easy to check that $G_{i,j}$ is $(n-1)$-connected. Then it follows from Theorem 4 that $d_{i,j} \geq n-3$ or $d_{i,j} = 0$. If $d_{1,2} = 0$ then $A_1 \cup A_2$ is faithful. Since for any $v \in V(L(K_n)) - V(G_{1,2})$, v is adjacent to at least four vertices of $G_{1,2}$, we can apply Lemma 3 inductively to imply that $H = L(K_n)$.

If $d_{1,2} = n-2$ and clearly $V(H_{1,2}) = n-1$ then $|V(H_{1,2})| = V(H_1)$.

Without loss of generality we may also identify vertices of H_1 with vertices of G_1 in such a way that an edge of G_1 has the same ends in G_1 and in H_1. Let e_k be the edge with ends $\{1,2\}$, $\{1,k\}$, and e_k' be the edge with ends $\{1,2\}$, $\{2,k\}$ in $G_{1,2}$ for $k=3,4,\ldots,n$. If $\{1,2\}$ is an end of e_k' in the $H_{1,2}$, let $\{1,j\}$ be the other end, then e_j, e_k' form a cycle in $H_{1,2}$. But e_j, e_k' form a path from $\{1,j\}$ to $\{2,k\}$ in $G_{1,2}$, contrary to Lemma 1. Therefore $\{1,2\}$ is not an end of e_k' in $H_{1,2}$ for any k. Since $e_3', e_4', \ldots, e_n'$ form a star in G_2, they must form a star in H_2. Therefore they form a star in $H_{1,2}$. According to the argument above, $\{1,2\}$ cannot be a vertex of the star. But $H_{1,2}$ has only $n-1$ vertices, and the star has also $n-1$ vertices: this is a contradiction. Therefore $d_{1,2}=n-3$, and (ii) of Theorem 5 is true for $G_{1,2}$ and $H_{1,2}$ and similarly for every $G_{i,j}$ and $H_{i,j}$.

Case 3.a. (Suppose (ii) (a) of Theorem 5 is true for every $G_{i,j}$, $i,j \in \{1,2,\ldots,n\}$). It is easy to verify that the edges with ends $\{i,k\}$, $\{k,j\}$ in $G_{i,j}$ are the only choices for the $e_1, e_2,\ldots,e_{n-2}$ in (ii) (a) of Theorem 5. Now we index the vertices of $H_{1,2}$ as shown in Figure 3.

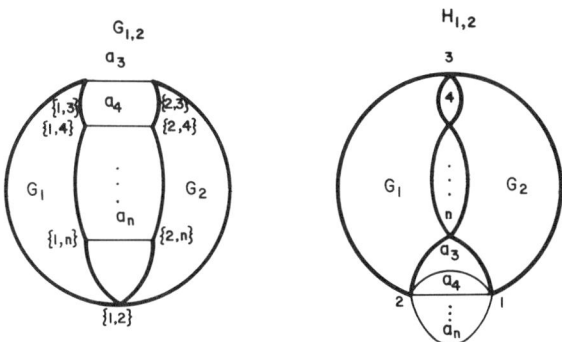

Figure 3

All edges of $G_{1,2}$ satisfy the rules of (ii) (a) of the main theorem. It is obvious that $E(G_{1,2}) \cap E(G_{1,3})$, which consists of $E(G_1)$, a_3 and the edge of G_2 with ends $\{1,2\}$, $\{2,3\}$, covers $V(H_{1,2})$. This implies that $V(H_{1,3})=V(H_{1,2})$. By applying (ii) (a) of Theorem 5 to $G_{1,3}$, it is straightforward to check that all edges of $G_{1,3}$ satisfy the rules of (ii) (a) of the main theorem. Similarly $V(H_{1,k}) = V(H_{1,2})$ and all edges of $G_{1,k}$ satisfy the rules of (ii) (a) of the main theorem for $k=4,\ldots,n$. Since $E(L(K_n)) = \bigcup_{k=2}^{n} E(G_{1,k})$, this implies that (ii) (a) is true.

Case 3.b. (Suppose (ii) (b) of Theorem 5 is true for some $G_{i,j}$, say $G_{1,3}$). It is easy to see that in $G_{1,3}$ all but one of the edges with ends $\{1,k\}$, $\{k,3\}$ are the only choices for the $e_1, e_2, \ldots, e_{n-3}$ in (ii) (b) of Theorem 5. We can index and construct $H_{1,3}$ as shown in Figure 4.

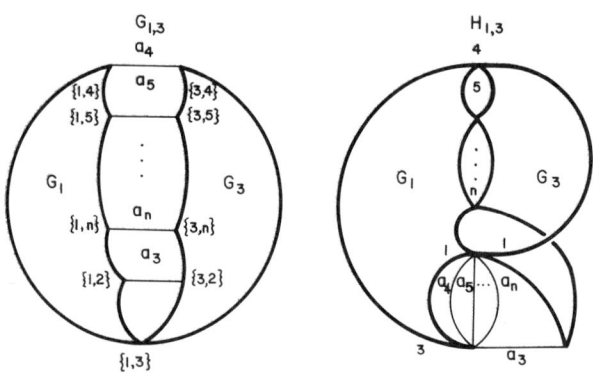

Figure 4

It is straightforward to check that all edges in $G_{1,3}$ satisfy the rules of (ii) (b) of the main theorem. Since $E(G_{2,3}) \cap E(G_{1,3})$ covers $V(H_{1,3})$, we have $V(H_{2,3}) = V(H_{1,3})$. It is not hard to show that the only way to apply (ii) of Theorem 5 to $G_{2,3}$ is so that all edges of $G_{2,3}$ satisfy the rules of (ii)(b) of the main theorem. Since $E(G_{1,k}) \cap (E(G_{1,3}) \cup E(G_{2,3}))$ covers $V(H_{1,3})$ (the edge with ends $\{1,2\}$, $\{2,k\}$ covers 2), we have $V(H_{1,k}) = V(H_{1,3})$ for $k=4,\ldots,n$. Again, it is not hard to show that the only way to apply (ii) of Theorem 5 to $G_{1,k}$ is so that all edges of $G_{1,k}$ satisfy the rules of (ii) (b) of the main theorem. Since $E(L(K_n)) = \bigcup_{k=3}^{n} E(G_{1,k}) \cup E(G_{2,3})$, this implies that (ii) (b) is true.

5. Epilogue.

Alternative (ii) of the main theorem shows examples for which G is (2n-4)-connected, and $K(H) \subseteq K(G)$ with codimension $-\frac{1}{2}n(n-3)$. In (ii) (b) we can add an edge with ends $\{1,2\}$, $\{i,j\}$ in $L(K_n)$ and add the same edge to $(n-2)K_n$ with ends i,j for any $i,j \in \{3,4,\ldots,n\}$. We now construct new examples. Let $G=L(K_n)+t$, $H=(n-1)K_n$ be such that the edges in $L(K_n)$ have their ends in H defined as in (ii) (a) of the main theorem and the edge with ends t, $\{i,j\}$ in G has ends i,j in H for $i,j \in \{1,2,\ldots,n\}$. This is an example for $K(H) \subseteq K(G)$, where the codimension is $\frac{1}{2}n(n-3)+1$, such that G is $(2n-3)$-connected. The question remains "Do there exists some theorem generalizing Theorem 5 and Conjecture 1 so that the above examples are the exceptional cases?".

References

[H] F. Harary, *Graph Theory*. Addison-Wesley Publishing Company (1969).

[J] F. Jaeger, "On graphic minimal spaces", Proc. of Montreal Conf. Combinatorics '79, Annals DM No. 8, pp. 123-126.

[S] C. H. Shih, "On Graphic Subspaces of Graphic Spaces", Doctoral Thesis, The Ohio State University (1982).

Extremal Graph Problems, Degenerate Extremal Problems, and Supersaturated Graphs

Miklós Simonovits

A Survey of Old and New Results

NOTATION. Given a graph, hypergraph G^n, . . . , the upper index always denotes the number of vertices, $e(G)$, $v(G)$ and $\chi(G)$ denote the number of edges, vertices and the chromatic number of G respectively. Given a family $\boldsymbol{L}$ of graphs, hypergraphs, $ex(n, \boldsymbol{L})$ denotes the maximum number of edges (hyperedges) a graph (hypergraph) G^n of order n can have without containing subgraphs (subhypergraphs) from $\boldsymbol{L}$. The problem of determining $ex(n, \boldsymbol{L})$ is called a *Turán-type extremal problem*. The graphs attaining the maximum will be called *extremal* and their family will be denoted by $EX(n, \boldsymbol{L})$.

1. Introduction

Let us restrict our consideration to ordinary graphs without loops and multiple edges. In 1940, P. Turán posed and solved the extremal problem of K_{p+1}, the complete graph on $p+1$ vertices [39,40]:

TURÁN THEOREM. If $T^{n,p}$ denotes the complete p-partite graph of order n having the maximum number of edges (or, in other words, the graph obtained by partitioning n vertices into p classes as equally as possible, and then joining two vertices iff they belong to different classes), then

(a) $T^{n,p}$ contains no K_{p+1}, and

(b) all the other graphs G^n of order n not containing K_{p+1} have less than $e(T^{n,p})$ edges.

Using a theorem of Erdős and Stone [24], Erdős and Simonovits [18] derived that if $\boldsymbol{L}$ is an arbitrary family of forbidden graphs, and

$$p(\boldsymbol{L}) = \min_{L \in \boldsymbol{L}} \chi(L) - 1, \qquad (1)$$

then

$$ex(n, \mathbb{L}) = (1 - \frac{1}{p})\binom{n}{2} + o(n^2). \tag{2}$$

They also conjectured, and later proved independently [13,14,30].

THEOREM 1. Given a family $\mathbb{L}$ of forbidden graphs with minimum chromatic number $p+1$ and an $L \in \mathbb{L}$ with chromatic number $p+1$, then for $c = 2 - 1/v(L)$,

$$ex(n, \mathbb{L}) = (1 - \frac{1}{p})\binom{n}{2} + O(n^c), \tag{2*}$$

and every extremal graph S^n can be obtained from a $T^{n,p}$ by deleting and adding $O(n^c)$ edges. Further, the minimum degree $\underline{d}(S^n) = (1 - \frac{1}{p})n + O(n^{c-1})$.

This means that the extremal numbers and the extremal structure depend very loosely on $\mathbb{L}$, they are asymptotically determined by the minimum chromatic number $p+1$. Further, (2) implies that $ex(n, \mathbb{L}) = o(n^2)$ iff $p=1$, that is, iff $\mathbb{L}$ contains at least one bipartite graph. This case will be called *degenerate*.

Recently, writing a survey on extremal graph theory [36], I came to realize that one of the most intriguing, most important and rather underdeveloped areas of extremal graph theory is the theory of degenerate extremal graph problems. In this case the structure of extremal graphs tends to become very complicated.

Of course, there are very many interesting results in this field, see e.g. [1,2,6,7,12,15,17,19,26,29,32]. Still, this underdevelopment was one of the reasons why the author with some other coauthors started investigating degenerate extremal graph problems more systematically. Some of the newer results achieved by the author and others can be found in [25,21,22,23,32]. Here I would like to give a survey which is almost completely devoted to degenerate extremal graph problems, degenerate extremal hypergraphs problems and the corresponding *theory of supersaturated graphs*. Some repetition compared to [36] is unavoidable, but I have tried to minimize it. Anyone wishing to find further literature on extremal graph theory is recommended to read, among others, Bollobás' excellent book [5], or my survey [36].

Before restricting our consideration to degenerate extremal problems, let us pose the following question:

2. Which General Questions Should be Asked in Extremal Graph Theory?

Obviously, such a question can be answered in many different ways, and most of the answers are biased in some way or other. Perhaps one fair answer could be:

(a) Try to find general questions, related to the *general* theory of extremal graphs, and

(b) try to ask specific questions, which cannot be answered by the general theory, which attack completely new areas, new phenomena, and therefore lead to completely new (and meaningful!) theories.

It is meaningless to speak too much of (b) in generalities. The reader, however, should remember that the whole of extremal graph theory (and many other theories) developed this way, inductively: through solving particular problems, proving conjectures which often seemed too particular at first glance, and still they often led to many other interesting and general results, rich theories. So let us get back to (a) and try to give a sketch of the general problems in extremal graph theory.

The problems in extremal graph theory can be classified according to their OBJECTS and their TYPES. This is illustrated (without much explanation) on the chart on the next page.

Below we shall see that many of the extremal graph theorems can quite well be described by these categories.

3. Degenerate Extremal Graph and Hypergraph Problems

Above we restricted our consideration to ordinary graphs, however, extremal graph problems naturally arise for r-uniform hypergraphs as well. Further, extremal graph theorems are known for multigraphs, multidigraphs, multihypergraphs as well, see e.g. [7,8.12,29,32]. However, in the "multi-" cases we have to fix an upper bound on the multiplicity of edges to get finite maxima. For the sake of simplicity, in this survey we shall restrict ourselves to ordinary graphs and uniform hypergraphs. The definitions of $ex(n, \mathcal{L})$ and $EX(n, \mathcal{L})$ are obvious.

DEFINITION 1. Let us consider r-uniform hypergraphs and let $\mathcal{L}$ be a family of forbidden hypergraphs. The extremal hypergraph problem of $\mathcal{L}$ will be called *degenerate* if $ex(n, \mathcal{L}) = o(n^r)$.

TWO EXAMPLES.

THEOREM 2. Kővári-T.Sós-Turán, [26]. Let $K_{p,q}$ denote the complete bipartite graph with p and q vertices in its color-classes. Then

OBJECTS	TYPES OF QUESTIONS
ordinary graphs	nondegenerate extremal problems
multigraphs	degenerate extremal problems
digraphs	—lower bounds using random graphs
hypergraphs	—lower bounds by finite geometrical constructions
....	—upper bounds by counting arguments
	—upper bounds by recursions
graphs with partitioned vertex set (subgraphs of $K(n_1,\ldots,n_r)$)	supersaturated graphs
	compactness results
	....
	Perturbation problems
	—degree perturbation
	—chromatic perturbation
	—Turán-Ramsey theorems (Ramsey perturbation)
APPLICATIONS	(α) application of other principles in extremal graph theory
	—algebraic methods, eigenvalue methods
	—random graph "constructions" (see above)
	(β) Application of extremal graph theory in combinatorial geometry, analysis, probability theory.
	(γ) continuous versions of extremal graph theorems.

$$ex(n,K_{p,q}) \leq \frac{1}{2}\sqrt[p]{q-1}\cdot n^{2-1/p} + O(n). \tag{3}$$

This was generalized by Erdős [12]:

THEOREM 3. Restrict ourselves to r-uniform hypergraphs. Let $K_r^{(r)}(m,\ldots,m)$ denote the r-uniform hypergraph obtained by fixing rm vertices x_{ij} ($i=1,\ldots,r$, $j=1,\ldots,m$) and taking all the m^r "transversals" $(x_{1,j_1}, x_{2,j_2}, \ldots, x_{r,j_r})$ as hyperedges. Then

$$ex(n, K_r^{(r)}(m,\ldots,m)) = O(n^{r-m^{-(r-1)}}). \tag{4}$$

REMARK 1. Theorem 3 is a very good example of what was stated of questions about type (b) in §1. Its proof is neither trivial nor too complicated, and one not too much acquainted with extremal graph theory will find it too particular. However, on the one hand it is a direct and elegant generalization of Theorem 2, (apart from the value of the constant), and on the other hand, it is widely applicable in extremal graph problems. One of its important applications is

PROPOSITION 1. Restrict our considerations to r-uniform hypergraphs. The problem of $\mathbb{L}$ is degenerate iff $\mathbb{L}$ contains an L^* the vertices of which can be colored by r colors so that each r-edge gets r different colors.

PROOF. The condition of the proposition is equivalent to the requirement that L^* be contained in some $K_r^{(r)}(m,\ldots,m)=L'$. By (4) $ex(n,L')=o(n^r)$. If $L^* \subseteq L'$, then $ex(n,L^*)=o(n^r)$, and consequently, $ex(n, \mathbb{L})=o(n^r)$. This proves one half of Proposition 1. The other half follows easily from the fact that if no $L \in \mathbb{L}$ is contained in any $K_r^{(r)}(m,\ldots,m)$, then $S^n := K_r^{(r)}(m,\ldots,m)$ for $m=n/r$ contains no prohibited subhypergraphs and has cn^r edges. Thus $\mathbb{L}$ is not degenerate. Further applications of (4) will be given later.

If we wish to give some other examples of degenerate extremal problems — as we do — then we should start with the simplest ones. The extremal problem for *paths* was solved by Erdős and Gallai [15], the problem for *even cycles* by Erdős (unpublished), and a generalization of it was given by Bondy and Simonovits [2]. Here we mention only the original result on $ex(n,C_{2t})$, (where C_{2t} denotes the cycle of $2t$ vertices).

THEOREM 4. $ex(n,C_{2t})=O(n^{1+1/t})$.

Turán asked the following question: if we fix one of the *regular polyhedra*, (tetrahedron, cube, octahedron, dodecahedron, icosahedron), and denote the graph formed by its vertices and edges by L, how large is $ex(n,L)$? His theorem answers the case of the tetrahedron, the others were settled by Simonovits (dodecahedron, icosahedron, [33], [34], and by Erdős and Simonovits, (octahedron, cube), see [19], [20]. Again, rather surprisingly, one of the simplest polyhedra, the cube turned out to be the most difficult case.

THEOREM 5. If Q denotes the graph determined by the vertices and edges of a cube, and Q^* denotes the graph obtained by joining two opposite vertices of the cube, then

$$ex(n,Q) \leq ex(n,Q^*) = O(n^{8/5}).$$

We conjecture that these results are sharp[1]. Unfortunately, no nontrivial lower bound is known for the cube problem. (In this sense

---

(1) This means the existence of a $c > 0$ such that $ex(n,Q) > cn^{8/5}$ ($n > 8$).

the word "settled" does not apply to the cube.)

4. Why are Degenerate Extremal Graph Problems Important?

Since the main topic of this survey is the theory of degenerate extremal graph problems, it is quite natural to ask the above question. There are always many possible answers to such questions. One of them could be that in some sense *many nondegenerate extremal graph problems can be reduced to degenerate extremal problems*. More precisely, if $\mathcal{L}$ is a given family of forbidden graphs, one defines $p = p(\mathcal{L})$ by (1) and then one can find (in many cases) p families $M_1, \ldots, M_p$ of forbidden graphs such that

(a) *If S^n is extremal for $\mathcal{L}$, then we can find p extremal graphs $T_i \in EX(n, M_i)$ of $n/p + o(n)$ vertices each, such that "S^n is the product of the T_i's"*, where the *product* of the T_i's is defined as follows: take vertex disjoint copies of these graphs and join every $x \in V(T_i)$ to every $y \in V(T_j)$ for every $1 \le i < j \le p$. (The product of many graphs will be denoted by XT_i, of two graphs G and H by $G \times H$.)

(b) *If T_i are now arbitrary graphs not containing subgraphs from M_i, then XT_i contains no $L \in \mathcal{L}$.*

(c) The extremal graph problem of M_i is degenerate.

Here I give only one illustration of such "reduction" theorems, namely the octahedron theorem. However, there are many other similar reduction theorems, e.g. the icosahedron and the dodecahedron theorems [34, 33].

THE OCTAHEDRON THEOREM. (Erdős, Simonovits [20]). If L denotes the octahedron graph ($=K_{2,2,2}$), then there exists an n_1 such that for $n > n_1$ every extremal graph $S^n \in EX(n, L)$ is the product of two other graphs: $S^n = U_1 \times U_2$, where U_1 is extremal for C_4, U_2 is extremal for P_3, $v(U_i) = n/2 + o(n)$ (where $v(G)$ denotes the number of vertices of G), and if U' and U'' are some other graphs and U' contains no C_4, U'' contains no P_3, then $U' \times U''$ contains no octahedron graph L.

Generally, theorems stating that the extremal graph for a nondegenerate extremal problem is the product of some extremal graphs for a degenerate problem, will be called REDUCTION THEOREMS. And now we can say that *degenerate extremal graph problems are interesting partly* in their own right, and partly *because nondegenerate extremal graph theorems are often reducible to degenerate extremal graph theorems*, see [35].

5. Some General Open Questions on Degenerate Extremal Graph Problems

To be sincere, many of the open problems on degenerate extremal problems seem to be completely hopeless because the extremal graphs conjectured have a fairly regular and yet very complicated structure. Still, it is worth formulating quite general and hopeless conjectures, since they often indicate very well in which direction we should go, or which types of simpler questions should be solved.

Below, we formulate four general questions.

4.1 The Problem of Rational Exponents

CONJECTURE 1. Let $\mathbb{L}$ be a family of bipartite graphs. Does there exist a rational $\alpha \in [1,2)$ such that $ex(n, \mathbb{L})/n^\alpha$ converges to a *positive* limit. Here the requirement that the limit should be positive, is important otherwise the conjecture would immediately follow from Kővári-T. Sós-Turán theorem. A weakening of this conjecture is

CONJECTURE 2. Given a family $\mathbb{L}$ of bipartite forbidden graphs, there exist constants, c and $c' > 0$ and $\alpha \in [1,2)$ for which

$$c \cdot n^\alpha \leq ex(n, \mathbb{L}) \leq c' n^\alpha$$

The main idea behind both conjectures is that while the structure of extremal graphs may be fairly complicated, it must be simple in some other sense. For 3-uniform hypergraphs, Conjecture 2 does not hold, as was shown by Ruzsa and Szemerédi [29].

4.2 The Problem of Strong Compactness

CONJECTURE 3. Let $\mathbb{L}$ be a *finite* family of bipartite graphs. Then there is an $L \in \mathbb{L}$ such that

$$1 \leq \frac{ex(n,L)}{ex(n,\mathbb{L})} = 0(1).$$

CONJECTURE 4. Let $\mathbb{L}$ be a *finite* family of bipartite graphs. Then there is an $L \in \mathbb{L}$ such that

$$\frac{ex(n,L)}{ex(n,\mathbb{L})} \to 1$$

Clearly, the second conjecture is much stronger. Both assert that if we prohibit a finite family of bipartite graphs, then there is always *one* among them, the exclusion of which has almost the same effect as if we excluded the whole of $\mathbb{L}$: one forbidden graph dominates the whole problem. Here *the finiteness of $\mathbb{L}$ is very important:* if $\mathbb{L}$ is the (infinite) family of all the cycles, then $ex(n, \mathbb{L}) = n-1$. However, for any finite subclass $\mathbb{L}^*$ of $\mathbb{L}$ there exists an $\alpha>0$ such that $ex(n,\mathbb{L}^*)>c\cdot n^{1+\alpha}$.

For non-degenerate extremal graph problems even the stronger Conjecture 4 immediately follows from the Erdős-Simonovits theorem, [18]. One final remark should be made: we stated Conjectures 3,4 in affirmative form, however some examples suggest that perhaps they are false. It would be interesting to find counterexamples.

4.3 The Problem of Everywhere Dense Exponents

CONJECTURE 5. For every $1\leq\alpha<\beta\leq 2$ there exists a *finite* family $\mathbb{L}$ of forbidden graphs such that

$$n^\alpha < ex(n,\mathbb{L}) < n^\beta$$

if n is large enough.

CONJECTURE 6. (Stronger). For every $1\leq\alpha<\beta\leq 2$, there exists a *finite* $\mathbb{L}$ and a $\gamma\in(\alpha,\beta)$ such that $\lim_{n\to\infty} ex(n,\mathbb{L})/n^\gamma$ exists and is positive. (Perhaps for every rational $\gamma\in(1,2)$ there exists such a finite $\mathbb{L}$.)

4.4 The Problem of Weak Vertices

In some reduction theorems the following property of L plays important role: *"One can delete a vertex v of L so that $ex(n,L-v)=o(ex(n,L))$"*. Vertices satisfying this condition will be called *weak*. The problem we would like to state is: *"Characterize those bipartite graphs which have weak vertices."*

Clearly, the trees have no weak vertices, C_{2k}, $K_{p,q}$ for $p=2,3$ do have (and probably $K_{p,q}$ does have weak vertices even for $p\geq 4$). The graph in Figure 1 (as it is easy to check) has no weak vertices. (For some literature, see [31].)

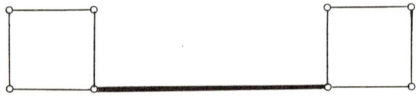

Figure 1.

PROOF METHODS: Obviously, to solve an extremal problem we need an upper and a lower bound on $ex(n,\mathbb{L})$. Both may be fairly difficult to find. However, at present, we know much less about how to get lower bounds. That is why we start with this question.

6. How to Get Lower Bounds?

In trying to find a lower bound for a degenerate Turán-type extremal problems we usually need a construction, but sometimes use the method of random graphs. However, the method of random graphs somehow turns out to be rather weak; I do not know of any case of a finite degenerate $\mathcal{L}$ when the method of random graphs gave a sharp lower bound. The explicit constructions are mostly based on FINITE GEOMETRICAL arguments. The method of using finite geometrical constructions seems to be fairly powerful, and many times gives sharp lower bounds. Still, we do not know enough about finite geometries, and perhaps this is why in many cases we only conjecture that an appropriate finite geometrical construction could help, but are unable to find it. I shall not go into detailed discussion of this method, it is described in [5] or [36], or in the original papers, e.g. [2,6].

7. How to Get Upper Bounds in Degenerate Extremal Problems?

Mostly we use one or two of the following methods (combined):

(a) *blowing up;*

(b) *recursion theorems;*

(c) application of *supersaturated theorems*

Since we often combine these methods, several of the illustrations given below will be "mixed" ones.

8. Blowing Up

First we sketch the method, then give some applications. In many extremal graph problems, we first apply some regularization theorem to the original graph, asserting that *in any graph G^n we can find a subgraph G^m the minimum degree of which is almost its average degree and such that G^m still has many edges*. One such theorem is

THEOREM 7. Let G^n be a graph with E edges and average degree $d = \frac{2E}{n}$. Then there exists a subgraph G^m with $\underline{d}(G^m) > \frac{d}{2}$, where $\underline{d}(G)$ denotes the minimum degree. (Clearly, if $d = \frac{2e(G^n)}{n} \to \infty$, then $m \to \infty$, too.)

This theorem is mostly applied with $e(G^n) = c \cdot n^{1+\alpha}$ where $\alpha > 0$. Sometimes we need the sharper result of Erdős and Simonovits [19]:

THEOREM 8. For every $\alpha > 0$ and $c_1 > 0$, there exist a $\beta > 0$ and $c_2, c_3 > 0$, such that if $e(G^n) \geq c_1 n^{1+\alpha}$, then G^n contains a subgraph G^m with $m > n^\beta$, and $\underline{d}(G^m) \geq c_2 m^\alpha$, $\bar{d}(G^m) \leq c_3 m^\beta$, where $\bar{d}(G)$ denotes the maximum degree. If $c_1 \to \infty$, then $c_2, c_3 \to \infty$.

Both results say that we may restrict our consideration primarily to "regular" graphs. The next lemma asserts that we may assume that G^n is bipartite:

LEMMA 1. *If G is an arbitrary graph and H is a bipartite subgraph of G having the maximum number of edges, then $d_H(x) \geq d_G(x)/2$ for every vertex, where $d_H(x)$ and $d_G(x)$ denote the degrees in the corresponding graphs.*

Assume now that we would like to prove that $ex(n, \mathit{L}) = O(n^{1+\alpha})$. Then we fix a graph G^n with at least $c_1 n^{1+\alpha}$ edges, choose a $G^m \subseteq G^n$ according to Theorem 7 and then a bipartite $H^m \subseteq G^m$ with $\underline{d}(H^m) \geq 1/2 c_2 m^\alpha$. After these two preparatory steps, we can carry out the actual "blowing up". We choose an arbitrary $x \in V(H^m)$ and denote by S_j the set of vertices having distance j from x. Then we prove that the condition "$L \not\subseteq H^m$" implies that, for some $r = r_L$ the size of S_j is much larger than the size of S_{j-1} $j = 1,2,\ldots,r$. We know that $|S_1| \geq \underline{d}(H^m) \geq c_2 m^\alpha$, and we known also that $|S_r| < m$. These two facts yield that for the considered exponent α, c_2 is bounded, and therefore c_1 is bounded as well. This yields that $ex(n, \mathit{L}) = O(n^{1+\alpha})$.

This method was used by Bondy and Simonovits to prove the *even cycle* theorem [2], and also by Faudree and Simonovits to prove a generalization of this result, see below [25].

Bondy and Simonovits conjectured the following generalization of Theorem 4:

DEFINITION. Let $C_{k,t}$ denote the graph obtained by joining two vertices x,y by k vertex-independent paths of t edges each.

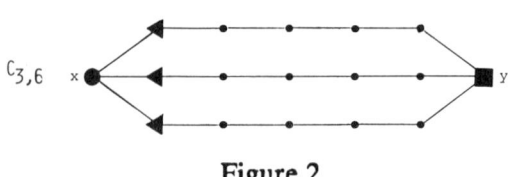

Figure 2

THEOREM 9. (Faudree, Simonovits [25].) $ex(n, C_{k,t}) = O(n^{1+1/t})$. Clearly, this generalizes Theorem 4. As a matter of fact, Faudree and Simonovits have proved a much more general theorem.

DEFINITION. Let L be a bipartite graph with a given 2-colouring c, say in red and blue. Let x be a new vertex and joint it to each blue vertex of L by vertex-independent paths of $t-1$ edges (Figure 3). The resulting graph will be denoted by L_t or $L_t(L,c)$.

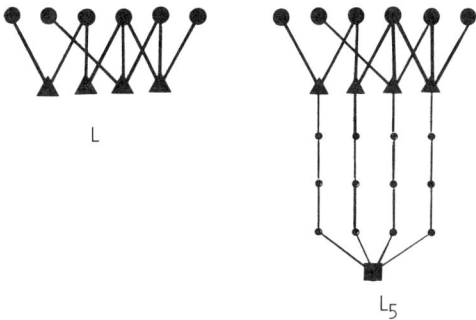

Figure 3.

Now we can formulate the Faudree-Simonovits *recursion theorem*.

THEOREM 10. If T is a tree or a forest with a given 2-colouring c, and $L^* = L_t(L,c)$, then $ex(n, L^*) = O(n^{1+1/t})$.

Clearly, starting with a star $T = K_{1,k}$ we get back to Theorem 9. In proving Theorem 10 we used again the *blowing up* method. Theorem 10 has many different applications and some generalization. Here, we have no space to go into the details.

9. Recursion Theorems

If we wish to prove general degenerate extremal graph theorems, (or supersaturated theorems, see below), one way to do this is to look for recursion theorems. These assert that if we have bounds for a graph L, then we can obtain other bounds for some more complicated graph L^*. Generally, this L^* is obtained from L by some given operation. Theorem 10 above is one illustration of such recursion theorems. Below, we shall give another recursion theorem, due to Erdős and Simonovits.

DEFINITION. Let L be a bipartite graph properly coloured by red and blue. Let $K_{t,t}$ be coloured by red and blue and join each red vertex of $K_{t,t}$ to each blue one of L, each blue vertex of $K_{t,t}$ to each red one of L, (see fig. 4.) The resulting graph will be denoted by $L(t)$.

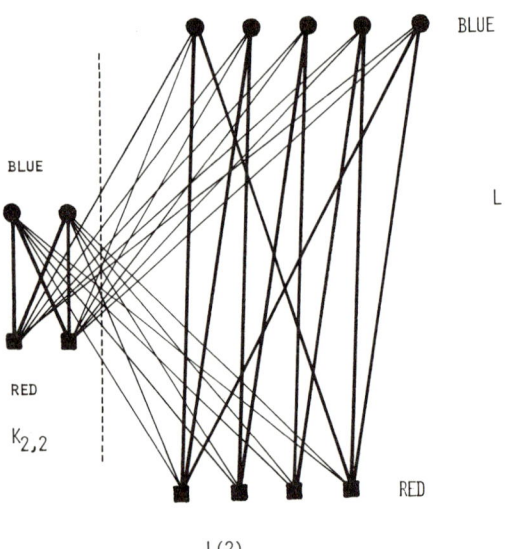

Figure 4

THEOREM 11. Assume that L is a given bipartite graph with a fixed 2-colouring, and that $ex(n,L)=O(n^{2-\alpha})$. Define β by

$$\frac{1}{\beta} - \frac{1}{\alpha} = t.$$

Then $ex(n,L(t))=O(n^\beta)$.

One application of this theorem is when $L=C_6$, hence $ex(n,L)=O(n^{4/3})$. Since $\beta=8/5$, and $L(1)=Q^*$ of Theorem 5, Theorem 11 immediately yields Theorem 5. As a matter of fact, Theorem 11 was obtained by trying to prove Theorem 5. However Theorem 11 has many other interesting consequences, too; for example,

THEOREM 12. Let L be a bipartite graph properly coloured by red and blue. If we may delete t red and t blue vertices so that the resulting graph is a tree, then $ex(n,L)=O(n^{2-1/(t+1)})$.

The proof of Theorem 12 from Theorem 11 is trivial. Theorem 12 contains a result of Erdős (a sharpening of the Kővári-T. Sós-Turán theorem) as a special case. Indeed, let $L=(K_{p+1,p+1}$-an edge). Then Theorem 12 can be applied: we may delete $p-1$ red and $p-1$ blue vertices of L so that the resulting graph is a path P_3. Thus $ex(n,L)=O(n^{2-1/p})$.

There are also recursion theorems of other types. For instance, Erdős and Simonovits [22] proved a result where they obtained upper bounds on $(r+1)$-uniform hypergraph problems from r-uniform hypergraph results: the recursion goes according to the edge-size of the hypergraph. This theorem implies Theorem 3 immediately.

10. Supersaturated Graphs

To describe how supersaturated graph theorems imply degenerate extremal graph results, we first explain what we mean by a supersaturated graph.

Let us consider the case when $\mathbb{L}$ is fixed and G^n has $E > ex(n, \mathbb{L})$ edges. Clearly, G^n must contain forbidden subgraphs $L \in \mathbb{L}$. As it turns out, in most cases G^n must contain not only one but very many prohibited $L \in \mathbb{L}$. Such graphs will be called *supersaturated* or *L-supersaturated*.

There exists a fairly extensive literature on the case when G^n is K_p-supersaturated. Without trying to be complete in any sense, we list the following references: [3,4,5,11,12,27,28]. For the case of L-supersaturated graphs in general, see [21,22,23,37].

Here we are interested primarily in supersaturated graphs where the corresponding extremal problem is degenerate. Nevertheless, let us start with a quite general theorem.

THEOREM 13. (Erdős, Simonovits, [22]) Let us consider r-uniform hypergraphs and let $\mathbb{L}$ be a given finite family of forbidden graphs. Assume (for the sake of simplicity) that all the graphs $L \in \mathbb{L}$ have v vertices. Then, for every $c > 0$, there exists a $c' > 0$ such that, if

$$e(G^n) > ex(n, \mathbb{L}) + cn^r,$$

then G^n contains at least $c' \cdot n^v$ copies of $L \in \mathbb{L}$.

Obviously, this result is sharp in the sense that G^n has at most $O(n^v)$ copies of $L \in \mathbb{L}$. We arrive at much deeper and more difficult problems if we consider the case of *"weakly supersaturated graphs"*: the case when $e(G^n) = ex(n, \mathbb{L}) + o(n^r)$. For the sake of simplicity, we shall restrict below our considerations to the case of ordinary graphs $(r=2)$. Further, we shall assume that $\mathbb{L}$ consists of bipartite graphs only. The theorems below state (in some sense) that if $e(G^n) > ex(n,L)$ is fixed, then G^n has the minimum number of copies of $L \in \mathbb{L}$ if G^n is a random graph. Observe first that if G^n is a random graph with E edges, then G^n contains on average

$$\binom{n}{v} \cdot \left(\frac{2E}{n^2}\right)^e \approx c_L \cdot \frac{E^e}{n^{2e-v}}$$

copies of $L \in \mathbb{L}$.

CONJECTURE 7. Let L be a bipartite forbidden graph such that $ex(n,L)=O(n^{2-\alpha})$, for some $\alpha \in (0,1)$. Then there exist two constants c and $c'>0$ such that if $E=e(G^n)>cn^{2-\alpha}$, then G^n contains at least $c' \cdot \frac{E^e}{n^{2e-v}}$ copies of L, where $e=e(L)$, $v=v(L)$.

A weakening of this conjecture is

CONJECTURE 8. For every L, there exists an α and $c,c'>0$ such that if $e(G^n)>cn^{2-\alpha}$, then G^n contains at least $c' \cdot \frac{E^e}{n^{2e-v}}$ copies of L, where $e=e(L)$ and $v=v(L)$.

In [37], Simonovits proved Conjecture 7 for C_{2t} with $2-\alpha=1+1/t$; in [21], Erdős and Simonovits proved it for P_t†; in [23] they proved it for many other cases, including the cube graph and $K_{p,q}$. Since [23] is published in this very volume, there is no reason to repeat the results listed either in the introduction of [23] or among its main results. The reader will find there, among other things, a conjecture stronger than Conjecture 7. Further, one of the main results on [23] is a "RECURSION THEOREM ON SUPERSATURATED GRAPHS":

THEOREM 14. If Conjecture 7 holds for L with some constant α, then it also holds for the β defined in Theorem 11 and the $L(t)$ defined in the preceding definition.

(Thus, for example, the fact that Conjecture 7 holds for C_6 implies that it also holds for the graph Q^* obtained from the cube graph Q by joining two opposite vertices.

11. How to Apply Theorems on Supersaturated Graphs to Prove Extremal Graph Theorems

The method of using supersaturated graph results to prove Turán type extremal theorems is as follows: We would like to prove that $ex(n,L)=O(n^{2-\alpha})$. Assume that we can find another bipartite graph L' such that

(i) if $e(G^n)>E=cn^{2-\alpha}$ for some large c, then G^n contains at

---

† For walks W^k of length k the analogous (or, more precisely, a sharper) inequality follows from some matrix theoretical inequalities.

least $N=N(L',E)$ copies of L';

(ii) if G^n contains at least N copies of L', then it contains a copy of L.

From (i) and (ii) we derive that $ex(n,L)<cn^{2-\alpha}$.

Of course, this very method is often used with a slightly different "ideology": we say that "we count the number of copies of L' in G^n". First we get an upper bound on this, using the fact that $L\not\subseteq G^n$, then we get a lower bound on the number of copies of L' in G^n, using the fact that it has many edges. Comparing the two estimates, we obtain that G^n cannot have too many edges.

Obviously, there is not much difference between the two viewpoints. Still, in this survey we choose the first one, which will turn out to be quite useful.

Let us give some illustrations. (1) The proof of the Kővári-T. Sós-Turán theorem is one good example: we calculate the number of copies $L'=K_{1,p}$ in a graph. Each vertex x of degree d yields $\binom{d}{p}$ such $K_{1,p}$'s. Hence one can easily find a (sharp) lower bound on the number of copies of $K_{1,p}$ in G^n in terms of $e(G^n)$ and n. On the other hand, if the number of these is larger than $(q-1)\binom{n}{p}$, then there exists a p-tuple $x_1,\ldots,x_p$ occurring in at least q stars $(y_j,x_1,\ldots,x_p)$. These vertices x_i and y_j form a $K_{p,q}$ in G^n.

The cube theorem (Theorem 5), and the more general Theorem 11 (the recursion theorem on $L(t)$) were proved also by using a supersaturated graph argument. There we counted the number of copies of C_4 in G^n. Further, the "supersaturated graph recursion theorem" of [23], that is, Theorem 14, was again proved by a similar method.

One interesting application of this method is to the case when the forbidden graphs are cycles. To describe this application we first formulate two conjectures [21].

CONJECTURE 9. $ex(n,C_{2t})=\dfrac{1}{2}\cdot n^{1+1/t}+o(n^{1+1/t})$.

CONJECTURE 10. For any t and $k\geq 2$ $ex(n,C_{2t},C_{2k-1})=\left(\dfrac{n}{2}\right)^{1+1/t}+o(n^{1+1/t})$.

In both cases the upper bound of the conjecture follows from the Bondy-Simonovits or the Erdős theorem, apart from the value of the multiplicative constants. The background to these conjectures is that one may reasonably believe that the extremal graphs are almost regular, and if we exclude only C_{2t}, then the degrees are around $^t\sqrt{n}$, while in the second case the extremal graphs are bipartite, with roughly $n/2$ vertices in each colour-class and the degrees are only around $^t\sqrt{n/2}$. Although many results are known on similar problems about even cycles, Conjecture 9 has been specified only for $t=2$. In [21], Erdős and Simonovits proved that

$$ex(n,\{C_4,C_5\})=\left(\frac{n}{2}\right)^{3/2}+o(n^{3/2}).$$

Their proof uses sharp supersaturated graph theorems on walks of length k in graphs. It follows from some well known inequalities on non-negative matrices [41] (see comments below) that

THEOREM 15. If $E=e(G^n)$ and $d=\dfrac{2E}{n}$ is the average degree, then G^n contains at least $(1/2)nd^k$ walks of length k (where we regard the walks $(x_0,\ldots,x_k)$ and $(x_k,\ldots,x_0)$ as identical). Further, if $d\to\infty$, then G^n contains at least $(1/2)nd^k+o(nd^k)$ paths of length P.

Theorem 15 is sharp: a regular graph contains exactly $(1/2)nd^k$ walks of length k and $(1/2)nd^k-o(nd^k)$ paths of length k if $d\to\infty$. This theorem was conjectured by Simonovits, proved by Godsil for walks of even length and by Faudree and McKay for $k=3^s$. Finally Godsil found some matrix inequalities [41] immediately implying it for all walks. Below, we sketch how one can use it to prove the upper bound in

THEOREM 16. $\left(\dfrac{n}{2}\right)^{3/2}-o(n^{3/2})\le ex(n,\{C_4,C_5\})\le\left(\dfrac{n}{2}\right)^{3/2}+O(n).$

To prove this theorem we take a graph of G^n with E edges, count how many walks of length three occur in it, and conclude that there exists an edge (x,y) contained in more than n paths of the form (x,y,z,t) or (y,x,z,t). Hence, we find two paths of the form (x,y,z,t) and (x,y,z',t) or two paths of the form (x,y,z,t) and (y,x,z',t). In the first case we find a C_4, in the second one a C_5. (In trying to give precise proofs for such theorems, the main difficulty is always to get rid of the "coincidences": above, for example, we have to ensure that $z\ne z'$, otherwise our C_4 or C_5 is degenerate.)

11. Final Remarks

Supersaturated extremal theorems can be used in many other cases, too. Thus, for example, we may use supersaturated theorems on some L to prove supersaturated theorems on other ones. This happened in the proof of the recursion theorem on supersaturated graphs

(Theorem 14). A similar situation occurred in [22], where we gave a recursion theorem yielding supersaturated hypergraph theorems. The recursion uses results on r-uniform hypergraphs to prove supersaturated theorems on $(r+1)$-uniform hypergraphs. Right now, one of the most intriguing supersaturated graph conjectures, which does not seem completely hopeless (though we cannot prove it), is Conjecture 8.

One more remark: applying supersaturated graph theorems to prove extremal graph theorems or other supersaturated graph results seems to be a fairly powerful method. The reason for this may be that, in these applications we may use rough averaging methods and convexity arguments, which are not greatly affected by the irregularities in the structure of our graph.

References

[1] C. Benson: Minimal regular graphs of girth eight and twelve, Canadian J. Math. *18*. (1966) 1091-1094.

[2] J. A. Bondy and M. Simonovits: Cycles of even length in graphs, J. Combin. Theory *16* (1974) 97-105.

[3] B. Bollobás: Relations between sets of complete subgraphs, Proc. Fifth British Combin. Conference, Aberdeen, 1975, 79-84.

[4] B. Bollobás: On complete subgraphs of different orders, Math. Proc. Cambridge Phil. Soc. *79* (1976) 19-24.

[5] B. Bollobás: Extremal Graph Theory, London Math. Soc. Monographs, N11. Academic Press. 1978.

[6] W. G. Brown: On graphs that do not contain a Thomsen graph, Canadian Math. Bull. *9* (1966) 281-285.

[7] W. G. Brown, P. Erdős and V. T. Sós: On the existence of triangulated spheres in 3-graphs and related problems, Periodica Math. Hung. *3* (3-4), (1973) 221-228.

[8] W. G. Brown, M. Simonovits: Digraph extremal problems, hypergraph extremal problems and densities of graph structures, Discrete Mathematics (to appear).

[9] P. Erdős: On sequences of integers, one of which divides the product of two others and some related problems, mitt. Forschungsins Math. u. Mech. Tomsk, *2* (1938) 74-82.

[10] P. Erdős: On the structure of linear graphs, Israel J. Math. *1*. (1963) 156-160.

[11] P. Erdős: On a theorem of Rademacher-Turán, Illinois J. Math. *6* (1962) 122-127.

[12] P. Erdős: On the number of complete subgraphs and circuits contained in graphs, Casopis pro Pěstovani Matem. *94* (1969) 290-296.

[13] P. Erdős: Some recent results on extremal problems in graph theory, Theory of Graphs, International Symp. Rome, 1966, 118-123.

[14] P. Erdős: On some new inequalities concerning extremal properties of graphs, Theory of Graphs, Proc. Coll. Tihany, (Hungary) 1966, 77-81.

[15] P. Erdős and T. Gallai: On maximal paths and circuits of graphs, Acta Math. Acad. Sci. Hungar. *10* (1959)33-356.

[16] P. Erdős and A. Rényi: On the evolution of random graphs, Magyar Tud. Akad. Mat. Kut. Int. Közl. *5* (1960) 17-61.

[17] P. Erdős, A. Rényi and V.T. Sós: On a problem of graph theory, Studia Sci. Math. Hungar. *1* (1966) 215-235.

[18] P. Erdős and M. Simonovits: A limit theorem in graph theory, Studia Sci. Math. Math. Hungary. *1* (1966) 51-57.

[19] P. Erdős and M. Simonovits: Some extremal problems in graph theory, Coll. Math. Soc. J. Bolyai *4* (1969) 377-390.

[20] P. Erdős and M. Simonovits: An extremal graph problem, Acta Math. Acad. Sci. Hung, 22 (3-4), (1971) 275-282.

[21] P. Erdős and M. Simonovits: Compactness results in extremal graph theory, Combinatorica, *2* (3) (1982) 275-288.

[22] P. Erdős and M. Simonovits: Supersaturated graphs and hypergraphs, Combinatorica, *3* (2) (1983).

[23] P. Erdős and M. Simonovits: Cube-supersaturated graphs and related problems, this very volume.

[24] P. Erdős and A. H. Stone: On the structure of linear graphs, Bull. Amer. Math. Soc. *52* (1946), 1089-1091.

[25] R. J. Faudree and M. Simonovits: On a class of degenerate extremal problems, Combinatorica, *3* (1) (1983) 97-107.

[26] T. Kővári, V.T. Sós and P. Turán: On a problem of Zarankiewicz, Coll. Math. *3* (1954) 50-57.

[27] L. Lovász and M. Simonovits: On the number of complete subgraphs of a graph, Proc. Fifth British Combin. Conference, Aberdeen, (1975) 431-442.

[28] L. Lovász and M. Simonovits, On the number of complete subgraphs of a graph, II. Studies in Pure Math. (dedicated to the memory of Paul Turán) Akadémiai Kiadó+Birkhauser Verlag, 1983 459-495.

[29] I. Ruzsa and E. Szemerédi: Triple systems with no six points carrying 3 triangles, Coll. Math. Soc. J. Bolyai, *18* Combinatorics, Keszthely, Hungary, 1976, 939-945.

[30] M. Simonovits: A method for solving extremal problems in graph theory, Theory of Graphs, Proc. Coll. Tihany (1966), 279-319.

[31] M. Simonovits: Extremal graph problems, Proc. Calgary Internat. Conf. on Combinatorial Structures and Their Applications (1969), 399-401.

[32] M. Simonovits: Note on a hypergraph extremal problem, Hypergraph Seminar Columbus, Ohio, USA (1972). Lecture notes in Mathematics 411. Springer Verlag Berlin, 1974, 147-151.

[33] M. Simonovits: Extremal graph problems with symmetrical extremal graphs, additional chromatic conditions, Discrete Math. *7* (1974) 349-376.

[34] M. Simonovits: The extremal graph problem of the icosahedron, J. Combin. Theory, *17* (B)1 (1974) 69-79.

[35] M. Simonovits: Extremal graph problems and graph products, Studies in Pure Math. (dedicated to the memory of Paul Turán), Akadémiai Kiadó+Birkhauser Verlag 1982.

[36] M. Simonovits: Extremal Graph Theory, Selected Topics in Graph Theory, II, edited by L. W. Beineke and R.J. Wilson, Academic Press, London, New York,

San Francisco.

[37] M. Simonovits: Cycle-supersaturated graphs, under publication.

[38] R. Singleton: on minimal graphs of maximum even girth, J. Combin. Theory *1* (1966), 306-332.

[39] P. Turán: On an extremal problem in graph theory, Mat. Fiz. Lapok *48* (1941), 436-452.

[40] P. Turán: On the theory of graphs, Colloq. Math. *3* (1954), 19-30.

[41] H. P. Mulholland and C. A. B. Smith: An inequality arising in genetical theory, Amer. Math. Monthly *66* (1959) 673-683, (among others).

Elements of a Decomposition Theory for Matroids*

Klaus Truemper

ABSTRACT

We present a new decomposition theory for matroids with several attractive properties which we have just begun to explore. Results obtained to-date include: (1) decomposition theorems for matroids which make precise the intuitive notion that one can decompose a matroid or remove a number of specified elements without destroying 3-connectivity, (2) characterizations of minimal violation matroids for certain inherited properties, and (3) an algorithm that recursively constructs decomposition theorems. These results imply a number of important known matroid theorems as well as several nontrivial new ones.

1. Introduction

In this paper we propose a new decomposition for matroids with several attractive properties: (1) the decomposition is defined for matroids of arbitrary connectivity, (2) if a k-connected matroid ($k \geq 3$) is decomposed, then the pieces are 3-connected minors, (3) the decomposition can be dualized, i.e., if a matroid M can be decomposed into M_1 and M_2, then M^* can be decomposed into M_1^* and M_2^*, (the asterisk indicates the dual), and (4) the decomposition is easily reversed if the pieces are representable over a given field. By now it is obvious that the decomposition has numerous uses in matroid theory. Several ideas have already been worked out in detail, and a few of them have already been written up in a series of papers [18,19,20]. This paper essentially summarizes and interprets those published results. We decided to omit all proofs to achieve a short and uncluttered

*This work was supported in part by the National Science Foundation under Grant No. MCS-8305462.

exposition; the reader interested in details should refer to the cited papers.

The paper is organized as follows. The remainder of this section introduces definitions and the decomposition. Section 2, which is based on [18], contains a decomposition theorem for matroids which makes precise the intuitive notion that one can either decompose a matroid or remove a number of specified elements in any order without destroying 3-connectivity. The actual situation is not quite as simple, but the exceptions can be well characterized. The theorem can be considerably strengthened when the matroid is binary. In that case roughly one of 4 situations must occur: (1) the matroid is decomposable, or (2) a number of specified elements can be removed without destroying 3-connectivity, or (3) the matroid belongs to one of 2 well-described classes, or (4) the matroid has 12 elements or less. The latter theorem is sufficiently complex that a version restricted to graphs still is of interest. We utilize the latter 2 theorems for binary and graphic matroids in Section 3 to characterize minimal violation matroids of certain inherited properties (a property P is *inherited* if its is exhibited by all minors of any matroid that has P). This material is based on [19]. Let M be a binary minimal violation matroid of an inherited property P, i.e., all proper minors of M have P but M does not. We then have the following theorem: Suppose P has certain composition and extension properties. Then M is a minor of one of the matroids referred to under (3) above, or it has at most 12 elements. Graphicness, planarity, and regularity have all or almost all of the desired composition and extension properties, and thus by routine arguments one produces the well-known minimal violation matroids. In Section 4, which relies on [20], we examine conditions for decomposability, in particular necessary and sufficient (and sometimes just sufficient) conditions under which a decomposition of a proper minor of a matroid M can be extended to a decomposition of M. These conditions are then utilized in an algorithm that recursively produces decomposition theorems. The algorithm is of sufficient simplicity to permit hand computations for cases of significant interest. As a demonstration of the power of the algorithm we apply it to certain classes of graphic and binary matroids, and thus produce a number of nontrivial new and old decomposition theorems with surprising ease.

The work described here is connected in numerous ways with prior efforts, especially by P. D. Seymour and W. T. Tutte. It seemed best that we point out the relationships as we go along, except for some general comments concerning the basic decomposition idea. Even a cursory search reveals that composition and decomposition results for matroids abound in the literature (see [1,3,6] and the references cited therein), but the schemes we are aware of address low connectivity (typically 2-separability, e.g., in [1,6]), or they apply to composition but not decomposition (e.g., in [3]), or they do not allow dualizing (e.g., in [13]). The reader should not be misled by the preceding observation, which is *not* meant to be critical of the excellent work of the cited references. Indeed, for quite a while we tried to work with known decompositions, but finally felt that the schemes were not suitable for our purposes. For 2-separable matroids our approach is effectively the same as those described in the above references. From then on, however, our decomposition is materially different from any method we know of.

Next we introduce relevant definitions. Let X be a base of a matroid M on a set S, and $Y = S - X$. Construct a $\{0,1\}$ matrix $\hat{B} = [I|B]$ as follows. S is to be the set of indices of the columns of $\hat{B}$; in particular X is to the index the columns of the identity I, say in the order of $x_1, x_2, \ldots, x_m$. Then we index the rows by $x_1, x_2, \ldots, x_m$ as well. Let $y \in Y$, and suppose $\bar{X}$ is the subset of X that forms a circuit with y. Then in the column of B with index y we set element B_{xy} equal to 1 if $x \in \bar{X}$, and equal to 0 otherwise. Any $\hat{B}$ that may be so constructed from M is a *partial representation* of M, and it is nothing but a matrix representation of the fundamental circuit set of H. Whitney [26]. Note that this construction can always be carried out unless M consists only of loops. In the latter case we may formally take $\hat{B}$ to be a matrix without rows (we call a matrix without rows or columns *empty*). We have utilized partial representations in prior work [15-17], and the arguments to follow rely on the matrix theory for such representations developed in [17]. Briefly, with each square submatrix $\bar{B}$ of $\hat{B}$, say of row and column index sets $\bar{X} \subseteq X$ and $\bar{Z} \subseteq X \cup Y$, respectively, we associate a *determinant*, $\det \bar{B}$, which is 1 if the set $(X - \bar{X}) \cup \bar{Z}$ is a base of M, and equal to 0 otherwise. There may be another submatrix of $\hat{B}$ that (possibly after row and/or column permutations) looks like $\bar{B}$. The related row and column index sets cannot both be identical with $\bar{X}$ and $\bar{Z}$, respectively, and for this reason we will consider such a matrix to be different from $\bar{B}$. Thus for mathematical exactness we could specify $\bar{B}$ of $\hat{B}$ by the triple $(\bar{X}, \bar{Z}, X)$. We avoid this cumbersome notation unless confusion of $\bar{B}$ with some other matrix is possible. We also use expressions like "$\bar{B}$ is singular (nonsingular)" with the obvious interpretation. If $\bar{B}$ is a not necessarily square submatrix of $\hat{B}$, we define the *rank* of $\bar{B}$ to be the order of the largest nonsingular submatrix of $\bar{B}$. The matrix $[B^t|I]$ is a partial representation of M^*, the dual of M, and any square submatrix $\bar{B}$ of B is nonsingular if and only if $(\bar{B})^t$ is nonsingular in B^t (in the triple notation mentioned above $(\bar{B})^t$ becomes $(\bar{Z}, \bar{X}, Y)$). If we delete a column with index $y \in Y$ from $\hat{B}$, we obtain a partial representation of $M\backslash y$, where "\" denotes deletion. (If e is an element, we write $M\backslash e$ and M/e instead of $M\backslash\{e\}$ and $M/\{e\}$ to unclutter the notation). The determinants of the square submatrices of the reduced matrix are unchanged under such a column deletion. By duality a deletion of a row and column of $\hat{B}$ with index $x \in X$ produces a partial representation of M/x, where "/" denotes contraction. Again, the determinants are not affected by this operation. A *reduction* is a deletion or contraction, while an *addition (expansion)* is the inverse of a deletion (contraction). An *extension* is an addition or an expansion.

The reader may wonder why we worked out and present here most results in terms of partial representations. Why not use standard matroid terminology? There are several reasons for this choice. (1) Any partial representation typically displays a number of bases, circuits, cobases, cocircuits, we well as numerous relationships among them (see, e.g., the matrices of (4.4) below). We have found this simultaneous display of such relationships most helpful in the development of new concepts, theorems, and proofs. (2) The transition to matroids representable over a given fields is almost trivial: the 1's of any partial representation are replaced by appropriate nonzeroes of the field. This fact often provides insight into and motivates a strengthening of results when one restricts attention to certain classes of representable matroids (see, e.g., theorem 4.2 below). (3) The use of partial representations often motivates a constructive proof of a theorem, and if one works a bit carefully, one frequently finds a related polynomial and practically useful testing algorithm as a byproduct. - On the other hand, in our experience certain parts of matroid theory apparently are not easily handled via partial representations. Thus such representations could be viewed as an additional and sometimes quite helpful tool. After this brief excursion we continue with additional definitions.

Let A be a matrix. Then $\hat{A}$ denotes $[I|A]$, where I is an identity of appropriate order. In the display of matrices unspecified entries are always to be taken as 0. We typically write the index sets of the columns above a matrix and those of the rows to the left of it, and for $\hat{A}$ the column indices of I are always the same as the row indices of $\hat{A}$. We define $G(A)$ to be the following bipartite graph. Each row and each column of A generates a node, and each nonzero A_{ij} leads to an edge connecting nodes i and j. We say that A is *connected* if $G(A)$ is connected. Partial representations allow a simple characterization of matroid connectivity as follows.

LEMMA 1.1 (W.H. Cunningham [4], S. Krogdahl [7]): Let $\hat{B}$ be a partial representation of a matroid M. Then M is connected if and only if $\hat{B}$ is connected.

Let M be a matroid with rank function $r(\cdot)$ on a set S. If 2 elements of S form a circuit in M (M^*), they are said to be *parallel* (*series*) elements. Any circuit of cardinality equal to 3 in M (M^*) is a *triangle* (*triad*). M is *k-separable* [23] if S can be partitioned into S_1 and S_2 such that $|S_1|, |S_2| \geq k$ and $r(S_1)+r(S_2) \leq r(S)+k-1$. The pair (S_1, S_2) is then a *k-separation* of M. M is *k-connected* if it has no l-separation, $l \leq k-1$; in the case of $k=2$, M is simply said to be *connected*. For a given $k \geq 2$, M is $(k+)$-*separable* if (1) M is ($\lceil k/2 \rceil + 1$)-connected, (2) both the rank and corank of M are at least k, and (3) M has a k-separation where the sets S_1 and S_2 satisfy $|S_1|, |S_2| \geq k+1$. Here $\lceil n \rceil$ denotes the smallest integer greater than or equal to n. A k-separation or $(k+)$-separation (S_1, S_2) is *exact* if $r(S_1)+r(S_2) = r(S)+k-1$. M is a *1-sum* if it is the disjoint union of 2 matroids M_1 and M_2. This situation is denoted by $M = M_1 \oplus M_2$. M is a *k-sum*, $k \geq 2$, if M has a partial representation $\hat{B}$ where

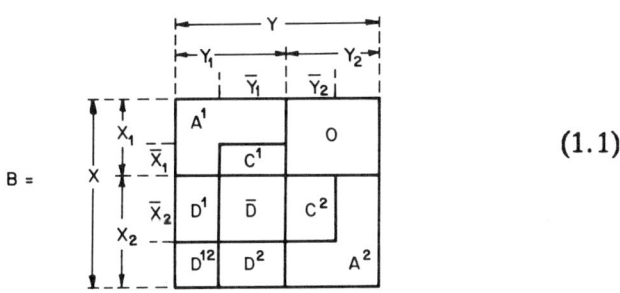
(1.1)

observes the following conditions.

(a) $C^1(C^2)$ is a connected nonempty proper submatrix of $A^1(A^2)$, and it has no nested rows (columns).

(b) $\overline{D}$ is a nonsingular matrix and rank $\overline{D}$ = rank $D = k-1$, where (1.2)

$$D = \begin{array}{|c|c|} \hline D^1 & \overline{D} \\ \hline D^{12} & D^2 \\ \hline \end{array}$$

We recall that 2 $\{0,1\}$ vectors c and d are *nested* if $c_j=1$ implies $d_j=1$, for all j, or $d_j=1$ implies $c_j=1$, for all j. By the previous observations the matrices $\hat{B}^1$ and $\hat{B}^2$ defined by

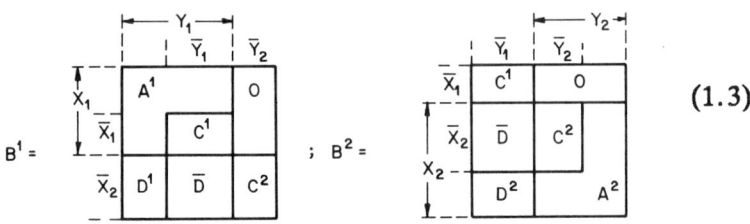 (1.3)

are partial representations of $M_1=M/(X_2-\bar{X}_2)\backslash(Y_2-\bar{Y}_2)$ and $M_2=M/(X_1-\bar{X}_1)\backslash(Y_1-\bar{Y}_1)$ respectively, and the determinants of their submatrices agree with those of $\hat{B}$. We call M_1 and M_2 the *components* of the k-sum and write $M=M_1\oplus_k M_2$. Note that $\oplus_k$ is not commutative, for all $k\geq 2$. It is trivial to verify that $M=M_1\oplus_k M_2$ if and only if $M^*=M_2^*\oplus_k M_1^*$. Note that the minor

$$\bar{M}=M/(X_1-\bar{X}_1)\cup(X_2-\bar{X}_2)\backslash(Y_1-\bar{Y}_1)\cup(Y_2-\bar{Y}_2)$$

is present in both M_1 and M_2. For $k\geq 3$ $\bar{M}$ is always 3-connected, and a comparison of (1.1) and (1.3) reveals that identification (in some sense) of the $\bar{M}$ minor of M_1 with the $\bar{M}$ minor of M_2 produces M from M_1 and M_2. For this reason we call $\bar{M}$ the *connecting matroid* of the decomposition. If M_1 and M_2 are representable over a given field, and if they have standard representation matrices $\hat{B}^1$ and $\hat{B}^2$ (not necessarily $\{0,1\}$), where B^1 and B^2 are given by (1.3) and satisfy (1.2) with $D^{12}=D^2\bar{\bar{D}}D^1$, $\bar{D}$ being the inverse of $\bar{D}$, then we can compose M_1 and M_2 to a matroid M (which we also denote by $M_1\oplus_k M_2$) by defining M to be the matroid represented by $\hat{B}$ with B of (1.1) over the field, where the submatrices of B are those of (1.3) and D^{12} is the matrix just specified. If both M_1 and M_2 are representable over 2 or more fields, this composition procedure may generate several matroids depending on the field over which B^1 and B^2 are expressed. Though the notation "$M_1\oplus_k M_2$" may thus be ambiguous when used for composition, this will not cause any difficulty since the underlying field will always be clear from the context.

A k-sum is *proper* if both submatrices A^1 and A^2 of B of (1.1) are connected; it is *semi-proper* if one of the following two situations prevails; either A^1 is connected and A^2 is equal to C^2 with one additional zero row adjoined to the bottom of C^2, or A^2 is connected and A^1 is equal to C^1 with one additional zero column adjoined to the left hand side of C^1.

To reduce confusion we derive the k-connectivity definitions, ($k \geq 2$) for graphs from those for matroids, so a graph is *k-connected*, $k \geq 2$, if that is true for the associated graphic matroid. However, when a graph is claimed to be *connected* we mean that every pair of nodes is joined by a path. This is *not* the same as saying that the (graphic) matroid is connected. Note that our definition of k-connectivity for graphs, $k \geq 2$, implies the customary one (defined via removal of nodes), and hence the existence of k internally node-disjoint paths between any two disjoint node sets. The exact relationships are nicely proved in [5].

Now and then we use the term "efficient algorithm". By this we mean an appropriate Turing machine which relies on an independence black box to decide dependence/independence of the sets of the given matroid, and whose total effort for producing the desired answer is bounded by a polynomial in the size of the groundset of the matroid.

Any other matroid terminology used later on may be found in the book by D. J. A. Welsh [25].

2. Decomposition Theorems

Intuitively one would expect that one can decompose any k-connected matroid, $k \geq 3$, or remove a number of specified elements in any order without destroying 3-connectivity, The following theorem makes this notion precise for general matroids.

THEOREM 2.1 (General Matroid): Every 3-connected matroid M observes at least one of the following conditions.

(1) M is a 3-sum.

(3) M has a (4+)-separation.

(3) M has at least one triangle or triad. For every triangle $\{e,f,g\}$: $M\backslash e$, $M\backslash f$, $M\backslash g$, $M/e\backslash g$, $M/f\backslash e$, $M/g\backslash f$, $M/\{e,f,g\}$ are all 3-connected, and $M\backslash\{e,f,g\}$ is connected. For every triad $\{e,f,g\}$: M/e, M/f, M/g, $M\backslash e/g$, $M\backslash f/e$, $M\backslash g/f$, $M\backslash\{e,f,g\}$ are all 3-connected, and $M/\{e,f,g\}$ is connected.

(4) M is 4-connected. For all pairs $\{e,f\}$ of distinct elements, $M\backslash\{e,f\}$ is 3-connected or has series elements, $M/\{e,f\}$ is 3-connected or has parallel elements, and $M\backslash e/f$ is 3-

connected. There exist 2 distinct elements e and f such that $M\backslash\{e,f\}$ or $M/\{e,f\}$ is 3-connected; for any 2 such elements there exists a third one, say $g\neq e,f$, such that at least one of the minors $M\backslash\{e,f,g\}$, $M\backslash\{e,f\}/g$, $M/\{e,f,g\}$, $M/\{e,f\}\backslash g$ as well as all extensions of that minor in M are 3-connected.

(5) M is 4-connected and has at most 12 elements. For all pairs $\{e,f\}$ of distinct elements, $M\backslash\{e,f\}$ has series elements, $M/\{e,f\}$ has parallel elements, and $M\backslash e/f$ is 3-connected.

(6) M has at most 9 elements, or the rank or corank of M is 3 or less.

On the surface (1) and (2) permit a wide range of possibilities but application of a simple *shifting algorithm* reduces that range to just a few well-described cases described next.

THEOREM 2.2: Let a matroid M be $(k+)$-separable, some $k\geq 2$. Then M has a partial representation $\hat{B}$ where B is the matrix

$$B = \begin{array}{c|c|c} & Y_1 & Y_2 \\ \hline X_1 & A^1 & 0 \\ \hline X_2 & D & A^2 \end{array} \qquad (2.1)$$

such that (1) the sets of X_i, Y_i, of (2.1) are nonempty, $i=1,2$, (2) the sets $X_i \cup Y_i$, $i=1,2$, define an $(l+)$-separation of M, where $\bar{k}=\lceil k/2 \rceil + 1 \leq l \leq k$, (3) A^1 of (2.1) is connected, or it is one of the matrices of (2.2) below,

$$\left. \begin{array}{l} \text{(i)} \quad [1\ldots 1 | 0 \ldots 0];\ \text{size } 1\times l;\ \bar{k}\leq p<l,\ \text{where } p \text{ is the number of 1's;} \\ \\ \text{(ii)} \quad \begin{bmatrix} 1\ldots 1 | 0\ldots 0 \\ 0\ldots 0 | 1\ldots 1 \end{bmatrix};\ \bar{k}\leq p_i<l,\ \text{where } p_i \text{ is the number of 1's of row } i,\ i=1,2; \end{array} \right\} \qquad (2.2)$$

and (4) A^2 of (2.1) is connected or is the transpose of one of the matrices of (2.2).

Much more can be said when the matroid is known to be binary. The improvements are largely due to the fact that any 2-separation (S_1,S_2) of a connected binary matroid without series or parallel elements must have $|S_i| \geq 5$, $i=1,2$. Note, however, that (3) and (3*) of theorem 2.3 below are weaker than (3) of theorem 2.1. In return we gain the interesting case (6) of theorem 2.3.

THEOREM 2.3 (Binary Matroids): Every 3-connected binary matroid M observes at least one of the following conditions.

(1) One of a), b) below applies.

   (a) M is a proper 3-sum with 3-connected components M_1 and M_2.

   (b) M is a semi-proper 3-sum, and the component M_i with connected A^i is 3-connected.

(2) M is a proper 4-sum with 3-connected components M_1 and M_2.

(3) M has a triangle $\{e,f,g\}$ such that $M\backslash e$, $M\backslash f$, $M\backslash g$, $M/e\backslash g$, $M/f\backslash e$, $M/g\backslash f$, $M/\{e,f,g\}$ are all 3-connected, and $M\backslash\{e,f,g\}$ is connected.

(3*) M is 3-connected and has a triad $\{e,f,g\}$ such that M/e, M/f, M/g, $M\backslash e/g$, $M\backslash f/e$, $M\backslash g/f$, $M\backslash\{e,f,g\}$ are all 3-connected, and $M/\{e,f,g\}$ is connected.

(4) (As in Theorem 2.1).

(5) M is 4-connected and has at most 12 elements. For all pairs $\{e,f\}$ of distinct elements $M\backslash\{e,f\}$ has series elements, $M/\{e,f\}$ has parallel elements, and $M\backslash e/f$ is 3-connected. If M is regular, then M is isomorphic to R_{10} (defined by (4.11) below).

(6) There exists a representation matrix $\hat{B}$ of M where B or B^t is one of the matrices below.

(i)

There are at least 3 blocks of type [11].

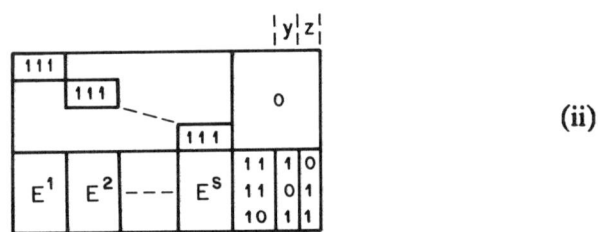

(ii)

Each E^i is nonsingular, and $s \geq 1$. If $s=1$ ($s \geq 2$), one (one or both) of the columns labelled y, z need not be present. If M is regular, then each E^i is the matrix $\begin{bmatrix} 1 & 0 & 0 \\ 1 & 1 & 0 \\ 1 & 1 & 1 \end{bmatrix}$ and y must be absent.

(7) M has at most 9 elements.

When the matroid is graphic and not just binary, additional refinements are possible. The resulting theorem can be nicely stated using graph terminology once we adopt the following conventions. To display any graph, say G, we utilize points for the nodes and line segments for the edges as usual. The drawing of G is frequently simplified by omission of most nodes and edges. Instead we create one or more connected closed areas using *heavy* lines. All nodes of G not explicitly shown are understood to be in the interior of these areas, and any edge not explicitly shown connects two nodes (explicitly shown or not) of the same closed area. If a closed area is not labelled, then its interior does not contain any nodes. Paths are indicated by dashed lines connecting two nodes. Any 2 such paths have disjoint internal nodes. Finally K_n ($K_{m,n}$) refers to the complete (complete bipartite) graph on n (m and n) nodes.

THEOREM 2.4 (Graphs) Every 3-connected graph G observes at least one of the following conditions.

(1) G is and a proper or semi-proper 3-sum, and may be decomposed in one of the three ways shown below.

(a) Proper 3-sum:

ELEMENTS OF A DECOMPOSITION THEORY

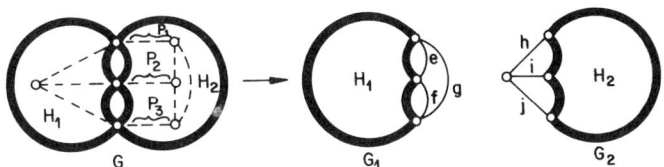

G_1 and G_2 are 3-connected, and $G_1/\{e,f,g\}$ and $G_2\backslash\{h,i,j\}$ are 2-connected. P_1, P_2, and P_3 may be null paths.

(b) Semi-proper 3-sum, case 1:

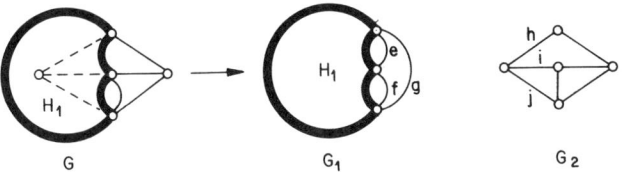

G_1 is 3-connected and $G_1/\{e,f,g\}$ is 2-connected.

(c) Semi-proper 3-sum, case 2:

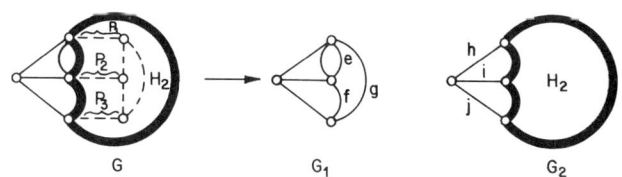

G_2 is 3-connected and $G_2\backslash\{h,i,j\}$ is 2-connected. P_1, P_2, P_3 may be null paths.

(2) G is a proper 4-sum, and may be decomposed as follows.

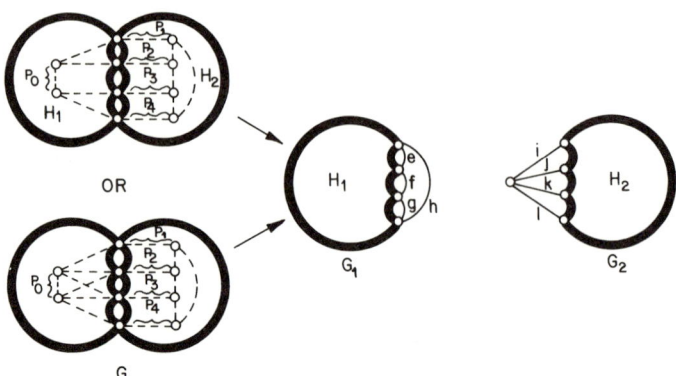

G_1 and G_2 are 3-connected, and $G_1/\{e,f,g,h\}$ and $G_2\backslash\{i,j,k,l\}$ are 2-connected. $P_0, P_1, \ldots, P_4$ may be null paths.

(3) (3*),(4) (Easily derived from Theorem 2.3. The triad of Theorem 2.3 (3*) is a star in G.)

(5) (cannot occur).

(6) G is one of the following graphs.

  (i) $K_{3,n}$; $n \geq 4$.

  (ii)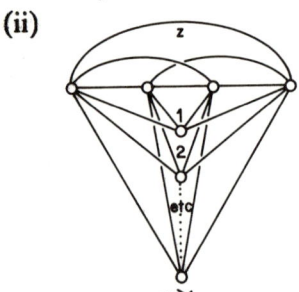

If $s \geq 2$, the edge labelled z may not be present.

(7) G is one of the following graphs: K_n, $1 \leq n \leq 4$; the wheel with four spokes; K_5 with one edge deleted; the (planar) dual of the latter graph; $K_{3,3}$.

The vacuous case (5) of Theorem 2.4 arises from the fact that the numbering in that theorem is in agreement with that of Theorem 2.3 to simplify comparisons.

Finally we should mention that an efficient algorithm exists for each of the Theorems 2.1, 2.3 and 2.4 that selects an applicable alternative of the theorem for a given matroid.

3. Minimal Violation Matroids.

The three theorems of Section 2, in particular Theorems 2.3 and 2.4 for binary and graphic matroids, respectively, appear to be useful in several ways. Here we demonstrate how one may utilize Theorems 2.3 and 2.4 to characterize minimal violation matroids of inherited properties. The basic idea can be outlined as follows. From the decomposition Theorem 2.3 we know that every binary matroid (1) is decomposable, or (2) has series/parallel elements or has a minor $\overline{M}$ with 3 elements less than M such that every minor of M containing $\overline{M}$ is 3-connected (there is an insignificant exception when triangles or triads are involved), or (3) M belongs to one of 2 well-specified classes, or (4) M has at most 12 elements. Now let P be an inherited property defined for a class **M** of binary matroids. We say P has the *composition property* **CP** if any decomposable $M \in \mathbf{M}$ has P provided the components of the decomposition have P. Furthermore P is said to have the *extension property* **EP** if $M \in \mathbf{M}$ has P provided all proper minors of M have P and case (2) above applies. The simplest theorem is then the following: Let P have the composition and extension properties. Then any minimal violation matroid $M \in \mathbf{M}$ of P belongs to one of the 2 classes of (3) above or has at most 12 elements. Though this theorem has some uses, its assumptions are quite restrictive, and one would want to weaken them to extend the range of applicability. This indeed can be done, and one obtains a list of potentially useful theorems; the most promising ones are included here. In the next section we extend the concept of composition property so that we may utilize decomposition theorems generated by the recursive algorithm of that section for the characterization of minimal violation matroids. Despite all this our treatment of minimal violation matroids is of course very much incomplete, and the reader should view the results below just as supporting evidence for the fundamental notion that characterizations of inherited properties are often obtainable by appropriate decomposition approaches. We felt that working out some significant examples would be helpful. The properties of graphicness, planarity, and regularity seemed sufficiently complex to test our ideas, so we apply our theorems to them to obtain the well-known minimal violation matroids by K. Kuratowski [8] and W. T. Tutte [21,22]. While proving these theorems we also obtain some results that are likely to be useful for other investigations.

We start with precise definitions of the concepts just mentioned. Let P be an inherited property defined on a class **M** of binary matroids where throughout we assume that **M** is closed under the taking of minors. Formally one could take P to be a subset of **M**, or a function from **M** to $\{0,1\}$. We only consider matroid properties that are invariant under any relabelling of the matroid elements. Therefore in this section we need not distinguish between isomorphic matroids, and for convenience we temporarily consider any two isomorphic matroids to be equal.

DEFINITION 3.1: P has the *composition property* $\mathbf{CP}_1, \mathbf{CP}_2, \mathbf{CP}_3, \mathbf{CP}_{3_s}, \mathbf{CP}_4$ if any $M \in \mathbf{M}$ has P provided M is a 1-sum, connected proper 2-sum, 3-connected proper 3-sum, 3-connected semi-proper 3-sum, 3-connected proper 4-sum, respectively, and the components of the decomposition have P. Finally P has the *composition property* **CP** it it has $\mathbf{CP}_1$, $\mathbf{CP}_2$, $\mathbf{CP}_3$, $\mathbf{CP}_{3_s}$, and $\mathbf{CP}_4$.

DEFINITION 3.2: P has the *extension property* $\mathbf{EP}_2$ if any series or parallel extension in **M** of a matroid $M \in \mathbf{M}$ has P provided M has that property. P has the *extension property* $\mathbf{EP}_3$, $\mathbf{EP}_{3^*}$, $\mathbf{EP}_4$, $\mathbf{EP}_{4+}$ if any $M \in \mathbf{M}$ which has 10 or more elements and which is not a proper 3- or 4-sum or a semi-proper 3-sum, has P provided all proper minors of M have P and M satisfies (3), (3^*), (4), (5), respectively, of Theorem 2.3. Finally P has the *extension property* **EP** if it has $\mathbf{EP}_2$, $\mathbf{EP}_3$, $\mathbf{EP}_{3^*}$, and $\mathbf{EP}_4$.

We denote by **N** the class of binary matroids specified by (6) of Theorem 2.3. With these definitions we have the following results.

THEOREM 3.1: Let P be an inherited property defined on a class **M** of binary matroids. Suppose P has the extension property **EP**, and assume $M \in \mathbf{M}$ is a minimal violation matroid of P.

(1) If P has $\mathbf{EP}_{4+}$ and **CP**, then $M \in \mathbf{N}$ or M has at most 9 elements.

(2) If P has **CP**, then M satisfies (5) of Theorem 2.3, or $M \in \mathbf{N}$, or M has at most 9 elements.

(3) If P has $\mathbf{CP}_1$, $\mathbf{CP}_2$, $\mathbf{CP}_3$, and $\mathbf{CP}_{3_s}$, then M is a 3-connected proper 4-sum with 3-connected components, or M satisfies (5) of theorem 2.3, or $M \in \mathbf{N}$, or M has at most 9 elements.

(4) If P has CP_1, CP_2, and CP_{3_s}, then M is a 3-connected proper 3- or 4-sum with 3-connected components, or M satisfies (5) of theorem 2.3, or $M \in N$, or M has at most 9 elements.

Suppose **M** is a subset of the class of regular matroids. If case (5) of Theorem 2.3 applies to M, then $M = R_{10}$; if $M \in N$, then M or M^* is represented by one of the graphs of (6) of Theorem 2.4. If **M** is a subset of the graphic or cographic matroids, then case (5) of Theorem 2.3 cannot occur.

One can also investigate sufficiency conditions for decomposition with the aid of Theorems 2.1, 2.3 and 2.4. It appears that the techniques of the next section are far more powerful, but a related minimal violation theorem may still be of interest. Define $Q_2(P)$ and $Q_3(P)$ to be the following properties for a given inherited property P on a class **M** of binary matroids: a matroid $M \in \mathbf{M}$ has $Q_2(P)$ if each minor of M is a 1-sum, or a proper 2-sum, or has series or parallel elements, or has P. In the case of $Q_3(P)$ we demand that each minor of M is a 1-sum, or a proper 2-sum, or a 3-connected proper 3-sum, or has series or parallel elements, or has P. By these definitions properties $Q_2(P)$ and $Q_3(P)$ are clearly inherited.

THEOREM 3.2: Let P be an inherited property on a class **M** of binary matroids, and assume that P has **EP** and CP_{3_s}.

(1) Every minimal violation matroid $M \in \mathbf{M}$ of $Q_2(P)$ is a 3-connected proper 3- or 4-sum with 3-connected components, or M satisfies (5) of Theorem 2.3, or $M \in N$, or M has at most 9 elements.

(2) If for every 3-connected proper 3-sum of **M** each component has at least 8 elements, then every minimal violation matroid $M \in \mathbf{M}$ of $Q_3(P)$ is a 3-connected proper 4-sum with 3-connected components, or M satisfies (5) of Theorem 2.3, or $M \in N$, or M has at most 9 elements.

The statements about **N** and (5) of Theorem 2.3 made in the last part of Theorem 3.1 apply here as well. If **M** is a subset of the class of regular matroids, then the condition in (2) is always satisfied.

Quite straightforward checking produces the next theorem where L, L_p, and R are the properties of graphicness, planarity, and regularity, defined on the class **M** of binary matroids.

THEOREM 3.3:

(a) L, L_p and R have **EP** and **EP**$_{4+}$.

(b) L and L_p have **CP**.

(c) A binary k-sum has R is this is so for the components, $1 \leq k \leq 3$. In particular R has **CP**$_k$, $1 \leq k \leq 3$, and **CP**$_{3s}$. However, R does not have **CP**$_4$.

The lack of **CP**$_4$ for R does not cause any complication here since a binary 3-connected proper 4-sum has R if all of its proper minors have R. This observation plus Theorems 3.1 and 3.3 and simple checking easily lead to the well-known characterizations of L, L_p, and R. Below we refer to the Fano matroid by F_7.

THEOREM 3.4 (W. T. Tutte [22]; A binary matroid has L (L_p) if and only if it has no F_7, F_7^*, $K_{3,3}^*$, or K_5^* (F_7, F_7^*, $K_{3,3}$, $K_{3,3}^*$, K_5, or K_5^*) minor.

COROLLARY 3.4.1 (K. Kuratowski [8]: A graph is planar if and only if it has no $K_{3,3}$ or K_5 minor.

THEOREM 3.5 (W. T. Tutte [21]: A binary matroid has R if and only if it has no F_7 or F_7^* minor.

P. D. Seymour has given a reasonably short proof of Theorem 3.4 in [12], and C. Thomassen [14] has published elegant proofs of Corollary 3.4.1 and numerous other characterizations of planarity while assuming **M** to be the class of graphic matroids. Theorem 3.5 has been proved via characterizations of $GF(3)$-representability by R. Reid (who did not publish his proof), R. E. Bixby [2], P. D. Seymour [11], and in [15]. However, the proof via Theorems 3.1 and 3.3 is much shorter than those cited (of course, we prove less, too).

4. A Recursive Construction of Decomposition Theorems

In this section we develop necessary and sufficient (and sometimes just sufficient) conditions for decomposability. First we present conditions under which a decomposition of a minor of a matroid M cannot be extended to a decomposition of M. Next these conditions are embedded into an algorithm that in each iteration derives new nondecomposable matroids as well as new proper minors of such matroids from the ones found so far. The scheme starts with a user-supplied list of matroids within a class **M**. It is so designed that either it proves all matroids of **M** to be decomposable save those of a finite subclass (specified by the algorithm), or it continues indefinitely. In the latter situation the output of the algorithm is still of interest since it gives decomposability results for a subclass of **M** that is characterized by excluded-minor conditions. Finally we apply the algorithm to classes of graphic and binary matroids, and produce a number of nontrivial new and old decomposition theorems with comparative ease.

Let M be a matroid on a set S, and N be a minor $M/X_3\backslash Y_3$ where X_3 contains no circuit, Y_3 contains no cocircuit, and $X_3 \cap Y_3 = \emptyset$. Suppose for some $k \geq 1$ we know of an exact k-separation (T_1, T_2) of $T = S - (X_3 \cup Y_3)$ of N. Below all k-separations and $(k+)$-separations are required to be exact, so as a matter of convenience we omit "exact" throughout this part. We then may ask whether or not one can extend this k-separation to one for M, or more precisely, whether or not one can assign the elements of $S-T$ to T_1 and T_2 so that the resulting sets S_1 and S_2 constitute a k-separation (S_1, S_2) of M. We say "the k-separation of N induces a k-separation of M" if such an assignment is possible. We first investigate 3 problems related to such induced separations.

P1: Efficiently decide whether or not a given k-separation of N induces a k-separation of M.

We describe an efficient scheme that solves P1 and give a simple and useful characterization of M when the k-separation of M does not exist. Suppose the latter situation prevails. Two seemingly difficult problems for which we have some partial solutions are as follows.

P2: Efficiently decide whether or not M has a proper minor $\overline{M}$ such that (i) N is a minor of $\overline{M}$, and (ii) the k-separation of N does not induce a k-separation of $\overline{M}$.

P3: Characterize the minimal minors $\overline{M}$ of M that have a minor N_1 isomorphic to N such that a k-separation of N_1 which corresponds to the k-separation of N under one of the isomorphisms between N_1 and N, does not induce a k-separation of $\overline{M}$.

Let's start with P1. The assumed k-separation may be viewed as follows. Choose a base $X_2 \subseteq T_2$, then add a set $X_1 \subseteq T_1$ so that $X_1 \cup X_2$ is a base of N. The partial representation $\hat{B}_N$ produced from this basis has

$$B_N = \begin{array}{c|cc} & Y_1 & Y_2 \\ \hline X_1 & A^1 & 0 \\ \hline X_2 & D & A^2 \end{array} \quad ; \quad X_i \cup Y_i = T_i \; ; \; i = 1,2 \tag{4.1}$$

with rank $D = k-1$. To simplify the exposition we assume throughout this section that $X_i, Y_i \neq \emptyset$, $i = 1,2$. The results described below are easily adapted to the situation where one or two of the four sets are empty. (The case with two empty sets is quite uninteresting, though). Since $N = M/X_3 \backslash Y_3$, M has a partial representation $\hat{B}_M$ with

$$B_M = \begin{array}{c|c|cc|c} & Y_1 & Y_{31} & Y_{32} & Y_2 \\ \hline X_1 & A^1 & \tilde{A}^1 & 0 & 0 \\ \hline X_3 \begin{array}{c} X_{31} \\ X_{32} \end{array} & & \tilde{D} & \tilde{A}^2 & \\ \hline X_2 & D & & & A^2 \end{array} \tag{4.2}$$

The submatrix composed of $A^1, 0, D, A^2$ has the same determinantal structure as B_N of (4.1). Then the k-separation of N induces one for M if and only if we can partition X_3 into X_{31}, X_{32} and Y_3 into Y_{31}, Y_{32} as shown in (4.2) such that

$$\text{rank } \tilde{D} = \text{rank } D \; (= k-1) \tag{4.3}$$

It is well known that the question "Can B_M of (4.2) be partitioned so that (4.3) is satisfied?" is equivalent to a matroid intersection problem since it can be restated as rank $D \stackrel{?}{=} \min_{X_{3i}, Y_{3i}} \{r_M(X_1 \cup Y_1 \cup X_{31} \cup Y_{31})$

$+ r_M(X_2 \cup Y_2 \cup X_{32} \cup Y_{32}) | X_{31} \cup X_{32} = X_3, Y_{31} \cup Y_{32} = Y_3\}$

$- r_M(S)$

which by J. Edmonds' matroid intersection theorem is equivalent to

rank $D \stackrel{?}{=} \max\{|J| \,|\, J$ independent in M_1 and $M_2\}$

$+ \sum_{i=1,2} r_M(X_i \cup Y_i) - r_M(s)$

where $r_M(.)$ is the rank function of M, $M_1 = M/X_1 \cup Y_1 \setminus X_2 \cup Y_2$, and $M_2 = M/X_2 \cup Y_2 \setminus X_1 \cup Y_1$.

Hence any one of the known efficient matroid intersection/partitioning algorithms (see [9] and [25]) could be employed to answer the question. However, the problem is much simpler and can be resolved by a quite straightforward theorem, which together with the subsequent Lemma 4.1 is roughly equivalent to the statements (8.1)-(8.10) of P. D. Seymour's paper [13]. The proof, though, is quite different from that of [13]. It is based on an utterly simple algorithm that either produces the desired partition of X_3 and Y_3 or creates one of the violation matrices of the theorem. The following definition will simplify the exposition. Let B^1, and $B^2 = [B^1 | C]$ or $B^2 = [\frac{B^1}{C}]$ be submatrices of a partial representation. We then say "B^1 spans C" if B^2 has the same rank as B^1.

THEOREM 4.1: B_M of (4.2) cannot be partitioned such that (4.3) holds if and only if one of the six matrices below may be derived from B_M by deletion of some rows of X_3 and some columns of Y_3. n always denotes a positive integer.

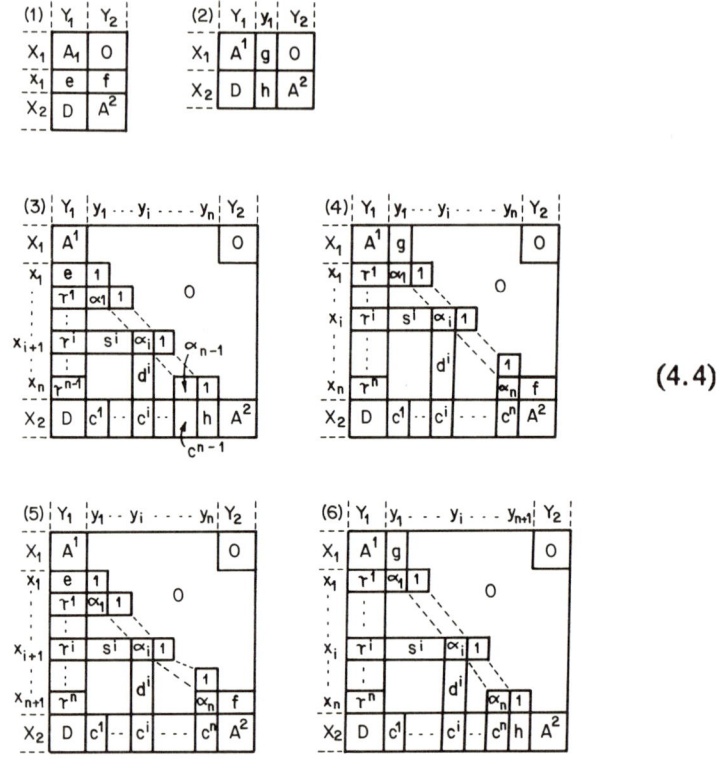

(4.4)

The following statements hold for (1)-(6):
(a) D does not span e or h, and f and g are nonzero.
(b) Let D^i be any submatrix of (3)-(6) defined by the column index of α_i and all columns to the left of it, and the row index of α_i and all rows below it. Then $\text{rank} D^i = \text{rank} D + 1$, but for any proper submatrix $\overline{D}$ of D^i that does not contain α_i, $\text{rank}\overline{D} \leq \text{rank} D$. In particular, $s^i = 0$ if $r^i = 0$, and $d^i = 0$ if $c^i = 0$.   (4.5)

Suppose B_M itself is one of the matrices (1)-(6). Then B_M is minimal in the following sense: B_M cannot be partitioned but deletion of any row of X_3 or any column of Y_3 makes a partition possible Note, however, that the minor N may be producible from M by a number of different reduction sequences, and that we may be able to find a smaller instance of (1)-(6) by searching through all the related partial representations. Problem P2 asks for an efficient test to decide the outcome of such a search. In general, P2 seems to be difficult, but we have been able to construct a polynomial (but not practically usable for $k \geq 3$) test for the special case when (i) **M** is a subset of the class of matroids representable over a given finite field **F**, (ii) each $M \in \mathbf{M}$ is specified by a standard representation matrix over **F**, (iii) if N has two standard representation matrices over **F** with same support, then one matrix is a scaled version of the other one, and (iv) k is bounded by some given constant. If the given field is $GF(2)$ of $GF(3)$, the requirements of (ii) and (iii) are trivial. Thus an efficient test is always possible if **M** is contained in the class of binary or ternary matroids and k is bounded by some constant. The proof of the above claims makes use of an easily proved, but very useful lemma by P. D. Seymour [13], here expressed in our matrix notation.

LEMMA 4.1 (=Statement (8.4) of [13]). Let $\hat{B}_M$ be a partial representation of a matroid M where B_M is one of the matrices (1)-(6) of (4.4) and satisfies (4.5), and define N to be the minor of M corresponding to the submatrix of B_M composed of A^1, D, 0 and A^2. Suppose the k-separation (T_1, T_2) of N induces a k-separation in every proper minor of M that in turn has N as a minor. Then $M\backslash x$ and M/y do not have N as a minor, for all $x \in X - (X_1 \cup X_2)$ and $y \in Y - (Y_1 \cup Y_2)$. In particular

$$\text{each } c^i \neq 0 \text{ unless it resides in the same column as g;} \quad (4.6)$$
$$\text{each } r^i \neq 0 \text{ unless it resides in the same row as f.}$$

The next theorem is the basis for the polynomial testing algorithm for P2 mentioned above.

THEOREM 4.2: Let M and N be the matroids of Lemma 4.1.

(a) If $k=2$, then M has at most 5 elements beyond those of N.

(b) If $k \geq 3$, and M is representable over a finite field **F**, then the number of elements M has beyond those of N, is bounded by a function **n** that depends only on k and the size m of **F**; for example, $\mathbf{n}(k,m) = 2^{m+2}k^{4m+1} + 4$ will do.

The bound of 5 in Theorem 4.2 is best, even for graphic matroids. Furthermore, the reader is most likely to wonder whether or not the substantial difference between the above bounds for $k=2$ and

$k \geq 3$ isn't simply a result of our approach. More specifically, one might ask whether or not there are bounds for general matroids and values of k greater than 2. Such bounds might be allowed to depend on k, or the size of N. However, there exists a class of examples that demonstrate that no such bounds exist in general for any $k \geq 3$. In each case, N has only $2(k+1)$ elements.

For P3 we have some partial answers that go beyond the results for P2, and that will be of considerable use for our algorithm. The next result implies, but is not implied by, Theorem (9.1) of the previously mentioned paper [13] by P. D. Seymour. We shall not bother to make a detailed comparison, but only remark that the condition on circuits in part (i) of Theorem (9.1) precludes application of theorem (9.1) to a number of interesting cases. For example, that theorem cannot be applied if D consist of a nonsingular matrix bordered by zero rows and columns.

LEMMA 4.2: Suppose that N with B_N of (4.1) is a minor of a matroid M, and that the k-separation (T_1, T_2) of N does not induce a k-separation for M. Furthermore assume that the following holds for every proper minor $\overline{M}$ of M that in turn has a minor N_1 isomorphic to N: every k-separation of N_1 corresponding to (T_1, T_2) of N under one of the isomorphisms between N_1 and N induces a k-separation of $\overline{M}$. Then any B_M of M of type (3), (4), (5), or (6), of (4.4) satisfies the following.

(a) If e or r^i is not in the same row as f, then that vector is not parallel to any row of A^1 and is not a unit vector with 1 in a column y, provided y is not a loop of N/T_2.

(b) Vector $\begin{bmatrix} g \\ \hline c^1 \end{bmatrix}$ is not parallel to a column y of $\begin{bmatrix} A^1 \\ \hline D \end{bmatrix}$ if y is not a coloop of $N\backslash T_2$, and it is not a unit vector with 1 in a row x if x is not a coloop in $N\backslash T_2$.

(4.7)

(c) If h or c^j is not in the same column as g, then that vector is not parallel to a column y of A^2 and is not a unit vector with 1 in a row x, provided x is not a coloop of $N\backslash T_1$.

(d) Vector $[r^n|f]$, $n \geq 1$, is not parallel to a row

x of $[D|A^2]$ if x is not a loop of N/T_1,
and it is not a unit vector with 1 in a column y
if y is not a loop of N/T_1.

The algorithm relies on one additional lemma stated below, and some 3-connectivity results (P. D. Seymour's Theorem (7.3) of [13] and parts of [17]).

LEMMA 4.3: Suppose a matroid N is a proper k-sum, $k \geq 2$, say with partial representation $\hat{B}$, where B is the matrix of (1.1). If the k-separation $(X^1 \cup Y^1, X^2 \cup Y^2)$ of N induces a k-separation of a connected matroid M that has N as a minor, then M is a proper k-sum, which has 3-connected components if $k \geq 3$ and M is 3-connected.

We now come to the second part of this section, which concerns the algorithm itself. A few definitions are needed. Differences between isomorphic matroids will no longer be of interest, so from now one we treat any 2 such matroids as if they were equal. Let $\bar{S}$ and $\bar{T}$ be subsets of the same cardinality of the groundsets of 2 equal (i.e., isomorphic) matroids N_1 and N_2 respectively. We say that $\bar{S}$ *corresponds to* $\bar{T}$ if there exists an isomorphism between N_1 and N_2 that maps $\bar{S}$ onto $\bar{T}$. Statements like "the k-separation (S_1, S_2) of N_1 corresponds to the k-separation (T_1, T_2) of N_2" are analogously interpreted. A *wheel* with $s \geq 3$ *spokes* is the graphic matroid of the following graph. Take a cycle (the *rim*) with s edges; introduce one additional node, and connect it with each node of the cycle. A *whirl* with $s \geq 3$ spokes is derived from a wheel with s spokes by declaring the rim to be independent. We consider U_4^2, the uniform matroid of rank 2 on 4 elements, to be a whirl with 2 spokes.

Suppose we have a class **M** of matroids that is closed under the taking of minors, and a nonempty subclass **V**$^\bullet$, where each matroid in **V**$^\bullet$ is 3-connected and has at least 4 elements. Further suppose that a matroid of **V**$^\bullet$, say N, has a k-separation (T_1, T_2) as in (4.1), for some $k \geq 3$. In principle we could identify all minimal matroids $M \in $ **M** such that (i) M has N as a minor, and (ii) there exists a minor N_1 of M isomorphic to N such that a k-separation of N_1 that corresponds to (T_1, T_2) of N, does not induce a k-separation of M. Suppose we now replace N in **V**$^\bullet$ by these minimal matroids, getting, say **V**. A moment's reflection convinces us that the following claim is valid: For every $M \in $ **M** at least one of the statements below holds: (1) M has no minor in **V**$^\bullet$; (2) M has a minor in **V**; (3) M has a minor equal to N; for every such minor, say N_1, every k-separation of N_1 corresponding to (T_1, T_2) of N induces a k-separation of M. We cold repeat the above process, this time starting with **V** instead of **V**$^\bullet$, and thus could generate another potentially interesting claim. Indeed, one could continue until **V** becomes empty or one becomes tired of carrying out the computations. We omit details of the rather obvious inductive arguments and instead examine the first iteration (involving N) more closely. A generally formidable task in that iteration is the identification of the minimal matroids to be added to $\mathbf{V}^\bullet$ upon removal of N. But we have a list of necessary conditions (4.4)-(4.7) that such minimal matroids must observe, so we could add to $\mathbf{V}^\bullet$ all matroids of **M** observing these conditions for the given N and its k-separation (T_1, T_2). Denote by $\overline{\mathbf{V}}$ the collection of these matroids. Clearly $\overline{\mathbf{V}}$ includes the minimal matroids, and the above claim remains valid when we use $\mathbf{V} = (\mathbf{V}^\bullet - \{N\}) \cup \overline{\mathbf{V}}$. With this change we still face a second problem. The number of minimal matroids, and hence the number of matroids in $\overline{\mathbf{V}}$, could be infinite or so large that just listing of the matroids cannot be carried out on any computer in reasonable time. A simple observation helps overcome this obstacle: the previous claim remains valid if we creat **V** by adding to $\mathbf{V}^\bullet - \{N\}$ some *minor* of V for every $V \in \overline{\mathbf{V}}$. There are, of course, many ways to generate a small set of such minors. We have found the following recursive procedure quite attractive since each iteration of the procedure produces a potentially interesting claim of the type mentioned above.

In the first iteration we place into **V** upon removal of N all matroids arising from (1) and (2) of (4.4), plus certain cases arising from (7)-(10) of (4.8) below with $n = 1$. (The numbering of (4.8) continues that of (4.4) for reasons to become apparent shortly). In (7)-(10) below the submatrix composed of A^1, O, D, and A^2 corresponds to the minor N analogous to (1)-(6) of (4.4).

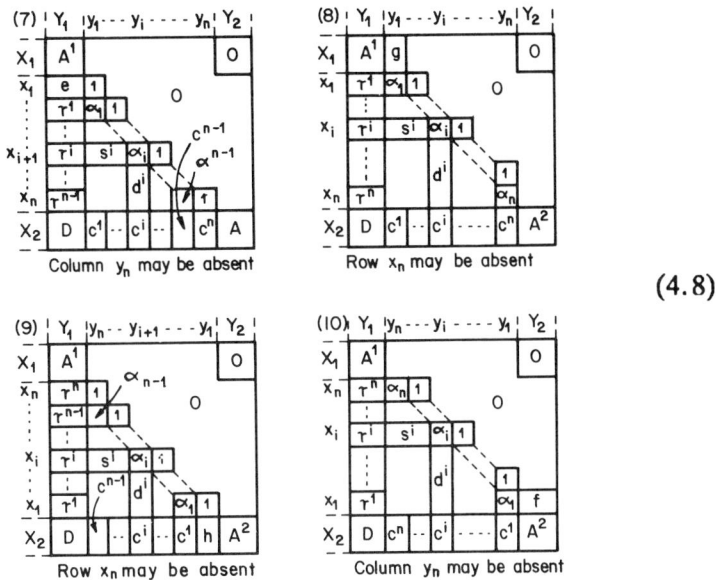

(4.8)

The precise rule is as follows. Select either all matroids of **M** corresponding to (7) without y_1, and to (8) without x_1, or all matroids of **M** corresponding to (9) without x_1, and to (10) without y_1. In all instances we demand (4.5)-(4.7). A comparison of (7) and (8), or of (9) and (10), with (3)-(6) of (4.4) confirms that **V** now contains a minor of each matroid of $\overline{\mathbf{V}}$. Thus the previous claim still holds for this **V**. The choice of (7), (8) versus (9), (10) may be based on any criterion as long as the following condition is satisfied. For (7), (8) we demand that (4.9) below holds, and for (9), (10) we require (4.10).

A^1 has no zero column (equivalently: N/T_2 has no loops) (4.9)

A^2 has no zero row (equivalently: $N\backslash T_1$ has no coloops) (4.10)

Why did we select such conditions? Space limitations prevent a detailed discussion, so suffice it to say that (4.7) becomes more effective when (4.9) or (4.10) holds and the above described rule concerning (7)-(10) is enforced. We must make sure, of course, that N is only accepted for processing when B_N of the k-separation satisfies (4.9) or (4.10).

Inductively assume that after one or more iterations **V** contains a minor of type (7) or (8) of (4.8) for each matroid of $\overline{\mathbf{V}} - \mathbf{V}$. (The case of (9), (10) is handled analogously, so we omit details). Remove such a minor from **V**, say V. If it is of type (7), add to **V** all 1-element extensions in **M** of type (3) or (5) of (4.4) and of type (7) or (4.8). If it is of type (8), instead add to **V** all 1-element extensions in **M** of type (4) or (6) of (4.4) and of type (8) of (4.8). In every instance the related matrix must observe (4.5)-(4.7). It is easy to see that the new **V** satisfies the inductive assumption, and thus may again be used in the previous claim.

In steps 1-4 of the decomposition algorithm below we have implemented the above procedures. To differentiate between matroids produced via (4.4) and (4.8) we store matroids using triples (V, a, b) with the following convention. $(V, \varnothing, \varnothing)$ corresponds to a matroid V of the original $\mathbf{V}^{\bullet}$ or one produced via (4.4), while (V, N, B_V) corresponds to a matroid V produced via (4.8). N of the latter triple is the matroid inducing the decomposition and B_V is the matrix of (4.8) giving rise to V. The collection of all triples is always called **W**. A few technical details of steps 3 and 4 require further discussion. The partitioning of $\mathbf{W}_N^2$ and the related redefinition of $\mathbf{W}_N^1$ and $\mathbf{W}_N^2$ essentially changes some triples of type (V, N, B_V) to $(V, \varnothing, \varnothing)$. It can be shown that the required absence of series and parallel elements of V assures V to be 3-connected. Indeed, one can prove that V of any $(V, \varnothing, \varnothing)$ produced by the algorithm must be 3-connected due to (4.4)-(4.10). An aspect of step 4 not touched upon so far concerns the elimination of isomorphic cases. The previous claim is not invalidated by the existence of isomorphic matroids in **V**, but computationally the presence of such matroids is quite undesirable. A bit of checking reveals that isomorphism of triples (V, N, B_V) requires careful definition so that decomposition claims remain valid upon removal of isomorphic cases from **W**. Thus we say that 2 triples $(V_1, \varnothing, \varnothing)$ and $(V_2, \varnothing, \varnothing)$ are *isomorphic* if V_1 and V_2 are isomorphic, and 2 triples (V_1, N, B_{V_1}) and (V_2, N, B_{V_2}) are *isomorphic* if there exists an isomorphism between V_1 and V_2 that maps

T_1 and T_2 of N in V_1 onto T_1 and T_2, respectively, of N in V_2, and that maps each x_i and y_i of B_{V_1} onto x_i and y_i, respectively, of B_{V_2}. Obviously an isomorphism of the latter type exists if and only if $\hat{B}_{V_1}$ is obtainable from $\hat{B}_{V_2}$ by pivots within A^1 or A^2, and by (possibly repeated) use of the following operation: exchange two rows or columns both of whose indices are either in T_1 or T_2. In part (a) of step 4 we then eliminate as many isomorphic triples as is computationally feasible or attractive. About part (b) of step 4 we only mention that an inductive argument establishes validity of that reduction process.

For a justification of step 5 of the algorithm let us return to the beginning of the discussion, where we assumed the existence of a k-separation of type (4.1) for N. (Indeed, we later on required B_N of such a k-separation to satisfy (4.9) or (4.10)). Suppose that N has no such k-separation, or that we cannot find one while expending a reasonable computational effort. We then replace $(N,\varnothing,\varnothing)$ of **W** by all $(V,\varnothing,\varnothing)$ such that V is a 3-connected 1-element extension of N in **M**. If N is a wheel (whirl), we also place $(V,\varnothing,\varnothing)$ into **W** where V is the next larger wheel (whirl), provided the latter matroid occurs in **M**. By a theorem of P.D. Seymour [13] (see also [17]) we then can claim the following: Any $M \in \mathbf{M}$ with N as a minor is 2-separable, or is equal to N, or has a minor in $\{V | (V,\varnothing,\varnothing) \in \mathbf{W}\}$. In step 5 we have implemented this process, thus assuring that the algorithm can always continue unless **W** becomes empty. Finally in **N** we collect all matroids for which a k-separation is determined in step 2, and in **S** the matroids N processed by step 5.

Next we state the complete algorithm.

Decomposition Algorithm:

0. (Initialize the matroid classes) Select a class **M** of matroids that is closed under the taking of minors, and a nonempty subclass $\mathbf{V}^\bullet \subseteq \mathbf{M}$ where each $V \in \mathbf{V}^\bullet$ is 3-connected and has at least 4 elements. Define $\mathbf{W} = \{(V,\varnothing,\varnothing) | V \in \mathbf{V}^\bullet\}$, $\mathbf{V} = \mathbf{V}^\bullet$, and $\mathbf{N} = \mathbf{S} = \varnothing$.

1. (Process another W of **W**) If $\mathbf{W} = \varnothing$, stop. Otherwise remove a $W = (V,a,b)$ from **W**. If $a = b = \varnothing$, go to 2. Otherwise (i.e., $W = (V,N,B_V)$ for some $N \in \mathbf{N}$) go to 3.

2. (Derive a k-separation for $N = V$) Define N to be equal to V. Attempt to find a B_N of (4.1) for N that satisfies (4.9) or (4.10), and where $X_i, Y_i \neq \varnothing$, $i = 1, 2$. If it is computationally unattractive or infeasible, or if it is simply impossible, to

find such a B_N, go to 5. Otherwise define $B_V = B_N$, add N to **N**, and change W to (V, N, B_V).

3. (Build new triples from $W = (V, N, B_V)$). Collect in sets $\mathbf{W}_N^1$ and $\mathbf{W}_N^2$ all triples $(\bar{V}, N, B_{\bar{V}})$ where $\bar{V}$ is in **M**, $\hat{B}_{\bar{V}}$ is a partial representation of $\bar{V}$, $B_{\bar{V}}$ can be obtained from B_V by adjoining either one row or one column, and where $B_{\bar{V}}$ satisfies (4.5)-(4.7) as well as the following conditions.
   (a) Triples for $\mathbf{W}_N^1$: $B_{\bar{V}}$ is one of the matrices (1)-(6) of (4.4).
   (b) Triples for $\mathbf{W}_N^2$:
       (ba) If $V \neq N$: $B_{\bar{V}}$ is one of the matrices (7)-(10) of (4.8), and is of the same type as B_V.
       (bb) If $V = N$: Either all $B_{\bar{V}}$ are of type (7), (8) of (4.8), or they are all of type (9), (10) of (4.8). In the former case N must satisfy (4.9), and in the latter (4.10). If N satisfies both (4.9) and (4.10), the choice may be based on any criterion.

   Partition $\mathbf{W}_N^2$ into $\mathbf{W}_N^{21}$ and $\mathbf{W}_N^{22}$ such that no $\bar{V}$ of the triples of $\mathbf{W}_N^{21}$ has series or parallel elements; any rule may be applied to decide the partition as long as it is in agreement with the latter condition. Redefine $\mathbf{W}_N^1$ by $\mathbf{W}_N^1 \cup \mathbf{W}_N^{21}$ and $\mathbf{W}_N^2$ by $\mathbf{W}_N^{22}$. Replace each triple $(\bar{V}, N, B_{\bar{V}})$ in $\mathbf{W}_N^1$ by $(\bar{V}, \emptyset, \emptyset)$.

4. (Add new cases to **W** and redefine **V**) Do (a) and (b) below to the extent it is computationally feasible or attractive.
   (a) Check the triples of $\mathbf{W}_N^1$ and $\mathbf{W}_N^2$ for isomorphism; remove from $\mathbf{W}_N^1$ and $\mathbf{W}_N^2$ all triples of each class of pairwise isomorphic triples except for one arbitrarily selected triple of each class.
   (b) Identify and remove from $\mathbf{W}_N^1$ and $\mathbf{W}_N^2$ triples whose matroid $\bar{V}$ has a minor in $\{\bar{V} | (\bar{V}, \emptyset, \emptyset) \in \mathbf{W}\}$.

   Augment **W** by $\mathbf{W}_N^1 \cup \mathbf{W}_N^2$, redefine **V** as $\{\bar{V} | (\bar{V}, \cdot, \cdot) \in \mathbf{W}\}$, and go to 1.

5. (N is to induce 2-separations) Add N to **S**. Let $\mathbf{W}_N^1$ be the set $\{(\bar{V}, \emptyset, \emptyset) | \bar{V}$ is a 3-connected 1-element extension of N in **M**$\}$. If N is a wheel (whirl), add $(\bar{V}, \emptyset, \emptyset)$ to $\mathbf{W}_N^1$ where $\bar{V}$ is the next larger wheel (whirl), provided the latter matroid occurs in **M**. Define $\mathbf{W}_N^2 = \emptyset$ and go to 4.

The next theorem summarizes and extends the above analysis of the algorithm. Its proof is based on an inductive examination of the various reductions and substitutions.

THEOREM 4.3: Suppose one has performed any number of iterations through the various steps of the decomposition algorithm, and that another iteration is about to start with step 1. The sets **N**, **S**, and **V** in existence at that time, together with the matrices B_N, $N \in \mathbf{N}$, previously found in step 2, may then be utilized to produce the following theorem and corollary.

THEOREM 4.4: Every $M \in \mathbf{M}$ obeys one of the conditions below.

(1) M has at most 3 elements.

(2) M is 2-separable.

(3) M has no minor in $\mathbf{V}^\bullet \cup \mathbf{N} \cup \mathbf{S}$.

(4) M has a minor in $\mathbf{V}$.

(5) M is equal to some $N \in \mathbf{S}$.

(6) M has a minor in **N**, say N, for which the following holds. (i) every k-separation of every minor of M equal to N induces a k-separation of M as long as the k-separation of the minor corresponds to the k-separation (T_1, T_2) of N defined via B_N; (ii) each such induced k-separation of M can be turned into a proper k-sum decomposition with 3-connected components provided B_N can be further subdivided to become a matrix of (1.1) that shows N to be a proper k-sum.

COROLLARY 4.4.1: The statements (7) and (8) below hold for every 3-connected $M \in \mathbf{M}$ on at least 4 elements and with a minor in **N** or **S**, and for every element v of such M.

(7) M has a 3-connected minor in $\mathbf{N} \cup \mathbf{N}_1 \cup \mathbf{S} \cup \mathbf{V}$ that contains v, where

$$\mathbf{N}_1 = \left\{ N_1 \in \mathbf{M}: \begin{array}{l} N_1 \text{ is a 3-connected 1-element extension of some } N \in \mathbf{N}, \\ \text{and the k-separation } (T_1, T_2) \text{ of } N \text{ as specified by } B_N \\ \text{induces a k-separation of } N_1 \end{array} \right\}$$

(8) If no $V \in \mathbf{V}$ has series or parallel elements, and if (6) of theorem 4.4. does not apply of M, then M has a 3-connected minor in $\mathbf{S} \cup \mathbf{V} \cup \mathbf{V}_1$ that contains v, where

$\mathbf{V}_1 = \{V_1 \in \mathbf{M} | V_1 \text{ is a 3-connected 1-element extension some } V \in \mathbf{V}\}$.

A particularly pleasant feature of the algorithm is the builtin flexibility that permits the user to explore numerous ideas and conjectures even when **M** and **V**° are held constant. So far we have identified two broad strategies that have been useful in our work with the algorithm.

The first strategy attempts to achieve decompositions in step 2 with small k (say $k \leq 4$ or 5). This approach produces theorems useful for the solution of many combinatorial optimization problems, and now and then amazing theorems of alternative.

The second strategy allows decompositions involving arbitrarily large values of k, but for $k \geq 4$ calls for proper k-sum decompositions that satisfy certain additional requirements. Particularly attractive appears to be the condition that the proper k-sum decomposition of N in step 2 has the connecting matroid in a stipulated subclass of **M**. Aim of the second strategy are theorems that say that all or almost all members of **M** are decomposable in a well-specified fashion. Such theorems are potentially useful for the determination of minimal violators of certain properties. For this we extend the concept of composition property introduced in Section 3. Let **H** be a collection of pairs $(H,(\bar{T}_1,\bar{T}_2))$, where H is in **M** and has groundset $\bar{T}_1 \cup \bar{T}_2$, such that the partition $(\bar{T}_1,\bar{T}_2)$ of the groundset allows H to be a connecting matroid of a k-sum, $k \geq 3$, with $\bar{X}_i \cup \bar{Y}_i = \bar{T}_i$, $i=1,2$. We then say that a property P defined on the class **M** of matroids has the *composition property* CP_k^H, $k \geq 3$, if the following statement is satisfied for all 3-connected $M \in \mathbf{M}$: If M is a proper k-sum, $k \geq 3$, whose connecting matroid H and sets $\bar{X}_i, \bar{Y}_i$, $i=1,2$, correspond to a pair $(H,(\bar{X}_1 \cup \bar{Y}_1, \bar{X}_2 \cup \bar{Y}_2)) \in \mathbf{H}$, then M has P if this is so for the components M_1 and M_2. P has CP_∞^H if it has CP_k^H for all $k \geq 3$. Suppose now almost all 3-connected members of **M** have a proper k-sum decomposition with $(H,(\bar{X}_1 \cup \bar{Y}_1, \bar{X}_2 \cup \bar{Y}_2)) \in \mathbf{H}$. If P has CP_∞^H, then the search for minimal violators may become trivial or at least greatly simplified.

Sometimes the sets **W** and **V** created in the iterations of the algorithm exhibit a certain pattern, and one can predict them by an inductive argument for any number of iterations. One thus can establish the sets **W** and **V** produced by an infinite sequence of such iterations, and then may continue the algorithm with the **W** and **V** sets so determined. It is easily checked that this extension of the algorithm does not affect the validity of Theorem 4.3. However, some other conclusions, e.g., about computational efficiency, may become invalid by this modification.

The usefulness of an algorithm is satisfactorily demonstrated once it has been shown to solve significant problems heretofore considered difficult. Here one would demand that the algorithm be capable of producing significant new theorems as well as difficult old ones. In the final part of this section it is demonstrated that our algorithm can indeed produce such theorems. We should emphasize that the capabilities of the algorithm (and schemes derived from it) go considerably beyond the results described here, and that more work has been, should be, and will be done.

First we use the algorithm to investigate graphic matroids. For efficiency of computations it is desirable to specialize the algorithm so that one manipulates the graphs directly instead of their representation matrices. Such a specialization is indeed possible and actually quite easily found. We omit a description due to space limitations, and instead list some of the theorems produced so far by the algorithm. First we define the graphs referred to in those theorems, along with T_1 or T_2 of a 3-sum decomposition where applicable.

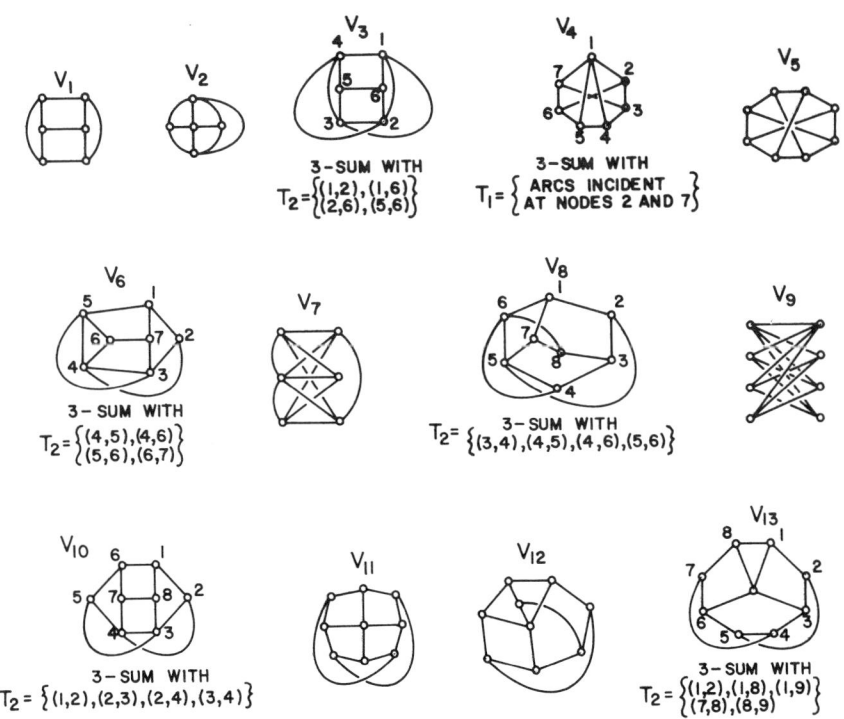

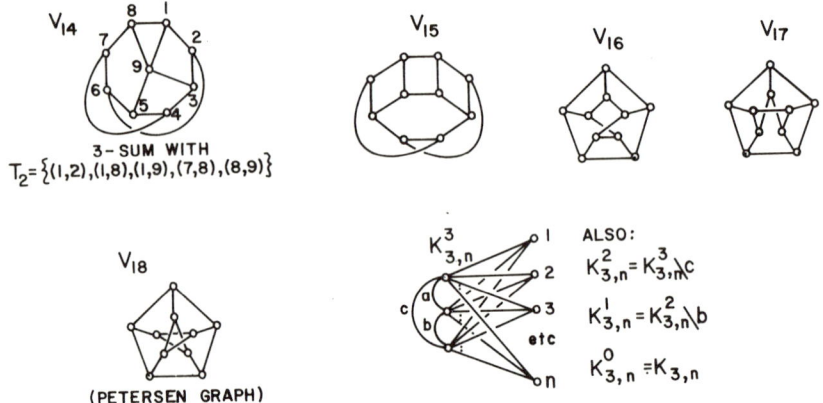

Figure 4.1: Graphs of Decomposition Theorems 4.5 - 4.8

Below, H_k is always the wheel with k spokes, $k \geq 3$.

THEOREM 4.5: Every 3-connected graph with 6 or more arcs and without minors in $\{K_{3,3}, K_5\}$ is equal to H_3, H_4, V_1 or V_2, or it has a proper k-sum decomposition, some $k \geq 3$, with 3-connected components, such that the connecting matroid is a wheel whose spokes form the set $\bar{X}_1 \cup \bar{Y}_1$ of the decomposition.

The exclusion of $K_{3,3}$ and K_5 in theorem 4.5 is, of course, equivalent to the requirement of planarity, by K. Kuratowski's theorem [8]. The latter theorem follows from the former one by most trivial arguments as follows.

LEMMA 4.4: The property of planarity has CP_2 and CP_∞^H where $H = \{(H, (\bar{T}_1, \bar{T}_2)) | H$ is a wheel and $\bar{T}_1$ is the set of spokes of $H\}$.

COROLLARY 4.5.1 (K. Kuratowski [8]): A graph is planar if and only if it does not have a minor in $\{K_{3,3}, K_5\}$.

THEOREM 4.6: Every 3-connected graph with 6 or more arcs and without minor in $\{V_4, K_5\}$ is planar or in $\{K_{3,j}^i | 0 \leq i \leq 3, j \geq 3\}$.

The next theorem is a strengthened version of a famous decomposition result by K. Wagner [24].

THEOREM 4.7: For any 3-connected graph M one of the statements below applies.
  (1) M is planar.
  (2) $M = V_5$ or K_5.
  (3) $M \in \{K_{3,j}^i | 0 \leq i \leq 3, j \geq 3\}$, and it is a 3-sum unless $M = K_{3,3}$.
  (4) M has V_4 has a minor, and the 3-sum decomposition of every such minor (as given in Figure 4.1) induces a 3-sum decomposition of M.
  (5) M has V_3 as a minor.

Finally, we have the following theorem.

THEOREM 4.8: For any 3-connected graph M one of the statements below applies.
  (1) M does not have V_3 as a minor.
  (2) M is one of the following graphs: $V_9, V_{11}, V_{12}, V_j, 15 \leq j \leq 18$.
  (3) M has a minor $\overline{M}$ equal to one of the graphs $V_3, V_6, V_8, V_{10}, V_{13}, V_{14}$; the 3-sum decomposition of every minor of M that is equal to $\overline{M}$ (as given in Figure 4.1) induces a 3-sum decomposition of M.
  (4) M has V_7 as a minor.

A quick glance at V_7 confirms it to be a 3-sum, so additional work should and will be done. The reader may wonder how much effort was needed to produce these theorems. The number of iterations is quite small, though manual isomorphism checking is tedious at times. Theorem 4.5 requires 8 iterations (including an induction step), Theorem 4.6 can be found in 7 iterations (again including an induction step), Theorem 4.7 is determined in 3-iterations, and finally Theorem 4.8 requires 13 iterations.

We now leave graphs and turn to the decomposition of certain binary matroids. Since P.D. Seymour's proof of his regular matroid decomposition [13] is at times complicated, it is tempting to look for significant simplifications. One can indeed find such simplifications, but space limitations do not permit a detailed presentation here. Suffice it to say that first a reasonably short proof produces the 2 minimal 3-connected matroids R_{10} and R_{12} that are regular but not graphic or cographic. The latter matroids have representation matrices $\hat{B}_{10}$ and $\hat{B}_{12}$, respectively, where

$$B_{10} = \begin{pmatrix} 1 & 0 & 0 & 1 & 1 \\ 1 & 1 & 0 & 0 & 1 \\ 0 & 1 & 1 & 0 & 1 \\ 0 & 0 & 1 & 1 & 1 \\ 1 & 1 & 1 & 1 & 1 \end{pmatrix} \qquad B_{12} = \begin{array}{c|cc|cc|cc} & \multicolumn{2}{c}{Y_1} & \multicolumn{2}{c}{Y_2} & & \\ \hline X_1 & 1 & 0 & 1 & 1 & 0 & 0 \\ & 0 & 1 & 1 & 1 & 0 & 0 \\ \hline & 1 & 0 & 1 & 0 & 1 & 1 \\ X_2 & 0 & 1 & 0 & 1 & 1 & 1 \\ \hline & 1 & 0 & 1 & 0 & 1 & 0 \\ & 0 & 1 & 0 & 1 & 0 & 1 \end{array} \qquad (4.11)$$

The partitioning of B_{12} indicates a proper 3-sum decomposition of R_{12}. With R_{10} and R_{12} definiting $\mathbf{V}^{\bullet}$, the decomposition algorithm produces a slightly stronger version (that apparently P.D. Seymour was aware of but did not include in [13]) of the desired theorem in 3 iterations. The improvement has pleasant implications for testing algorithms for matroid regularity.

THEOREM 4.9 (P.D. Seymour [13]): Every regular matroid M can be decomposed into graphic and cographic matroids and copies of R_{10} by repeated 1-sum, proper 2-sum, and proper 3-sum decompositions. Specifically, if M is 3-connected and not graphic or cographic, then $M = R_{10}$ or M has an R_{12} minor, and the proper 3-sum decomposition of every such R_{12} minor as given in (4.11) induces a proper 3-sum decomposition of M into 3-connected matroids M_1 and M_2. Conversely, every binary matroid produced from graphic and cographic matroids and copies of R_{10} by repeated 1-sum, proper 2-sum, and proper 3-sum composition is regular.

Since the pleasant behavior of regular matroids is induced by the absence of U_4^2, F_7 and F_7^*, one might want to move from regularity to somewhat weaker properties and see whether or not attractive decompositions are still possible. In the most straightforward modification one restricts oneself to binary matroids and rules out F_7^* but not F_7 (or vice versa). P.D. Seymour [13] looked into this and found that any binary matroid with an F_7 minor but without F_7^* is 2-separable or equal to F_7. (Our decomposition algorithm yields this result in 1 iteration). Thus one would want to impose a somewhat different condition. For example, suppose we know that a specified element, say l, is not contained in any F_7^* minor of a binary matroid M. Such matroids M play an important role for matroid-generated clutters (see P.D. Seymour [10]). One quickly determines that M is quite a special matroid, but its structural properties seem complicated at the outset. In joint work with

F. T. Tseng, we have applied the decomposition algorithm to this situation. Here we only outline the decomposition result since details and related clutter theorems will be published elsewhere. In contrast to the regular case the algorithm goes on for quite a while (20 iterations) before it stops. We need a few definitions to state the theorem so obtained. Two matroids are *l-isomorphic* if both matroids contain an element l, and if there is an isomorphism that maps l of one matroid onto l of the second one. Define $N_1, N_2, \ldots, N_6$ to be the binary matroids represented by $\hat{B}_1, \hat{B}_2, \ldots, \hat{B}_6$ where

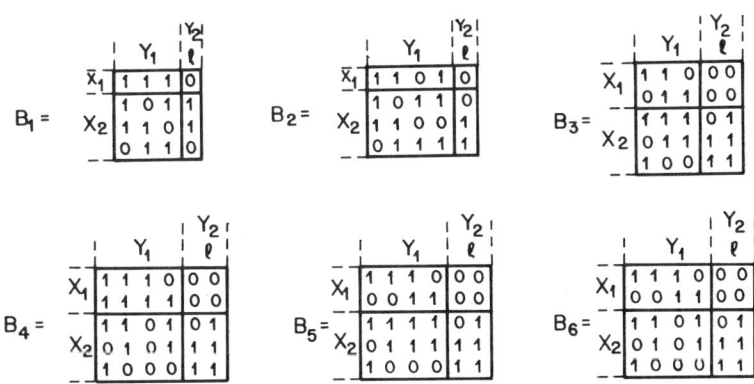

In each case the indicated partition leads to a 3-sum decomposition.

THEOREM 4.10: (with F.T. Tseng). Let M be a 3-connected nonregular binary matroid, and suppose a specified element l of M is not an element of any F_7^* minor of M. Then $M = F_7$, or M has one of $N_1, N_2, \ldots, N_6$, say N, as a minor such that every 3-sum decomposition of every minor $\overline{M}$ of M that is l-isomorphic to N, induces a 3-sum decomposition of M, provided the 3-sum decomposition of $\overline{M}$ corresponds to the one of N under one of the l-isomorphisms.

Theorem 4.10 may be applied recursively. Details and related composition results will be included in the aforementioned joint paper with F.T. Tseng.

References

1. R. E. Bixby, "Composition and Decomposition of Matroids and Related Topics", Thesis, Cornell University (1972).
2. R. E. Bixby, "On Reid's Characterization of the Ternary Matroids", *J. Comb. Theory (B)* 26 (1979) 174-204.
3. T. Brylawski, "Modular Constructions for Combinatorial Geometries ", *Trans. Am. Math. Soc.* 203 (1975) 1-44.
4. W. H. Cunningham, "A Combinatorial Decomposition Theory", *Thesis,* University of Waterloo (1973).
5. W. H. Cunningham, "On Matroid Connectivity", *J. Comb. Theory (B)* 30 (1981) 94-99.
6. W. H. Cunningham and J. Edmonds, "A Combinatorial Decomposition Theory", *Can. J. Math.* 32 (1980) 734-765.
7. S. Krogdahl, "The Dependence Graph for Bases in a Matroid", *Discrete Math.* 19 (1977) 47-59.
8. K. Kuratowski, "Sur le problème de courbes gauche en topologie", *Fund. Math.* 15 (1930) 271-283.
9. E. Lawler, *Combinatorial Optimization,* Holt, Rinehart and Winston, New York, 1976.
10. P. D. Seymour, "The Matroids with the Max-Flow Min-Cut Property", *J. Comb. Theory (B)* 23 (1977) 189-222.
11. P. D. Seymour, "Matroid Representation over GF(3)", *J. Comb. Theory (B)* 26 (1979) 159-173.
12. P. D. Seymour, "On Tutte's Characterization of Graphic Matroids", *Ann. Disc. Math.* 8 (1980) 83-90.
13. P. D. Seymour, "Decomposition of Regular Matroids", *J. Comb. Theory (B)* 28 (1980) 305-359.
14. C. Thomassen, "Planarity and Duality of Finite and Infinite Graphs", *J. Comb. Theory (B)* 29 (1980) 244-271.
15. K. Truemper, "Alpha-Balanced Graphs and Matrices and GF (3)-Representability of Matroids", *J. Comb. Theory (B)* 32 (1982) 112-139.
16. K. Truemper, "On the Efficiency of Representability Test for Matroids", *Europ. J. Comb.* 3 (1982) 275-291.
17. K. Truemper, "Partial Matroid Representations," Working Paper, University of Texas at Dallas, 1981.
18. K. Truemper, "A Decomposition Theory for Matroids, I: General Results", Working Paper, University of Texas at Dallas, 1982.
19. K. Truemper, "A Decomposition Theory for Matroids, II: Minimal Violation Matroids", Working Paper, University of Texas at Dallas, 1982.
20. K. Truemper, "A Decomposition Theory for Matroids, III: Decomposition Conditions", Working Paper, University of Texas at Dallas, 1983.
21. W. T. Tutte, "A Homotopy Theory for Matroids", *Trans. Am. Math. Soc.* 88 (1958) 144-174.
22. W. T. Tutte, "Matroids and Graphs", *Trans. Am. Math. Soc.* 90 (1959) 527-552.

23. W. T. Tutte, "Connectivity in Matroids", *Can. J. Math.* 18 (1966) 1301-1324.
24. K. Wagner, "Über eine Eigenschaft der ebenen Komplexe", *Math Annal.* 114 (1937) 570-590.
25. D. J. A. Welsh, *Matroid Theory*. Academic Press, New York, 1976.
26. H. Whitney, "On the Abstract Properties of Linear Dependence", *Am. J. Math.* 57 (1935) 509-533.

Map-Colourings and Differential Equations

W.T. Tutte

ABSTRACT

The differential equation $H''(2t+5H-3tH')=48t$ has a solution

$$H = t^2(1 + \sum_{n=1}^{\infty} h_n t^n)$$

as a power series in t, wherein h_n is the number of 4-coloured rooted triangulations of the sphere with $2n$ faces.

The maps of this paper are triangulations of the sphere. Each face is a triangle in the sense that it is bounded by a simple closed curve made up of three edges. Such a map can have no loop. It may include a digon, or circuit of two edges.

But both residual domains of such a circuit must be triangulated.

A map is "rooted" when a vertex, face and edge, mutually incident, are specified as the "root" elements.

Let k be a positive integer. A k-*colouring* of a triangulation T is a colouring of its vertices with k given colours, so that no two vertices joined by an edge are given the same colour. Not all the k given colours need be used. Now each T has an associated *chromatic* polynomial $P(T, \lambda)$ in a complex variable λ. The degree of the chromatic polynomial is the number of vertices of T, and its value at $\lambda = k$ is the number of k-colourings of T.

We shall be concerned with the number

$$h_n(\lambda) = \sum_{T \in S(n)} P(T, \lambda) \qquad (1)$$

where $S(n)$ is a complete set of combinatorially distinct rooted maps of $2n$ faces. We can call $h_n(k)$ the number of combinatorially distinct k-coloured rooted maps. But we shall discuss the determination of $h_n(\lambda)$

for arbitrary values of λ.

The numbers $h_n(\lambda)$, for a given λ, can conveniently be regarded as coefficients in a formal power series

$$h(t, \lambda) = 1 + \sum_{n=1}^{\infty} h_n(\lambda) t^n \qquad (2)$$

in an indeterminate t.

In papers on "Chromatic Sums" I have given some equations from which $h_n(\lambda)$ can be calculated for small values of n. Unfortunately they involve more complicated functions than $h(t, \lambda)$, and the coefficients in these functions have to be calculated. In [1] for example an equation is found by a straightforward generalization of the methods that have proved successful in rooted planar enumeration without coloration. To get an equation it is found necessary to introduce a power series

$$g = g(x, y, z, \lambda) \qquad (3)$$

in three variables x, y and z. (We treat λ as a constant in any one investigation.) g is an enumerating series concerned with maps somewhat more general than triangulations. Details can be found in [1]. An auxiliary series $q = q(x, z, \lambda)$ is derived from g by writing $y = 1$, and another $l = l(y, z, \lambda)$ by taking the coefficient of x^2 in g. It is found that only even powers of z occur in l. We are to identify z^2 with the variable t of Equation (2).

The series $h = h(t, \lambda)$ is related to q and l as follows. It is obtained from q by taking the coefficient of x^2, and from l by putting $y = 1$.

The final equation involves g, q and l. It is

$$xg = x^3 y \lambda (\lambda - 1) + \lambda^{-1} yzgq + yz(g - x^2 l) - x^2 y^2 z \frac{g - q}{y - 1}. \qquad (4)$$

We can solve it in the sense that we can calculate from it the coefficients of successive powers of z in g, and therefore in g, l and h. D. H. Younger has gone in this way up to the 14th power of z. His results yield coefficients in $h(t, \lambda)$ going up to the sixth power of t. They are exhibited in the following formula, in which $\mu = \lambda - 2$.

$$\begin{aligned} h = \lambda(\mu + 1)\{&1 + \mu t + (4\mu^2 - \mu)t^2 \\ &+ (24\mu^3 - 15\mu^2 + 3\mu)t^3 \end{aligned} \qquad (5)$$

$$+ (176\mu^4 - 189\mu^3 + 83\mu^2 - 14\mu)t^4$$
$$+ (1456\mu^5 - 2287\mu^4 + 1607\mu^3 - 572\mu^2 + 84\mu)t^5$$
$$+ (13056\mu^6 - 27429\mu^5 + 26946\mu^4 - 14943\mu^3 + 4548\mu^2$$
$$- 594\mu) \cdots \}.$$

There are special values of λ for which the determination of $h(t, \lambda)$ simplifies. Obvious examples are 0, 1 and 2. For these $h_n(\lambda)$ vanishes for each positive n, since a triangulation cannot be coloured in fewer than 3 colours. In other examples peculiar identities hold between chromatic polynomials, and these can be exploited to find chromatic sums. Thus in [2] the case $\lambda = \tau^2$, where $\tau = (1 + \sqrt{5})/2$, is discussed and the following equation is obtained.

$$\tau^3(1 + \tau y)(l - \tau^2 y) = y^2 t l^2 + \tau^2 y^3 t \frac{l - h}{y - 1}. \quad (6)$$

In [3], for $\lambda = 3$, we find that

$$6(1 + y)(l - 6y) = (1 + y)ytl^2 - 6y^2 tl + 6y^3 t \frac{l - h}{y - 1}. \quad (7)$$

In each of these papers an explicit formula is obtained for the general coefficient in h.

It has been conjectured by S. Beraha that the numbers

$$B_n = 2 + 2\cos(2\pi/n), \quad (8)$$

n being a positive integer, are of special importance in the theory of chromatic polynomials. Accordingly B_n is often called the "nth Beraha number" by combinatorialists. The first six Beraha numbers are 4, 0, 1, 2, τ^2, 3. Of these the first is notoriously important in map-colouring theory, and the other five are the special values of λ we have just noted. In [4] it is found that when $\lambda = B_n$ and $n > 4$ the number of variables required reduces from 3, as in Equation (4), to 2, as in (6) and (7).

One feels that (6) and (7) ought to be special cases of (4), but it is very difficult to see them as such. "Chromatic Sums V" is mainly an account of a struggle to exhibit (6) and (7) as logical consequences of (4), though it is also concerned with generalizing them to other Beraha numbers. ([4]). Some simplifications and clarifications are made in [5]. They involve a change of notation. First y is replaced by a variable V related to it as follows.

where
$$y^{-1} = 1 + V - v, \tag{9}$$

$$v = (2 - \mu)^{-1} = (4 - \lambda)^{-1}. \tag{10}$$

Then l is replaced by the function
$$K = \lambda^{-1} tl - yt(1 - y)^{-1} + y^{-2} - y^{-1} + v, \tag{11}$$

that is
$$K = \lambda^{-1} tl + t(v - V)^{-1} + (v - V)^2 + V. \tag{12}$$

Another useful symbol is
$$s = \left(\frac{K}{4} - \mu^2\right)^{1/2}. \tag{13}$$

Our equation for l, or K, in the case $\lambda = B_n$ depends to some extent on whether n is even or odd. Let us write $n = 2M$ in the first case, and $n = 2M - 1$ in the second. Let us also write $\theta = 0$ or 1 according as n is even or odd. Then the equation can be written thus.

$$\cos n \cos^{-1}(V/2s) - \cos n \cos^{-1}(v/2s) = (-1)^M \sum_{i=2}^{M} q_i s^{-2i+\theta}. \tag{14}$$

Here the q_i are power series in t alone, and they are initially unknown.

The left side of (14) reduces to a polynomial in s^{-1} by way of the identity
$$\cos n \cos^{-1}(x/2) = \frac{1}{2n}(-1)^{M+\theta} \sum_{r=0}^{M-\theta} \frac{(-1)^r (M + r - 1)!}{(2r+\theta)!(M - r - \theta)!} x^{2r+\theta}. \tag{15}$$

Formula (14) is thus equivalent to an equation of degree $M - 1$ for s^{-2}, with coefficients involving V and $M - 1$ unknown power series q_i in t.

Some information about the q_i can be got by considering the behaviour of the two sides of (14) when y approaches 1, that is when V approaches v. In [6] it is found in this way that

$$q_2 = M^2 v \frac{t}{4 - \mu^2} \tag{16}$$

when n is even, and that

$$q_2 = \frac{(2M-1)t}{2(4-\mu^2)} \tag{17}$$

when n is odd. The number of unknown functions q_i is thus reduced to $M-2$.

Further work along these lines relates q_3 to h. In the even case

$$q_3 = \frac{M^2v^2t(\lambda^{-1}th+v)}{(4-\mu^2)^2} - \frac{M^2t^2}{2(4-\mu^2)^2} - \frac{(M+1)M^2(M-1)v^3t}{6(4-\mu^2)} \tag{18}$$

and in the odd case

$$q_3 = \frac{(2M-1)t(\lambda^{-1}th+v)}{2(4-\mu^2)^2} - \frac{(2M-1)M(M-1)v^2t}{4(4-\mu^2)}. \tag{19}$$

The other q_i can be expressed in terms of the derivatives of l with respect to y, evaluated at $y=1$. (h is the zeroth derivative of this kind.) However no such expressions, for $i > 3$, are obtained in [6]. It is however shown that the constant term in each q_i is zero.

We note that to determine h it is only necessary to find what q_3 is. So it may not be necessary to get a complete solution of (14) for s in terms of V and t.

As for the proof of (14) it can be said here only that in [4] and [5] the equation is deduced, with difficulty, from (4), and that (6) and (7) are found to be special cases of it.

The theory is not applicable for $n=1$ and $n-2$, and it is utterly trivial when $n=3$ or 4. In the cases $n=5$ and 6, corresponding to Equations (6) and (7) respectively, we have a quadratic equation for l, with coefficients involving only one truly unknown power series q_3 in t. Such quadratics are common-place in the theory of planar enumeration, and there is a standard procedure for solving them. It is found that there exists a power series ξ in t whose substitution for V makes both the discriminant of the quadratic and its V-derivative vanish. Their vanishing gives two equations that can be solved for the two unknowns q_3 and ξ.

In [6] we find an extension of this method to the equations of higher degree, with more unknown functions q_i, that are the special cases of (14). It is found convenient to transfer the expression $\cos n \cos^{-1}(v/2s)$ to the right side of (14). By (15) we then get the equation

$$(-1)^M \cos n \cos^{-1}\left(\frac{V}{2s}\right) = \sum_{i=0}^{M-\theta} p_i s^{-2i+\theta}, \qquad (20)$$

where p_0 and p_1 are known and each other p_j can be expressed in terms of the corresponding q_j. It is now p_3 that we need to determine.

In [6] we again find some parametric equations. In general the parameter ξ of the quadratic case is replaced by $M - 2$ distinct parameters ψ_m, $(m = 2, 3, \cdots M - 1)$. These parameters also have distinct squares. Each ψ_m satisfies the two equations

$$(-1)^{M+m} = \sum_{r=\theta}^{M} p_r \psi_m^{2r-\theta}, \qquad (21)$$

$$0 = \sum_{r=\theta}^{M} (2r - \theta) p_r \psi_m^{2r}. \qquad (22)$$

Here we have $2M - 4$ equations for the $2M - 4$ unknowns ψ_m and $p_r(r > 2)$. By (16) and (17) q_2 is determined, and therefore p_2 is known. The cases $n = 7$ and $n = 8$ are indeed solved in [6], in the sense that t, p_2 and p_3 are expressed in terms of a single parameter.

Another equation of interest can be obtained by differentiating (21) with respect to t and combining the result with (22). It is

$$\sum_{r=2}^{M} p_r' \psi_m^{2r} = 0 \qquad (23)$$

Here and in what follows a prime denotes differentiation by t.

Some relations between the p_i can be obtained by comparing equations (22) and (23). In [7] the comparison is made in terms of a polynomial

$$U = \sum_{r=\theta}^{M} p_r w^{2r-\theta} \qquad (24)$$

in an indeterminate w. It is found convenient to write also

$$u = w^2 \qquad (25)$$

From (23) we can infer that the roots of $\partial U/\partial t = 0$ as an equation in u, and after division by the appropriate power of w, are the squares ψ_m^2. From (22) we find similarly that these squares are also roots of $\partial U/\partial u = 0$. Since the degree of the u-polynomial is one higher in the second case we can infer an identity

$$(\alpha + \beta u)\partial U/\partial t = 2u^2 \partial U/\partial u, \tag{26}$$

where α and β are independent of u, but may involve t.

Using the known values of p_1 and p_2 we discover by a comparison of coefficients in (26) that

$$\alpha = -\lambda. \tag{27}$$

Equation (26) does not determine β directly. Instead it gives the following recursion formula for the p_i.

$$\lambda p'_{r+1} = \beta p'_r - (2r - \theta)p_r \quad (r \geq 2). \tag{28}$$

In the case $r = 2$ this equation relates β to p_3, and therefore to h. Perhaps the result is best expressed in terms of the t-integral γ of β having a zero constant term. It is

$$\gamma = \lambda^{-1} v t^2 h + v^2 t + 3t^2/2. \tag{29}$$

It is noteworthy that Equations (27) and (29) are valid in both the even and the odd cases.

In [7] there is further discussion of the squared parameters as zeros of polynomial equations in u, and a second partial differential equation involving U results. It is

$$u^4(1 - U^2) = -Y(\partial U/\partial t)^2. \tag{30}$$

The polynomial Y, introduced like $(\alpha + \beta u)$ to balance degrees, is of the fourth degree. Since it is found to divide by u we can write it as

$$Y = u(A + Bu + Cu^2 + Du^3) \tag{31}$$

where A, B, C and D are independent of u but may involve t.

It is possible to eliminate U from the two equations (26) and (30). This is done in [7], with the emergence of the following partial differential equation for Y.

$$(8 - 2\beta')uY - 2u^2 \partial Y/\partial u + (\alpha + \beta u)\partial Y/\partial t = 0 \tag{32}$$

In view of (31) the last equation is equivalent to the following system of five simultaneous differential equations for A, B, C and D.

$$-\lambda A' = 0, \tag{33}$$

$$(6 - 2\beta')A - \lambda B' + \beta A' = 0, \tag{34}$$

$$(4 - 2\beta')B - \lambda C' + \beta B' = 0, \tag{35}$$

$$(2 - 2\beta')C - \lambda D' + \beta C' = 0. \tag{36}$$

$$-2\beta'D + \beta D' = 0 \tag{37}$$

By eliminating A, B, C and D from these equations we arrive in [7] at the following differential equation for γ.

$$(1 - \gamma)(-v^2 t + 12t^2 + 10\gamma) + 6t\gamma'\gamma'' = 0 \tag{38}$$

The constants of integration here, and the boundary conditions for our partial differential equations, are easily obtained from the theorem that the constant term in each q_i is zero.. The constant term in β is found from (28) to be v^2, and that in γ is by definition zero. We can use (38) to determine coefficients of successive powers of t in γ, β and h and so recover, and perhaps continue, Formula (5). We now have an equation giving h directly, in terms of one variable only.

Equation (38) can of course be written in terms of h rather than γ. Equation (29) suggests that

$$H = t^2 h \tag{39}$$

may be more convenient. In terms of H the differential equation is as follows.

$$\lambda^{-1} H''\{t + 9v^{-1}t^2 + 10\lambda^{-1}H - 6\lambda^{-1}tH'\} \tag{40}$$
$$= -2v^{-1}t + 6t - 20\lambda^{-1}v^{-1}H + 18^{-1}v^{-1}tH'.$$

Equation (38) was obtained on the assumption that λ is a Beraha number. But it is shown in [7] that it extends to all complex numbers λ, with the exception of 4. Putting $\lambda = 4$ makes v infinite and so invalidates some of the steps in the argument.

To find an equation for the important case $\lambda = 4$ we take limits in (40) as λ tends to 4 and v^{-1} therefore tends to zero. We find that

$$H''\{2t + 5H - 3tH'\} = 48t. \tag{41}$$

Using this equation B. Richter has evaluated $h_n(4)$, the number of combinatorially distinct 4-coloured rooted triangulations, as far as $n = 55$.

References

[1]. W.T. Tutte, *Chromatic sums for rooted planar triangulations: the cases $\lambda = 1$ and $\lambda = 2$*. Can. J. Math., 25 (1973), 426-447.

[2]. W.T. Tutte, *Chromatic sums for rooted planar triangulations: the case $\lambda = \tau + 1$*, Can J. Math., 25 (1973), 657-671.

[3]. W.T. Tutte, *Chromatic sums for rooted planar triangulations: the case $\lambda = 3$*, Can J. Math., 25 (1973), 780-790.

[4]. W.T. Tutte, *Chromatic sums for rooted planar triangulations: special equations*, Can J. Math., 26 (1974), 893-907.

[5]. W.T. Tutte, *On a pair of functional equations of combinatorial interest*, Aequationes Math., 17 (1978), 121-140.

[6]. W.T. Tutte, *Chromatic Solutions*, Waterloo Research Report CORR 81-12.

[7]. W.T. Tutte, *Chromatic Solutions II*, Waterloo Research Report CORR 81-18.

Combinatorial Classification of the Regular Polytopes

Andrew Vince

ABSTRACT

A generalization of a polytope, called a graph encoded map (GEM), is defined graph theoretically. We classify the set of symmetric GEMs whose underlying topological space is a closed simply connected manifold. In particular, this yields a combinatorial proof of the well known classification of the regular polytopes. This proof has the advantage of avoiding metric technicalities.

1. Introduction.

A polytope is the convex hull of a finite set of points in Euclidean space E^n. A supporting hyperplane of a polytope P is a hyperplane that intersects P in such a way that P lies in one of the closed half spaces determined by the hyperplane. Intersections of a polytope P with supporting hyperplanes are polytopes. With the exception of P itself, these are called faces. The set of all faces of an n-dimensional polytope P form a cell complex of dimension $n-1$, called the boundary complex of P. The polytopes of dimension 2 and 3 are polygons and polyhedra, respectively.

An automorphism of an n-dimensional polytope P in E^n is an isometry of E^n leaving P as a whole fixed. A polytope is regular if is automorphism group is sufficiently transitive. This will be made precise in Section 2. The regular polytopes have been completely classified [2] and are listed in Table 1.

The set of faces of a polytope is partially ordered by inclusion. In [9] McMullen defines combinatorial equivalence and combinatorial regularity of polytopes in terms of the poset of faces. McMullen then proves that the set of combinatorially regular polytopes essentially coincides with the set of metrically regular polytopes.

Polytope	Dimension	Schläfli symbol
n-gon	2	$\{n\}$
simplex	≥ 2	$\{3,3,\ldots,3\}$
hypercube	≥ 2	$\{4,3,\ldots,3\}$
cross polytope	≥ 2	$\{3,\ldots,3,4\}$
dodecahedron	3	$\{5,3\}$
icosahedron	3	$\{3,5\}$
25-cell	4	$\{3,4,3\}$
120-cell	4	$\{5,3,3\}$
600-cell	4	$\{3,3,5\}$

Table 1. Regular Polytopes

THEOREM (McMullen). *Every combinatorially regular polytope is combinatorially equivalent to a metrically regular polytope.*

Although McMullen takes a combinatorial point of view, equivalence and regularity defined without reference to the metric, the objects under investigation are still polytopes in Euclidean space. Here we formulate a purely combinatorial generalization of a polytope. This generalization originally appeared in the author's work [12,13] under the name combinatorial map. A related concept, called a crystallization, was independently investigated in a topological setting by Ferri and Gagliardi [5,6]. It was rediscovered by Lins [8], where it is called a graph-encoded map (GEM). We adopt this suggestive terminology. The concept is actually implicit in and preceded by the more general concepts of chamber complex and chamber system due to Tits [10,11].

Let $I=\{0,1,2,\ldots,n-1\}$ be fixed throughout this paper. A *graph-encoded map (GEM)* is a connected graph G, regular of degree n, whose lines are I-colored such that no two incident lines are the same color. To motivate this definition, consider the boundary complex $B(Q)$ of the cube Q. The dual graph G of the barycentric subdivision of $B(Q)$ is shown in Figure 1. The lines of this graph are colored in $I=\{0,1,2\}$ as follows: Each vertex in the barycentric subdivision represents either a 0-face (vertex), 1-face (edge) or 2-face (face) of the cube. A line $\{u,v\}$ in G is colored i if and only if the simplexes corresponding to points u and v in the barycentric subdivision differ only by an i-dimensional vertex. The same procedure can be carried out for the boundary complex of any polytope P. Let $G(P)$ denote the *GEM associated with the polytope* P.

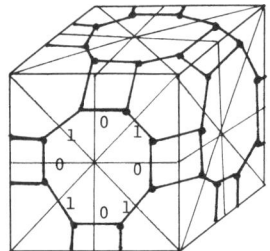 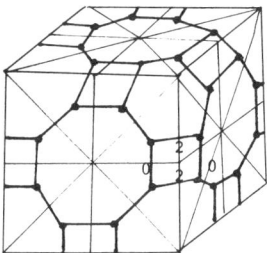

Figure 1. Faces of the graph-encoded map associated with the cube.

The *dimension* of a GEM over I is defined as $n=|I|$. An *isomorphism* of two GEMs is a color preserving graph isomorphism. For $J \subseteq I$ let G_J be the subgraph of G obtained by deleting all lines not colored in J. Each connected component of G_J is called a *residue of type J*. The residues of type $I-\{i\}$ are called *i-faces* of G. Two distinct faces f and f' are called *incident* if $f \cap f' \neq \emptyset$. Let X denote the set of faces of G and * the incidence relation on faces. Then the pair $S(G)=(X,*)$ is referred to as the *incidence structure* of G. Figure 1 shows a 2-face and a 1-face of the dimension 3 GEM associated with the cube. Intuitively we have in mind the front face of the cube and its right edge. As expected this 1-face and 2-face are incident.

To any GEM G over I is associated an $(n-1)$-dimensional simplicial complex ΔG as follows. For each point v in the point set $V(G)$ of G, let Δv be a simplex of dimension $n-1$. Arbitrarily assign to each vertex of Δv a distinct element of I. Call the set of elements assigned to a face s of Δv, the type of s. Let K be the disjoint union of the set $\{\Delta v | v \in V(G)\}$. In K identify two simplexes $s \subseteq \Delta v$ and $s' \subseteq \Delta v'$ of the same type J if and only if v and v' are joined by a path in G colored in $I-J$. If $\sim$ denotes this identification, take $\Delta G = K/\sim$. Intuitively ΔG can be thought of as being built from $(n-1)$-simplexes, one for each point of G, such that two $(n-1)$-simplexes share a common codimension 1 face if the corresponding points are adjacent in G. The space $|G|:=|\Delta G|$ is called the *underlying topological space of G*. R If $|G|$ is *a manifold, then we say that G is a GEM on a manifold*. In the example of Figure 1, $|G|$ is a 2-sphere.

Besides polytopes, graph-encoded maps offer a scheme for representing maps on surfaces, both orientable and non-orientable, higher dimensional analogues of maps on surfaces, tessellations, the hypermaps of Walsh [14] and certain toroidal complexes of Coxeter, Shephard and Gruünbaum [4,7]. See [13] for details concerning these applications.

2. Symmetric Graph-Encoded Maps.

Let G be a GEM over I and let $\Gamma(G)$ denote its group of automorphisms. If $\Gamma(G)$ acts transitively on the set of points of G, then we say that G is *symmetric*. This is a strong requirement. For a polytope P each point of $G(P)$ corresponds to an n-tuple $\{f_0, f_1, \ldots, f_{n-1}\}$ consisting of pairwise incident faces, one for each dimension. Hence for the GEM $G(P)$ to be symmetric, it is not only necessary that the automorphism group act transitively on faces of dimension i for all i, but it must act transitively on the set of all such n-tuples. This is usually referred to as flag transitivity. The GEMs associated with the polytopes listed in Table 1 are all symmetric. Other familiar examples of symmetric GEMs come from regular tessellations of the plane (Euclidean or hyperbolic) and regular maps on surfaces [13]. Note that the Cayley graph of any group generated by involutions yields a symmetric GEM. Let W be a group generated by involutions $\{r_i | i \in I\}$. Recall that the Cayley graph $G(W)$ is an I-colored graph defined as follows. The points of $G(W)$ are the elements of W and two points w and w' are joined by a line colored i if and only if $w' = wr_i$. Since W acts on $G(W)$ by left multiplication, $G(W)$ is a symmetric GEM.

Let G be a symmetric GEM over I and R any residue of type $\{i,j\}$. If R is finite, then it is a cycle in G consisting of lines alternately colored i and j. Let p_{ij} be half the length of this cycle. If R is infinite, take $p_{ij} = \infty$. Since G is symmetric, p_{ij} does not depend on which residue of type $\{i,j\}$ is considered. The *diagram* $D(G)$ of G is obtained by representing each $i \in I$ as a node labeled i and connecting nodes i and j by a line labeled p_{ij}. By convention the line is omitted when $p_{ij} = 2$ and the line label is omitted when $p_{ij} = 3$. If $G = G(P)$ is associated with a polytope P, then $D(G)$ is just the Schläfli symbol of the polytope P. For example, the Schläfli symbol of a cube Q is $\{4,3\}$ indicating that each face is a 4-gon and 3 faces surround each vertex. The diagram of $G(Q)$ is shown in Figure 2.

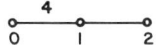

Figure 2

In general, different GEMs may have the same diagram. The relationship between the diagram of a GEM and the diagram of a Coxeter group is taken up in the next section.

There is a straightforward way of constructing a symmetric GEM of dimension (n_1+n_2) from symmetric GEMs of dimensions n_1 and n_2. Let G_1 and G_2 be GEMs over $I_1=\{0,1,\ldots,n_1-1\}$ and $I_2=\{n_1,n_1+1,\ldots,n_1+n_2-1\}$ with point sets $V(G_1)$ and $V(G_2)$. The *join* G_1*G_2 is a GEM over $I_1\cup I_2$ with point set $V(G_1)\times V(G_2)$. Two points (u_1,u_2) and (v_1,v_2) are joined by a line colored i whenever $[u_1=v_1$ and u_2 is joined to v_2 by a line colored $i]$ or $[u_2=v_2$ and u_1 is joined to v_1 by a line colored $i]$. This is the standard join construction for graphs, together with the appropriate line coloring. An example is shown in Figure 3. A GEM is called *reducible* if it is isomorphic to the join of two other GEMs. Otherwise it is *irreducible*. Note that $D(G_1*G_2)=D(G_1)\cup D(G_2)$, where the union is disjoint.

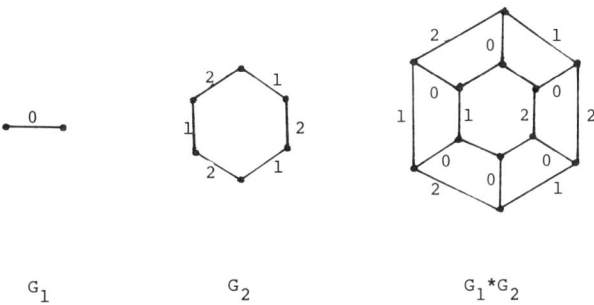

Figure 3. A reducible GEM.

3. Symmetric GEMs on Closed Simply Connected Manifolds.

In this section we classify the symmetric GEMs whose underlying topological space is a closed simply connected manifold. This result will generalize the classification of regular polytopes and also McMullen's theorem stated in the introduction. Recall that a Coxeter group over I is a group generated by involutions with presentation

$$<r_i,\ i\in I|\ (r_ir_j)^{p_{ij}}=1,\ p_{ii}=1,\ p_{ij}\geq 2\ i\neq j> \tag{3.1}$$

We do not eliminate the possibility that $p_{ij}=\infty$, in which case the relation $(r_ir_j)^{p_{ij}}=1$ is absent. By abuse of language we use the term Coxeter group to refer to the presentation (3.1). The *diagram* for a Coxeter group is constructed by representing each r_i by a node labelled i and connected nodes i and j by a line labelled p_{ij}. By convention the line is omitted when $p_{ij}=2$ and the line label is omitted when $p_{ij}=3$. A Coxeter group is uniquely determined by its diagram. If the diagram D of a Coxeter group W is disconnected, then the r_i fall into disjoint sets corresponding to the components of D. Consider the subgroups of W generated by these subsets. Any two elements in distinct subgroups commute, so that W is the direct product of these subgroups. Hence a Coxeter groups is said to be *reducible* if its diagram is disconnected. Otherwise it is *irreducible*. Coxeter [1] classified all finite irreducible groups with presentation (3.1) and identified some as the automorphism groups of the regular polytopes. The finite irreducible Coxeter groups are listed in Table 2.

Group	Diagram	Order	Automorphism group of
A_n	•—•—•—⋯—•	$(n+1)!$	simplex
B_n	•—•—•—⋯—•—^{4}—•	$2^n n!$	hypercube cross polytope
D_n	•—•—•—⋯—•<	$2^{n-1} n!$	
E_6	(diagram)	51840	
E_7	(diagram)	2903040	
E_8	(diagram)	696729600	
F_4	•—^{4}—•—•	1152	24-cell
H_3	•—^{5}—•	120	dodecahedron icosahedron
H_4	•—•—^{5}—•	1440	120-cell 600-cell
I_2^r	•—^{r}—•	$2r$	r-gon

Table 2.
Finite Irreducible Coxeter Groups

As pointed out in the previous section, the Cayley graph $G(W)$ is a symmetric GEM when W is a Coxeter group. For W a finite irreducible Coxeter group, we shall call a GEM of the form $G(W)$ a *Coxeter map*. Among the Coxeter maps are all the regular polytopes. However, the Coxeter maps corresponding to E_6, E_7, E_8 and D_n, $n \geq 4$ are not associated with any polytope.

THEOREM. The symmetric irreducible graph-encoded maps on closed simply connected manifolds are exactly the Coxeter maps.

The proof will proceed by a series of lemmas. If W is a group generated by involutions $\{r_i | i \in I\}$ and H is a subgroup of W, then the Schreier coset graph $G(W,H)$ is a generalization of the Cayley graph. The points of $G(W,H)$ are the right cosets of W/H and two points c and c' are joined by a line colored i if and only if $c' = cr_i$. When H is the trivial subgroup, the Schreier coset graph is the Cayley graph of W.

LEMMA 1. *If G is a GEM, then $G \approx G(W,H)$ for some group W generated by involutions and subgroup $H \leq W$.*

PROOF. Let G be a GEM over I. For each $i \in I$ define a permutation r_i of $V(G)$ by $r_i u = u'$ if and only if $\{u,u'\}$ is colored i. Let W be the permutation group on $V(G)$ generated by the $\{r_i\}$ and let H be the stabilizer of a point u of G. The function $G(W,H) \to G$ given by $Hw \mapsto w^{-1}(u)$ is an isomorphism.

LEMMA 2. *Let H be a subgroup of a group W generated by involutions. The graph-encoded map $G(W,H)$ is symmetric if and only if H is normal in W.*

PROOF. Assume H is normal in W and $a,b \in W$. Then the map $Hw \mapsto H(ba^{-1})w$ induces an automorphism of $G(W,H)$ taking Ha to Hb. Conversely if $G(W,H)$ is symmetric and $a \in W$, then there exists an automorphism g such that $g(H) = Ha$. This implies that $g(Hw) = Haw$ for all $w \in W$, and therefore $Hah = g(Hh) = g(H) = Ha$ for all $h \in H$.

In the next two lemmas W_* denotes the Coxeter group with presentation $\langle r_i, i \in I | r_i^2 = 1 \rangle$ and H_* denotes a normal subgroup. For $J \subseteq I$ let W_J be the subgroup of W_* generated by $\{r_i | i \in J\}$ and let H_k be the subgroup of H normally generated by the subgroups $\{H \cap W_J | |J| = k\}$. That is, H_k is the smallest normal subgroup of W_* containing these subgroups.

LEMMA 3. *If $|G(W_*,H_*)|$ is simply connected, then $H_* = H_{n-1}$.*

PROOF. A path in a GEM is defined to be a sequence of points such that consecutive points in the sequence are adjacent. If α and β are paths such that the terminal point of α is the initial point of β then the product $\alpha\beta$ is the concatenated path. The inverse α^{-1} is the path obtained by listing the points of α in the reverse order. Two paths $\beta\gamma\delta$ and $\beta\gamma'\delta$ with the same initial and terminal points are called elementary k-homotopic if $\delta\delta^{-1}$ is contained in some residue of type J where $|J| = k$. Two paths are k-homotopic if there is a sequence of elementary k-homotopies connecting them. This notion is due to J. Tits [11]. The set of k-homotopy classes of closed paths based at some fixed point forms a group in the usual way. Denote this group by $\pi(G,k)$. In [12 Theorem 6.2] it is shown that this group and the fundamental group of the underlying topological space of G are related as follows:

$\pi(|G|) \approx \pi(G, n-1)$. Also [12 Theorem 8.8] if $G = G(W_*, H_*)$, then $\pi(G, k) \approx H_*/H_k$. The lemma is a consequence of these two facts.

LEMMA 4. If $|G(W_*, H_*)|$ is a manifold, then $H_{n-1} = H_2$.

PROOF. Since $|G(W_*, H_*)|$ is a manifold, each open star st(s), where s is a simplex of condimension > 2 in $\Delta G(W_*, H_*)$, is contractible, hence simply connected. Lemma 4 is proved in [12 Corollary 6.3] by applying the arguments of Lemma 3 to the residues of type J, $|J| > 2$ in $G(W_*, H_*)$.

LEMMA 5. If W is a reducible Coxeter group, then $G(W)$ is a reducible GEM.

PROOF. If W is reducible, then W is the direct product of two other Coxeter groups $W = W_1 \times W_2$. Now

$$G(W) \approx G(W_1 \times W_2) \approx G(W_1) * G(W_2),$$

so that $G(W)$ is reducible.

PROOF OF THE THEOREM. Assume that G is a symmetric irreducible GEM over I on a closed simply connected manifold. By Lemma 1, $G \approx G(W, H)$ for some group W generated by involutions and some subgroup H. Let $\phi: W_* \to W$ be the canonical homomorphism, taking the ith generator of W_* to the ith generator of W, and let $H_* = \phi^{-1}(H)$. Then $G \approx G(W, H) \approx G(W_*, H_*)$. Since G is symmetric H_* is normal in W_*. Lemmas 3 and 4, together with the assumption that $|G|$ is a simply connected manifold imply $H_* = H_2$. Thus $G \approx G(W_*, H_*) \approx G(W_*, H_2) \approx G(W_*/H_2)$. But $W' = W_*/H_2$ is also a Coxeter group. By Lemma 5, W' is irreducible because $G \approx G(W')$ is assumed irreducible. Finally W' is finite, because $|G|$ closed implies ΔG has finitely many simplexes, and $G(W')$ is the dual graph of ΔG.

References

1. H.S.M. Coxeter, The complete enumeration of finite groups of the form $R_i^2 = (R_i R_j)^{k_{ij}} = 1$, J. London Math. Soc. 10 (1935), 21-25.
2. ———, "Regular Polytopes," The Macmillan Company, New York, 1972.
3. H.S.M. Coxeter and W.O. Moser, "Generators and Relations for Discrete Groups," Springer-Verlag, New York, 1965.
4. H.S.M. Coxeter and G.C. Shephard, Regular 3-complexes with toroidal cells, J. Comb. Theory B 22 (1977), 131-138.
5. M. Ferri, Crystallizations of 2-fold branched coverings fo S^3, Proc. Amer. Math. Soc. 73 (1979), 271-276.
6. C. Gagliardi, A combinatorial characterization of 3-manifold crystallization, Boll. Un. Mat. Ital. 16A (1979), 441-449.

7. B. Grünbaum, Regularity of graphs, complexes and designs, in Colloques internationaux C.N.R.S., N°260 - Problèmes Combinatoire et Théorie des Graphes, Orsay (1976), 191-197.
8. S. Lins, Graph-encoded maps, J. Comb. Theory B 32 (1982), 171-181.
9. P. McMullen, Combinatorially regular polytopes, Mathematika 14 (1967), 142-150.
10. J. Tits, "buildings of Spherical Type and Finite BN-Pairs," Springer-Verlag, New York, 1974.
11. _____, A local approach to buildings, preprint.
12. A. Vince, Combinatorial maps, J. Comb. Theory B, to appear.
13. _____, Regular combinatorial maps, J. Comb. Theory B, submitted.
14. T.R.S. Walsh, Hypermaps versus bipartite maps, J. Comb. Theory B 19 (1975), 155-163.

Products of Highly Regular Graphs

Paul M. Weichsel

ABSTRACT

If A is the adjacency matrix of a graph Δ then A_i, the ith distance matrix, is defined by: $(A_i)_{rs} = 1$ when α_r and α_s are vertices a distance i apart and 0 otherwise. A graph is *distance-regular* if and only if A_i is a polynomial of degree i in A for $i = 1,\ldots,d$, with d the diameter of Δ. A graph is *distance-polynomial* if A_i is a polynomial in A with no condition on the degree. We consider cartesian products of distance-regular and distance-polynomial graphs and derive conditions on the spectra (the sequence of *distinct* eigenvalues) which imply that the products are distance-regular or distance-polynomial. We show, for example, that the cartesian product of a distance-regular graph with itself is distance-regular if and only if it has the same intersection array as a cartesian power of a certain complete graph K_r.

1. Introduction.

A graph is usually referred to as being highly symmetric if its automorphism group acts transitively on certain subobjects of the graph. Important examples are vertex-transitive and edge-transitive graphs, graphs whose groups act transitively on paths of some given length and graphs whose groups act transitively on each set of pairs of vertices a given distance apart. This latter type of graph is called *distance-transitive*.

Each type of transitivity induces a local regularity condition, e.g., a vertex-transitive graph must be regular in the usual sense. In this note we will study the local regularity induced by distance- transitivity and some variations of it. Such graphs are called *distance-regular* and have been studied from several different points of view. [See, for example, [1] Chapters 20 and 21]. They can be defined in terms of sets of integers a_i, b_i, c_i as follows:

If α and β are vertices of the graph, Γ, of diameter d, whose distance apart is i, denoted by $\partial(\alpha, \beta) = i$, then

$a_i = |\{\omega \in V\Gamma | \partial(\omega, \alpha) = 1 \text{ and } \partial(\omega, \beta) = i\}| i = 0,\ldots,d$

$b_i = |\{\omega \in V\Gamma | \partial(\omega, \alpha) = 1 \text{ and } \partial(\omega, \beta) = i + 1\}| i = 0,\ldots,d - 1$

$c_i = |\{\omega \in V\Gamma | \partial(\omega, \alpha) = 1 \text{ and } \partial(\omega, \beta) = i - 1\}| i = 1,\ldots,d.$

The graph Γ is called *distance-regular* if it is connected and if the a_i, b_i and c_i depend only on i and not on the choice of α and β. Clearly $b_0 = k$, the valency of Γ, $c_1 = 1$ and $k = a_i + b_i + c_i$ for all $i = 1,\ldots,d - 1$ with $k = a_d + c_d$. The pair of sequences $\{k, b_1,\ldots,b_{d-1}; 1, c_2,\ldots,c_d\}$ is called the *intersection array* of a distance-regular graph and contains all of the essential information about the graph, but does not determine the graph up to isomorphism. For more background information about distance-regular graphs the reader should refer to [1, Chapters 20 and 21].

Distance-regular graphs can also be characterized in terms of their distance matrices A_i which are defined as follows. Let Γ be a graph with n vertices. A_i is an $n \times n$ matrix which has a 1 in position (r, s) if $\partial(\alpha_r, \alpha_s) = i$ and 0 otherwise. Thus $A = A_1$, the adjacency matrix of Γ. It can be shown that if Γ is a distance-regular graph, then A_i is a polynomial of degree i in A [1, Theorem 20.7]. Moreover, the converse also holds [2, Theorem 2.1], and there exists a recurrence defining these polynomials in terms of integers a_i, b_i, c_i. Thus if $v_i(x)$ is the polynomial defined by $A_i = v_i(A)$ with degree $v_i(x) = i$, then

$c_{i+1} v_{i+1}(x) + (a_i - x)v_i(x) + b_{i-1}v_{i-1}(x) = 0$ with $v_0(x) = 1$, $v_1(x) = x$, $i = 1,\ldots,d - 1$. Moreover, the minimum polynomial of A is

$$(x - k)(1 + x + v_2(x) + \cdots + v_d(x)).$$

There are connected, regular graphs, however, for which A_i is a polynomial in A, $i = 1,\ldots,d$ but not necessarily of degree i, for every i. We have named these graphs *distance-polynomial* [2]. A further weakening occurs if we simply require that $A_i A_j = A_j A_i$ for all i and j, and we call such graphs *distance-commuting*. This last property is easily seen to be a consequence of the condition that for each pair of vertices α, β of Γ, there is an automorphism that interchanges them.

In this note we will consider cartesian products of such graphs. Our purpose is to investigate regularity conditions which survive the stress of cartesian products. Our results will be in terms of the intersection array and the set of eigenvalues of the graphs in question.

In section III we will derive a necessary and sufficient condition for the cartesian product of two distance-polynomial graphs to be distance-polynomial (Theorem 3.1.). In section IV we characterize those distance-regular graphs whose cartesian square is also distance-regular (Corollary 4.7).

Will denote graphs by capital Greek letters and matrices by capital latin letters. The vertex set of the graph Γ will be denoted by $V\Gamma$. The cartesian product of graphs Γ and Δ is $\Gamma \times \Delta$. If K_r is the complete graph on r vertices, then $K_r \times \cdots \times K_r$ (t times) will be denoted by K_r^t. All graphs considered in this note are finite without loops or multiple edges.

2. Cartesian Products.

In this section we collect some elementary results about cartesian products of graphs.

If Γ and Δ are graphs, then $\Gamma \times \Delta$, the *cartesian product* of Γ and Δ is a graph whose vertex set is $V\Gamma \times V\Delta$ such that if $(\alpha_1, \beta_1), (\alpha_2, \beta_2) \in V\Gamma \times V\Delta$, then (α_1, β_1) is adjacent to (α_2, β_2) if either $\alpha_1 = \alpha_2$ and β_1 is adjacent to β_2 in Δ or α_1 is adjacent to α_2 in Γ and $\beta_1 = \beta_2$.

If A and B are the adjacency matrices of Γ and Δ, then the adjacency matrix of $\Gamma \times \Delta$ is denoted by $A \times B$ and is easily seen to satisfy:

$$A \times B = A \otimes I_s + I_r \otimes B \qquad (1)$$

where $U \otimes V$ is the Kronecker product of matrices, $r = |V\Gamma|$, $s = |V\Delta|$ and I_r and I_s are the identity matrices of orders r and s.

If $(\alpha_1, \beta_1), (\alpha_2, \beta_2) \in V(\Gamma \times \Delta)$, then it is easy to see that $\partial((\alpha_1, \beta_1), (\alpha_2, \beta_2)) = r$ if and only if $\partial(\alpha_1, \alpha_2) = r_1$ in Γ and $\partial(\beta_1, \beta_2) = r_2$ in Δ with $r_1 + r_2 = r$. Thus it follows that if C_i is the ith distance matrix of the graph with adjacency matrix C, then

$$(A \times B)_i = A_i \otimes I_s + A_{i-1} \otimes B + A_{i-2} \otimes B_2 \qquad (2)$$
$$+ \cdots + A \otimes B_{i-1} + I_r \otimes B_i$$

where A is an $r \times r$ matrix and B is an $s \times s$ matrix. It is understood that if the diameter of Γ is d, then $A_r = 0$ when $r > d$. It is immediate that diameter of $(\Gamma \times \Delta)$ = (diameter of Γ) + (diameter of Δ). (3)

If Γ and Δ are regular graphs with adjacency matrices A and B, then

valency of $(\Gamma \times \Delta)$ = (valency of Γ) + (valency of Δ). (4)

Much of our attention in the sequel will be focused on the sequence of distinct eigenvalues of a matrix A in descending order denoted by

$$\operatorname{spec} A = (\lambda_1, \lambda_2, \ldots, \lambda_p).$$

Since the adjacency matrix of a graph is a real symmetric matrix, all of the eigenvalues are real.

If A and B are square matrices of orders r and s with eigenvectors $\vec{u}$ and $\vec{v}$ and eigenvalues λ and μ, then $A \otimes B$ acts on $\vec{u} \otimes \vec{v}$ by the rule: $(A \otimes B)(\vec{u} \otimes \vec{v}) = \lambda\mu(\vec{u} \otimes \vec{v})$. Therefore,

$$(A \times B)(\vec{u} \otimes \vec{v}) = (\lambda + \mu)(\vec{u} \otimes \vec{v}). \quad (5)$$

Hence if $\operatorname{spec} A = (\lambda_1, \ldots, \lambda_p)$ and $\operatorname{spec} B = (\mu_1, \ldots, \mu_q)$, then

$$\operatorname{spec}(A \times B) = (\lambda_i + \mu_j \mid i = 1,\ldots,p; j = 1,\ldots,q) \quad (6)$$

with duplicate sums deleted and the remaining sums ordered appropriately.

We conclude this section with the following:

THEOREM 2.1 Let Γ and Δ be distance-commuting graphs. Then $\Gamma \times \Delta$ is distance-commuting.

PROOF. Let A and B be the adjacency matrices of Γ and Δ. Then $A_i A_j = A_j A_i$ and $B_i B_j = B_j B_i$ for all appropriate i and j. Now since Kronecker product distributes over matrix addition, it follows from (2) that $(A \times B)_i (A \times B)_j = (A \times B)_j (A \times B)_i$.

3. Distance-Polynomial Graphs.

We now consider cartesian products of distance-polynomial graphs.

Let A and B be the adjacency matrices of the distance-polynomial graphs Γ and Δ. Then there are polynomials $f_i(x)$, $g_j(y)$ with real coefficients such that $A_i = f_i(A)$ for $i = 0,\ldots,d_1$ and $B_j = g_j(B)$ for $j = 0,\ldots,d_2$ with d_1 the diameter of Γ and d_2 the diameter of Δ. Clearly $f_0(A) = I$ and $f_1(A) = A$. Thus it follows from (2) that

$$(A \times B)_i = \sum_{j=0}^{i} f_{i-j}(A) \otimes g_j(B)$$

with the convention that $f_r(A) = g_s(B) = 0$ if $r > d_1$ and $s > d_2$. We therefore have defined a class of polynomials

$$u_i(x, y) = \sum_{j=0}^{i} f_{i-j}(x) g_j(y) \tag{7}$$

for $i = 0, \ldots, d_1 + d_2$.

If $\vec{v}$ is an eigenvector of A with eigenvalue λ and $\vec{w}$ is an eigenvector of B with eigenvalue μ, then $\vec{v} \otimes \vec{w}$ is an eigenvector of $(A \otimes B)_i$ for all i and

$$(A \otimes B)_i(\vec{v} \otimes \vec{w}) = u_i(\lambda, \mu)(\vec{v} \otimes \vec{w}).$$

We are now in a position to state the main theorem of this section.

THEOREM 3.1 Let A and B be the adjacency matrices of the distance-polynomial graphs Γ and Δ. Let spec $A = (\lambda_1, \ldots, \lambda_p)$ and spec $B = (\mu_1, \ldots, \mu_q)$. Then $\Gamma \times \Delta$ is distance-polynomial if and only if whenever $\lambda_i + \mu_j = \lambda_k + \mu_l$, then $u_r(\lambda_i, \mu_j) = u_r(\lambda_k, \mu_l)$ for all $r = 1, \ldots, d$ with d the diameter of $\Gamma \times \Delta$.

PROOF. Since Γ and Δ are distance-polynomial, they are distance-commuting. Hence by 2.1 $\Gamma \times \Delta$ is distance-commuting. Thus $A \times B = (A \times B)_1, (A \times B)_2, \ldots, (A \times B)_d$ commute in pairs. Since they are all real symmetric matrices they can be simultaneously diagonalized with the same basis of eigenvectors. Thus it follows that the condition for $(A \times B)_i$ to be expressible as a polynomial in $A \times B$ is that whenever a pair of elements (eigenvalues) on the diagonal of $A \times B$ (diagonalized) are equal, then the pair of elements in the same positions of $(A \times B)_i$ (diagonalized) also be equal. Since $(A \times B)(\vec{v} \otimes \vec{w}) = (\lambda + \mu)(\vec{v} \otimes \vec{w})$ if $\vec{v}$ is an eigenvector of A with value λ and $\vec{w}$ is an eigenvector of B with values μ and $(A \times B)_i(\vec{v} \otimes \vec{w}) = u_i(\lambda, \mu)(\vec{v} \otimes \vec{w})$, the theorem follows.

COROLLARY 3.2 Let Γ and Δ be distance-polynomial graphs with adjacency matrices A and B. Let spec $A = (\lambda_1, \ldots, \lambda_p)$ and spec $B = (\mu_1, \ldots, \mu_q)$. If $\lambda_i + \mu_j \ne \lambda_k + \mu_l$ for all $i, k = 1, \ldots, p$ and all $j, l = 1, \ldots, q$, then $\Gamma \times \Delta$ is distance-polynomial.

COROLLARY 3.3. If Γ is a distance-regular graph of diameter 2, then $\Gamma \times \Gamma$ is distance-polynomial.

PROOF. A distance-regular graph of diameter 2 has exactly 3 distinct eigenvalues.

We will now show that, unlike the class of distance-commuting graphs, distance-polynomial graphs are not closed under cartesian product. In fact, the graphs in the example below are even distance-regular.

EXAMPLE. Let Γ be the 8-cycle and Δ the complete graph on two elements, K_2. Then Γ has diameter 4, Δ has diameter 1 and the polynomials that define the distance matrices of Γ are:
$$v_2(x) = x^2 - 2, \quad v_3(x) = x^3 - 3x \quad \text{and} \quad v_4(x) = \frac{1}{2}(x^4 - 4x^2 + 2).$$
Thus the polynomials $u_i(x, y)$ given in (7) are:

$u_2(x, y) = v_2(x) + xy$ (since Δ has diameter 1)
$u_3(x, y) = v_3(x) + v_2(x)y$
$u_4(x, y) = v_4(x) + v_3(x)y$
$u_5(x, y) = v_4(x)y.$

Now let A and B be the associated adjacency matrices of Γ and Δ and we have:

spec $A = (2, \sqrt{2}, 0, -\sqrt{2}, -2)$ and
spec $B = (1, -1)$.

Thus $2 + (-1) = 0 + 1 = 1$ and in order for $\Gamma \times \Delta$ to be distance-polynomial, we must have $u_i(2, -1) = u_i(0, 1)$ for $i = 2, 3, 4, 5$. But $u_2(2, -1) = 0$ and $u_2(0, 1) = -2$ and thus $\Gamma \times \Delta$ is not distance-polynomial.

One can also show, using the same procedure, that if $\Gamma = \Delta =$ the 6-cycle, then $\Gamma \times \Gamma$ is not distance-polynomial.

The operation of forming cartesian products affords the possibility of constructing many examples of distance-polynomial graphs.

COROLLARY 3.4. Let Γ be a distance-polynomial graph. Then for any positive integers a and b, there is a distance-polynomial graph Δ such that $\Gamma \times \Delta$ is distance-polynomial, the diameter of $\Gamma \times \Delta$ is greater than a and the valency of $\Gamma \times \Delta$ is greater than b.

PROOF. Construct Δ as follows. Choose a positive integer r which is greater than every difference of pairs of distinct eigenvalues of Γ. Now the graphs K_r, $K_r \times K_r$, $K_r \times K_r \times K_r$,... are all distance-regular and for each one, only the multiples of r occur as differences of distinct eigenvalues. Thus $\Gamma \times K_r^t$ is distance-polynomial for each t. Since diameter of $(\Gamma \times K_r^t) = $ (diameter of Γ) $+ t$ and valency of

$(\Gamma \times K_r^t) = $ (valency of Γ) $+ t(r - 1)$, it follows that by choosing t sufficiently large, $\Gamma \times \Delta$, with $\Delta = K_r^t$, satisfies the conditions of the Corollary.

Thus any distance-polynomial graph can be embedded in a distance-polynomial graph of arbitrarily large diameter and valency.

4. Distance-Regular Graphs.

If Γ is a distance-regular graph of diameter d, then Γ has $d + 1$ distinct eigenvalues - the smallest number possible. That property, however, is not enough to guarantee distance-regularity. If we know that Γ is distance-polynomial, then having precisely $d + 1$ distinct eigenvalues does imply distance-regularity. We first need to prove a lemma.

LEMMA 4.1. *Let Γ be a distance-polynomial graph of diameter d and adjacency matrix A. Suppose that $A_j = f(A)$ and the degree of the polynomial $f(x)$ is greater than j. Then the degree of $f(x)$ is greater than d.*

PROOF. Suppose that $\deg f(x) = r$, $j < r \le d$. Then

$$A_j = f(A) = l_0 I + l_1 A + \cdots + l_r A^r \text{ and } l_r \ne 0.$$

Consider a pair of vertices α, β of Γ with $\partial(\alpha, \beta) = r$ and assume that α and β are listed in rows t and s of A. Thus $(A_j)_{ts} = 0$ since $j < r$ while $(A^i)_{ts} = 0$ for $i = 0,\ldots,r - 1$, and $(A^r)_{ts} \ne 0$. Hence $l_r(A^r)_{ts} = 0$ and so $l_r = 0$, a contradiction. Thus $r > d$.

THEOREM 4.2. *Let Γ be a graph of diameter d. Then Γ is distance-regular if and only if it is a distance-polynomial graph with exactly $d + 1$ distinct eigenvalues.*

PROOF. Let A be the adjacency matrix of Γ and assume that Γ is distance-polynomial of diameter d with exactly $d + 1$ distinct eigenvalues. Then the minimal polynomial of A has degree $d + 1$. Thus every polynomial in A is equal to a polynomial of degree $\le d$. Hence if $A_j = f_j(A)$, then $\deg f_j(x) \le d$. Thus according to the lemma above, $\deg f_j(x) \le j$. But clearly $\deg f_j(x) \not< j$ and so $\deg f_j(x) = j$ for $j = 1, 2,\ldots,d$. Hence Γ is distance-regular. The converse is obvious.

We can now establish a necessary and sufficient condition for the cartesian square of a distance-regular graph to be distance-regular. First a combinatorial lemma.

LEMMA 4.3. *Let $\{\alpha_1, \alpha_2, \ldots, \alpha_p\}$ and $\{\beta_1, \ldots, \beta_q\}$ be sets of p and q real numbers with $p, q \ge 2$. Then $|\{\alpha_i + \beta_j|$*

$i = 1,\ldots,p$ and $j = 1,\ldots,q\}| = p + q - 1$ if and only if both sets form arithmetic progressions with the same common difference.

PROOF. Assume that both sets have been arranged in decreasing order: $\alpha_1 > \alpha_2 > \cdots > \alpha_p$, $\beta_1 > \beta_2 > \cdots > \beta_p$. Then $\{\alpha_1 + \beta_1, \alpha_2 + \beta_1, \ldots, \alpha_p + \beta_1, \alpha_p + \beta_2, \ldots, \alpha_p + \beta_q\} = S$ is a set of $p + q - 1$ distinct numbers.

Now suppose that both sets are arithmetic progressions with common difference γ. Since the difference between any pair of adjacent elements of S is $\pm\gamma$ and $\alpha_1 + \beta_1$ is the largest possible sum while $\alpha_p + \beta_q$ is the smallest, it follows that $\alpha_i + \beta_j \in S$ for all $i = 1,\ldots,p$ and $j = 1,\ldots,q$.

Conversely, let $T = \{\alpha_i + \beta_j | i = 1,\ldots,p \text{ and } j = 1,\ldots,q\}$ and assume that $S = T$. We now induct on p with q fixed. Thus let $p = 2$ and $S = \{\alpha_1 + \beta_1, \alpha_2 + \beta_1, \alpha_2 + \beta_2, \cdots, \alpha_2 + \beta_q\}$. Then $\alpha_1 + \beta_2 = \alpha_2 + \beta_{i(1)}$, $\alpha_1 + \beta_3 = \alpha_2 + \beta_{i(2)}, \ldots, \alpha_1 + \beta_q = \alpha_2 + \beta_{i(q-1)}$. Hence $1 \leq i(1) < i(2) < \cdots < i(q-1) \leq q - 1$ and therefore $i(1) = 1, i(2) = 2,\ldots,i(q-1) = q - 1$. Thus $\alpha_1 + \beta_2 = \alpha_2 + \beta_1$ and so $\alpha_1 - \alpha_2 = \beta_1 - \beta_2 = \beta_2 - \beta_3 = \cdots = \beta_q - \beta_{q-1}$, and the lemma holds for the case $p = 2$. Now suppose that the lemma holds for $p - 1, q$. Thus let $T = S$ and as above, $S = \{\alpha_1 + \beta_1, \alpha_2 + \beta_1, \ldots, \alpha_p + \beta_1, \alpha_p + \beta_2, \ldots, \alpha_p + \beta_q\}$. If we delete α_1 from the first set, then the set $S^1 = \{\alpha_2 + \beta_1, \ldots, \alpha_p + \beta_1, \alpha_p + \beta_2, \ldots, \alpha_p + \beta_q\}$ has $(p - 1) + q - 1$ elements and by induction $\beta_i - \beta_{i+1} = \gamma = \alpha_j - \alpha_{j+1}$ for $i = 1,\ldots,q$ and $j = 2,\ldots,p$. We must prove that $\alpha_1 - \alpha_2 = \gamma$. By hypothesis $\alpha_1 + \beta_2 \in S$ and thus $\alpha_1 + \beta_2 = \alpha_i + \beta_1$. Now if $i > 2$, we have $\alpha_i + \beta_1 = \alpha_i + \gamma + \beta_1 - \gamma = \alpha_{i-1} + \beta_2$. Therefore, $\alpha_1 = \alpha_{i-1}$ and since all α_i's are distinct it follows that $\alpha_1 + \beta_2 = \alpha_2 + \beta_1$ and hence that $\alpha_1 - \alpha_2 = \beta_1 - \beta_2 = \gamma$, which proves the lemma.

NOTATION: As we have already remarked, a pair of distance-regular graphs, say Γ and Γ', with the same intersection array need not be isomorphic. We will use the notation $\Gamma \sim \Gamma'$ when Γ and Γ' have the same array but may not be isomorphic.

LEMMA 4.4. Let $\Gamma, \Gamma', \Delta, \Delta'$ be distance-regular graphs with $\Gamma \sim \Gamma'$ and $\Delta \sim \Delta'$. If $\Gamma \times \Delta$ is distance-polynomial, then $\Gamma' \times \Delta'$ is also distance-polynomial.

PROOF. Recall that since Γ and Δ are distance-regular there are polynomials $v_i(x)$ and $w_i(x)$ $i = 0, 1,\ldots,d$, with d the diameter of Γ and

Δ such that $\Gamma_i = v_i(\Gamma)$ and $\Delta_i = w_i(\Delta)$. Now if $u_i(x, y) = \sum_{j=0}^{i} v_{i-j}(x)w_j(y)$ and spec $(\Gamma) = (\lambda_1, \ldots, \lambda_r)$ and spec $(\Delta) = (\mu_1, \ldots, \mu_s)$, then $\Gamma \times \Delta$ is distance-polynomial if and only if $u_i(\lambda_a, \mu_b) = u_i(\lambda_c, \mu_d)$ whenever $\lambda_a + \mu_b = \lambda_c + \mu_d$.

Thus the condition that $\Gamma \times \Delta$ be distance-polynomial depends strictly on the polynomials $u_i(x, y)$. But these polynomials are determined by the intersection arrays of Γ and Δ as are the spectra of Γ and Δ. Since the spectra of Γ' and Δ' as well as the arrays of Γ' and Δ' are the same as those of Γ and Δ, then if $\Gamma \times \Delta$ is distance-polynomial, so is $\Gamma' \times \Delta'$.

We will now determine the intersection arrays of these graphs by examining the polynomials $u_i(x, y)$. We first need a lemma about the coefficients of the polynomials $v_i(x)$.

LEMMA 4.5. Let $v_i(x)$ be defined recursively by: $v_0(x) = 1$, $v_1(x) = x$, $c_{i+1}v_{i+1}(x) + (a_i - x)v_i(x) + b_{i-1}v_{i-1}(x) = 0$ where a_i, b_i and c_i are given integers. Let $v_r(x) = \beta_r^r x^r + \beta_{r-1}^r x^{r-1} + \cdots + \beta_0^r$. Then $\beta_r^r = \dfrac{1}{c_r \cdots c_2}$ and

$$\beta_{r-1}^r = -\dfrac{(a_1 + \cdots + a_{r-1})}{c_r \cdots c_2}, \text{ for } r = 2, 3, \ldots.$$

PROOF. If $r = 2$, then $c_2 v_2(x) + (a_1 - x)x + b_0 = 0$. Thus $\beta_2^2 = \dfrac{1}{c_2}$ and $\beta_1^2 = -\dfrac{a_1}{c_2}$. Now assume the lemma for r and consider

$$\beta_{r+1}^{r+1} = \dfrac{\beta_r^r}{c_{r+1}} = \dfrac{1}{c_{r+1}c_r \cdots c_2}$$

and

$$\beta_r^{r+1} = \dfrac{-a_r \beta_r^r}{c_{r+1}} + \dfrac{\beta_{r-1}^r}{c_{r+1}} = \dfrac{-a_r}{c_{r+1}c_r \cdots c_2} - \dfrac{(a_1 + \cdots + a_{r-1})}{c_{r+1}c_r \cdots c_2} = -\dfrac{(a_1 \cdots + a_r)}{c_{r+1} \cdots c_r}.$$

THEOREM 4.6. Let Γ be a distance-regular graph of diameter d and valency k such that $\Gamma \times \Gamma$ is also distance-regular. Then Γ has the same intersection array as K_s^t where $s = \dfrac{k}{d} + 1$ and $t = d$.

PROOF. Let A be the adjacency matrix of Γ. Since $\Gamma \times \Gamma$ is distance-regular, there is for each $r = 1, 2, \ldots, 2d$ a polynomial $u_r(x)$ of degree r satisfying

$$(A \times A)_r = u_r(A \times A) = \sum_{i=0}^{r} \alpha_i^r (A \times A)^i.$$

On the other hand,

$$(A \times A)_r = \sum_{i=0}^{r} v_{r-i}(A) \otimes v_i(A)$$

where $v_i(x)$ is defined by $A_i = v_i(A)$ (see section 1). Now if these matrices act on the eigenvector $\vec{u} \otimes \vec{w}$ with x and y the eigenvalues of $\vec{u}$ and $\vec{w}$ under A, then we get the polynomial formula:

$$\sum_{i=0}^{r} \alpha_i^r (x+y)^i = u_r(x, y) = \sum_{i=0}^{r} v_{r-i}(x) v_i(y). \tag{8}$$

This formula is an identity for $r = 1, \ldots, d$ since it is satisfied by $(d+1)^2$ pairs of eigenvalues of A.

We obtain our result by comparing the coefficients of x^r, $x^{r-1}y$, x^{r-1} and $x^{r-2}y$ on both sides of (8). On the right we have:

$$\frac{x^r}{c_r \cdots c_2} + \frac{x^{r-1}y}{c_{r-1} \cdots c_2} - \frac{(a_1 + \cdots + a_{r-1})}{c_r \cdots c_2} x^{r-1}$$

$$- \left(\frac{a_1 + \cdots + a_{r-2}}{c_{r-1} \cdots c_2} + \frac{a_1}{c_{r-2} \cdots c_2 c_2} \right) x^{r-2} y + \cdots.$$

On the left we have: $\alpha_r^r x^r + \alpha_r^r r x^{r-1} y + \alpha_{r-1}^{r-1} x^{r-1} + \alpha_{r-1}^{r-1}(r-1) x^{r-2} y + \cdots$. Hence $\alpha_r^r = \dfrac{1}{c_r \cdots c_2}$ and $r \alpha_r^r = \dfrac{1}{c_{r-1} \cdots c_2}$, whence $c_r \cdots c_2 = r c_{r-1} \cdots c_2$ and so $c_r = r$. Therefore, $c_r \cdots c_2 = r!$.

Next $\alpha_{r-1}^{r-1} = -\dfrac{(a_1 + \cdots + a_{r-1})}{r!}$ and $(r-1)\alpha_{r-1}^{r-1} = -\left(\dfrac{a_1 + \cdots + a_{r-2}}{(r-1)!} + \dfrac{a_1}{(r-2)!2} \right)$. Thus $\dfrac{a_1 + \cdots + a_{r-1}}{r!} = \dfrac{a_1 + \cdots + a_{r-2}}{(r-1)(r-1)!} + \dfrac{a_1}{(r-1)!2}$ from which it follows by a simple induction that $a_r = r a_1$.

Now since Γ has valency k, $k = a_d + c_d = d a_1 + d$. Let $\dfrac{k}{d} = t$ and $b_r = k - a_r - c_r = k - r(a_1 + 1) = k - rt$. Thus the intersection array of Γ is

$$\{k, k-t, k-2t, \ldots, k-(d-1)t; 1, 2, 3, \ldots, d\}.$$

But it is easy to see that this is precisely the intersection array of K_s^t with $t = d$ and $k = st - 1$.

REMARK. This theorem can also be proved somewhat more directly by considering the elements of $\Gamma \times \Gamma$ and computing the values of a_r and c_r of $\Gamma \times \Gamma$ in terms of a_r and c_r of Γ.

COROLLARY 4.7. Let Γ be a distance-regular graph. Then $\Gamma \times \Gamma$ is distance-regular if and only if $\Gamma \sim K_s^t$ with t the diameter of Γ and $s = (\text{valency of } \Gamma)/t + 1$.

PROOF. If $\Gamma \times \Gamma$ is distance-regular, then the result is Theorem 4.6.

If $\Gamma \sim K_s^t$, then by Lemma 4.4, $\Gamma \times \Gamma$ is distance-polynomial. But if d is the diameter of K_s^t, then the number of distinct eigenvalues of $\Gamma \times \Gamma$ is $2d + 1$. Thus $\Gamma \times \Gamma$ is distance-regular by Theorem 4.2.

REMARK. Graphs with the same intersection array as K_s^t have been characterized by Y. Egawa [2].

References

[1] Norman Biggs, Algebraic Graph Theory, (Cambridge University Press, London) 1974.

[2] Yoshimi Egawa, Characterization of $H(n,q)$ by the parameters, J. Comb. Theory (A), *31* (1981), 108-125.

[3] Paul M. Weichsel, On distance-regularity in graphs, J. Comb. Theory (B), *32* (1982), 156-161.

The Number of Complete Subgraphs In Graphs with Non-Majorizable Degree Sequences

Douglas B. West

ABSTRACT

Erdős proved that any graph that has no K_{r+1} as a subgraph can be degree-majorized by a complete r-partite graph. More generally, let $f_r(n, \Delta)$ be the smallest number of copies of K_{r+1} in a graph with n vertices and maximum vertex degree Δ that is not degree-majorizable by an r-partite graph. We show constructively that $f_r(n, \Delta) = 1$ when $\Delta \geq \lceil n - \frac{n-1}{r} + \frac{r-1}{2} \rceil$. Trivially, $f_r(n, \Delta) = \infty$ when $\Delta \leq \lfloor n - \frac{n}{r} \rfloor$. The construction generalizes to give $f_r(n, \Delta) \leq (n - \Delta)^t$, where t is the smallest integer such that $n - 1 \geq (n - \Delta)r + \binom{r-t}{2}$. We prove that this construction is optimal for $r = 2$ and for $r = 3$, $n \leq 7$. This yields $f_2(n, \frac{n+1}{2}) = \frac{n-1}{2}$, for which we characterize the extremal graphs.

1. Introduction

In 1941, Turán [17] proved that the graph on n vertices with most edges that has no K_{r+1} as a subgraph is the r-partite graph with the most edges, which occurs when the part-sizes differ by at most one. This has been proved and extended in many ways. To facilitate discussion, we call the complete graph K_r an $r-clique$, without requiring that an r-clique be a maximal complete subgraph of a graph. A number of papers have examined the number of $(r + 1)$-cliques that must arise as the number of edges increases, particularly for the case $r = 2$. Moon and Moser [15] proved that a graph on n vertices with m edges has at least $\frac{m}{3n}(4m - n^2)$ triangles. Bollobás [2] extended these results by obtaining lower bounds for the minimum number of r-cliques in a graph on n vertices that has at least x p-cliques. See also [4], [7], [8], [13], [14], [16]. A good discussion of many of these results appears in [3].

Erdős [11] gave a proof of Turán's Theorem by characterizing the maximal degree sequences among graphs with no $(r + 1)$-clique. In particular, he proved that any graph G containing no $(r + 1)$-clique is "degree-majorized" by some complete r-partite graph H. Letting the degree sequence $d(G)$ of a graph be the non-increasing order of its vertex degrees, we say that H *degree–majorizes* G if $d_i(G) \leq d_i(H)$ for all i. If a graph is degree-majorized by an r-partite graph, we say it (or its degree sequence) is r-*majorizable*. We wish to extend the theorem of Erdős to an extremal theorem analogous to those mentioned above. How many $(r + 1)$-cliques must appear in a graph on n vertices if it is not r-majorizable? It is easy to find a graph with maximum vertex degree $\Delta = n - 1$ that has only one K_{r+1} but is not r-majorizable. To make the problem non-trivial, we restrict the value of $\Delta(G)$. Bounds on vertex degrees have been examined before for this type of question. Zarankiewicz [18] first noted that graphs with no $(r + 1)$-clique must have some vertex with degree at most $\frac{r-1}{r}n$. This was improved slightly in [1] under the assumption that the graph is also known not to be r-colorable (i.e., not r-partite). See also [6], [12] for related extremal results on vertex degrees.

To facilitate discussion, we refer to a graph on n vertices with maximum degree Δ as an (n, Δ)-*graph*. Let $f_r(n, \Delta)$ be the minimum number of $(r + 1)$-cliques in an (n, Δ)-graph that is not r-majorizable. The purpose of this paper is to determine $f_r(n, \Delta)$ completely when $r = 2$ and to obtain a uniform upper bound for all cases, in terms of r, n, and Δ.

Any (n, Δ)-graph with $\Delta \leq \lfloor n - \frac{n}{r} \rfloor$ is r-majorized by the complete r-partite graph in which the part sizes differ by no more than 1. By convention, we say $f_r(n, \Delta) = \infty$ in that case. In Section 2 we provide a construction to show that $f_r(n, \Delta) = 1$ whenever $\Delta \geq \lceil n - \frac{n-1}{2} + \frac{r-1}{2} \rceil$. To settle the only remaining case on $r = 2$, we show in Section 3 that $f_2(n, \frac{n+1}{2}) = \frac{n-1}{2}$. The graph achieving this is unique except for $\frac{1}{2}(n - 3)$ optional edges. In Section 4, we generalize these constructions to show that $f_r(n, \Delta) \leq (n - \Delta)^t$, where $t = t(r, n, \Delta)$ is the smallest integer such that $n - 1 \geq (n - \Delta)r + \binom{r-t}{2}$.

We conjecture that the bound of Section 4 is optimal. We have proved this for $r = 2$, for $r = 3$, $n \leq 7$, and for all (r, n) when Δ is sufficiently large. We note that this is similar to an old conjecture of Erdős and Stone [13] analogous to a theorem of Rademacher; Erdős [10] later proved it for $r = 3$. (See also [14].) They conjectured that any graph with rn vertices and at least $\binom{r}{2} n^2 + 1$ edges contains at least n^{r-1} $(r + 1)$-cliques. By the pigeonhole principle, such a graph always has a vertex of degree at least $(r - 1)n + 1$, but need not have one of higher degree. If it has none of higher degree, it cannot be r-majorizable. Since $t(r, rn, (r-1)n + 1) = r - 2$, the present conjecture would say such a graph has at least $(n - 1)^{r-2} (r + 1)$-cliques. Their lower bound is higher because they require more edges than needed to prevent a $(rn, (r-1)n + 1)$-graph from being r-majorizable.

2. The Construction for Large Δ

THEOREM 1. If $\Delta \geq \lceil n - \frac{n-1}{r} + \frac{r-1}{2} \rceil$, then $f_r(n, \Delta) = 1$.

PROOF: Erdős' Theorem gives the lower bound. For the upper bound, we need only construct a non-majorizable (n, Δ)- graph G having only one $(r + 1)$-clique. We begin with a complete r-partite graph on $n - 1$ vertices and then form G by adding an additional vertex z joined to one vertex in each part. Let the parts be A_i, and let the neighbor of z in A_i be x_i. The only $(r + 1)$-clique in G consists of z and $\{x_i\}$. The problem is to determine when (and how) the part-sizes $\{|A_i|\}$ of the initial r-partite graph can be chosen so that the resulting graph has maximum degree Δ and is not r-majorizable.

We claim it suffices to let the sizes of the parts be $\{k_i\}$ where $\{k_i\}$ are r integers such that $n - \Delta = k_1 < k_2 < \cdots < k_r$ and $\Sigma k_i = n - 1$. We also claim that this can be done precisely when $\Delta \geq \lceil n - \frac{n-1}{r} + \frac{r-1}{2} \rceil$ to get a degree sequence with the desired properties. An example of such a graph G with $(r, n, \Delta) = (3, 10, 8)$ and $\langle k_i \rangle = (2, 3, 4)$ is shown in Figure 1.

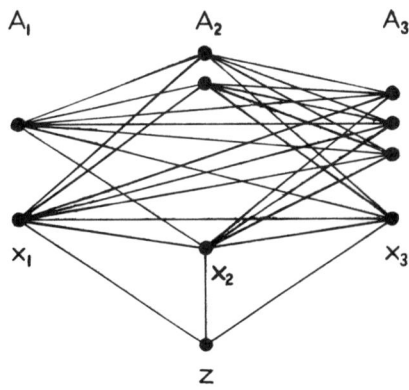

Figure 1

To show that a graph G with the k_i so chosen is not r-majorizable, we begin by computing the degree sequence of G. The degree of x_i is $n - k_i$. The degrees of the other vertices in A_i are $n - k_i - 1$. The degree of z is r. In the case $r = 1$ and $\Delta = n - 1$, the resulting sequence is $(n - 1, 2, 2, 1, 1, 1, \ldots)$, which is not 2-majorizable. In all other cases, the requirement $\Sigma k_i = n - 1$ implies $n - k_i > r$. In addition, the k_i form a strictly increasing sequence, so $n - k_i - 1 \geq n - k_{i+1}$. Therefore, the degree sequence of G is

$$(n-k_1, n-k_1-1, \ldots \mid n-k_2, n-k_2-1, \ldots \mid \cdots \mid n-k_r, n-k_r-1, \ldots \mid r)$$

where the ith segment of this sequence has k_i terms, $1 \leq i \leq r$.

We claim that G is not r-majorizable. A graph is r-majorizable if and only if it has a vertex partition into r parts such that each vertex in a part of size q has degree at most $n - q$. This can be tested by the "greedy algorithm". Partition the vertices by repeating the following two steps r times.

1) Place the remaining vertex of highest degree d ($=n - k_1$ at first) in a new block.
2) Fill the rest of that block with the vertices having the next $n - d - 1$ highest degrees.

G is r-majorizable if and only if this procedure places *all* the vertices into the r blocks.

We have applied this procedure above to the degree sequence of

G, placing vertical bars as far to the right as possible to indicate when a part must end. Since the resulting r parts do not exhaust the vertices, the sequence is not r-majorizable.

It remains only to determine for what values of k_1 this construction can be performed. To make k_1 as large as possible and satisfy the constraints on $\{k_i\}$, these parameters should be a set of consecutive integers summing to $n - 1$. In particular, we have

$$n - 1 = \sum k_i \geq \sum_{i=1}^{r} k_1 + i - 1 = rk_1 + \frac{r(r-1)}{2} \qquad (*)$$

From the two ends of this we get

$$n - \Delta \leq \frac{n-1}{r} - \frac{r-1}{2} \quad \text{or} \quad \Delta \geq \lceil n - \frac{n-1}{r} + \frac{r-1}{2} \rceil$$

which gives the lower bound on Δ. To perform this construction for larger values of Δ, simply increase the value of $k_r, k_{r-1}, \ldots,$ as needed to compensate for the decrease in k_1. This is always possible, except when $r = 2$ and $\Delta = n - 1$, which can be handled as described earlier. □

3. The Intermediate Case when $r = 2$

When $r = 2$, the only case remaining is $f_2(n, \frac{n+1}{2})$ for odd values of n. First we give a construction for an upper bound.

LEMMA 1. $f_2(n, \frac{n+1}{2}) \leq \frac{n-1}{2}$.

PROOF. Partition n vertices into four vertex sets, A, B, C, and D, having sizes $\frac{1}{2}(n - 3)$, $\frac{1}{2}(n - 3)$, 2, and 1, respectively. Form G by joining each vertex of A to each of B, each of B to each of C, and placing a triangle on $C \cup D$. The construction is illustrated in Figure 2.

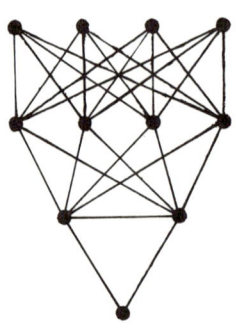

class	multiplicity	degree
A	$\frac{n-3}{2}$	$\frac{n-3}{2}$
B	$\frac{n-3}{2}$	$\frac{n+1}{2}$
C	2	$\frac{n+1}{2}$
D	1	2

Figure 2

The resulting maximum degree is $\frac{1}{2}(n+1)$ and there are $\frac{1}{2}(n-1)$ triangles. The degree sequence is not 2-majorizable, since there are $\frac{1}{2}(n+1)$ vertices of degree $\frac{1}{2}(n+1)$ — each would have to appear in a part with less than half the vertices, but no single such part can contain $\frac{1}{2}(n+1)$ vertices. With any fewer vertices of degree $\frac{1}{2}(n+1)$, no problem arises. □

The construction could also be modified by letting the vertex in D be joined to any subset of the $\frac{1}{2}(n-3)$ vertices in A. Let W be the collection of all such graphs. We will see that the graphs in W are the only graphs that achieve the minimum. In order to prove this, we need two numerical lemmas about the number of complete subgraphs that must be present when a graph contains other complete subgraphs of "high degree". Let the *degree* of a complete subgraph H of a graph be the sum of its vertex degrees. We write this as $d(H)$. Results have long been known about when high degree of an r-clique forces its inclusion in some $(r+1)$-clique. One of the most general appears in [9]. We phrase it for $r+1$ instead of the usual r.

LEMMA 2. *If H is any $(r+1)$-clique in G and $d(H) > rn$, then there exist in G at least $d(H) - rn$ distinct $(r+2)$-cliques each containing H.*

PROOF. We need only show that at least $d(H) - rn = s$ vertices not in H adjoin every vertex of H. There are $d(H) - r(r+1) = s + r(n - r - 1)$ edges between H and the $n - r - 1$ vertices of $G - H$. Since no vertex can be joined to more than $r + 1$ vertices of

H, the pigeonhole principle implies that at least s must be joined to exactly $r + 1$. □

The most well-known case of this is $r = 1$; two adjacent points whose degrees sum to at least $n + k$ lie together in at least k triangles.

Since we are interested in the number of $(r + 1)$-cliques, we will be more interested in the following variation.

LEMMA 3. *If H is an $(r + 1)$-clique in G and $(r - 1)n + r < d(H) \leq rn$, then G contains at least $d(H) - (r - 1)n - r$ $(r + 1)$-cliques that each contain at least r vertices of H. If $d(H) > rn$, then G contains at least $r[d(H) - rn] + n - r$ such $(r + 1)$-cliques.*

PROOF. Like the previous lemma, this is a simple application of the pigeonhole principle. We may assume there are no edges among the verties of $G - H$, since they could only cause additional K_{r+1}'s. The number of $(r + 1)$-cliques involving vertices of H is minimized by adding the other edges greedily. In other words, add edges so that all the vertices in $G - H$ have $r - 1$ neighbors in H before any have r neighbors, and all have r neighbors in H before any have $r + 1$. To prove that this works, let t be the number of vertices in $G - H$ neighboring r vertices in H, and let s be the number neighboring $r + 1$ vertices of H. Then the number of $(r + 1)$-cliques in G is $t + (r + 1)s + 1$, counting one for H itself. There are $d(H) - r(r + 1)$ edges between H and $G - H$. In the lower range mentioned, after distributing $r - 1$ of these to each of the $n - r - 1$ vertices in $G - H$, we have $t = d(H) - (r - 1)n - r - 1$ and $s = 0$. In the upper range, $s = d(H) - rn$, as in Lemma 1, and $t = n - r - 1 - s$. In either range, computing $t + (r + 1)s + 1$ gives the lower bound claimed. □

Of course, both of these lemmas give best-possible bounds, achieved by letting $G - H$ be an independent set. Together, they give a short proof of the lower bound.

THEOREM 2. $f_2(n, \frac{n+1}{2}) = \frac{n-1}{2}$. *Furthermore, the only $(n, \frac{n+1}{2})$-graphs that are not 2-majorizable yet have only $\frac{1}{2}(n - 1)$ triangles are those in the set W described in Lemma 1.*

PROOF. Lemma 1 gave the upper bound. Let $I = I(G)$ be the set of vertices of degree $\frac{1}{2}(n + 1)$ in an $(n, \frac{n+1}{2})$-graph G. Note that G is

2-majorizable if and only if I has less than $½(n + 1)$ vertices, as mentioned earlier. If G is non-majorizable, we will show that either there is a triangle among the vertices of I, or there are more than $½(n - 1)$ triangles elsewhere in the graph. In the former case, the degree-sum in this triangle is $3(n + 1)/2$. Since $3(n + 1)/2 \geq n + 3$ for $n \geq 3$, we can apply Lemma 3 with $r = 2$. It implies there will be at least $3(n + 1)/2 - n - 2 = ½(n - 1)$ distinct triangles in the graph, each of which contains at least two of these three vertices in I.

For any edge in I, the degrees of the endpoints sum to $n + 1$, so by Lemma 2 any such edge must appear at least one triangle. Also note that, since I consists of at least $\frac{n+1}{2}$ vertices of degree $\frac{n+1}{2}$, each must neighbor another; hence every vertex in I belongs to an edge in I and thus to a triangle in G.

Suppose there is no triangle in I. By the above, in this case it will suffice to show there must be more than $½(n - 1)$ edges in I. Let e be the number of edges in I. The vertices not in I have degree at most $½(n - 1)$, so they can absorb at most $½(n - 1)(n - |I|)$ of the degree sum from I. The remainder is twice e. We have

$$2e \geq (½(n + 1))|I| - (n - |I|)(½(n - 1))$$
$$= n(|I| - (½(n - 1))) \geq n,$$

where the last inequality follows from $|I| \geq \frac{1}{2}(n+1)$.

This means $e > ½(n - 1)$, which is the desired result.

This is also the first step toward uniqueness; not having a triangle in I forces *more* than $½(n - 1)$ triangles in G. Therefore, for a graph achieving the minimum we can assume there is a triangle T in I. We have seen there are at least $½(n - 1)$ triangles in G containing at least two vertices of T. If these $½(n - 1)$ triangles are the only triangles in G, then each vertex of I belongs to one of them, since every vertex in I lies on a triangle. Any such triangle other than T is also a triangle in I and also has $½(n - 1)$ triangles involving at least two of its vertices. If there are at most $½(n - 1)$ triangles altogether, these must be the same triangles, and hence they all share an edge.

Thus, minimality requires that all $½(n - 1)$ triangles share a common edge in I, and that all the other vertices in I appear as apexes of these triangles. To show G is in W, let the set C be the common vertices of the triangles. The apexes of the triangles are a set of $½(n - 1)$ vertices, of which at least $½(n - 3)$ have degree $½(n + 1)$; let this be the set $B \cup D$. There can be no edges between these vertices without creating more triangles, so if the vertices of B are to have degree

½(n + 1), they must all be joined to all the remaining vertices, which form the set A. Now no further edges can be added to the graph without creating more triangles (except between D and A), so G lies in W. □

4. The Uniform Upper Bound

The above constructions generalize to give an upper bound valid for all cases. The bound appears to be optimal, but we have not yet succeeded in showing this.

THEOREM 3. $f_r(n, \Delta) \le (n - \Delta)^t$, where t is the smallest integer such that

$$r(n - \Delta) + \binom{r-t}{2} \le n - 1.$$

If there is no such integer, we take $t = \infty$.

PROOF. Note that there is no such integer if and only if $\Delta \le \lfloor n - \frac{n}{r} \rfloor$. On the other hand, we get $t = 0$ if and only if $\Delta \ge \lceil n - \frac{r-1}{r} + \frac{r-1}{2} \rceil$. This agrees with the result of Section 2.

We proceed as in the proof of Theorem 1. Again, we begin with an r-partite graph on $n - 1$ vertices. However, it no longer suffices to join the remaining vertex to only one vertex in each part, because there will not be enough vertices of sufficiently high degree to make the degree sequence non-majorizable.

Again, let the parts be A_i, with sizes k_i. This time we make the small parts the same size. We choose the k_i so that $n - \Delta = k_1 = \cdots = k_t = k_{t+1} < \cdots < k_r$ and $\Sigma k_i = n - 1$. Note that by the definition of t we have $t < r$ when it is finite. To construct the graph, we join the last vertex z to every vertex in A_i when $i \le t$ and to a single vertex x_i from each part A_i with $i > t$. The graph is schematically illustrated in Figure 3 for the case $t = 2$, $r = 5$.

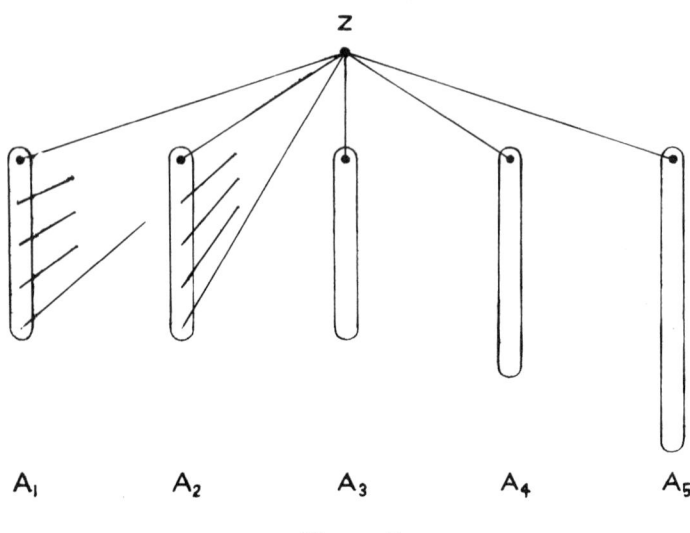

Figure 3

The equality among part-sizes and joining of z to all vertices in A_i "propagates" the needed high degree. Each neighbor of z in A_i has degree $n - k_i$; each non-neighbor has degree $n - k_i - 1$. Thus the degree sequence is

$$(n-k_1,\ldots | n-k_1,\ldots | \cdots | n-k_{t+1}, n-k_{t+1}-1,\ldots | \cdots$$
$$\cdots | n-k_r, n-k_r - 1,\ldots | tk_1 + r-t)$$

where each of the first $t+1$ segments of this sequence has k_1 terms, and the ith segment has k_i terms, for $t+2 \le i \le r$.

All $(r + 1)$-cliques in this graph consist of z and one neighbor of z from each A_i. The number of these is $(n - \Delta)^t$.

As in section 2, this degree sequence is non-majorizable by construction, even if the degree of z is large enough to require a different ordering of the degrees. That reordering will be needed, for example, if $\Delta = n - (n - 1)/r$ and $t = r - 1$, in which case the degree of z is also Δ. For $r = 2$, this is precisely the case discussed in section 3, and the graph we have constructed is the graph in W with the most edges. The edges that can be deleted without losing non-majorizability are those joining the vertices of A_r other than x_r to some single vertex not in $\{z\} \cup A_r$.

To get the best upper bound from this construction, we need only determine how small t can be made when k_1 is as large as required and Σk_i is as small as required. To make Σk_i small, we again let the part-sizes increase singly after k_{t+1}. We get

$$n - 1 = \Sigma k_i \geq tk_1 + \sum_{i=0}^{r-t-1} k_t + i = (n - \Delta)r + \binom{r-t}{2},$$

which completes the proof.

5. Further Remarks

To motivate the conjecture, we present a short proof that $f_3(7,5) = 4$. This is the next undetermined case.

THEOREM 4. $f_3(7,5) = 4$. Furthermore, there are only two $(7,5)$-graphs with four 4-cliques that are not 3-majorizable, namely those constructed in the proof of Theorem 3.

PROOF. A $(7,5)$-graph G is not 3-majorizable only if it has at least five vertices of degree 5 and at most two vertices of smaller degree. Let H be the subgraph of G induced by vertices of degree 5. We may assume by discarding edges that H has exactly five vertices and $G - H$ consists of isolated vertices. Otherwise G would be K_6 plus an isolated vertex, in which case it would have twenty triangles.

Each vertex in H has degree at least three in H. Every edge in H has degree 10 in G and thus belongs to at least three triangles in G. Since there are only two vertices in $G - H$, at least one of these triangles belongs entirely to H. Every triangle in H has degree 15 in G; by Lemma 2, it belongs to 4-clique in G.

There are at least eight edges in H, since each vertex of H neighbors at least three others. By the theorem of Moon and Moser, there are at least $\lceil \frac{8}{15}(32 - 25) \rceil = 4$ triangles in H. If there is no 4-clique in H, we are done, since the 4-cliques containing each of the triangles in H must then be different. So, assume H has a 4-clique. The remaining vertex of H neighbors at least three of those vertices, creating another 4-clique in H.

Let I be the three vertices shared by the two 4-cliques in H. The vertices of $H - I$ cannot be joined, else G contains K_5 with five 4-cliques. Thus, they have degree 3 in H and must each be joined to the two vertices in $G - H$. The vertices of I must each be joined to one vertex in $G - H$. Whenever two of them are joined to the same vertex in $G - H$, which must happen at least once, two additional 4-cliques are created, by including either of the vertices in $H - I$.

The restrictions obtained in proving this imply that the only (7,5)-graphs with four 4-cliques that are not 3-majorizable are those constructed in the proof of Theorem 3. □

This type of argument can be pushed farther; we have done so to obtain the appropriate lower bound for $f_3(3k + 2, 2k + 2)$. However, the argument becomes increasingly tedious as r increases. Therefore, we omit the proof of the lower bound for $f_3(3k + 2, 2k + 2)$ in the expectation that a more elegant inductive proof exists.

Acknowledgement

The author thanks Scott Smith for stimulating discussions on this problem.

References

[1] B. Andrásfai, P. Erdös, and V.T. Sós, On the connection between chromatic number, maximal clique, and minimal degree of a graph, *Discrete Math.*, 8 (1974), 205-208.

[2] B. Bollobás, On complete subgraphs of different orders, *Math. Proc. Camb. Phil. Soc.* 79 (1976), 19-24.

[3] B. Bollobás, Extremal problems in graph theory, *J. Graph Theory* 1 (1977), 117-123.

[4] B. Bollobás, Relations between sets of complete subgraphs, *Proc. 5th Brit. Comb. Conf. (Aberdeen) 1975*, Utilitas Math., Winnipeg (1976), 79-84.

[5] B. Bollobás, P. Erdös, and E. Szemerédi, On complete subgraphs of r-chromatic graphs, *Discrete Math.* 13 (1975), 97-107.

[6] J.A. Bondy, Dense neighborhoods and Turán's theorem, Univ. of Waterloo Technical Report, 1980.

[7] G.A. Dirac, Extensions of Turán's theorem on graphs, *Acta Math. Sci. Hungar.* 14 (1963), 418-422.

[8] C.S. Edwards, Triangles in simple graphs and some related results, *Proc 5th Brit. Comb. Conf.* (1975), Utilitas Math., Winnipeg (1976).

[9] C.S. Edwards, The largest number of triangles with a common edge in a graph, in *Problemes Combinatoires et Theorie des Graphes*, C.N.R.S. Colloq., Paris (1976), 123-128.

[10] P. Erdös, On a theorem of Rademacher-Turán, *Illinois J. Math.* 6 (1962), 122-127.

[11] P. Erdös, On the graph theorem of Turán (in Hungarian), *Mat. Lapok* 21 (1970), 249-251.

[12] P. Erdös and M. Simonovits, On a valence problem in extremal graph theory, *Discrete Math* 5 (1973), 323-334.

[13] P. Erdős and A.H. Stone, On the structure of linear graphs, *Bull. Amer. Math. Soc.* 52 (1946), 1087-1091.

[14] L. Lovász and M. Simonovits, On the number of complete subgraphs of a graph, *Proc. 5th Brit. Comb. Conf. (1975)*, Utilitas Math., Winnipeg (1976), 431-442. (further results in memorial volume to Turán).

[15] J.W. Moon and L. Moser, On a problem of Turán, *Publ. Math. Inst. Hungar. Acad. Sci.*, 7 (1962), 283-286.

[16] S. Roman, The maximum number of q-cliques in a graph with no p-clique, *Disc. Math.* 14 (1976), 365-371.

[17] P. Turán, On an extremal problem in graph theory (in Hungarian), *Mat. Fiz. Lapok* 48 (1941), 436-452. See also: On the theory of graphs, *Colloq. Math.* 3 (1954), 19-30.

[18] K. Zarankiewicz, On a problem of Turán concerning graphs, *Fund. Math.* 41 (1954), 137-145.

UNSOLVED PROBLEMS

Must a p-Chromatic Graph Contain a p-Scheme?

Pierre Duchet
Henri Meyniel

A p-scheme in a graph G is constituted by p vertices $v_1,\ldots,v_p$ of G and by $\binom{p}{2}$ subsets C_{ij} ($1 \leq i \neq j \leq p$; $C_{ij}=C_{ji}$) with the following properties.

(1) C_{ij} is connected and contains v_i and v_j.
(2) $C_{ij} \cap C_{kl} = \emptyset$ if i, j, k, l are distinct.

CONJECTURE S - *A p-chromatic graph contains a p-scheme.*

The problem is certainly difficult but may be tractable. We offer 1000 Canadian dollars or 5000 French F. for a proof.

OBSERVATION 1. It is quite easy to see that conjecture implies the four color property, since a planar graph cannot contain a 5-scheme. (cf[13], prop.5.])

OBSERVATION 2. The conjecture S is weaker than the Hadwiger's conjecture [10] *"a p-chromatic graph is contractible onto K_p, the complete graph with p vertices"*. (cf. conjecture RS below). Thus, the conjecture S is true for $p \leq 4$, and for $p=5$ via the 4-colour theorem [1, 2] and the Wagner's equivalence result [14,15 for a short proof]: the conjecture is also true for almost every graph [4].

Conversally, we don't know what is the largest integer $g(p)$ such that every p-scheme is contractible onto $K_{g(p)}$. In [7], it is shown that $g(p) \geq [\sqrt{2p}]$ for $p>1$ and $g(p) \leq \frac{17}{27}p$ for p sufficiently large.

OBSERVATION 3. If a p-scheme (v_i, C_{ij}) in a graph G satisfies the additional condition: (3) $C_{ij} \cap C_{ik} = \{v_i\}$ if i,j,k are distinct. Then, G contains a *subdivided K_p* ($=p$ vertices that are pairwise joined by disjoint paths). So, the conjecture S corresponds to an idea similar to those Hajòs had when he proposed his conjecture: *"every p-chromatic graph contains a subdivided K_p"*. Since Catlin [5] has discovered counterexamples to Hajòs conjecture, Erdös and Fajtlowicz [9] proved that almost every graph is a counterexample (see [3] for a more precise result).

OBSERVATION 4. The condition (2) is a very natural condition when we consider *Kempe exchanges* in a coloured graph. Given a p-colouring C of a graph G, if C_{ij} denotes the set of vertices coloured in i or j, then the C_{ij}'s satisfy obviously the condition (2). A connected component of the subgraph induced by C_{ij} is usually called a bicoloured *(C,i,j)-component*. To each bicoloured component B, corresponds a "Kempe exchange": the interchange of colours i and j in B; remark that the resulting colouring may have one colour less (compare to [12] for details).

Very often, the consideration of Kempe exchange is the only way the chromatologist knows for obtaining results: for instance see [2]. As in the 4-colour problem, one might try to prove conjecture S via Kempe exchange. But this seems hopeless considering the following example: the simplest known counterexample to Hajòs' conjecture is the graph with vertex-set $\mathbb{Z}5 \times \mathbb{Z}3$ where (x,y) is joined by (x',y') iff $x=x'$ or $x-x'=\pm 1$ (=lexicographical product $C_5[C_3]$). One can verify that this graph is 8-chromatic; moreover every q-colouring for $q \geq 9$ can be reduced to a 8-colouring by a sequence of Kempe exchanges (i.e. the Kempe chromatic number [12] is 8). In this graph, no subdivided K_8 exists and, furthermore, no 8-colouring exists with a "coloured 8-scheme": 8 vertices coloured with different colours that are pairwise connected by bicoloured components. (For a proof, show that only four 8-colourings exist, up to isomorphism). However, there exist a 9-colouring with a coloured 9-scheme!

OBSERVATION 5. The condition (2) above appears also as a natural property in the context of *abstract convexities*. For an introduction to this subject we refer to [6,11]; an (abstract) convexity on a connected graph G with vertex set V is a collection **C** of *connected* subsets of V which is stable under intersection. The members of **C** are called the convex sets; $\emptyset$ and V are convex sets. The convex hull of a subset $A \subset V$ is defined as the intersection of the convex sets containing A. The Helly number and the Radon number for convexities are parameters usually studied. Concerning convexities on a graph, we noted in [8] that very close relationships exist between these parameters and the Hadwiger number of the graph (= the size of the largest complete graph in which the graph is contractible); the Radon property with order p is:

(R_p) Every subset with p elements contains two disjoint subsets whose convex hull have a common element

If a set $A = \{a_1, \ldots, a_p\}$ of vertices contradicts (R_p) for some convexity on a graph, the condition (2) is satisfied by the sets $C_{ij}=$ convex hull of $\{a_i, a_j\}$. A stronger form of conjecture S follows:

CONJECTURE RS. *Suppose that every convexity on a connected graph G satisfies the Radon property with order p: then G admits a colouring with p-1 colours.*

It turns out that the conjecture RS is still weaker than the Hadwiger's conjecture. To see this fact, remark that, for a connected graph G which is contractible onto K_p, the vertex-set admits a partition into p disjoint connected subsets $V_1, \ldots, V_p$ that are pairwise adjacent. So, defining convex sets as all the possible unions of V_i's, we obtain a convexity contradicting (R_p).

We offer 1000 Canadian dollars for a counterexample.

References

[1] K. Appel, W. Haken, Every planar map is four-colourable, Part I: discharging, *Illinois J. Math. 21 (1977), 429-490.*

[2] K. Appel, W. Haken, J. Koch, Every planar map is four-colorable, Part II: reducibility, *Illinois J. Math. 21 (1977), 491-567.*

[3] B. Bollobas, P.A. Catlin, Topological Cliques of Random Graphs, *J. Comb. Th. B 30 (1981), 224-227.*

[4] B. Bollobas, P.A. Catlin, P. Erdos, Hadwiger's conjecture is true for almost every graph, *Eur. J. Comb. 1 No. 3 (1980), 195-199.*

[5] R.A. Catlin, Hajós graph-colouring conjecture: variations and counterexamples, *J. Comb. Th. B 6(1979), 268-274.*

[6] M. Cochand, P. Duchet, Sous les pavés ..., "Proceedings Int. Conf. Marseille-Luminy (1981), Theorie des graphes et Problèmes combinatoires". Ann. of Disc. Math. 17 (1983), 191-202.

[7] P. Duchet, Schemes of graphs, in Graph Theory and Combinatorics, B. Bollobas Ed., Ac. Press 1984, 145-153.

[8] P. Duchet, H. Meyniel, Ensembles convexes dans les graphes, I: nombres de Helly et de Radon pur graphes et surfaces, *Eur. J. Comb., 4 (1983), 127-132.*

[9] P. Erdős, S. Fajtlowicz, On the conjecture of Hajós, Combinatorica 1 (1981) 141-143.

[10] H. Hadwiger, Ungelöste Problem, Element Math. 13 (1958), 127-128.

[11] D. Kay, E.W. Womble, Axiomatic convexity theory and relationships between the caratheodory, Helly and Radon numbers, *Pac. J. Math. 38 (1971), 471-485.*

[12] M. Las Vergnas, H. Meyniel, Kempe classes and the hadwinger Conjecture, *J. Comb. Th. B 31 (1981), 95-104.*

[13] W.T. Tutte, Toward a theory of crossing numbers, *J. Comb. Th. 8 (1970), 45-53.*

[14] K. Wagner, Beweis einer Abschwachung der Hadwiger-Vermutung, Math. Am. 153 (1964), 139-141.

[15] H.P. Young, A quick proof of Wagner's equivalence theorem, J. London Math. Soc. (2) 3 (1971), 661-664.

Selected Problems

P. Erdős

1. Szemerédi's Extremal Graph Problem

Let $G(n;e)$ be a graph of n vertices and e edges. Szemerédi conjectured (a few days ago) that to every ϵ and k there is an n_0 such that if $e > n^{1+\epsilon}$ and $n > n_0$ then there are two disjoint sets, S_1 and S_2, of vertices of our graph, $|S_1| = |S_2| = k$, for which, if $A \subset S_1, B \subset S_2, |A| \geq \frac{k}{2}, |B| \geq \frac{k}{2}$, then there is at least one edge joining A and B.

Szemerédi tells me that this, if true, would imply my following old conjecture: Let $X_1, \ldots, X_n$ be n distinct points in the plane, then the number of pairs X_i and X_j whose distance is 1 is less than $c_\epsilon n^{1+\epsilon}$. He also informs me that, since in my problem $n^{1+\epsilon}$ can not be replaced by $n^{\frac{1+c}{\log\log n}}$, the same is true for his conjecture. Thus, if true, his conjecture must be quite delicate.

2. Erdős-Farber-Lovász Conjecture

Here is our old conjecture with Farber and Lovász. Let $|A_i| = n, 1 \leq i \leq n$, and $|A_i \cap A_j| \leq 1, 1 \leq i < j \leq n$. Is it then true that one can color the elements of $\bigcup_{i=1}^{n} A_i$ with n colors so that every A_i contains all the colors?

Here is an equivalent graph theoretic formulation: Let G be a graph which is the union of n edge disjoint complete graphs of size n. Is it true that the chromatic number of G is n?.

This conjecture is now 10 years old. It seems easy but is probably difficult and I offer 500 dollars for a proof or disproof.

3. A Problem of Simonovits and Myself

Let L be a graph, $f(n;L)$ be the smallest integer for which every $G(n;f(n;L))$ (i.e. a graph of order n and $f(n;L)$ edges) contains L as a subgraph. Denote by $A(n; L)$ the largest integer for which every $G(n; f(n; L))$ contains at least $A(n; L)$ copies of L. Determine or estimate $A(n; L)$ as accurately as possible. For which graphs L does $A(n; L)$ tend to infinity? This is known if L is complete. The simplest unsolved case is if L is C_4.

4. Ramsey Problems

Let $r(n_1,\ldots,n_k)$ be the smallest integer for which, if we color the edges of a complete graph of $r(n_1,\ldots,n_k)$ vertices with k colors, then for at least one i the i-th color contains a K_{n_i}. V.T. Sós and I conjectured that

$$r(3, 3, n) / r(n, n) \to \infty \tag{1}$$

I offer 100 dollars for a proof or disproof of (1). I also offer 100 dollars for the proof that

$$\lim_{n\to\infty} r(n, n)^{1/n} \tag{2}$$

exists and I offer 250 dollars for the value. (I am sure that the limit exists and I offer 10^4 dollars for a disproof.)

It must be clear to any right thinking person that

$$\lim_{n\to\infty} \frac{r(n + 1, n)}{r(n, n)}$$

exists and is greater than 1. We (Burr, Faudree, Rousseau, Schelp and I) thought that we had a proof that

$$r(n + 1, n) - r(n, n) > cn^2. \tag{3}$$

We have not been able to reconstruct our proof, but hope to do so soon. The difference in (3) must tend to infinity faster than any power of n, but we certainly have no proof of this.

One of the difficulties in proving these conjectures seems to be that all the good lower bounds are obtained by probabilistic methods and do not seem to be adaptable to give results about local properties.

5. On Pancyclic Graphs

I want to call attention to an old and forgotten problem of Bondy: Let $f(n)$ be the smallest integer for which there is a $G(n; f(n))$ which contains C_k (a circuit of k edges) for every k, $3 \le k \le n$. Bondy proved

$$n + \frac{\log n}{\log 2} < f(n) < n + \frac{\log n}{\log 2} + L(n) \qquad (4)$$

where $L(n)$ is the smallest r with $\log_r(n) < 2$ and $\log_r(n)$ is the r times iterated logarithm. Is it true that

$$f(n) - n - \frac{\log n}{\log 2} \to \infty$$

or is the lower bound in (4) essentially best possible?

6. Extremal Problems on Hypergraphs - The Conjecture of Turán

Let $L^{(k)}$ be a k-uniform hypergraph, and $f(n; L^{(k)})$ the smallest integer for which every $G^{(k)}(n; f(n; L^{(k)}))$ contains $L^{(k)}$ as a subgraph. The problems are much harder for $k > 2$ than for $k = 2$. Turán considered the case when $L^{(k)} = K^{(k)}(r)$ is the complete k-uniform hypergraph on r vertices. He has plausible conjectures for $k = 3$ and $r = 4$, $r = 5$, i.e. for $K^{(3)}(4)$ and $K^{(3)}(5)$. His conjectures would imply

$$\lim_{n \to \infty} \frac{f(n; K^{(3)}(4))}{\binom{n}{3}} = \frac{5}{9} \qquad (5)$$

and

$$\lim_{n \to \infty} \frac{f(n; K^{(3)}(5))}{\binom{n}{3}} = \frac{8}{9}.$$

I offer 500 dollars for the proof or disproof of (5) and 1,000 dollars for the determination of $f(n; K^{(k)}(r))$.

Here is my conjecture on the so-called jumping constants: Is it true that there is an increasing well ordered transfinite sequence

$$\alpha_1^{(k)} < \alpha_2^{(k)} < \cdots, \qquad \alpha_1^{(k)} = \frac{k!}{k^k}$$

such that for every $L^{(k)}$

$$\lim_{n\to\infty} \frac{f(n; L^{(k)})}{\binom{n}{k}}$$

is always one of the α's? I offer 1,000 dollars for a proof or disproof.

$\alpha_1^{(k)} = \frac{k!}{k^k}$ is an old theorem of mine. The fact that, for $k = 2$, $\alpha_i^{(2)} = \frac{1}{2}(\frac{i-2}{i-1})$ is the well known theorem of A. Stone, Simonovits and myself.

Thus the general conjecture is true for $k = 2$. I offer 500 dollars for a proof that $\alpha_2^{(3)} > \alpha_1^{(3)} = \frac{2}{9}$. See my paper Discrete Math. Vol. 1 for references.

7. On Triangle-Free Graphs

Let $G(n)$ be a regular graph of degree k_n. Assume that G is triangle free and the maximum number of independent vertices is k_n. (It can not be less since the star of a vertex must be independent.) Determine or estimate the smallest value of k_n for which such a graph can exist. In J.C.T. Vol. 1, I have an example with $k_n \sim n^{1-c}$, $c > 0$, but c is small. Is it true that for every $\epsilon > 0$ there is such a $G(n)$ with $k_n < n^{1/2+\epsilon}$? I offer 100 dollars for a proof or disproof.

On the Chessmaster Problem

P. Erdös

R.L. Hemminger

D.A. Holton

B.D. McKay

To illustrate the use of the Pigeonhole Principle, Brualdi [1, pp 16,22] considers the following problem. A chessmaster who has 11 weeks to prepare for a tournament decides to play at least one game every day, but in order not to tire himself he decides not to play more than 12 games during any one week. Show that, for any k with $1 \le k < 21$, there corresponds a succession of days during which the chessmaster will have played *exactly* k games.

If we let a_i be the number of games played during the first i days, (and take $a_0 = 0$) then the above problem translates into the following:

CMP: If $1 \le k \le 21$ and $0 = a_1 < a_1 < a_2 < \ldots < a_{77}$ such that $a_{i+7} \le a_i + 12 \; \forall i (0 \le i \le 70)$, then $\exists i, j (0 \le i < j \le 77)$ such that $a_j = a_i + k$.

An alternate interpretation could be obtained by replacing $(0 \le i \le 70)$ with $(i = 0, 7, 14, \ldots, 70)$ but the generalization we will make seems to be much more chaotic with this interpretation.

It turns out that the implication in CMP holds for all k with $1 \le k \le 77$ but does not hold for any $k > 77$ (just take $a_i = i, \forall i$). This moved one of the authors to raise the following general question.

GCMP: Let k, n, b and c be positive integers with $b \le n, c$ and let $A = \{a_i\}_{i=0}^{n}$ be a sequence of integers. Consider the followig conditions on A:

(1) $0 = a_0 < a_1 < a_2 < \ldots < a_n$,

(2) $a_{i+b} \le a_i + c, \; \forall i (0 \le i \le n-b)$ and

(3) $\exists i, j \; (0 \le i < j \le n)$ such that $a_j = a_i + k$.

For what values of k, n, b and c does $(1) \wedge (2) \Rightarrow (3)$?

The case $n = b$, $k \le 2b - c - 1$ is an Olympiad problem (see [3]) which has been analyzed for other values of k by Hutchinson and Trow [2]; also see [4] for an answer to a question they raised. We have shown that (1) $\wedge$ (2) $\wedge$ $(n=b)$ $\Rightarrow$ (3) precisely for those k with $k \lfloor b/k \rfloor > c-b$.

We have already noted that the implication does not hold for $k>n$ so hereafter we will assume that $k \leq n$. The just-stated results suggest that $c=2b$ is somehow limiting. The following examples confirm that impression since (1) $\wedge$ (2) hold in them, but (3) fails to hold.

EXAMPLE 1: For $c>2b$ and k even, define A by taking

$$a_i = \begin{cases} 2i & \text{if } \lfloor 2i/k \rfloor \text{ is even} \\ 2i+1 & \text{otherwise} \end{cases}.$$

EXAMPLE 2: For $c \geq 2b$ and k odd; indeed for k/d odd where d is any divisor of (k,b), the g.c.d. of k and b, define A by taking

$$a_{md+r} = 2md+r \quad (0 \leq r \leq d-1, \ 0 \leq md+r \leq n).$$

We observe that for $i<j$ in Example 1, $a_j - a_i$ is one of the numbers $2(j-i)$, $2(j-i)-1$ or $2(j-i)+1$, and so $a_{i+b} \leq a_i + (2b+1) \leq a_i + c$. On the other hand one easily sees that (3) fails to hold since k is even.

In Example 2 we note that $a_{md+r} + k = 2md+r+hd$ (with h odd) $= (2m+h)d+r$ which is not in A since $2m+h$ is odd. Moreover, noting that $a_t = 2t - r_t$ where r_t denote the remainder when t is divided by d, we see that $a_{t+b} - a_t = 2(t+b) - r_t$ where r_t denotes the remainder when t is divided by d, we see that $a_{t+b} - a_t = 2(t+b) - r_{t+b} - (2t - r_t) = r_t - r_{t+b} + 2b = 2b$ since d divides b. Thus $a_{t+b} = a_b + 2b \leq a_t + c$.

The following result is useful in attempting a proof by contradiction.

LEMMA 1: If (1) and the negation of (3) hold, then $a_{i+k} \geq a_i + 2k$, $\forall i (0 \leq i \leq n-k)$.

PROOF: From the negation of (3), we see that for each i, the numbers $a_{i+1}, a_{i+2}, \ldots, a_n$ include at most one each from each of the $k-1$ sets $\{a_i+1, a_i+k+1\}$, $\{a_i+2, a_i+k+2\}, \ldots, \{a_i+k-1, a_i+2k-1\}$. Likewise none of them equals $a_i + k$ and so the lemma follows by (1) and the Pigeonhole Principle.

The following two results, due to the first author, confirms most of the conjectures put forward by the other three authors at the Silver Jubilee conference. As in Lemma 1, the Pigeonhole Principle is a vital ingredient.

LEMMA 2: If $c \le 2b-1, 1 \le k \le n$ and r is the number such that $a_r < k \le a_{r+1}$, then $r+b \le n =>$ (3) holds.

PROOF: Since $a_0 = 0 < k$ and $a_k \ge k$ by (1), such an r exists. But then the $2b+1$ numbers $a_0+k, a_1+k, \ldots, a_b+k, a_{r+1}, a_{r+2}, a_{r+b}$ all lie in the set $\{k, k+1, \ldots, k+c\}$ by (2), and hence (by the Pigeonole Principle) two of them are equal since $c+1 \le 2b$.

COROLLARY. If $c \le n+1$, then (3) holds.

PROOF. If $r+b > n$, then we have

$$a_k \le a_n - (n-k) \text{ by (1)}$$
$$\le a_{n-b} + c - (n-k) \text{ by (2)}$$
$$< a_r + c - (n-k) \text{ since } r > n-b$$
$$\le a_r + c - (c-1-k) \text{ since } c \le n+1$$
$$= a_r + k + 1$$

Thus $a_k < 2k$ since $a_r < k$. The corollary follows from the two lemmas.

We note in passing that this establishes the claim that the CMP is true for all k with $1 \le k \le 77$.

The situation for $n < c-1$ is confusing at best. For example, for $n = k = c-2 = 2b-3$, there are always examples satisfying (1) and (2) but not (3); in the particular case $b=5, c=9, n=k=7$,

$$0, 3, 5, 6, 8, 9, 11, 14$$

is such a sequence. In fact, computer generated information suggested that the Corollary was sharp on some intervals of values of k; for example, we conjectured that for $c-1 = 2b-2$, there are k-sequences satisfying (1) and (2) but not (3) for each k with $c - \lfloor b/2 \rfloor \le k < c-1$. We note that this is indeed correct and is consequence of (f) in the Theorem to follow since $c - \lfloor b/2 \rfloor = 2b - 1 - \lfloor b/2 \rfloor > 3b/2$.

Some semblance of order was restored when we discovered a generalization of the Corollary that gave sharp results. It is (f) in the Theorem to follow. However, in using the computer we shifted emphasis slightly and we need a difinition to explain the shift.

DEFINITION: We call an n-sequence *acceptable* if it satisfies (1) and (2). Moreover, for $c \ge b$, we define the integer $N(b,c,k)$ by $N(b,c,k) = \min\{n \mid (3) \text{ holds for all acceptable } n\text{-sequences}\}$. Thus, for each $n \ge N(b,c,k)$, (3) holds for every acceptable n-sequence, while, for each $n < N(b,c,k)$, there exist an acceptable n-sequence for which (3) fails. Of course we always have $N(b,c,k) \ge b$ (otherwise (2) holds vacuously and acceptable N-sequences are easily constructed for which (3) fails) and $N(b,c,k) \geq k$ (otherwise $a_i = i$ gives an acceptable N-sequence for which (3) fails). In fact, we can characterize when equality holds in these two statements as well as when $N(b,c,k) = \infty$. We now summarize these and other results on $N(b,c,k)$. As most of these results are due to two of the authors who feel they are on the verge of even better results we will not include proofs here. However, we should note that (f), which generalizes the Corollary, admits a proof similar to that of the Corollary.

THEOREM: We have the following (recall that $c \geq b$).

(a) If $c > 2b$, then $N(b,c,k) = \infty$ for all k.
(b) If $c = 2b$ and $2^m | k => 2^m | b$, then $N(b,c,k) = \infty$ for all k.
(c) If $c = 2b$ and there is a positive integer m such that $2^m | k$ but $2^m \nmid b$, then $N(b,c,k) \leq b + k - (b,k)$.
(d) If $c < 2b$, then $N(b,c,k) \leq b + k - (b,k)$.
(e) If $c < 2b$ and $k \leq b$, then $N(b,c,k) = b$ if and only if $k \lfloor b/k \rfloor > c - b$.
(f) If $c < 2b$, then $N(b,c,k) = k$ if and only if either $k = b$ or $k \geq \max(b, 2(c-b))$.
(g) If $c < 2b$, then $N(b,c,c-b) = \max(b, 2(c-b))$.
(h) For each positive integer m $N(mb, mc, mk) \geq mN(b,c,k)$.

The computer generated information was very helpful in guessing what was to be proved in several of the above. Values of $N(b,c,k)$ were generated in the case of $c = 2b$ for all k, $b \leq 100$ and in the case $c = 2b - i$, $1 \leq i \leq 10$ for k and b with $1 \leq b \leq 50, 1 \leq k \leq 2b$. Based on that information we propose the following conjectures and questions.

CONJECTURES AND QUESTIONS:

(i) Equality holds in (c) above. We do know that equality often does not hold in (d).

(ii) We suggest the following generalization of the Corollary (and of (f) in the case of $c = 2b - 1$).

$N(b, 2b-1, k) = mk$ if $b = mk$,

$N(b, 2b-1, k) = mk$ for $m \geq 2$ if $\lfloor \dfrac{2b+2m-4}{2m-1} \rfloor \leq k < \lfloor \dfrac{2b+2m-6}{2m-3} \rfloor$,

$N(b, 2b-1, k) > mk$ otherwise.

(iii) Equality holds in (h).
(iv) Between (e) and (f) of the Theorem, we have quite a bit of information. Some of it is illustrated in the two figures below which

correspond to the cases $2(c-b) \leq b$ and $2(c-b) > b$. Note that $c-b < b$ since we are assuming we have $c < 2b$. The endpoint case $k = c-b$ is covered by (g).

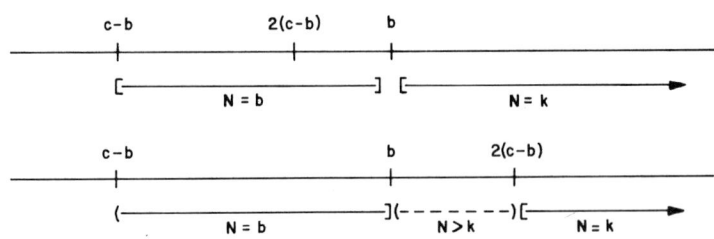

(v) If $c < 2b$, then $N(b+k, c+2k, k) = N(b,c,k) + k$.

(vi) In case (b) of the Theorem, the only known infinite acceptable sequences which fail (3) have a periodic stucture. Are there any which are *non-periodic* ?

In order that the reader can get some feel for these results and conjectures the table to follow gives the computer generated information for the case $b = 18, c = 35$. Each entry is of the form $k:N(b,c,k)$.

1:18	2:18	3:18	4:20	5:20	6:18	7:21	8:24	9:18	10:22
11:23	12:24	13:26	14:28	15:30	16:32	17:34	18:18	19:24	20:26
21:27	22:29	23:28	24:30	25:32	26:34	27:34	28:34	29:34	30:34
31:34	32:34	33:34	34:34	35:35	36:36	37:37	38:38	39:39	40:40

and $k:k$ for $k \geq 2(c-b) = 34$.

References

[1] R.A. Brualdi, *Introductory Combinatorics*. New York, Elsevier North-Holland, Inc., 1977.

[2] J.P. Hutchinson and P.B. Trow, Some pigeonhold principle results extended, *Amer. Math. Monthly* 87 (1980), 648-651. MR82g:0516.

[3] D.O. Shklarsk, N.N. Chentzov and I.M. Yaglom, *The USSR Olympiad Problem Book*. W.H. Freeman, San Francisco, 1962.

[4] W.D. Wei, *Math. Reviews* (July 1982) #82g:0516.

The Number Of Acyclic Orientations Of The n-Cube

H. Fleischner

Let W_n be the graph of the n-dimensional cube having 2^n vertices and $n2^{n-1}$ edges. Denote by d_n the number of all orientations of the edges of W_n, and by k_n, f_n respectively, the corresponding numbers of cyclic, acyclic orientations. Obviously, $2^{n2^{n-1}} = d_n = f_n + k_n$. The following results have been obtained, so far, by the author (and are easily verified).

1) $k_n > d_n(1-(\frac{7}{8})^{n2^{n-3}})$ if n is even

2) $2^{c2^{n-1}} \leq f_n < 2^{(1-\epsilon)n2^{n-1}}$ for any $n \geq 3$ and $c \in (2,3)$, and some small ϵ.

We write $f_n = 2^{g(n)2^{n-1}}$ and ask for the behaviour of g(n). Is it bounded above? If not, is it growing linearly or, say, logarithmically? It is easy to see that g(n) is strictly monotonically increasing. One could also tackle this problem by asking if there exists some $\alpha > 1$ such that $k_n < d_n(1-(\frac{1}{2})^{n2^{n-\alpha}})$.

Let furthermore $f_{n,r}$ denote the number of acyclic orientations of W_n having precisely r sinks. Clearly, $f_n = \sum_{r=1}^{2^{n-1}} f_{n,r}$. However, how do these functions $f_{n,r}$ behave for fixed r? What kind of relations exist between these functions for fixed n?

It should be noted that these questions stem from modelling problems in genetics.

Uniquely 3-colorable Planar Graphs

Wen-Lian Hsu

Conjecture. In a uniquely 3-colorable planar graph G, if there does not exist two triangles with a common edge, then G must have an odd cycle with no chords.

T-Free Arrangements of Hyperplanes

Richard Stanley

Let $X = \{H_1, \ldots, H_\nu\}$ be a finite set of hyperplanes in $\mathbb{C}^n$. Let L be the set of all intersections of the H_i's, ordered by reverse inclusion (including the void intersection $\mathbb{C}^n$). Suppose for simplicity $H_1 \cap \cdots \cap H_\nu = \{0\}$. L is a geometric lattice studies extensively by Zaslavsky (e.g., Memoirs of the A.M.S., #154). Let Ω be the set of all polynomial functions $p: \mathbb{C}^n \to \mathbb{C}^n$ satisfying $p(H_i) \subset H_i$ for $1 \leq i \leq \nu$. (In other words, $p = (p_1, \ldots, p_n)$ where $p_i \in \mathbb{C}[x_1, \ldots, x_n]$, and if $x \in H_i$ then $(p_1(x), \ldots, p_n(x)) \in H_i$.) Clearly Ω is a module over the polynomial ring $R = \mathbb{C}[x_1, \ldots, x_n]$. If Ω is a *free* R-module then we call the arrangement X *free in the sense of Terao* or *T-free*. It can then easily be shown that Ω has a basis consisting of exactly n elements $p^{(1)}, \ldots, p^{(n)}$, where all the components of $p^{(j)}$ are homogeneous polynomials of the same degree d_j. A remarkably theorem of Terao (Inv. math. 63 (1981), 159-179) is equivalent to the statement that the characteristic polynomial $\sum_{t \in L} \mu(0,t) \lambda^{\dim t}$ of L is given by $\prod_{j=1}^{n}(\lambda - d_j)$.

Problem. Find a purely geometric or matroid-theoretic characterization of T-free arrangements and the numbers $d_1, \ldots, d_n$. In particular, does the property of being T-free depend only on L? It can be shown, using Terao's Removal Theorem (J. Fac. Sci. Tokyo 27 (1980), 293-312, esp. pp. 303-304) that if L is supersolvable (as defined in Alg. Univ. 2 (1972), 197-217), then X is T-free. However, there exist non-supersolvable T-free arrangements.